MICROBIOLOGY: THE LABORATORY EXPERIENCE

STEVE KEATING PENNSYLVANIA STATE UNIVERSITY

W. W. NORTON & COMPANY

NEW YORK · LONDON

W. W. Norton & Company has been independent since its founding in 1923, when William Warder Norton and Mary D. Herter Norton first published lectures delivered at the People's Institute, the adult education division of New York City's Cooper Union. The firm soon expanded its program beyond the Institute, publishing books by celebrated academics from America and abroad. By mid-century, the two major pillars of Norton's publishing program—trade books and college texts—were firmly established. In the 1950s, the Norton family transferred control of the company to its employees, and today—with a staff of four hundred and a comparable number of trade, college, and professional titles published each year—W. W. Norton & Company stands as the largest and oldest publishing house owned wholly by its employees.

Editor: Betsy Twitchell
Associate Editor: Katie Callahan
Developmental Editor: Michael Zierler
Senior Project Editor: Thomas Foley
Associate Production Director: Benjamin Reynolds
Copy Editor: Norma Sims Roche
Managing Editor, College: Marian Johnson
Marketing Manager, Biology: Jake Schindel
Designer: Anna Reich
Photo Editor: Trish Marx
Photo Researcher: Elyse Reider
Permissions Manager: Megan Jackson
Permissions Clearer: Elizabeth Trammell
Editorial Assistant: Taylere Peterson
Composition and page layout by codeMantra
Illustrations by codeMantra
Project Manager at codeMantra: Ellora Sengupta
Manufacturing by: Quad/Graphics—Versailles KY

ISBN: **978-0-393-92364-3**

W. W. Norton & Company, Inc., 500 Fifth Avenue, New York, NY 10110

ww.norton.com

W. W. Norton & Company Ltd., Castle House, 75/76 Wells Street, London W1T 3QT

1 2 3 4 5 6 7 8 9 0

DEDICATION

This book is dedicated to the memory of my father, Thomas Keating, Jr., a ramblin' wreck from Georgia Tech, and one hell of an engineer.

CONTENTS

Microbiology is one of those subjects that can interest everyone because microbes affect our lives in so many different ways, both positively and negatively. In my microbiology lecture, I begin the course with photos from 19th-century cemeteries that I took while on vacation in Maine. Similar to any cemetery from the 1800s, these photos show that many of the graves are of children who died of infectious disease. I use these photos to illustrate that we live in a very different world today, a world in which diagnostic methods and antimicrobial drugs have resulted in a dramatic decrease in childhood mortality, thanks to microbiology.

In my lab course, I tell my students that one does not have to be planning to be a microbiologist to appreciate that the methods of the microbiology lab have given us a greater understanding of the biological world, created useful products, and have saved countless lives. Students taking a microbiology lab will hopefully grow to understand and value the way in which discoveries have been made in this field of science and how they might make discoveries of their own in the future. My personal curiosity about the subject, and the fun of watching others develop their own interests in microbiology, while they gain proficiency with the field's methods, have made teaching these labs, and writing this manual, very enjoyable indeed.

My interest in microbiology dates back to the seventh grade when I first used a microscope in a biology class. I was already interested in biology, and the microscope opened up a whole new world of organisms which were fascinating, ubiquitous, and very significant in terms of their impact on us. As an undergraduate, I majored in microbiology and was also drawn to the sights and smells of the microbiology lab. There were so many media colors, so many kinds of microbes to observe, and everything was alive and growing. This was a type of biology in which there was no need to dissect dead, stiff, and colorless frogs and fetal pigs reeking of formaldehyde. Best of all, unlike the chemistry and physics labs, the microbiology labs actually worked.

After working with insect viruses in graduate school, I began teaching microbiology and ecology at a small college where I was able to be in the lab as the students worked. It was a real pleasure to watch students' confidence increase as they learned the basic techniques of the microbiology lab, and as their growing interest led them to ask questions that went beyond the immediate task at hand. In this environment, it was possible for me to directly explain the lab, observe student errors, provide verbal directions, and answer questions as they arose. This gave me ample practice in explaining the principles and concepts behind microbiology labs, and led me to anticipate certain common student errors in lab technique. These experiences are reflected in this manual, both in the introductory material and in the numerous notes to students in the protocols.

This changed when I moved on to work at a large university and began teaching through graduate teaching assistants, who were new to teaching, and in some cases, had not had a microbiology lab course as undergraduates. At this point, the lab manual became much more important as I was teaching primarily through the manual, but many of these manuals did not provide enough material to get the students interested in the labs or to thoroughly explain the labs, and more information was needed to reduce student errors. This motivated me to write a microbiology lab manual as a means of speaking directly to the students to increase their interest and understanding of the subject and to decrease the number of mistakes.

In writing this microbiology lab manual, I tried to keep the following things in mind:

- It can be challenging to get students to perform the essential task of reading lab manuals. If it's possible to avoid reading the manual, some students will act accordingly. So, I prepared a manual written in a more conversational style with a little humor and a few digressions to encourage students to read the manual and read it closely. Writing a manual is not exciting work, and I would often put in items that entered my mind by free association, in part, to make the work less tedious for me. However, I found that the students responded very positively to these mental wanderings, so I kept them in the manual to facilitate student interest.

- Most existing manuals are not as complete in terms of background information as they could be, so they are not as useful for lab courses taught by graduate TAs, who typically teach these courses at the larger research institutions. Manuals can leave out certain types of information if veteran microbiology instructors are in front of the classroom, because the instructors can fill in the gaps. However, graduate students are not experienced teachers and usually lack the depth of knowledge that is typical of more experienced microbiologists. This can be a problem when, for example, students raise questions that are not answered by the manual. As a lab course coordinator, I wanted to be sure that the students would get all of the information that they needed, regardless of who was at the front of the classroom.

- Students will also do and interpret things in unexpected ways. This is not surprising, given that most students in an introductory microbiology lab are completely new to the subject. In writing the manual, I tried to apply lessons learned from watching how students actually do the labs and from the observations relayed by TAs to anticipate how students will interpret and apply the information provided by a manual. Everyone is unhappy when a lab doesn't work, and so many steps in the protocols have additional notes designed to prevent some of the common errors that lead to disappointment. In contrast, success in lab leads to more interest, more enthusiasm, and a much better classroom atmosphere.

Microbiology: The Laboratory Experience is based on a manual that I prepared for teaching a multiple section introductory microbiology lab course at Penn State. We used that manual for several years, and it has been well-received by the students. Many students have commented that they like the extra quotes, analogies, and digressions, and that these additions made this one of the few manuals that they actually read. More importantly, the students have been able to follow the protocols and succeed in the lab, which has given them confidence and a very positive view of microbiology. Some students have even changed their majors to microbiology as a result of their experiences in this introductory lab course. We have also modified the protocols over the years, to address problems reported to me by our TAs. These changes have been intended to reduce confusion and common student errors. So, this manual has been well-tested and constantly improved, with the sole purpose being to promote the success of both the students and the TAs in the lab. Finally, it is my hope that lab instructors and students everywhere will find this manual to be a rich and useful resource for learning about microbiology.

FEATURES AND RESOURCES

The tendency in many microbiology lab manuals is to simply list steps for students to work through in lab protocols, resulting in their mechanical execution of lab work rather than promoting real learning. *Microbiology: The Laboratory Experience* avoids this common problem by using distinctive features, which promote genuine understanding of the microbiology behind each lab. These features include:

- **Extensive introductions** to each lab that explain concepts clearly and accessibly. This ensures that students—and inexperienced TAs—come to the lab session with the necessary background to not just go through the motions, but to actually learn. Most competing manuals simply ask students to follow a protocol and report results, but students who don't understand the composition of media, why a reagent produces a given reaction, or how a given method accomplishes a particular goal are just painting by numbers. These thorough introductions provide an antidote by explaining why procedures are performed in a particular way, providing context that enables students to meaningfully interpret the results of an activity. They also help students to understand the reasons behind the method, thereby preparing them to identify and correct errors that commonly lead to bad results.

- **Lab protocol sections which provide thorough explanations** for why certain procedures are performed in a particular way. This helps students to develop an understanding of the tools and techniques used in microbiology lab work. Because microbiology lab work can be overwhelming to undergraduates, the inclusion of **frequent hints and tips** that address common problems students encounter when doing each lab, also helps them to avoid potential mistakes.

- **Post-lab thought questions** that encourage students to consider how procedures and results can be applied beyond the classroom. These questions, which have been used by thousands of students at Penn State over the years, include:

  - "If you were a doctor, describe two cases or situations in which you might order a lab to perform an acid-fast stain."

  - "Assume that *Arthrobacter* cells do not secrete toxins into food, and consider what you've learned from lab results. Given that, would you be worried about the possibility of *Arthrobacter* multiplying in your body if your latest meal had been full of *Arthrobacter*? Why or why not?"

  - "Grape juice has a concentration of up to 25% sucrose. Assume, for the moment, that *E. coli*, *S. epidermidis*, and *S. cerevisiae* all fermented sucrose to produce ethanol and carbon dioxide. If all of these cell types produced the same fermentation products, would there be any reason to use *S. cerevisiae* to make wine instead of the other two species? Cite evidence from the lab to back up your answer."

  - "If you lost the labels on your *E. coli* and *P. vulgaris* stock culture tubes, how could you figure out which tube contained the *E. coli* cells?"

- **Dynamic visual content** which distills complex concepts and demonstrates correct lab procedures. In every lab, **impressive, up-to-date illustrations** created specifically for the manual, and others adapted from Norton's microbiology textbooks *Microbiology: An Evolving Science* and *Microbiology: The Human Experience,* present the visual content in an engaging manner. Showing students how to execute lab protocols through **photographs of lab procedures** is an important component of any microbiology lab manual; in this manual photographs were taken exclusively to match lab protocols in the manual exactly. Finally, it's important for students to see micrographs of the microorganisms they're studying, so the manual includes dozens of **meticulously researched micrographs**.

- An **engaging authorial voice**, using frequent anecdotes from popular culture and the history of science, draws students into the material. The extensive lab introductions are written in a conversational style, which motivates students to read before just jumping into the procedures. By including short histories that illustrate the process of scientific discovery and application, descriptions of the real world applications of lab methods, quotes from diverse sources such as The Beatles, Stephen Sondheim, Monty Python, and more, the author engages student interest, promotes careful reading of the manual, and most importantly facilitates actual learning.

Make no bones about it, adopting a new microbiology lab manual is a lot of work. But, to make the process easier, and perhaps even fun, Steve Keating has written a companion guide for instructors, *Teaching Microbiology Lab, A Guide to Microbiology: The Laboratory Experience.* This guide includes tips and tools for teaching successful, engaging courses, so it is also intended for lab coordinators who oversee large programs to distribute to their undergraduate and graduate TAs. The first half of the guide includes helpful suggestions for how to design a course and syllabus, create effective assignments and quizzes, interact with students or TAs, and much more. The second half provides a guide for teaching each individual lab with common student questions and mistakes noted, recommended sources for materials, instructions for media and culture preparation, suggestions for reducing costs and preparation time, answers to questions in the lab exercises, and ways to modify or expand on each lab.

ACKNOWLEDGMENTS

The process of creating a lab manual, especially a first edition, involves the hard work of dozens of contributors to whom I am very thankful. First and foremost, I'd like to acknowledge the unflagging support of the book's editors Betsy Twitchell and Katie Callahan. Without their attention to detail, and commitment to the schedule, you would not be holding this manual in your hands right now. This book's developmental editor, Michael Zierler, truly has no equal. His experience as a microbiology lab teacher himself, a sharp eye for clarity, and his commitment to helping me execute my vision at all costs have made the finished product better in innumerable ways. Editorial assistant Taylere Peterson rounds out the editorial team at Norton. She made sure files were where they were supposed to be and that everyone on the team had access to the correct information; for that she has my gratitude.

First-edition lab manuals involve a dizzying array of moving parts, yet senior project editor Thom Foley managed to stay organized and precise through multiple production passes. He is absolutely a master at his craft. Copy editor Norma Sims Roche provided an essential check to ensure that the published manual you hold in your hands is 100% correct in grammar, punctuation, and style. And the proofreader, Janel Mosley, did an excellent job of providing a last line of defense against errors and typos. Photo manager Trish Marx was also the photographer of the lab procedure images you see in this book. She brought an artistic, yet practical, eye to several shoots and did a wonderful job making the photos ready for the manual. Photo researcher Elyse Reider tracked down the rest of the photos you see in the manual from sources near and far. Permissions manager Megan Jackson and permissions clearer Elizabeth Trammell provided helpful guidance throughout the production of this book. Anna Reich is to thank for the manual's beautiful design, and Anne DeMarinis did an amazing job designing the cover. Last but not least, production manager Ben Reynolds has my thanks for juggling multiple vendors to produce a book with absolutely stunning art, a beautiful layout, professional printing and binding, and more.

My thanks also extend to codeMantra, who were not only the compositor for this title, but also created the stunning new art that enhances all of the labs. Ellora Sengupta was an invaluable resource as our main contact and coordinator of all of our interactions with codeMantra. Thank you.

I'd also like to thank the many microbiology lab instructors who contributed and provided feedback on various aspects of this manual. Maureen Morrow at SUNY New Paltz and her student Katherine Betuel have my deep gratitude for giving us access to a microbiology lab in which to shoot photos of lab procedures. My gratitude also extends to our six amazing accuracy checkers who read every single lab in this manual, looking for opportunities to make them not only more correct, but also more clear, engaging, and accessible. They are:

Blaise Boles, University of Iowa
Robert E. Carey, Lebanon Valley College
Mette P. Ibba, The Ohio State University
Sherry Meeks, University of Central Oklahoma
Dorothy Scholl, University of New Mexico
Uma Singh, Valencia College

And a huge thank you to the following reviewers, who provided invaluable feedback on the draft chapters. This manual is a better teaching and learning tool because of your contribution:

Lois C. Anderson, Minnesota State University, Mankato
Lisa Antoniacci, Marywood University
Daniel Aruscavage, Kutztown University
Jonathan Awaya, University of Hawaii at Hilo
Melody Bell, Vernon College
Kathleen A. Bobbitt, Wagner College
Lisa G. Bryant, Arkansas State University - Beebe
Kristin M. Burkholder, University of New England
Valerie Campbell, Arkansas State University - Newport
Alaina Campbell, Virginia Commonwealth University
Cassy Cozine, Coe College
Angus Dawe, New Mexico State University
Xin Fan, West Chester University of Pennsylvania
Mary B. Farone, Middle Tennessee State University
Gregory Frederick, LeTourneau University
Sandra Gibbons, University of Illinois at Chicago
Donald Glassman, Des Moines Area Community College
Melanie C. Griffin, Kennesaw State University
Ernest M. Hannig, University of Texas at Dallas
Robert J. Kearns, University of Dayton
Cynthia Keler, Delaware Valley College
Dana Kolibachuk, Rhode Island College
Susan Koval, University of Western Ontario
John Lee, City College of New York
Claudia Lemper, Iowa State University
Todd C. Lorenz, University of California, Los Angeles
Kari L. Murad, The College of Saint Rose
Pam Rich, The University of Akron
Veronica Riha, Madonna University
Jason A. Rosenzweig, Texas Southern University

Kathleen Sandman, The Ohio State University
Gene M. Scalarone, Idaho State University
Gary E. Schultz, Jr., Marshall University
Isdore Chola Shamputa, National Institutes of Health, United States
Cristina Takacs-Vesbach, University of New Mexico
Janyce Woodard, Little Priest Tribal College

I'd also like to recognize and thank the many students, teaching assistants, prep room staff people and instructors at Penn State who contributed to this manual. Dr. Carl Sillman, a true master of microbiology, provided invaluable assistance in the development of the microbial ecology and antibiotic producers from soils labs. Dr. Meredith Defelice patiently showed me the way during the development of the PCR lab and witnessed my astonishment when it actually worked. Dr. James McDonel authored a lab manual that provided a framework for the writing of the first version of my Penn State manual. Ms. Linda Price has been reliably making our media and preparing all of the materials for our labs for several years, and she provided many of the media formulas used in this manual; teaching a microbiology lab is easy when you have such great prep room people.

In addition, I'd like to note my deep appreciation to all of the dozens of teaching assistants who have served in the classroom trenches and done the actual lab instruction over the last fifteen years, while I tried to follow Monty Python's advice on how not to be seen. In particular, Ms. Melissa Mason has served as a microbiology lab teaching assistant for many years; she has been a vital source of feedback as to what works and what doesn't, and she caught and corrected numerous errors before the manual first reached W. W. Norton & Company, saving me much embarrassment.

Most of all, I'd like to express my gratitude to my family. Thanks to my parents, Tom and Jennie Keating, who gave me my first microscope for Christmas. I still have it. Despite my decision to enter the disreputable field of biology, instead of engineering, they were always behind me. Finally, many thanks to my wife, Chris, whose constant support and reassurance kept me going over what proved to be a long process of reviewing, editing, rewriting, rinsing and repeating. Without her unfailing encouragement, the manual would have remained "in the bottom of a locked filing cabinet stuck in a disused lavatory with a sign on the door saying 'Beware of the Leopard.'"

Sincerely,
Steve Keating

ABOUT THE AUTHOR

Steve Keating has been teaching and coordinating microbiology labs for 25 years. Over that time, he has acquired both a mastery of lab techniques and a catalog of information on the history and applications of microbiology lab methods. Steve earned his BS in Microbiology at the University of Maryland and his PhD in Entomology at Pennsylvania State University. He began his teaching career at St. Francis College (now St. Francis University), where he directly taught his students these microbiology laboratory techniques. For the past 15 years, Steve has taught microbiology at Pennsylvania State University, where he both lectures and coordinates the introductory lab courses in microbiology. Steve has twice won the Biochemistry and Molecular Biology Department's Althouse Outstanding Instructor Teaching Award.

SAFETY AND WASTE DISPOSAL PROCEDURES

Life is not without risks. In the microbiology laboratory, students are exposed to living microbial organisms, flames, chemical agents, sharp objects, and breakable materials. Each of these things, if improperly handled, poses potential health risks to the student. However, through instruction and self-awareness, attention and care, these risks can be minimized. The intent of this discussion, although it is not all-inclusive, is to inform you of potential risks, instill in you the importance of good safety practices, provide proper handling procedures, provide corrective measures should an incident occur, and serve as an aid to minimize risks.

BASIC SAFETY RULES

1. **Do not eat, drink, or chew gum in the laboratory at any time.**
2. Do <u>not</u> apply lip balm or makeup, or handle contact lenses and contact lens solutions, in the laboratory.
3. All electronic devices should be kept in backpacks or laptop bags and, if at all possible, should be stored away from your work area. In some labs, cell phones may be used to photograph results; otherwise, phones should be stored with other electronic devices.
4. <u>Wipe down</u> your bench and work area with disinfectant before each and every lab.
5. <u>Keep benches free of nonessential items</u> such as coats, backpacks, books from other courses, laptops, tablets, and so on. Do not place them on the floor because items on the floor can be contaminated by spills.
6. **Tie long hair behind your head and avoid wearing clothing with loose long sleeves.**
7. Open-toe and open-heel footwear <u>is prohibited</u>; there may be broken glass and spilled bacterial cultures on the floor, so wear closed-toe shoes to lab.
8. Lab coats may or may not be required by your instructor, but keep in mind that the stains used in lab will stain <u>everything</u>, including clothing. In other words, this is not the venue for showing off your finest and most expensive attire.
9. Your instructor may also require that you wear clothing that covers your legs,; that is, shorts and short skirts may be prohibited.
10. **Mouth pipetting is strictly prohibited.** Always pipette liquids with the aid of a mechanical pipetting device.
11. In cases in which we are working with <u>biosafety level 2 organisms</u>, you will be required to use disposable <u>gloves and glasses</u> (See the table to the right for examples of biosafety level 2 organisms that we may encounter in this course).
12. <u>Know the locations</u> of first aid and eyewash stations, safety showers, fire extinguishers, and if available, fire safety blankets. Be familiar with the use of this equipment. Report all accidental cuts and burns to your instructor.
13. <u>Wipe down</u> your bench and work area with disinfectant at the end of each and every lab.
14. Wash your hands with soap and water before leaving the lab for any reason. **Wash your hands, wash your hands, wash your hands!**

BIOSAFETY LEVEL 2 ORGANISMS
Proteus vulgaris
Pseudomonas aeruginosa
Salmonella typhimurium
Staphylococcus aureus
Streptococcus mitis
Streptococcus pyogenes

WASTE DISPOSAL

1. While details for disposal of lab materials will vary from lab to lab, the most important rule is simple: **Do not dispose of any materials contaminated with live bacteria or fungi in the regular trash at any time for any reason.** Such materials include used plastic pipettes, used plastic petri dishes, used test tubes, and used swabs.

2. After all masking tape labels have been removed, used glassware (test tubes, beakers, and so forth) should be placed in the area designated by your instructor so that they can be autoclaved later.

3. Used plastic petri dishes and plastic pipettes should be placed in autoclave bags so that they can be autoclaved later. You do not have to remove labels from disposable plastics because these items will be autoclaved and discarded.

4. Glass slides that you have prepared and stained should be discarded in sharps containers. Commercial and demonstration slides should be returned to slide storage boxes.

BUNSEN BURNERS

1. **Bunsen burners should always be turned off when not in use. Be sure to turn off the burners before you leave the lab.**

2. **Keep long hair pulled back or tied behind your head.**

3. **Secure any loose clothing.** The most common fire incident involves burner-ignited long sleeves.

4. While burners should be off when not in use, **don't ever assume that a burner is off**. Do not put any part of your body or clothing over a burner at any time.

MICROBIAL SPILLS

Treat all microbes used in the lab as if they are pathogenic (disease-causing).

Spills on skin

If you spill a live culture on your skin, go immediately to the nearest sink and rinse the affected area of the skin with disinfectant, then wash thoroughly with soap and water.

Benchtop and floor spills

1. Do not touch the spill.

2. Alert your instructor; the instructor will help you properly clean up and dispose of the material. Cover the spill with paper towels and saturate the paper towels with disinfectant for at least 10 minutes.

3. Dispose of any broken glassware in a sharps container and place the remaining materials in autoclave bags.

BROKEN GLASS

1. Alert the instructor, who will properly dispose of the material.
2. If broken glass is contaminated, flood the glass with disinfectant before handling and disposing of it.
3. All broken glass should be placed in a sharps container.

FIRE PROCEDURES

Know the location of fire extinguishers and fire alarm stations.

In the event of a very small fire:

1. **Alert your instructor immediately** and quickly put out the fire with a fire extinguisher. After the fire is extinguished, **immediately turn off the gas to all of the Bunsen burners in the lab** to prevent unburned gas from filling the room.
2. If you accidently ignite a beaker filled with alcohol, put it out by covering the beaker with a larger beaker. If this does not immediately extinguish the flame, then use a fire extinguisher. Again, after using the extinguisher, immediately turn off the gas to all of the Bunsen burners in the lab.

In the event of a larger fire:

1. Go to the nearest fire alarm pull station and activate the building alarm system.
2. **When the fire alarm sounds, evacuate the building immediately.**
3. Once you have left the building, do not reenter until the area is declared safe by fire, police, or environmental health and safety personnel.
4. Do not try to put out a fire with a fire extinguisher if there is any doubt that it can be done safely. Your safety is more important. Attempts to extinguish a fire should not be made until the alarm has been activated.

I have received and carefully read all of the information in the above Safety Procedures. I understand the procedures, have had my questions answered to the best of the instructor's ability, and intend to implement the policies listed in these procedures.

_____

Date

_____

Printed Name

_____

Signature

The ability to see microorganisms with the aid of a microscope and more recent imaging technologies revolutionized biology and medicine. This scanning electron micrograph shows deadly methicillin-resistant *Staphylococcus aureus* bacteria, commonly referred to by the acronym, MRSA. This species of bacteria is a scourge for hospitals. The photograph has been false-colored.

Learning Objectives

- Define and understand key terms related to microscopy, including resolution, magnification, working distance, and depth of field.

- Explain limits of light microscopes, especially with respect to types of microbes that can and cannot be seen with these microscopes.

- Understand the challenges created by increasing magnification and describe methods used to respond to these challenges.

- Use the oil immersion lens and understand how immersion oil reduces light scattering and improves image quality, a particularly important property when viewing cells at high magnification.

LAB 1

The Microscope

Things look bad enough without having to look closer at them.
—Bender B. Rodríguez in *Möbius Dick* (Futurama)

O h no, not "The Microscope" again. Yes, we're going to start this microbiology lab course with the microscope. While by now you have probably already worked with microscopes, you may not have used the oil immersion lens (100× lens, or 1,000× total magnification) or looked at many examples of bacterial cells. Use of high magnification and examination of cells that are much smaller than human cells can be a bit of a challenge, so this lab will give you a chance to both use the oil immersion lens and look at various types of bacteria. A typical compound microscope is shown in **Figure 1.1**.

Let's begin by defining some important terms in microscopy.

RESOLUTION

Resolution is (1) the ability to see objects that are close together as multiple, distinct, and separate objects; or (2) the minimum distance between two objects that reveals them as separate objects.

One way to think of resolution is to consider what happens when you look at someone holding up two fingers. When you are close to the person, you can

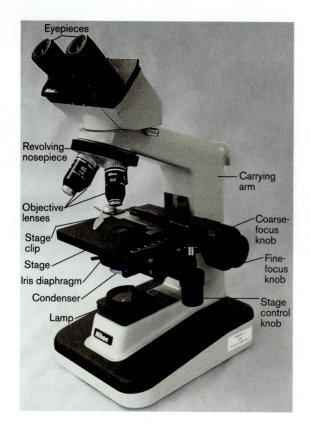

FIGURE 1.1 A typical compound light microscope.

clearly see two separate fingers. But as you move farther away from the person, it becomes increasingly difficult to see or resolve the fingers as separate, distinct objects. Eventually, it becomes impossible to see separate fingers using only the unaided eye. At this point, to resolve the individual fingers, you would need binoculars to magnify the image.

Turning this around, you can use your ability to resolve a given object of a consistent size as a rough measure of distance. The eighteenth-century military command "Don't shoot until you see the whites of their eyes" was a technique for persuading the troops to hold their fire until the opposing army was close enough for their highly inaccurate smoothbore muskets to be reasonably effective. When you can just barely resolve the whites of the eyes in the approaching faces, the faces are close enough to hit—at a reasonable rate—with musket fire.

Resolving power is a function of the wavelength of light that forms the image and the numerical aperture of the lens. **Numerical aperture** is a measure of the lens's ability to capture light and is a product of the type and quality of the lens. Even with the highest-quality lenses, resolution is limited to approximately the wavelength of light (λ) divided by 2 (or $\lambda \times 0.5$). Since the wavelengths of visible light range from about 0.4 micrometers (violet-blue) to 0.7 micrometers (red), visible light microscopes can successfully resolve objects separated by a minimum distance of 0.2–0.3 μm **(Figure 1.2A)**. (One micrometer, abbreviated μm, is equal to 1×10^{-6} meters.) However, if the objects are separated by a distance of less than about 0.2 μm, then the two separate items will begin to appear as if they were smeared together as one object **(Figure 1.2B)**.

In addition, when light is transmitted through the slide, most of the light will pass by objects smaller than about 0.1–0.2 μm without interacting with them. So, in the case of these tiny objects, resolution will be very low, and there will be little contrast between the objects and the brightly lit background. Since prokaryotic cells are typically about 0.5–3.0 μm in size, most individual bacterial cells can be resolved and observed with the light microscope. In contrast, most viruses are 0.010–0.100 μm and are thus far too small to be clearly resolved with visible light microscopy. Therefore, with the exception of a few very large viruses, bacteria are the smallest living things that can be seen with

A.

FIGURE 1.2 Resolving power.
A. Two objects can be clearly resolved by light microscopy if they are at least 0.3 μm apart.
B. Light waves of 0.4–0.7 μm cannot resolve two objects separated by a distance of less than about 0.2–0.3 μm. Instead, they blur together. In this example, two round objects of equal size appear as one dumbbell shape. If these two objects could be resolved when separated by only 0.15 μm, then they would appear like the panel on the right.

B.

a standard light microscope. To observe viruses, we need to use electron microscopes which were invented in the late 1930s. These microscopes use electromagnetic radiation with much, much shorter wavelengths than visible light.

MAGNIFICATION

Magnification is the relative enlargement of a specimen when it is seen through the microscope. In other words, through the use of lenses, an object's size appears to increase, so cells that are too small to be seen by the unaided eye appear to be much larger. When an object is sufficiently enlarged, in a virtual sense, then our eyes can observe it. Remember that the object is enlarged only in appearance, not in physical size.

However, magnifying objects does not change the limits to resolution imposed by the wavelengths of visible light. Therefore, the highest useful magnification will be that magnification that increases the apparent size of the smallest object resolvable by visible light to a point that we can see that object. The unaided eye can usually see objects that are 0.2 mm or larger, and since wavelengths of visible light limit resolution to 0.2 μm, or 1/1,000 of 0.2 mm, the highest useful magnification is about 1,000$\times$. At this magnification, the smallest objects that can be resolved appear to be about 0.2 μm $\times$ 1,000, or 0.2 mm, in size.

Total magnification

In the compound microscopes that we use in this lab course, **total magnification** is the product of the magnification of the **eyepiece (ocular) lenses** times the magnification of the **objective lenses**. Ocular lens magnification is always 10$\times$, and the objective lenses are typically (1) 4$\times$, (2) 10$\times$, (3) 40$\times$ or 45$\times$, and (4) 100$\times$. So the total magnifications available with typical light microscopes are those shown in **Table 1.1**.

Need for oil with high magnification

Here's a fact worth knowing: As the capacity of lenses to magnify increases, their diameter generally decreases, so an objective lens capable of 100$\times$ magnification (1,000$\times$ total magnification) has a very small diameter indeed. When the diameter of a lens is very small, then obviously, the light-gathering area of the lens is also very small.

Table 1.1 | MAGNIFICATIONS AVAILABLE WITH A COMPOUND MICROSCOPE

Ocular Lens	Objective Lens	Total Magnification
10$\times$	4$\times$	40$\times$
10$\times$	10$\times$	100$\times$
10$\times$	40 or 45$\times$	400 or 450$\times$
10$\times$	100$\times$	1,000$\times$

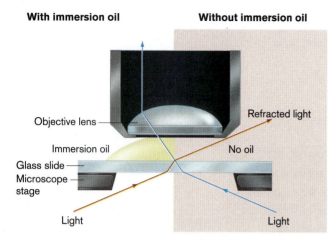

With immersion oil **Without immersion oil**

Objective lens

Immersion oil

Glass slide

Microscope stage

Light

Refracted light

No oil

Light

FIGURE 1.3 Oil immersion lens.
Oil with the same refractive index as glass reduces the bending and scattering of light as the rays pass through the slide and into the space between the slide and the lens.

Who cares? Well, this matters because when a light ray passes from the glass slide into the air between the slide and the lens, the ray is bent due to the differences in **refractive index** between glass and air. When two media (such as glass and air) differ in their refractive indexes, light waves will bend as they pass from one medium into another. If the ray exiting the glass is bent a lot (this depends on the entry angle), then it will miss the light-gathering surface of the lens. When the light-gathering surface is relatively large, as it is with a lower-magnification lens, then plenty of the light rays pass across the surface of the lens and contribute to the formation of the image. When the lens diameter is very small, however, a larger proportion of the rays miss the lens, and the quality of the image declines significantly.

To reduce this problem of "light scattering" when using the 100× objective lens, we put a layer of **immersion oil** between the slide and the lens. The oil has a refractive index similar to that of glass, so the bending and scattering of light is greatly reduced, and the quality of the image is significantly improved **(Figure 1.3)**. Remember that the oil is only to be used on the 100× oil immersion lens and never on any other lens. Furthermore, the oil must be removed from the lens with lens paper after every use.

Here are a few more terms and some practical considerations to keep in mind as you step up in magnification from 40× or 100× to 1,000× total magnification:

Working distance

Working distance is related to focal length or focal distance. Working distance is defined as the distance between the front edge of the lens and the specimen (focal length is measured from the center of the lens). Here's the important bit: Working distance decreases with increasing magnification. When the 100× oil immersion objective lens is used, the lens is only about 0.1 mm above the slide, and it's easy to end up grinding the lens into the slide. So, at higher magnifications, adjust the focus with the fine-focus knob only.

Depth of field

As **Figure 1.4** illustrates, **depth of field** is (1) the thickness of the slice of a specimen that will be in focus with a given lens, (2) the distance above and below the subject that appears to be in focus, or (3) the distance between the highest point and the lowest point of a three-dimensional object that is in sharp focus.

Why does this matter? The critical point is that **depth of field decreases with increasing magnification** (compare Fig. 1.4A and 1.4B). So, when working with the 100× oil immersion lens (1,000× total magnification), always remember that the slice that is in focus at any time is very thin or shallow. And that means that it's very easy for the specimen to get out of focus when you move the lens up and down.

Here's the solution to the problem of shallow depth of field:

When looking at bacterial cells, which are usually only 1–2 μm thick, always **focus on the image at low magnification first**. Then, when you switch to the 100× oil immersion lens, the lens will be at almost exactly the right height above the slide to get the cells into focus. This is because the lenses are **parfocal**, which means that the different objective lenses with different focal lengths are set in the rotating nosepiece at distances above the stage such that the image should remain nearly in focus when you switch from one lens to another. After switching to the 100× oil immersion lens,

A. Low-magnification lens.   **B.** High-magnification lens.

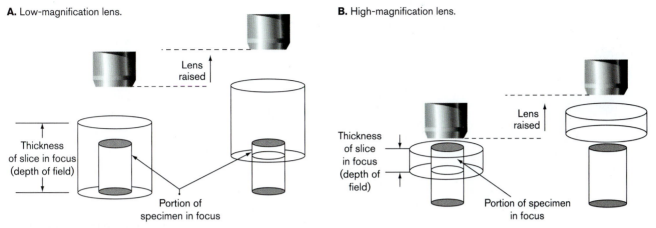

FIGURE 1.4 Depth of field.
A. Low-magnification lens. At lower magnifications, depth of field is deeper, and the lens can be moved up and down greater distances before the cells are completely out of focus. In this case, both lens positions have at least part of the specimen in focus.
B. High-magnification lens. At higher magnifications, depth of field is shallower, and very small changes in the height of the lens can result in loss of focus. In this case, only the lens on the left has a part of the specimen in focus.

you should use only the fine-focus knob, so that the thin slice that is in focus moves only a short distance up or down, until you find the point at which the cells are sharply in focus. This makes it less likely that you will "overshoot" the point where the shallow field in focus corresponds to the plane where the cells are located.

Amount of light required at higher magnification

As magnification increases, the diameter of the lens decreases, reducing the amount of light that passes through the lens and through the cells on the slide. To compensate, as you increase magnification, you must also take steps to increase the amount of light passing through the specimen. On most light microscopes, there are several ways to do this.

Light intensity dial On most microscopes, there is a dial or knob that can be used to adjust the lamp voltage, and thus the intensity of the light leaving the lamp. At higher magnifications, the dial should be adjusted so that the lamp produces more light.

Iris diaphragm The **iris diaphragm** controls the amount of light passing from the lamp through the bottom of the slide. At lower magnifications, the diameter of the iris opening can be decreased to increase contrast between the cell and the slide. At higher magnifications, the diameter of the opening should be increased to allow more light to pass through the cells.

Condenser The **condenser** is a lens system that focuses light coming up from the lamp below onto objects on the slide. At lower magnifications, it may help to lower the condenser to help increase the contrast between the cells and the slide, especially if the cells are unstained. At higher magnifications, moving the condenser up toward the bottom of slide will focus the light rays on a point on the slide that is directly below the $100\times$ oil immersion lens. The effect is to increase the intensity of the light passing through the cells that are in the field of view.

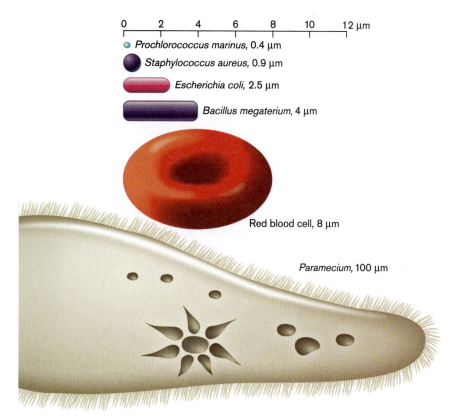

FIGURE 1.5 Relative sizes of various cells.
Paramecium and red blood cells are eukaryotic cells, while the remaining cells (bacterial cells) are prokaryotic cells.

Eukaryotic and prokaryotic cell sizes

All cells, except those of bacteria and archaea, are **eukaryotic cells**; these cells include protozoan, fungal, plant, and animal cells. Eukaryotic cells contain nuclei and are typically 10–20 μm (though there is great variation in their size) **(Figure 1.5)**. Cells of this size are not difficult to see with the light microscope, so you will look at prepared slides of *Paramecium* (protozoan) cells and *Saccharomyces* (yeast, fungal) cells to "warm up" on the microscope.

By contrast, all bacterial cells are **prokaryotic cells**. These cells lack nuclei, and they are much smaller in size (0.5–3.0 μm). Consequently, it is more challenging to observe these cells, and you will have to use the 100× oil immersion lens (1,000× total magnification) to resolve the images. This is the main goal for the day: Can you use the oil lens and see bacteria using the light microscope?

SUMMARY

	LOWER MAGNIFICATIONS	HIGHER MAGNIFICATIONS
Working distance	Longer Can use coarse-focus knob	Shorter Very short at 1,000× Risk of grinding lens Use fine-focus knob only
Depth of field	Deeper Easier to focus Can use coarse-focus knob	Shallow Easy to lose focus Use fine-focus knob only
Iris diaphragm	Use less light Close diaphragm down to increase contrast	Need more light Open diaphragm up to maximize light
Condenser	Lowering may improve contrast	Raise to base of stage to increase light focus
Microscope oil	Never use oil	Use immersion oil to reduce light scatter Only for 100× oil immersion lens (1,000× total magnification)

PROTOCOL

Materials

- Compound microscope with 100× oil immersion lens 95% ethanol
- Immersion oil
- Lens paper
- Kimwipes (lab tissues)
- Slides of *Saccharomyces* and *Paramecium* (or other eukaryotic microorganisms)
- Slides of various bacteria including bacillus (rod-shaped), coccus (spherical), and spiral or spirillum cells

Basic Steps for Using a Compound Light Microscope

Take another look at Figure 1.1 as you read and use this lab manual. It will help you use the microscope properly.

1. Carry a microscope to your bench by grasping the carrying arm with one hand and placing your other hand under the base **(Figure 1.6)**. Always carry the microscope with two hands!

FIGURE 1.6 The proper way to carry a microscope.

2. Plug in the microscope and turn on the lamp (the lamp switch will be in different locations on different microscopes).

3. Adjust the lamp to a medium setting; you will need to increase the amount of light at higher magnifications.

4. Place one of the slides on the microscope's stage so that the cells are on the upper surface of the slide.

   Eukaryotic cells are larger than prokaryotic cells, so the first time you use your microscope, the cells will be a little easier to find and get in focus if you start with a *Saccharomyces* or *Paramecium* slide.

   The "cell side" will be obvious when you are using prepared slides with coverslips, but most slides in this lab course are viewed without coverslips. If the slide is accidentally positioned cell side down, you will not be able to get the cells into focus at 1,000× magnification. At this high magnification, the working distance is very short and is less than the thickness of the slide. Placing slides on the stage with the cell side down ("slidis invertus") is a common cause of failure to focus.

5. Once the slide is on the stage, rotate the objective lenses so that the 10× lens is positioned directly over the cells on the slide (100× total magnification).

6. Look through the ocular lenses with both eyes and adjust the distance between the ocular lenses to match the distance between your eyes. When the lenses are properly adjusted, you should see a single, unified image with both eyes open. That is, the images from the two ocular lenses should be perfectly superimposed on each other.

7. Use the **coarse-focus** knob to bring the lens as close to the slide as possible, and then, while looking through the ocular lenses, rotate the coarse-focus knob so that the distance between the slide and the stage is increased as the stage is slowly lowered away from the 10× objective.

8. As the cells on the slide start to come into focus, begin using the **fine-focus** knob to move the objective lens up and down until the image is in sharp focus.

9. Once the cells appear to be in sharp focus, close your left eye, and if the cells are now no longer in focus, then again use the fine-focus knob to bring them into focus for your right eye only. If the cells are still in focus after you initially close your left eye, leave the fine-focus knob alone.

10. Now open your left eye and close your right eye. If the cells are no longer in focus, then rotate the left ocular lens until the cells are in focus. While the right ocular lens is set at a fixed height, the left ocular lens can be turned to raise or lower the lens. This allows you to adjust the ocular lenses to compensate for differences in vision in your two eyes. If the cells are still in focus after you initially close your right eye, leave the left ocular lens alone.

Do not touch the fine- or coarse-focus knob while adjusting the ocular lens. The goal is to adjust the microscope so that the image is in focus for both eyes at the same time. We do this by adjusting the height of the left ocular lens after the image is in focus for the right ocular lens.

11. Be certain that the object you are looking at is well centered. Now rotate the set of objective lenses so that the 40× or 45× objective lens is above the cells (all microscopes have either a 40× or a 45× lens).

    Remember that while focal length changes with changes in magnification, the lenses in these microscopes are <u>parfocal,</u> so the image should remain approximately in focus when you switch from one lens to another.

12. While looking at cells on the slide, **rotate** the **fine-focus knob only** to move the objective lens up and down until the image is in sharp focus. At 400/450× total magnification, the depth of field is quite shallow, so if you use the coarse-focus knob, just a little bit of rotation is enough to completely lose focus.

    The prepared slides may have been used by other students before you, and not all students are as careful as you will be. As a result, the 40/45× lens may become covered with oil, and these lenses are not made to be covered with oil. Oil on the lens will make it very difficult to see anything, even when the lens is at the proper distance for sharp focus. If you can't get anything in focus with the 40/45× lens, use the lens paper available in the lab to give the 40/45× lens a good rubdown (cleaning). If all else fails, you can try going directly from the 10× objective lens to the 100× oil immersion objective lens.

13. Once the cells are in focus at a total magnification of 400/450×, rotate the objective lenses so that slide is in between the 40/45× and 100× oil immersion lenses.

14. Now, place a drop of immersion oil on top of the coverslip, or if there is no coverslip, directly on top of the cells. The drop should be placed in the center of the slide so that when the 100× oil immersion lens is rotated into position, a seal will be formed between the bottom of the lens and the slide **(Figure 1.7)**.

    Once oil has been added to the slide, do **not** rotate the lens in such a way that the 40/45×

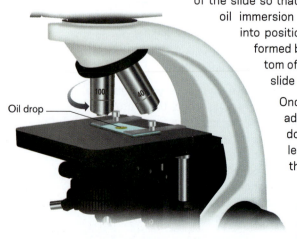

Oil drop

FIGURE 1.7 Using oil and the oil immersion lens.

lens passes back over the slide. Given the height of the 40/45× lens, the lens will inevitably pick up oil, and this is bad, bad, bad. Oil is only to be used on the 100× oil immersion lens and never on any other lens.

15. Rotate the 100× oil immersion lens into position over the cells, and while looking through the ocular lenses, **rotate the fine-focus knob only** to move the objective lens up and down until the image is in sharp focus.

    At 1,000× total magnification, depth of field is very shallow, so it is very important that you use the **fine**-focus knob **only**. If the **coarse**-focus knob is used, just the slightest rotation is enough to completely lose focus. In addition, working distance is very short, so if the **coarse**-focus knob is used, it's very easy to accidentally grind the lens into the slide. And that's bad.

16. When you are finished, slide the slide directly forward without raising it off the stage until it is well clear of the 100× lens. Remember that the distance between the slide and the 100× lens is very small, so keep the slide pressed to the stage as you pull it forward.

17. **After the slide has been removed, use the lens paper to remove as much oil as possible from the 100× oil lens.**

18. When finished for the day, rotate the nosepiece so that the 10× lens is clicked into position directly above the lamp (for use by the next student), turn off the lamp, unplug the microscope, replace the dust cover, and carry the microscope with two hands to the storage cabinet.

Examining Microorganisms through a Microscope

You will be using commercially prepared slides to examine five different types of cells. Remember that prokaryotic cells are much smaller than eukaryotic cells, so your challenge will be to get the prokaryotic cells into focus at 1,000× total magnification.

1. Before using each slide, gently wipe away any oil left on the slide from previous use with a paper towel or Kimwipe. If ethanol is available, dampen the towel or Kimwipe with a little bit of ethanol to enhance the removal of the oil.

2. For each slide or cell type, use the procedures given above to view the cells at 400/450× and 1,000× total magnification.

3. Sketch what you see at each magnification in the Results section. Cells should be drawn in such a way as to reflect differences in cell size.

4. When you have finished with a slide, use a Kimwipe and ethanol to wipe all oil off the slide. Examine the microorganisms on the next slide.

Name: _____   Section: _____

Course: _____   Date: _____

EUKARYOTIC CELLS

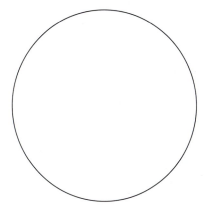

Saccharomyces - 400/450×

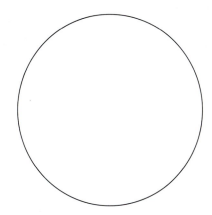

Saccharomyces - 1000×

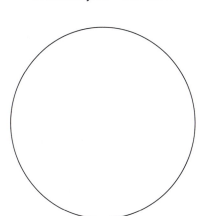

Paramecium - 400/450×

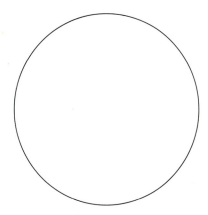

Paramecium - 1000×

PROKARYOTIC CELLS

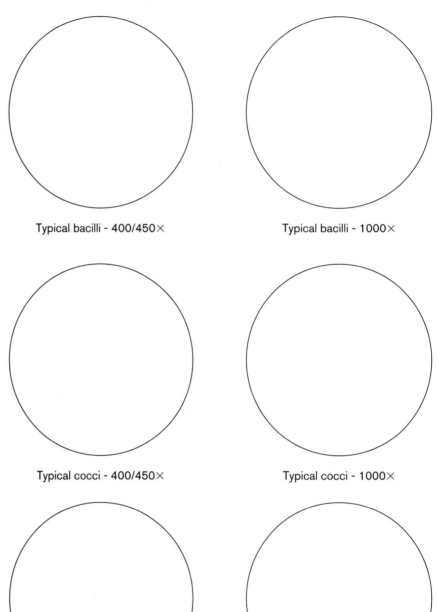

Typical bacilli - 400/450×

Typical bacilli - 1000×

Typical cocci - 400/450×

Typical cocci - 1000×

Typical spirilla - 400/450×

Typical spirilla - 1000×

Name: _____   Section: _____

Course: _____   Date: _____

1. Why won't we be looking at viruses under the microscope in this course? How is your answer related to the wavelengths of light and the maximum magnification of a light microscope?

2. What is the difference between "high resolution" and "high magnification"?

3. If the ocular lens magnification is 5× and the objective lens magnification is 40×, what is the total magnification?

4. When using higher magnification, you need more light. Which parts of the microscope can be adjusted to provide more light, and how would you adjust those parts to increase the light?

5. What is the relationship between changes in depth of field and the need to adjust the focus using only the fine-focus knob?

6.   a. When is it necessary to use a layer of oil between the slide and the lens?

   b. Why is it necessary to use a layer of oil?

   c. What is the oil doing to improve the function of the microscope?

7. Prokaryotic or bacterial cells must be viewed at 1,000× total magnification, so when working with bacteria, you must use the 100× objective lens. Describe two specific things that make it more challenging to use the 100× objective lens than the 10× objective lens. (Your answers should be unrelated to the need for oil because oil issues are covered under Question 6.)

8. Bacteria were first viewed with microscopes in the 1600s, but viruses have been directly observed with microscopes only since the 1930s. Still, microbiologists have known of the existence of viruses since the late 1800s. How did microbiologists know that viruses existed before the 1930s? (You will probably have to do a little research to answer this question.)

LAB 2

Learning Objectives

- Explain the uses, advantages, and disadvantages of broth media and agar media.

- Define the term "sterile" and be able to divide the world around you into "sterile" and "nonsterile."

- Define and describe the concept of "pure cultures" and appreciate the value of such cultures.

- Understand the principles and apply the methods used to transfer and maintain pure cultures.

Working with Culture Media and Pure-Culture Transfer

CULTURE MEDIA

In this course, you will be working with two forms of **media** or nutritive substances: broths and agars.

Broths

Broths are nutrient-containing liquids, and in liquids, cells can grow in three dimensions. Since the cells can grow in all directions and throughout the volume of the liquid, broths are good for several purposes:

1. Producing large numbers of cells

2. Observing the arrangements of cells (pairs, tetrads, chains, clusters, etc.) created as the cells divide in three dimensions

3. Observing and analyzing anaerobic ("no air") metabolic processes, such as fermentation, when cells grow at the bottom of the test tube where there is little or no dissolved oxygen

FIGURE 2.1 Agarobiose.
Agarobiose, a repeating disaccharide subunit in agar, is composed of D-galactose and 3,6-anhydro-L-galactose.

Agars

Agars are nutrient-containing media solidified with **agar**, a complex polysaccharide extracted from a multicellular red-purple algal species (often described as being a type of "seaweed"). The polysaccharide polymer is composed of two types of galactose subunits, D-galactose and 3,6-anhydro-L-galactose. These two types of sugars are joined to form a repeating disaccharide ("two-sugar") subunit called agarobiose (**Figure 2.1**). In contrast, most of the better-known polysaccharides, such as starch and glycogen, are built from glucose subunits. Agars are good for isolating individual bacterial colonies, separating mixtures of bacteria, and investigating colony morphology. Examples of broths and agars are shown in **Figure 2.2**.

So what's so great about agar? This polymer has several properties that make it an ideal gelling agent for the microbiology lab:

1. It is nontoxic to most microbes.

2. It is stable at temperatures required for sterilization (autoclaves typically reach 120°C) and 15 pounds per square inch of pressure.

3. It is physiologically inert and can't be digested or degraded by the vast majority of microbes growing in or on it.

4. Once solid, it doesn't melt or liquefy until heated to about 90°C. As a result, there is no risk that agar-based media will soften or melt when incubated at 37°C, the temperature used to grow most human-associated bacteria.

5. Once melted, agar doesn't solidify until it is cooled to about 40°C–45°C, so live bacteria can be mixed into and throughout the melted agar at temperatures that the bacteria can tolerate (45°C–50°C) until the agar further cools and solidifies.

6. Agar-based media can be poured into petri dishes and test tubes of various sizes. If a tube is tilted while the agar cools, the tilting creates an angled agar plane (called a **slant**) with a greater surface area than in a tube that wasn't tilted. More surface area means more room for microbial growth.

7. Most media use agar at a concentration of 1.5% (15 grams per liter, or per 1,000 ml), but if a softer, semisolid medium is needed, all you have to do is reduce the agar concentration to achieve the desired firmness.

FIGURE 2.2 Various types of microbiological media.

A Solution from the Kitchen

Robert Koch

Lina Hesse

Microbiologists in the late 1800s recognized that physically separating cells on a sterile solid surface was a necessary step in the creation of "pure cultures" (cultures containing just a single microbial species). Dr. Robert Koch, a German country doctor and microbiology pioneer, was using protein-based gelatins to solidify media, but this "Jell-O" technique left much to be desired. Among other problems, (a) the gelatin melted at 37°C, the ideal temperature for growing most human pathogens, and (b) many microbes produced proteases that digested and liquefied the gelatin, converting its protein threads into an amino acid soup.

The solution was provided by Frannie Hesse, the wife of Walther Hesse, one of Koch's lab assistants. Herr Hesse noticed that Frau Hesse was able to create solid jellies and puddings in her kitchen, even in warm weather. He asked how she did it and learned that his wife was using a gelling agent called "agar-agar." This agent, extracted from "seaweed," had been used for centuries in Asian cooking, and by the late nineteenth century, knowledge of the material had spread to some European kitchens, including Frau Hesse's.

Koch tried agar in his lab and found that he could produce a variety of media that remained solid at 37°C. With his sterile solid media in hand, he was able to create pure cultures of many different bacterial species, and he and his assistants went on to make numerous significant discoveries. Koch is remembered as one of the great names in the history of microbiology. But let's not forget Frau Hesse.

As an aside, the petri dish was invented in 1887 by J. R. Petri, another assistant of Robert Koch.

TRANSFERRING CULTURES

The ability to transfer pure-culture material without contamination from one location to another—that is, the ability to inoculate sterile media with pure-culture material—is one of the most important skills that you will learn in an introductory microbiology lab. If you can master this skill—and it's not that hard—then almost everything you do for the rest of the course will work well.

Sterility

The key to success in transferring cultures rests in understanding that the world is divided into "sterile" and "nonsterile." And as you work in the lab, you must be aware at all times of what is sterile and what is not. So what does "sterile" mean?

Sterile means that there are no viable organisms of any kind or in any form living in or on a given object or medium. This includes organisms that are currently inactive (such as a spore, a cyst, or a viral particle outside of a host cell) but which might become active later (by processes such as germination of a spore).

To paraphrase Monty Python's famous Dead Parrot Sketch:

> These cells have passed on! These cells are no more! They have ceased to be! They've expired and gone to meet their maker! They're stiffs! Bereft of life, they rest in peace! Their metabolic processes are now history! They're off the twig! They've kicked the bucket, they've shuffled off this mortal coil, run down the curtain and joined the bleedin' choir invisible!! These are EX-CELLS!!

So what is <u>not</u> sterile? Usually, almost anything you can see and touch in the lab is nonsterile. Benchtops, lab manuals, test-tube racks, the outsides of test tubes, the outsides of petri dishes, all the external surfaces of your body, and even the air are nonsterile. And here's one of the most important rules in microbiology:

> **If a sterile object touches anything that is not sterile, then the object that was once sterile is assumed to no longer be sterile. Do <u>not</u> use this object if sterility is required for success.**

For example, if you flame-sterilize your inoculating loop, but the loop then touches the bench or your hands or the outside of a test tube, or you lick it, then it's not sterile anymore. Flame it again before using it to transfer pure-culture material.

Pure-culture transfer

Throughout the transfer process, the goal is to keep pure cultures pure. Obsessing over what is sterile and what is not will help ensure that you successfully transfer pure-culture material from a tube or plate to a sterile medium where cells will continue to grow as a pure culture. So what is a pure culture?

A **pure culture** is a collection of cells of a single strain or species growing in an environment free from contamination by any other living forms. That is, it is a culture composed of a single species of microorganism growing in isolation from any other species. Almost everything we do in the microbiology lab requires pure cultures, so contaminated or mixed-species cultures are usually worthless. You want to do well in microbiology? Then follow this rule when transferring pure-culture material:

> **During the transfer, if pure-culture material touches anything nonsterile, then the pure culture is assumed to be contaminated. It is no longer pure. If this happens, start over. <u>Do not use contaminated material.</u>**

In this course, you'll learn several ways to transfer pure cultures and keep them from getting contaminated. In this particular lab, you'll use flame-sterilized, metal wire **inoculating loops** to do the transfers, and if you can correctly handle a loop in the microbiology lab, then you will go far. So, pay very close attention to the text description and photos included below in the "General protocol" section of the lab procedures.

DESCRIBING BACTERIAL GROWTH

Once you have a pure culture of a given species, you can observe the characteristics of that species. These characteristics include patterns of growth, which can be observed in liquid broths, and colony morphology, which can be observed on solid media.

Some useful terms for characterizing growth in liquid broths are given in the following table and illustrated in **Figure 2.3**.

Terms	Occurs When
Turbid	Cells are evenly suspended to produce a uniform cloudiness
Flocculent	Cells are floating in visible, separated clumps called "flocs"
Pellicle	Cells form a thick layer on the top of the broth; often due to high lipid content
Sediment	Cells sink to the bottom of the tube to form a loose pellet

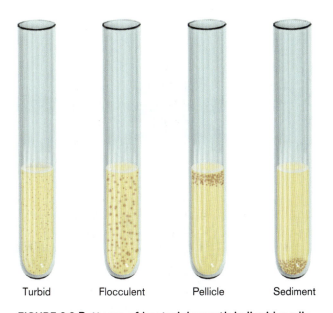

| Turbid | Flocculent | Pellicle | Sediment |

FIGURE 2.3 Patterns of bacterial growth in liquid media.

Colony morphology refers to the characteristics of the visible growth of expanding populations of microbes on solid and semisolid media. These traits, which can be seen without a microscope, include the color, size, shape, elevation, and surface features of the isolated colonies of a given species.

Surfaces of microbial colonies may be described as:

Smooth
Glistening
Rough or matte
Wrinkled
Dry or powdery

Profiles of various colony morphologies are identified in **Figure 2.4**.

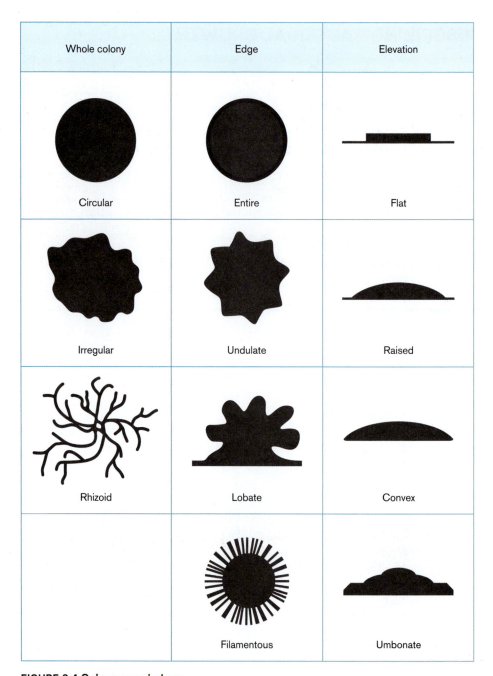

Whole colony	Edge	Elevation
Circular	Entire	Flat
Irregular	Undulate	Raised
Rhizoid	Lobate	Convex
	Filamentous	Umbonate

FIGURE 2.4 Colony morphology.
Terms to be used to describe colony morphology.

PROTOCOL

General protocol for transferring pure cultures from a test tube to sterile tube media

Follow the steps below whenever you need to transfer pure-culture materials from a test tube to any type of sterile tube medium, including during today's lab. These techniques are absolutely essential to success in any microbiology lab course, and once you've mastered these techniques, things should generally work well in this lab course. Learn to use the loop properly, and...

> *The road will rise to meet you,*
> *The wind will be always at your back,*
> *The sun will shine warm upon your face,*
> *The rains will fall soft upon your fields.*
>
> —Traditional Irish blessing

Figures 2.5–2.10 illustrate how to perform steps 1–10.

1. Use the striker to light the Bunsen burner. Pick up the inoculating loop between your thumb and first two fingers, and flame the loop until all parts of the entire length of the wire have briefly glowed red at some point in the flaming process **(Figure 2.5)**. This will take less time if you hold the loop at an angle close to parallel to the flame, but do not use an angle so close to parallel that it will inflict third-degree burns on your finger.

   Once the loop has been flamed, all parts of the wire that glowed red are now sterile. Now don't forget the rule: If a sterile object touches anything nonsterile, then the once sterile object is assumed to be nonsterile. Do not use the object if sterility is required for success. So, at this point, do not allow the loop to touch anything; don't put it down on the bench, don't touch the wire with your fingers, and don't lick the loop.

2. While still holding the freshly sterilized loop between your thumb and first two fingers (don't put the loop down!), use your nondominant hand to pick up the tube with the pure-culture material, and using your dominant hand, remove the cap with your pinkie finger so that the opening of the cap is pointing down **(Figure 2.6)**.

   Remember that the air in the lab is nonsterile, so keeping the cap opening pointing down reduces contamination, especially contamination from airborne fungal spores. Also, remember that the lab bench is crawling with "wee beasties," so instead of putting the cap down on the bench, you should do all of the following steps while holding the cap

FIGURE 2.5
The inoculating loop is sterilized by heating it in the flame until it glows red. Note the angle at which the loop is held; emulate this angle during loop sterilization and be sure to sterilize the entire length of the wire before use.

FIGURE 2.6
Caps should be removed from the tubes with your pinky finger and held with the open end of the cap pointing downward. Continue to hold the cap in the crook of your pinky until it is time to put the cap back on the tube.

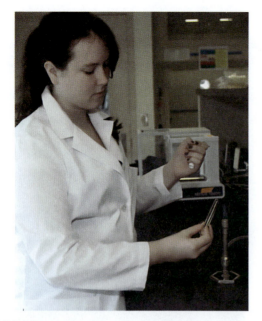

FIGURE 2.7
Sterilize the mouth of the test tube by briefly passing it through the flame. This also helps to prevent entry of airborne microbes.

FIGURE 2.8
Insert the loop into the culture to retrieve a sample of bacterial cells. In this step, the loop should touch only the culture and the inside walls of the test tube.

in your pinkie until it's time to put it back on the tube.

3. Briefly flame the lip of the glass tube **(Figure 2.7)**. This will kill any microbes that may have been transferred to the lip when you removed the cap, and it will create air currents that help to keep out airborne bugs (remember, hot air rises).

4. The sterile loop can now be inserted into the pure-culture tube. If the culture is in a broth, the loop can be immediately inserted into the broth, where it will cool quickly **(Figure 2.8)**. If the tube contains cells growing on agar, give the sterile loop about 15 seconds to cool before you stab the loop into the agar to complete the cooling process.

5. Once the loop has been inserted into the pure-culture tube, the next step is to get cells on the loop.

   In the case of broths, moving the loop into a cloudy part of the tube should do the job. If the cells are well mixed or suspended throughout the broth, any part of the broth will do.

   If cells are on agar slants, touch the loop to the cell layer and plenty of cells should be transferred to the loop. Keep in mind that cells near a stab made by a hot loop may have been killed by the heat of the loop, so touch a part of the slant that is away from the stab.

6. After picking up cells from the tube containing the pure culture, carefully withdraw the loop in such a way that it does not touch anything except the inside of the tube.

The inner wall of the tube will either be cell-free or will contain the same type of cells as the culture to be transferred, so it's OK to touch it with cell-filled loops. But any surface outside of the tube must be assumed to be contaminated, and if the loop touches anything other than the inner wall of the tube, the pure-culture material is no longer pure, and you should go back to step 3.

7. Once the loop is out of the pure-culture tube, while **holding the loop in such a way as to avoid contact with any surface**, flame the lip of the pure-culture tube and put the cap back on the tube, freeing your pinkie for new and exciting things.

8. Now, pick up the tube of sterile broth or agar that is the destination for your transferred cells, remove the cap with your talented pinkie **(Figure 2.9A)**, and briefly flame the lip of this second tube **(Figure 2.9B)**.

9. To transfer cells from the loop to new media, insert the cell-covered loop into the tube of sterile broth or agar. Broths are inoculated by briefly swirling the loop in the broth to wash off some cells **(Figure 2.10A)**. If just one cell is transferred into the broth, within 24 hours, you will have billions of cells of the given type to play with, thanks to the exponential nature of cell division. Slants are inoculated by gently brushing the surface of the agar for the length of the slant.

10. Once the new medium is inoculated, withdraw the loop, briefly flame the lip of the tube, place the cap back on the tube, thank your pinkie, and flame the loop until every part of the entire length of the wire glows red for at least a brief moment to kill remaining cells. You're done.

A.

B.

FIGURE 2.9
A. After cells have been transferred to your loop, replace the cap on the culture test tube.
B. Open a tube of sterile broth following the same technique that you used to open the tube with the culture. Sterilize the mouth of this test tube, too, by briefly passing it through the flame.

FIGURE 2.10
Insert the bacteria-covered loop into the test tube without touching any of its exterior surfaces, and transfer cells by touching the sterile broth with the loop.

Day 1
Materials

- Melted brain-heart infusion (BHI) deeps (3 test tubes per student)
- Glucose nutrient broths (GNB) (4 test tubes per student)
- Bunsen burner
- Striker
- Inoculating loop
- Masking tape or label tape
- Pens or pencils for labeling tubes
- Test-tube racks
- 47°C water bath

Cultures:
- *Escherichia coli*
- *Micrococcus luteus*
- Soil sample

TRANSFERRING PURE CULTURES TO SLANTS

1. Remove three test tubes filled with melted brain-heart infusion agar from the 47°C water bath (the water bath keeps the agar in a melted state until you're ready to use the tubes). As soon as the tubes are removed, place them in test-tube racks at an angle.

   You can hold the tubes at an angle by putting them through two offset tube rack holes or by tilting the whole rack against a fixed object. The angle should be sufficient to create a slanted surface of about 2 inches in length, but not so great as to spill the melted agar.

2. Allow about 15–20 minutes for the agar to cool and solidify. While you are waiting, you can work on other assigned tasks. The tubes should not be inoculated until the agar is firm and the surface remains slanted when you hold the tube in a completely upright position.

3. When the slants are ready, label each slant. Labels should include your name, the date, and a description of the contents of the tube (*E. coli*, *M. luteus*, or uninoculated).

4. Using the techniques described above, transfer a pure culture of *E. coli* to the first tube and a pure culture of *M. luteus* to the second tube in such a way as to keep the pure cultures pure. The third BHI tube should not be inoculated. This tube is an uninoculated control and should be incubated along with the *E. coli* and *M. luteus*–inoculated slants.

5. Place all tubes in the racks designated for incubation of class material. Tubes will be incubated at 28°C–30°C for 48 hours.

TRANSFERRING PURE CULTURES TO BROTHS

1. Label four tubes of glucose nutrient broth. Labels should include your name, the date, and a description of the tube's contents:

   a. *E. coli*
   b. *M. luteus*
   c. Soil
   d. Uninoculated control

2. Using the techniques described above, transfer a pure culture of *E. coli* to the first tube and a pure culture of *M. luteus* to the second tube in such a way as to keep the pure cultures pure. The third glucose nutrient broth tube should be inoculated with about a cubic centimeter (about 0.40 cubic inches) of soil. The fourth tube should not be inoculated; this tube is an uninoculated control.

3. Place all tubes in the racks designated for incubation of class material. Tubes will be incubated at 28°C–30°C for 48 hours.

Day 2
Examine all slants and broths from day 1 and record your observations in the Results section. Descriptions can and should use the terms given in the introductory section under "Broths" and "Agars" in addition to any other terms you choose to use, such as "yellow."

RESULTS

Name: _____   Section: _____

Course: _____   Date: _____

DESCRIPTION OF GROWTH ON AGAR SLANTS

1. *E. coli*

2. *M. luteus*

3. Uninoculated control

DESCRIPTION OF GROWTH IN BROTHS

1. *E. coli*

2. *M. luteus*

3. Soil

4. Uninoculated control

Name: _____   Section: _____

Course: _____   Date: _____

1. In this course, you need to be aware of what is sterile and what is not sterile at all times! For each of the following, note whether the object or surface is sterile or nonsterile.

    a. Benchtop

    b. Test-tube rack

    c. Hands and fingers

    d. Air in the microbiology lab

    e. Inner surfaces of uninoculated test tube

    f. Notebook and lab manuals

    g. Clothing

    h. Flamed loop

    i. Outer surfaces of uninoculated test tube

    j. Tongues

    k. Pens and pencils

    l. Inner surface of your nose and resident nose goblins

2. How could you tell if your *E. coli* slant was contaminated by *M. luteus*?

3. Would you describe the contents of the soil-inoculated broth as being a "pure culture"? Why or why not?

4. How did the uninoculated broth differ in appearance from the broths inoculated with *E. coli* and *M. luteus*? Based on your observations, how could you tell if a supposedly sterile, uninoculated broth was contaminated?

5. Later in this course, we will use a medium called phenol red glucose broth. When *E. coli* grows in this broth, the medium turns yellow, while *M. luteus* turns the medium red. With this in mind, explain the following outcome:

   A student used a loop to transfer *E. coli* cells from a broth culture to a tube of sterile phenol red glucose broth. The student then flamed the bottom half of the loop wire and used the loop to transfer *M. luteus* cells from a broth culture to a second tube of sterile phenol red glucose broth. A day later, both phenol red glucose broths were yellow in color. What happened here? What lesson can you learn from this?

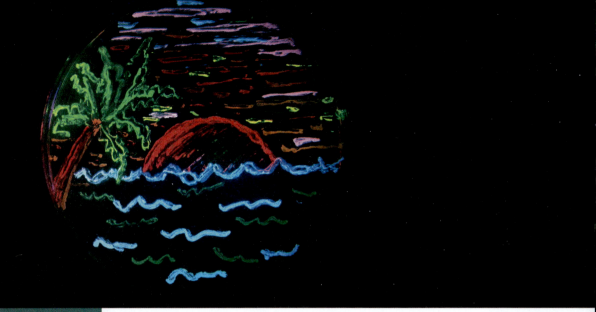

LAB 3

Creating and Storing Pure Cultures

Learning Objectives

- Understand that physical separation of cells is the critical step in creating pure cultures.

- Know and apply the concepts and methods involved in creating pure cultures by streak and pour plate procedures.

- Consider the advantages and disadvantages of streak plating and pour plating.

- Describe and understand the methods used to preserve a pure culture for months or years.

CREATING PURE CULTURES WITH AGAR-BASED MEDIA

In the real world, microbiologists often analyze environmental and medical samples that contain a mixture of microbial species. For example, when identifying the cause of an intestinal illness, one might start with fecal material that contains a **mixed culture** of dozens of human-associated bacterial and fungal species. How do we find a single-species "needle" in the microbial "haystack"? In short, we do this by creating pure cultures using the techniques described below.

Pure cultures are enormously, stupendously useful in microbiological procedures such as the identification and characterization of bacterial species. Therefore, many procedures begin with the creation of pure cultures, **a process that usually requires the physical separation of cells on a solid medium**. Once separated, if properly "fed and watered," a single cell will divide and divide and divide until there are millions of cells piled together to form a visible colony. Since each cell was derived from the original, isolated cell, this process generates a visible colony of genetically identical clones. In other words, as long as there is no contact with other colonies, the colony itself is a "pure culture" or "pure strain," a culture containing just one species. If cells from that colony can be transferred to separate sterile media without contact with any nonsterile surfaces, including other colonies, then the cells can continue to multiply in their new home to create a pure-culture stock.

That's all there is to creating pure cultures. All you really need is solid medium, separated cells, and a steady hand to transfer the cells. While there are a number of ways to isolate cells on solid media, all the methods are variations on the theme of physically separating cells. In this lab, you will use two methods, streak plating and pour plating, to isolate cells that will grow to produce isolated colonies and pure cultures.

Streak plating

In **streak plating**, you will use your inoculating loop to drag and physically separate cells on the surface of an agar plate. If all goes well, you will finish with cells that are separated by at least several millimeters. Then, when these cells divide, they will produce well-separated colonies that can be used to create pure cultures.

Success in this lab course is critically dependent on your ability to master streak plating techniques. A well-streaked plate will bring you much joy, but a poorly streaked plate leads only to despair.

1. Touch a flame-sterilized and cooled loop to a source of cells. Use the loop to transfer the cells to the surface of an agar plate by gently making three or four streaks on the plate to make a first set of streaks **(Figure 3.1)**. The loop should remain in contact with the agar surface as you move it back-and-forth. The streaks should form the chord of an arc that is about 75 degrees of the 360-degree perimeter of the plate and should cover a maximum of about 10%–15% of the total surface area of the plate. Note that the loop should be positioned at a shallow angle to reduce the chances of digging into the agar. While agar is solid, the stuff ain't exactly rock-hard.

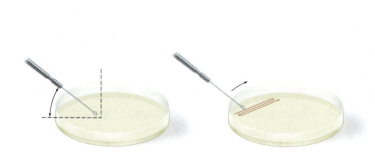

FIGURE 3.1 Streaking a plate to isolate bacteria: first streak.
Pay particular attention to the shallow angle of the loop as it is gently moved across the agar.

2. After finishing the first streaks, flame the loop to kill all the cells remaining on the loop. The first set of streaks will have deposited plenty of cells on the plate, and remember, our goal is to reach a point in the streaking where there are so few cells at a given place on the plate that these cells are several millimeters apart from one another. Since you do not want to add more cells to the plate, do not go back to the original source of cells for more loopfuls. Don't even think about it.

3. Once the loop has cooled, make a second set of three or four more back-and-forth streaks on the plate as shown in **Figure 3.2**. These streaks should cross through the ends of the previous streaks. This will pull some of the cells from the first set of streaks into the second set of streaks. But since you are only dragging a fraction of the cells from the first set of streaks into the second set of streaks, the cells should be increasingly farther apart from each other.

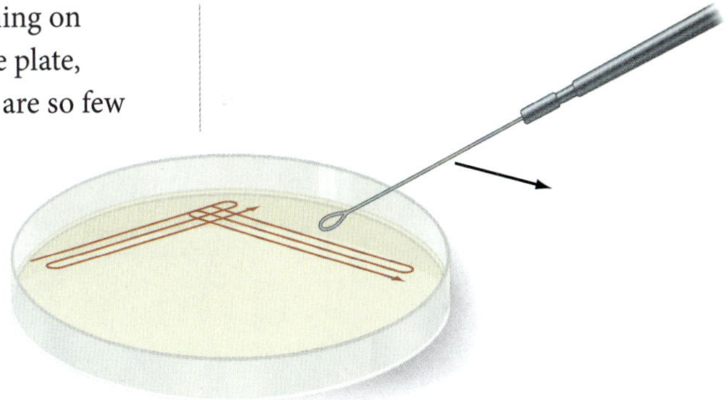

4. After finishing the second set of streaks (step 3), flame the loop again to kill any cells picked up by the loop. Again, the goal is to spread fewer and fewer cells farther and farther apart. So kill the cells on the loop, let the loop cool, and make a third set of back-and-forth streaks by dragging the loop through the ends of the second set of streaks **(Figure 3.3)**. This will pull some of the cells from the second set of streaks into the third set of streaks.

5. After finishing the third set of streaks (step 4), flame the loop again, let the loop cool, and make a fourth set of back-and-forth streaks by dragging the loop through the ends of the third set of streaks **(Figure 3.4)**.

6. After finishing the fourth set of streaks, flame the loop again, let it cool, and finish streaking with a single squiggle running from the ends of the fourth set of streaks into the unused middle of the plate **(Figure 3.5)**. Incubate the plate and hope for the best.

So what have you done? If you could magnify the plate about 1,000 times, you would see that the cells are very close together at the beginning of a streak and farther apart at the end of the streak **(Figure 3.6)**. Without magnification, you can't tell where along the streaks you will find individual cells separated by at least several millimeters, but we hope that you will find that this point exists somewhere along your streaks. Once the colonies appear, you'll know if you succeeded in separating the cells **(Figure 3.7)**.

7. With 18 to 24 hours of incubation, the cells will have divided enough to produce visible colonies. At points along the streaks where the cells were still very close together, there will be no separate colonies, just solid sheets of slime. But at points where the cells were well separated by streaking, there will be clearly isolated colonies composed of a single clone or single type of cell. These colonies are pure cultures, and if sampled with care (without contacting other colonies), cells from a separated colony can be transferred to tube or plate media to create a pure-culture stock.

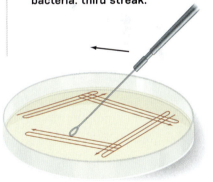

FIGURE 3.2 Streaking a plate to isolate bacteria: second streak.
Don't forget to sterilize the loop between the first and second streaks.

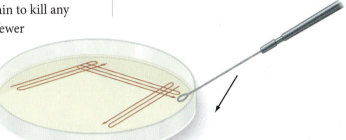

FIGURE 3.3 Streaking a plate to isolate bacteria: third streak.

FIGURE 3.4 Streaking a plate to isolate bacteria: fourth streak.

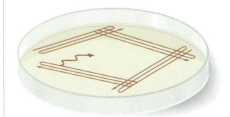

FIGURE 3.5 Streaking a plate to isolate bacteria: the final step.

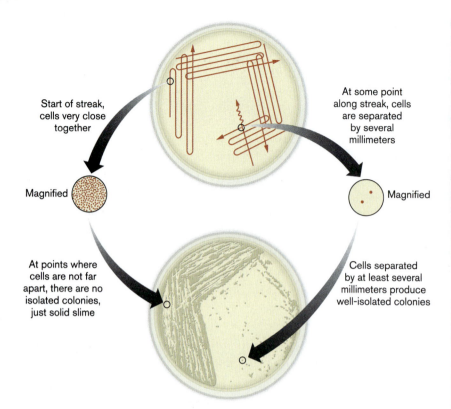

Start of streak, cells very close together

Magnified

At some point along streak, cells are separated by several millimeters

Magnified

At points where cells are not far apart, there are no isolated colonies, just solid slime

Cells separated by at least several millimeters produce well-isolated colonies

FIGURE 3.6 Streaking a plate to isolate bacteria: effects.

FIGURE 3.7 Streaked plates with isolated colonies.

Pour plating

Pour plating is another way to physically separate cells to produce isolated colonies. In this case, cells will be added to melted agar, the agar will be poured into a plate, and colonies will appear at points where the cells were trapped in the agar. Pour plating is more time-consuming, but this method has some advantages over streak plating. For example, pour plating usually produces a greater number of well-isolated colonies than streak plating. In addition, later in the course, we'll see that pour plating can also be used to determine the concentration of viable cells in a unit volume of liquid, such as the number of cells in a milliliter of milk.

The trick to using pour plating to create pure cultures is to find a dilution with the right number of cells in a given volume of melted agar. If you mix in too many cells, the cells won't be far enough apart when the agar solidifies. And if you don't mix in enough cells, you might not find any colonies at all. This is a Goldilocks problem: You don't want too many cells mixed in and you don't want too few. You want a number that is just right.

Of course, when you look at a tube full of cells, you cannot see the individual cells without magnification. So how can you tell how many cells are in a given tube of melted agar prior to pouring the agar into a plate?

The solution to this problem is **serial dilution**, a stepwise dilution procedure in which a given dilution in the series is used as the source of cells for the next dilution in the series. You may not know how many cells you have to begin with, but if you make several dilutions of cells, then each subsequent dilution will have a different and decreasing number of cells. If you make enough dilutions, then somewhere in that dilution series, you should find a tube with the right number of cells to produce a plate with tens to hundreds of isolated colonies per plate.

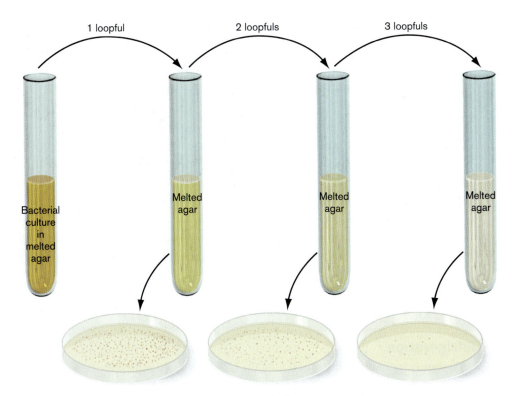

1 loopful   2 loopfuls   3 loopfuls

Bacterial culture in melted agar

Melted agar

Melted agar

Melted agar

FIGURE 3.8 Pour plate method: loop dilution technique.

There are several different ways to do a dilution series, but we will start with one of the simplest, the **loop dilution technique (Figure 3.8)**. In this case, the dilutions are created by transferring one to two loopfuls of cells from one tube of melted agar to the next in a series of three or four tubes. Each subsequent tube in the series will have a lower density of cells than the previous tube. After dilution, the melted agar in each tube is poured into plates, and if we're lucky, one of the plates will have tens to hundreds of isolated colonies.

When using the loop dilution technique, there are a couple of things that you should pay close attention to as you follow the procedure:

1. Melted agar is kept in a water bath at 47°C, a temperature that's cool enough for the cells to survive, but not so cool that the agar solidifies before it can be poured. When the tubes are removed from the bath, there will be drops of nonsterile water on the outsides of the tubes. Be certain to wipe the outsides of the tubes dry before pouring their contents into a plate, or drops of nonsterile water may drip into the plates along with the agar. Drops of nonsterile water in your plates would be bad.

2. At 47°C, the agar is melted, but it's close to the temperature at which it will solidify. When you remove the tubes from the bath, the agar will immediately begin to cool, and if you don't move fast, there will be chunks of solid agar in the tube when you pour it into your plate. This creates "extra chunky" or "home style soup" plates. Not aesthetically pleasing. So, before removing the tubes from the water bath, read all the steps in the protocol a couple of times to be sure you understand what you are doing, be certain that all plates are labeled, and don't just stand there as the agar cools in the tubes.

To be clear, while not ideal, chunky plates will still do the job of separating cells. If you produce a lumpy plate, do not repeat the loop dilution protocol. A limited number of melted agar tubes will be prepared for this lab, and repeating the procedure will exhaust the supply of tubes before every student in the lab has a chance to do the experiment.

PRESERVING BACTERIAL CULTURES

Once a pure culture has been created, there are often numerous reasons for wanting to preserve that culture for an extended time. Almost all preservation methods are designed to slow or stop cell metabolism and cell division. That is, the goal is to achieve a state of **bacteriostasis**, a condition in which the cells are still alive, but inactive. Growing cells cannot be stored for long periods because active cells produce toxic wastes and eventually exhaust their supply of nutrients. So, if growth continues, all the cells will eventually die. And a dead culture has limited value. There are two basic approaches to slowing the death rate of bacterial cultures. The first is to lower temperature, thereby decreasing metabolic functions. The second option is to remove water from the culture.

Refrigeration

Refrigeration uses temperatures of 0°C–4°C (32°F–40°F) to significantly slow cell activity, and many cultures can be stored at these temperatures for a month or two. However, these temperatures may not stop growth entirely, and cultures of some species will die off long before 2 months have passed.

Strict aerobes (species that require oxygen) are usually stored in the refrigerator on the surfaces of agar slants. Facultative anaerobes can use oxygen, but don't require it, so these organisms can be stored by stabbing an agar slant to push the cells deep into the agar; cells embedded in agar are much less likely to suffer death by drying. Strict anaerobes are killed by the presence of oxygen, and these cells must be covered by agar, oil, or some other material that reduces their contact with oxygen. As we will see in Lab 14, reducing agents such as thioglycollate are often added to media used for strict anaerobes in order to create an environment with little or no free oxygen (O_2).

Freezing

Effective **freezing** requires temperatures lower than −20°C, and many labs use freezers set at −70° to −80°C for long-term culture storage. Temperatures this low will stop all metabolic activity, and many species can be held in a form of suspended animation for many years. Though metabolically inactive, the cells are still viable and will grow again upon thawing. On the downside, freezing can be fatal to cells if the formation of large ice crystals rips apart delicate cell structures such as cell membranes. So cultures are routinely mixed with cryoprotective agents such as glycerol and litmus milk to reduce damage to the cells.

Lyophilizing

Lyophilizing, or "freeze-drying," is a dehydration process that requires freezing of the material to be preserved, followed by the removal of water by sublimation under low atmospheric pressure. Cells may be suspended in milk or blood serum to reduce mechanical damage from the freeze-drying process. **It's the removal of the water that is the key**, because almost all biological and metabolic chemistry of the cell occurs in the presence of water. So no water, no metabolism. After the water is removed, the cells can be stored for many, many years until they are reanimated by the addition of water. While lyophilized cultures are often stored at cold temperatures, this isn't essential, because it's the dehydration that counts.

PROTOCOL

Day 1
Materials

- Brain-heart infusion (BHI) agar plates (3 per student)
- Bunsen burner
- Striker
- Inoculating loop
- Wax pencils or Sharpies for labeling plates (or masking tape and pens)
- Mixed culture of *Escherichia coli*, *Micrococcus luteus*, and *Serratia marcescens*

STREAK PLATES

1. Using the streak plating technique described in the lab introduction, label and streak three BHI agar plates with the mixed culture. Again, use the mixed culture, which contains three bacterial species, for all of the streak plates.

   Remember, your goal is to separate the cells of the three bacterial species. The most common mistakes students make with this technique are (1) failing to flame the loop between the sets of streaks and (2) failing to go back through the end of the previous streaks when doing the next set of streaks.

2. Invert the plates and incubate them at 28°C–30°C for 48 hours.

   Agar plates should always be incubated upside down (agar side on top, lid on the bottom). If plates are not inverted, water vapor will condense on the inner surface of the empty upper lid. This water can drip onto the surface of the agar, and then bacteria from the separate colonies will mix together via the water drop bridge. Mixing means no pure, isolated colonies. By contrast, if the plate is incubated with the agar side on top, much of the water vapor within the plate will be absorbed by the agar. Drops of water are less likely to form, and, obviously, they can't drip down onto the agar surface if that surface is on top.

Day 2
Materials

- Streaked BHI plates from day 1
- Glucose nutrient broths in test tubes (3 per student)
- Melted brain-heart infusion agar in test tubes at 50°C (3 per student)
- Sterile petri dishes (3 per student)
- Bunsen burner
- Striker
- Inoculating loop
- Wax pencils or Sharpies for labeling plates
- Masking tape or label tape
- Pens or pencils for labeling tubes
- Mixed culture of *E. coli*, *M. luteus*, and *S. marcescens*

STREAK PLATES

1. Observe the BHI plates for isolated colonies and record your observations in the Results section. Your lab instructor will be looking at your plates, and he or she may make suggestions for increasing the number of isolated colonies.

   E. coli colonies are white to pale tan in color, *M. luteus* colonies are light yellow, and *S. marcescens* colonies are red. Depending on the relative number of *E. coli*, *M. luteus*, and *S. marcescens* cells in the initial mix, you may not see three different, distinct types of colonies, but you should have at least some colonies that are clearly, physically separated from the others.

2. Each isolated colony contains a single type of cell. To create pure cultures, select a total of three different, well-isolated colonies from any of your plates for transfer to sterile broths following the steps given below. Colonies can come from any plate, but if available, pick colonies of three different types (tan *E. coli*, yellow *M. luteus*, red *S. marcescens*). Record a description of each selected colony in the Results section.

3. To transfer each isolated colony, start by flaming an inoculating loop. Give the loop 15–20 seconds to cool, and then gently stab the loop into an area of the agar plate where there are no colonies nearby to finish the cooling. It's acceptable to cool a loop by stabbing into agar, but this must be done without any contact with any of the bacterial colonies on the plate, because after contact, the loop would no longer be sterile, right?

4. With your sterile, cool loop, gently touch one of the isolated colonies to transfer cells to the loop. The loop should touch one and only one isolated colony. Don't forget, each colony contains only one cell type and thus is effectively a tiny pure culture. But if you touch more than one colony, you can't be sure that you have only one cell type on the loop. The

skills required here are similar to those needed to play the kid's game "Operation."

5. Inoculate a glucose nutrient broth with the cells on your loop using the techniques described in Lab 2 for transfer of pure-culture material to broths (remove cap with pinky, flame lip of tube, insert loop without contacting nonsterile surfaces, swirl loop in broth, replace cap, and flame loop). Label the inoculated broth tube with your name, date, and the source or color of the colony.

6. Repeat steps 3 to 5 a total of three times, transferring cells from a total of three isolated colonies to separate tubes of glucose nutrient broth.

7. Incubate the tubes at 28°C–30°C for 48 hours.

POUR PLATES

1. Use a wax pencil or Sharpie permanent marker, or masking tape and pens, to label the bottoms of each of three empty petri dishes with your name, the date, and the word "Mixed." Then label one plate "Dilution 1," another plate "Dilution 2," and the third plate "Dilution 3."

2. Label one piece of tape "Dilution 1," another piece of tape "Dilution 2," and a third piece of tape "Dilution 3."

3. Review the loop dilution technique described under "Pour plates" in the introduction to this lab. Note that melted agar cools and solidifies rapidly, so be sure that you understand the procedure before you remove the melted agar from the water bath.

4. When you're ready, remove three melted brain-heart infusion agar tubes from the water bath. Use the labels from step 2 to quickly mark these tubes "Dilution 1," "Dilution 2," and "Dilution 3" with the pre-labeled tape.

5. Using a flame-sterilized loop, transfer one loopful of cells from the mixed culture broth to the melted agar tube labeled Dilution 1.

6. Use the loop to mix the cells throughout the agar by briefly stirring the melted agar with the loop. Be sure to move the loop up and down from the top to bottom of the melted agar to distribute the cells throughout the medium.

7. Transfer two loopfuls of cells from the Dilution 1 melted agar tube to the Dilution 2 melted agar tube and mix briefly as in step 6.

8. Transfer three loopfuls of cells from the Dilution 2 melted agar tube to the Dilution 3 melted agar tube and mix briefly as in step 6.

9. When the transfers are done, remember to flame the loop before putting it down on the benchtop.

10. Pour the contents of each tube into the appropriate empty sterile petri dishes, that is, pour the Dilution 1 tube into the Dilution 1 dish, and so on.

The inner surfaces of both halves of the petri dishes are sterile; so don't touch either inner surface with fingers or with anything else that is nonsterile. Do not put the lid on the benchtop when you pour the agar into the plate. Replace the lid as soon as the agar is poured to reduce the chances of contamination by microbes in the air.

11. Let the agar cool for 10–15 minutes or until solid. Make sure the agar is fully solidified before you move the plates.

12. Incubate the plates in an inverted position at 28°C–30°C for 48 hours.

Day 3
Materials

- Pour plates from day 2

- Broths inoculated on day 2 (pure cultures from streak plates)

- Nutrient agar or tryptic soy agar (TSA) plates (3 per student)

STREAK PLATES

1. Using the broths inoculated on day 2 with colonies from the day 1 streak plates, make one streak plate for each broth. To prepare these streak plates, use the same techniques you used to prepare the streak plates on day 1 with a goal of obtaining isolated colonies.

2. Label the plates and incubate them in an inverted position at 28°C–30°C for 48 hours.

POUR PLATES

Observe the day 2 pour plates and record your observations in the Results section. Note differences in form between colonies growing on the surface of the agar and colonies embedded in the agar. Note differences in the relative numbers of colonies at the different loop dilutions, and compare the number of isolated colonies found on pour plates and on streak plates.

Day 4
STREAK PLATES

Observe the day 3 streak plates and record your observations in the Results section. Compare these streak plates with the ones prepared on day 1. Be sure to note the change in the number of different colony types on these two sets of plates.

Name: _____   Section: _____

Course: _____   Date: _____

1. STREAK PLATES

a. Mixed-culture streak plates (streaked on day 1, observed on day 2)

Plate 1

Approximate number of isolated colonies:

Description of different colony types:
(May not see all colony types on given plate)

Colony type 1: _____

Colony type 2: _____

Colony type 3: _____

Plate 2

Approximate number of isolated colonies:

Description of different colony types:
(May not see all colony types on given plate)

Colony type 1: _____

Colony type 2: _____

Colony type 3: _____

Approximate number of isolated colonies:

Description of different colony types:
(May not see all colony types on given plate)

Colony type 1: _____

Colony type 2: _____

Colony type 3: _____

Plate 3

b. Description of colonies selected to be transferred from the streak plates to broths (observed on day 2)

Colony type 1: _____

Colony type 2: _____

Colony type 3: _____

c. Streak plates from pure cultures in broths (streaked on day 3, observed on day 4)

Approximate number of isolated colonies: _____

Description of different colony types:

Plate 1

Approximate number of isolated colonies: _____

Description of different colony types:

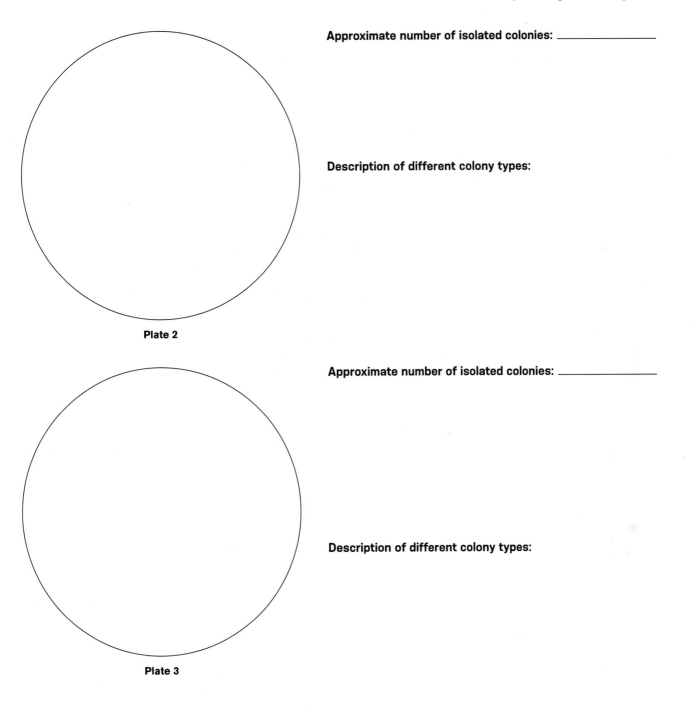

Plate 2

Approximate number of isolated colonies: _____

Description of different colony types:

Plate 3

2. POUR PLATES: MIXED CULTURE (POURED ON DAY 2, OBSERVED ON DAY 3)

How does the appearance of <u>embedded</u> colonies differ from that of <u>surface</u> colonies?

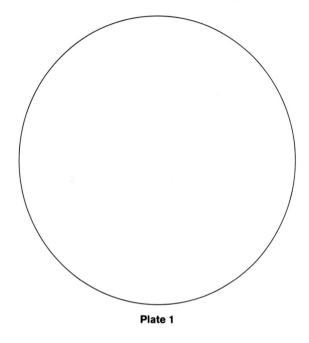

Plate 1

Approximate number of isolated colonies: _____

Description of different colony types:
(May not see all colony types on given plate)

Colony type 1: _____

Colony type 2: _____

Colony type 3: _____

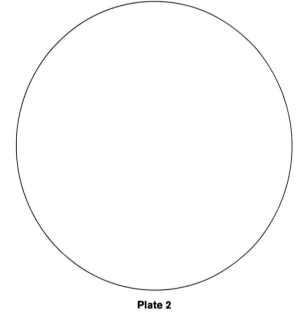

Plate 2

Approximate number of isolated colonies: _____

Description of different colony types:
(May not see all colony types on given plate)

Colony type 1: _____

Colony type 2: _____

Colony type 3: _____

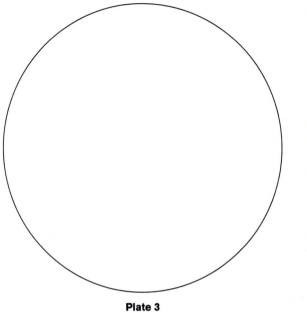

Plate 3

Approximate number of isolated colonies: _____

Description of different colony types:
(May not see all colony types on given plate)

Colony type 1: _____

Colony type 2: _____

Colony type 3: _____

Name: _____   Section: _____

Course: _____   Date: _____

1. Both streak plating and pour plating produce isolated colonies. What is the underlying explanation for why both methods work; that is, what are both methods doing with respect to the bacterial cells?

2. No doubt your streak plates produced many isolated colonies, but if a streak plate <u>fails</u> to produce isolated colonies, describe two things that you could do to improve your chances of generating isolated colonies.

3. Why do we care so much about producing isolated colonies? What is an isolated colony composed of? What can you do with an isolated colony?

4. How did the mixed-culture streak plates (streaked on day 1) and broth-culture streak plates (streaked on day 3) differ from each other in terms of the **number of colony types?** Explain why there is a difference.

5. Based on the differences in appearance between mixed-culture streak plates and broth-culture streak plates, what would lead you to believe that you created a pure culture when you inoculated the broths (on day 2)?

6. When you used the loop dilution technique to produce pour plates, did all the plates produce isolated colonies? If your goal was to ultimately create pure cultures of *E. coli*, *M. luteus*, and *S. marcescens*, which plate or which dilution would you use, and why would you use this plate?

7. What are the advantages and disadvantages of streak plating compared with pour plating? (Think about ease of use, number of isolated colonies, etc.)

LAB 4

Quantitative Plate Counts

Learning Objectives

- Compare different methods for determining bacterial cell populations.

- Understand the basic principles and concepts related to serial dilutions.

- Use and understand the techniques required to calculate bacterial cell counts.

- Consider the sources of error in quantitative spread plate counts.

SURVEY OF CELL COUNTING METHODS

There are numerous reasons for determining the density of bacterial populations or the number of cells per unit volume. Many methods have been developed to meet this need, some of which are described below.

Centrifugation

Cells in suspension can be spun down into a pellet in a centrifuge tube. The volume or dry weight of the pellet is proportional to the number of cells per unit volume in the original suspension.

Turbidity

A quicker way to estimate bacterial cell densities is to measure the turbidity or cloudiness of a suspension of cells by measuring the absorbance of the suspension with a spectrophotometer (Figure 4.1). A **spectrophotometer** sends light beams through a standard-sized tube of liquid, then measures the amount of light that reaches a detector on the other side of the tube. Because cells deflect, reflect, and absorb the light passing through a tube containing a cell

FIGURE 4.1 Spectrophotometer.

suspension, absorbance is roughly equal to, and can be correlated with, cell density. This method is used in Lab 13, where we examine the effects of pH changes on bacterial growth.

Direct microscopic counts

Cell density can also be determined by counting the number of cells within grid lines marked on a **hemocytometer slide** (a type of microscope slide) **(Figure 4.2)**. Since we know the length and width of the squares formed by the grid lines and the depth or distance between the coverslip and the slide, we have the information that we need to calculate the volume of the "box" within which the cells are counted. The number of cells within the box divided by the volume of the box in milliliters yields the number of cells per milliliter.

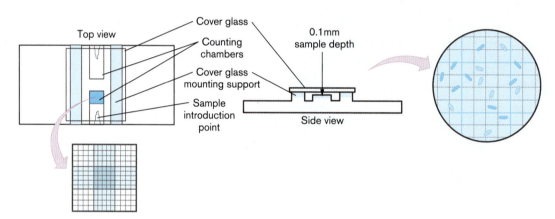

FIGURE 4.2 Hemocytometer slide.

Most probable numbers (MPN)

The **most probable numbers** method works by diluting a cell suspension to the point where there are no more live cells in a sample. The degree to which a suspension must be diluted before a sample will not have any cells in it is directly related to the cell density: Higher-density suspensions require more diluting before there are no cells in a diluted sample. The MPN method is used in Lab 30, "Water Analysis," to determine the numbers of intestinal bacteria in environmental water samples.

Quantitative plate counts

In **quantitative plate counting**, the method used in this lab, cells from different dilutions are either mixed with agar and poured into plates (see Lab 3) or spread onto the surfaces of various media, and the number of resulting colonies is used to calculate a **viable cell count**, **cell concentration**, or **cell** or **population density**; these are all estimates of the number of live cells per unit volume. Unlike some of the other methods, quantitative plate counting counts only live cells, because dead cells do not produce colonies.

QUANTITATIVE PLATE COUNT CONCEPTS AND PROCEDURES

Pour plate counts

In **pour plate counting**, a starting sample is serially diluted, and then measured amounts of the different diluted cell suspensions are mixed with melted agar and poured into sterile plates. Cell division by viable cells produces visible colonies in 18 to 24 hours, so the number of colonies reflects the number of live cells in the diluted samples mixed with the agar. Multiplying colony counts by the dilution factor for a counted sample yields the cell density of the starting sample. Pour plating is used in Lab 29, "Milk Microbiology," to measure bacterial cell densities in milk and other dairy products.

Spread plate counts

Spread plate counting is similar to pour plate counting in that a starting cell suspension is serially diluted, and then the number of viable cells in each dilution is determined by counting colonies on a plate. However, in spread plate counting, a measured amount of a cell suspension is spread on the surface of an agar media plate instead of being mixed with melted agar. Spread plate counts are used in this lab.

Colony-forming units (CFU)

Both the spread plating and pour plating techniques have the net result of separating individual cells in or on a solid medium. Each separated viable cell can divide and divide again, and eventually produces a visible colony. So, in theory, the number of colonies on the plate should be equal to the number of viable cells applied to the plate. And as we'll see below, this information is quite valuable, because if we know the number of cells applied to a plate, then we can calculate the cell density of an undiluted starting sample.

But there's a catch. The "one colony equals one cell theory" is generally accurate for species in which cells cleanly separate from each other at the end of the division process. For example, when an *E. coli* suspension is spread on a plate or mixed with agar, relatively few of the *E. coli* cells are stuck to each other, so most colonies really are the product of a single starting cell. In other species, however, the cells do not immediately separate from each other after cell division; instead, they may remain clustered in tetrads (*Micrococcus* species), clusters (*Staphylococcus* species), or chains (*Streptococcus* and *Bacillus* species). When cells of these species are spread onto a plate or mixed with agar, the individual colonies that we see are actually the product of a single cluster or single clump of several cells.

So, in practice, we cannot conclude that "one colony equals one cell," and as a result, our final calculation does not produce a value or cell count expressed as "cells per milliliter." Instead, our estimate of population density is reported in **colony-forming units per milliliter**, or **CFU/ml**. Now, if you're plating *E. coli*, then CFU/ml is probably pretty close to cells/ml, but if you're counting *Micrococcus* cells, then CFU/ml is probably closer to $4\times$ cells/ml. And if you are working with the irregular cell clusters of *Staphylococcus*, it's going to be very difficult to equate a particular CFU/ml value with any specific cells/ml value. So, as a rule, you should stick with CFU/ml, and all results should be reported using this unit of density.

Countable plates

To determine the CFU/ml value for a sample with an unknown cell density, we need plates with colonies growing from cells that were deposited in or on the media by spreading or pouring. But if there are too many or too few colonies per plate, then our estimate of CFU/ml will not be accurate. We need plates with a "Goldilocks" number of colonies: we need the number of colonies to be "just right." Fortunately, in the far distant past, a statistician sat down and figured out what this means for plate counting, so here are the rules for **countable plates**:

Spread plates: Countable plates are those with 20 to 200 colonies.

> As it might have been put in *Monty Python and the Holy Grail:*
>
> Countable spread plates shalt be those plates with 20 to 200 colonies. 20 to 200 shalt be the number that thou shalt count, and the number of the counting shalt be 20 to 200 colonies. 201 colonies shalt thou not count, nor either count 19 colonies, excepting that thou then proceed to 20. And 202 colonies is right out.

Pour plates: Countable plates are those with 30 to 300 colonies.
The numbers are higher for pour plates because the cells can grow throughout the depth of the agar and not just on the surface of the plate.

> Countable pour plates shalt be those plates with 30 to 300 colonies. 30 to 300 shalt be the number that thou shalt count, and the number of the counting shalt be 30 to 300 colonies. 301 colonies shalt thou not count, nor either count 29 colonies, excepting that thou then proceed to 30. And 302 colonies is right out.
>
> Once the number 200, being the 200th number, be reached for spread plates and the number 300, being the 300th number, be reached for pour plates, then lobbest thou the Holy Hand Grenade of Antioch towards thy foe, who, being naughty in my sight, shall snuff it.

Dilution series

When you are confronted with a test tube or flask full of bacteria and are asked to determine a CFU/ml for the cells in that flask, you obviously don't know the answer before you've done a plate count. This is what you might call a "known unknown": You know that you don't know the unknown CFU/ml value for the sample, and you'd like to find out what it is. To find the answer, you will need to apply cells to the plate in such a way as to produce a "countable plate," but how are you going to do this when you don't know the initial cell density of the sample? Given how little you know at this point, you could very easily put too many or too few cells on a plate. So what's the solution?

The answer is **serial dilution** (see Lab 3). Quantitative plate counting to determine the CFU/ml requires a series of dilutions of the original sample because, with enough dilutions, we will eventually hit upon a dilution that will give us countable plates. To make life a little easier, these dilution series are typically done using only tenfold and hundredfold dilutions, as opposed to the mathematically trickier twofold or fourfold dilutions. The tenfold and hundredfold dilutions are described in the table below.

Dilution	Description	Examples of Dilutions	Notation
Tenfold	1 part cell suspension to 9 parts water for 10 parts total	1 ml cell suspension + 9 ml water <u>or</u> 0.5 ml cell suspension + 4.5 ml water	1:10 10^{-1} 0.1
Hundredfold	1 part cell suspension to 99 parts water for 100 parts total	1 ml cell suspension + 99 ml water <u>or</u> 0.5 ml cell suspension + 49.5 ml water	1:100 10^{-2} 0.01

The key to doing tenfold and hundredfold dilutions is remembering that total dilution at any point in the process is equal to the sum of all the negative exponents of all the individual dilutions up to that point. This is equivalent to multiplying all of the decimal fractions of the individual dilutions together.

An example is shown in **Figure 4.3** and the accompanying table:

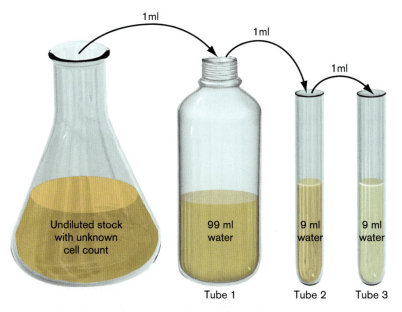

FIGURE 4.3 An example of a tenfold serial dilution.

	Stock	Tube 1	Tube 2	Tube 3
Individual dilution	$10^{-0} = 0$	10^{-2} (1 ml + 99 ml)	10^{-1} (1 ml + 9 ml)	10^{-1} (1 ml + 9 ml)
Total dilution	$10^{-0} = 0$	10^{-2}	10^{-3} $(-2) + (-1) = -3$	10^{-4} $(-2 + -1 + -1 = -4)$
Dilution factor	$10^{0} = 1$	10^{2}	10^{3}	10^{4}

For example, tube 2 was produced by making a 10^{-2} dilution to make tube 1, followed by a 10^{-1} dilution of tube 1 to make tube 2. The total dilution for tube 2 can be determined by adding the negative exponents; $^{-2} + ^{-1} = ^{-3}$, so the total dilution is 10^{-3}; or the two dilutions can be multiplied together: $10^{-2} \times 10^{-1} = 10^{-3}$, so the total dilution is 10^{-3}.

You will need to know the total dilution for a given tube because that information is used to determine the total dilution factor for that tube: a value that is a part of the CFU/ml formula. As long as we stick to multiples of ten in our dilutions, the total dilution factor for a given tube can be found by converting the negative exponent of the total dilution for that tube to a positive value. So, for example, if the total dilution is 10^{-3}, then the dilution factor is 10^3.

Applying cells to the plates

After the serial dilutions have been made, the cells must be applied to plates containing a growth medium to determine the number of viable cells per milliliter for each of the dilutions. In spread plating, 0.1 ml of each diluted cell suspension is spread across the surface of an agar plate with a sterile glass rod. If more than 0.1 ml is applied to a solid agar plate, it will take too long to dry. In pour plating, for each dilution, either 0.1 ml or 1.0 ml of cell suspension is mixed with melted agar and poured into a sterile petri dish to harden.

If 0.1 ml of cell suspension is applied, regardless of whether by spread plating or pour plating, this is effectively another tenfold dilution, and you must remember this when doing the final calculation. Remember that the goal is to produce a CFU per milliliter value, and if one-tenth of a milliliter is applied to the plate, then the number of colonies on that plate will be one-tenth of the CFU/ml of the cell suspension. For example, if 0.1 ml of a cell suspension containing 1,000 cells per milliliter is drawn from a test tube, then that 0.1 ml aliquot, or 0.1 ml portion of the whole volume, will contain 100 cells and will yield 100 colonies when applied to a plate. So, to convert the number of colonies on the plate to CFU/ml in the test tube, we'll need to multiply by 10: 100 colonies $\times$ 10 = 1,000 CFU/ml in the cell suspension. Keep this in mind as we look at the equations for converting colonies per plate into CFU/ml in the undiluted stock cell suspension.

Final calculation of CFU/ml

The final calculation of CFU/ml converts the number of colonies per plate into a CFU/ml value for an undiluted cell suspension with an originally unknown cell density. To do the calculation, you'll need to count the colonies on a countable plate as defined above. Occasionally, two different dilutions will produce countable plates, or counts within the limits of accuracy for a particular method (spread or pour plate). In this case, use the dilution with the count that is farthest from the upper and lower limits for a given method. For example, if 0.1 ml of a 10^{-4} dilution produces 145 colonies on a spread plate, and 0.1 ml of a 10^{-5} dilution produces 25 colonies on another spread plate, then use the counts from the 10^{-4} dilution, because 145 colonies is far from the upper limit of 200 colonies, while 25 colonies is close to the lower limit of 20 colonies.

Rounding rules

After the colonies have been counted and before the number can be used in the equation below, the counts should be rounded off according to the guidelines given in the following table:

Spread Plate	Pour Plate	Round To
20 to 50	30 to 50	Do not round. Use actual count
50 to 100	50 to 100	Nearest 5 or 10
100 to 200	100 to 300	Nearest 10

Equations

When 0.1 ml of cell suspension produces countable plates, then the CFU/ml of the original, undiluted cell suspension equals

$$10 \times \text{number of colonies} \times \text{dilution factor}$$

- We multiply by 10 because the number of colonies is actually colonies per 0.1 ml and the final value must be in CFU per 1.0 ml.
- "Number of colonies" is the number of colonies on countable plates.
- "Dilution factor" is the total dilution factor for the tube that produced countable plates.

When 1.0 ml of cell suspension produces countable plates, then the CFU/ml of the original, undiluted cell suspension equals

$$\text{Number of colonies} \times \text{dilution factor}$$

where number of colonies and dilution factor mean the same as for the previous equation.

Reporting CFU/ml and scientific notation

After you have completed the calculation of CFU/ml, there is one more task to be completed. The final value must be converted to scientific notation so that there is one digit to the left of the decimal point and one digit to the right of the decimal point. The reported value should look like this: $A.B \times 10^{C}$. If you need to adjust your calculated value to fit this format, remember that when moving the decimal point to the left, you must add one unit to the exponent for every digit "passed" in the shift. For example, to convert 120×10^{5} CFU/ml to the proper final form, you need to move the decimal point to the left two places and add two units to the exponent for a final form of 1.2×10^{7} CFU/ml. If the decimal point must be moved to the right, then subtract one unit from the exponent for every digit "passed" in the shift.

PUTTING IT ALL TOGETHER : A SUMMARIZING EXAMPLE

Here, again, is the dilution series used as an example above.

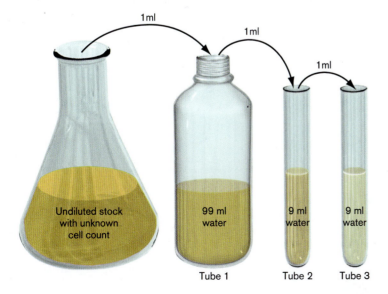

| Undiluted stock with unknown cell count | 1ml → | Tube 1 99 ml water | 1ml → | Tube 2 9 ml water | 1ml → | Tube 3 9 ml water |

After the dilutions were made, 0.1 ml of cell suspension was transferred from each tube to nutrient agar plates and spread with a glass rod. The various plates produced colony counts as shown in the table below.

	Individual Dilution	Total Dilution	Dilution Factor	Colonies per Plate
Tube 1	10^{-2}	10^{-2}	10^2	TNTC*
Tube 2	10^{-1}	10^{-3}	10^3	102
Tube 3	10^{-1}	10^{-4}	10^4	13

*If there are more than 300 colonies, then this may be reported as "TNTC," or "too numerous to count."

The concentration of cells in tube 1 was too high, and the concentration in tube 3 was too low, to produce countable plates, but the 0.1 ml aliquot from tube 2 contained more than 20 cells and fewer than 200 cells, and it produced a countable plate with 102 colonies on the plate. Using this plate, and following the rounding rules given above,

CFU/ml of undiluted stock $= 10 \times 100$ colonies $\times 10^3 = 1{,}000 \times 10^3$ CFU/ml

Converting this answer, $1{,}000 \times 10^3$ CFU/ml, to the proper reporting format requires moving the decimal point three places to the left, so three units are added to the exponent ($3 + 3 = 6$), yielding the final answer:

CFU/ml of undiluted stock $= 1.0 \times 10^6$ CFU/ml

Figures 4.4 to **4.6** illustrate what is happening in this example.

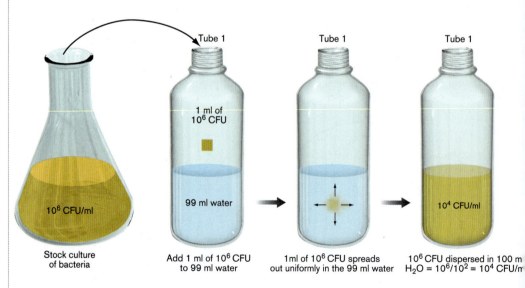

FIGURE 4.4 Dilution of stock bacteria.
The three bottles represent Tube1 at three different points in time. From left to right: a 1-ml sample of bacteria is added to 99 ml of water; the bacteria disperse in the water; after some time, the bacteria are evenly spaced throughout the water.

In the first step or first dilution, 1.0 ml of stock culture containing 10^6, or 1,000,000, CFU/ml is transferred to tube 1, which contains 99 ml of water (a 99-ml water blank), producing a 1:100 or 10^{-2} dilution (Figure 4.4). The final volume of tube 1 is 100 ml, and when 10^6 (or 1,000,000) CFU are dispersed or spread out in a volume of 10^2 (or 100) ml of water, it yields a final concentration of 10^4 (or 10,000) CFU/ml in tube 1.

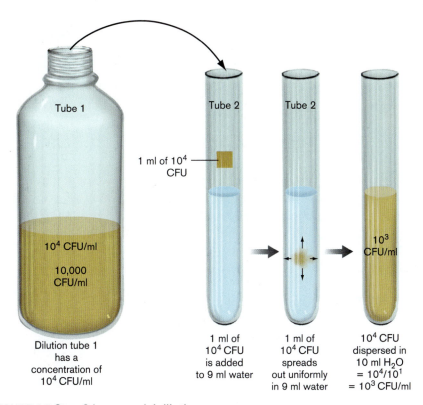

FIGURE 4.5 Step 2 in our serial dilution.
The three test tubes represent Tube 2 at three different points in time. From left to right: a 1-ml sample of bacteria is added to 9 ml of water; the bacteria disperse in the water; after some time, the bacteria are evenly distributed throughout the water.

In the second step or second dilution, 1.0 ml of the suspension containing 10^4, or 10,000, CFU/ml is transferred to tube 2, which contains 9 ml of water (a 9-ml water blank), producing a 1:10 or 10^{-1} dilution (Fig. 4.5). The final volume of tube 2 is 10 ml, and when 10^4 (or 10,000) CFU are dispersed or spread out in a volume of 10^1 (or 10) ml of water, it yields a final concentration of cells of 10^3 (or 1,000) CFU/ml in tube 2.

The sample from tube 2 resulted in 100 CFU on the nutrient agar plate (Fig. 4.6). Since 0.1 ml was applied to the plate, if you multiply 100 CFU per plate by 10, you'll know how many CFU/ml are present in the suspension in tube 2. In this case, after planting, we know tube 2 contains 1,000 CFU/ml.

Tube 2, the tube that produced countable plates, contains a thousand-fold (1:1,000, or 10^{-3}) dilution of the original suspension. So the CFU/ml is 1/1,000 the concentration in the undiluted stock. So to determine the concentration in the original stock, you would multiply by the dilution factor of 10^3, or 1,000, and this yields an answer of 1.0×10^6, or 1,000,000 CFU/ml, in the original stock.

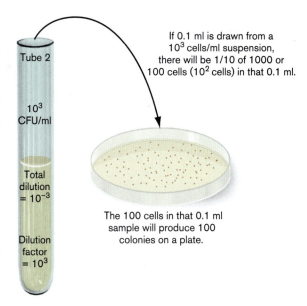

If 0.1 ml is drawn from a 10^3 cells/ml suspension, there will be 1/10 of 1000 or 100 cells (10^2 cells) in that 0.1 ml.

The 100 cells in that 0.1 ml sample will produce 100 colonies on a plate.

FIGURE 4.6 Countable colonies on a plate.

A summary of the spread plate dilution method and sample results are presented in **Figure 4.7** below.

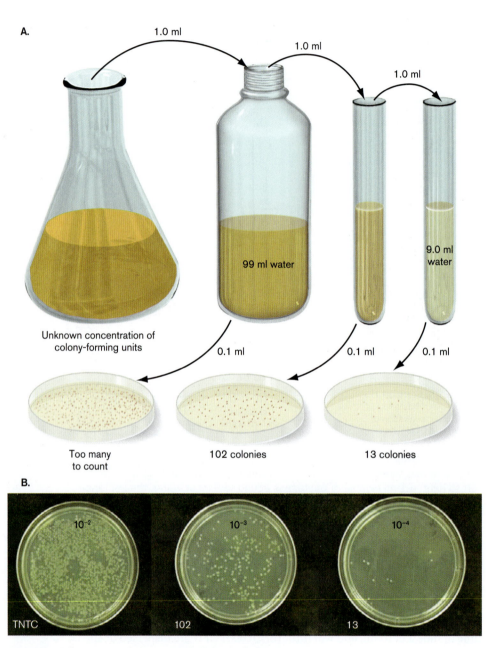

FIGURE 4.7 Summary of the spread plate technique.

PROTOCOL

Day 1
Materials
- 8 Glucose nutrient agar plates
- Beaker of alcohol
- 2-ml (blue-barrel) pipette pump
- Glass spreading rod[1]
- 4 1.0-ml sterile pipettes

- 99-ml water blank bottle
- 39-ml water blank test tubes

[1]An alternative to flaming alcohol-soaked glass spreaders is the much more sedate use of disposable plastic spreaders that come individually wrapped and sterilized. Although perhaps not as exciting as the spreader flambé, sterile plastic spreaders are perfectly suitable and significantly safer. Dispose of used spreaders as you would any contaminated waste.

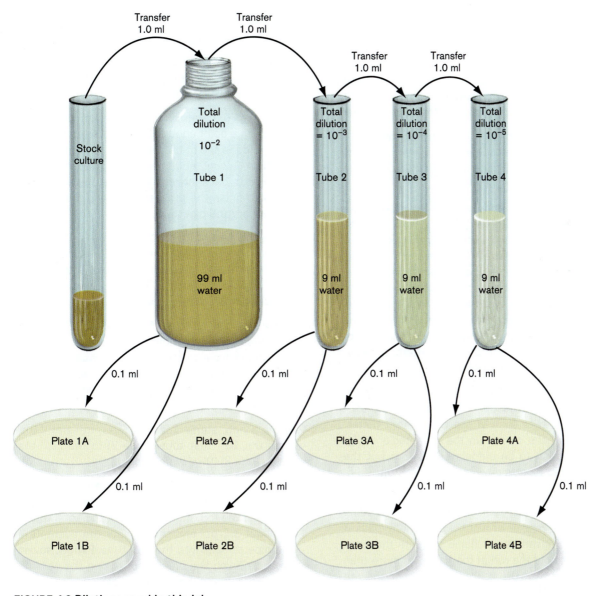

FIGURE 4.8 Dilutions used in this lab.

Culture:

- *Escherichia coli* stock of unknown CFU/ml

PROCEDURES

The dilutions in this lab are illustrated in **Figure 4.8**.

1. Label a 99-ml water blank bottle with your name and the label 10^{-2} (Tube 1).

2. Label three 9-ml water blank test tubes with your name. In addition, label the first tube 10^{-3} (Tube 2), the second tube 10^{-4} (Tube 3), and the third tube 10^{-5} (Tube 4).

3. Label eight glucose nutrient agar plates with your name. The first two plates should be labeled "Plate 1A" and "Plate 1B," the second two should be labeled "Plate 2A" and "Plate 2B," the third two should be labeled "Plate 3A" and "Plate 3B," and the last two should be labeled "Plate 4A" and "Plate 4B."

4. Pick up one tube of the *E. coli* stock of unknown CFU/ml, and be sure to record the identification number on your tube. Your lab instructor will need this ID number when comparing your estimate of CFU/ml with the actual density of the stock culture.

5. Gently swirl the stock culture to evenly distribute the cells in the stock. Then use a sterile 1.0 ml pipette to transfer 1.0 ml of stock to the 10^{-2} bottle (Tube 1) (99-ml water blank).

6. Use the pipette to gently stir the solution in the 99-ml water blank bottle (Tube 1) to distribute the cells throughout the water. Carefully draw the water up into the pipette to a point about halfway between the 1.0 ml mark and the top of the pipette, or to a total volume of about 1.2 ml, and then empty the contents of the pipette back into the bottle. Repeat several times to finish the process of mixing the cells into the water. Do not discard the pipette at this point, and do not let the pipette touch anything except the bottle's contents.

7. Using the same pipette used in steps 5 and 6, transfer 0.1 ml of suspended cells from the 10^{-2} bottle (Tube 1) to each of the two Plate 1 plates. Now discard the pipette in the receptacle designated for trash to be autoclaved. Do not put the contaminated pipette down on the benchtop.

8. After observing an instructor demonstration of the following technique, remove the glass spreading rod from the beaker of alcohol and briefly pass it through a Bunsen burner flame to ignite the alcohol. Do not hold the rod in the flame for more than a second or two; the goal is not to heat the rod to sterilize it, because the alcohol will have already killed any bacteria on the rod. Instead, the flame is simply used to remove the excess alcohol quickly by burning it away. In addition, **once the alcohol is ignited, keep the spreading end of the rod pointed downward** so that the flaming alcohol does not run down the rod and onto your hand.

9. Once the alcohol has been removed, following the techniques demonstrated by your lab instructor and shown in **Figure 4.9A–D**, use the alcohol-sterilized spreading rod to spread the 0.1 ml drops to all areas on each of the two plates. You do not have to sterilize the spreader between the two plates when the drops are from the same dilution, because the cell densities in the two drops are equal to each other. However, when you are done spreading cells on both plates, return the spreader to the beaker of alcohol.

Before touching the spreader to the 0.1 ml drop of cells, it's a good idea to touch the spreader to a sterile, drop-free part of the agar in the plate for a few seconds to ensure that it has been sufficiently cooled after the alcohol has been removed by flaming.

10. Using a new, sterile 1.0 ml pipette, transfer 1.0 ml from the 10^{-2} bottle (Tube 1) to the 10^{-3} test tube (Tube 2) (9-ml water blank).

11. As in step 6, carefully draw the water up into the pipette to a point about halfway between the 1.0 ml mark and the top of the pipette, or to a total volume of about 1.2 ml, and then empty the contents of the pipette back into the test tube. Repeat several times to finish the process of mixing the cells. Do not discard the pipette at this point.

12. Using the same pipette used in steps 10 and 11, transfer 0.1 ml of suspended cells from the 10^{-3} test tube (Tube 2) to each of the two Plate 2 plates. Now discard the pipette in the receptacle designated for trash to be autoclaved.

13. Use a sterilized spreader to spread the 0.1 ml drops to all areas on each of the two Plate 2 plates. Again, you do not have to sterilize the spreader between the two plates with drops from the same source. When you are done spreading cells on both plates, return the spreader to the beaker of alcohol.

14. Using a new, sterile 1.0 ml pipette, transfer 1.0 ml from the 10^{-3} test tube (Tube 2) to the 10^{-4} test tube (Tube 3) (9-ml water blank).

15. As in step 6, carefully draw the water up into the pipette to a point about halfway between the 1.0 ml mark and the top of the pipette, or to a total volume of about 1.2 ml, and then empty the contents of the pipette back into the test tube. Repeat several times to finish the process of mixing the cells. Do not discard the pipette at this point.

16. Using the same pipette used in steps 14 and 15, transfer 0.1 ml of suspended cells from the 10^{-4}

A.

B.

C.

D.

FIGURE 4.9 Spreading bacteria onto the surface of agar media.

test tube (Tube 3) to each of the two Plate 3 plates. Now discard the pipette in the receptacle designated for trash to be autoclaved.

17. Use a sterilized spreader to spread the 0.1 ml drops to all areas on each of the two Plate 3 plates. Again, you do not have to sterilize the spreader between the two plates with drops from the same source. When you are done spreading cells on both plates, return the spreader to the beaker of alcohol.

18. Using a new, sterile 1.0 ml pipette, transfer 1.0 ml from the 10^{-4} test tube (Tube 3) to the 10^{-5} test tube (Tube 4) (9-ml water blank).

19. As in step 6, carefully draw the water up into the pipette to a point about halfway between the 1.0 ml mark and the top of the pipette, or to a total volume of about 1.2 ml, and then empty the contents of the pipette back into the test tube. Repeat several times to finish the process of mixing the cells. Do not discard the pipette at this point.

20. Using the same pipette used in steps 18 and 19, transfer 0.1 ml of suspended cells from the 10^{-5} test tube (Tube 4) to each of the two Plate 4 plates. Now discard the pipette in the receptacle designated for trash to be autoclaved.

21. Use a sterilized spreader to spread the 0.1 ml drops to all areas on each of the two Plate 4 plates. Again, you do not have to sterilize the spreader between the two plates with drops from the same source. When you are done spreading cells on both plates, return the spreader to the beaker of alcohol.

22. When finished with all the plates, give them about 5 minutes to dry, then invert all the plates and incubate them at 37°C for 24 hours.

Day 2

1. Complete the Results section to determine the CFU/ml for your stock culture. Begin by counting the colonies on each of the eight plates from day 1 and record the counts on the appropriate lines in the Results section.

If in the process of counting the colonies on a plate, you reach a count of about 300 colonies or more, then you can stop counting that plate at that point, and you can write "TNTC" ("too numerous to count") on the appropriate line in the Results section. There is no need to count all the colonies on plates with more than 300 colonies per plate because you won't be using those plates to do your calculations.

Go ahead and record results if there are between 200 and 300 colonies on a plate. Strictly speaking, a spread plate with 200 to 300 colonies is not a countable plate. However, it's possible that the next dilution will produce plates with fewer than 20 colonies. In that case, if there are no plates with counts between 20 and 200, then you should use the plates with 200 to 300 colonies to do your calculations.

I read the news today oh, boy,
Four thousand holes in Blackburn, Lancashire,
And though the holes were rather small,
They had to count them all,
Now they know how many holes it takes to fill the Albert Hall.

—Lennon and McCartney

2. Once counts are completed, calculate the averages for each of the pairs of plates at a given dilution, find the dilution that produced an average count between 20 and 200, and then use that dilution and the average number of colonies at that dilution to calculate the CFU/ml of the original stock solution. As noted above, if there are no plates with counts between 20 and 200, then you should use the average value that is between 200 to 300 colonies to do your calculations.

Name: _____   Section: _____

Course: _____   Date: _____

STOCK ID NUMBER:

Dilution factors:

	INDIVIDUAL TUBE DILUTION*	TOTAL DILUTION†	TOTAL DILUTION FACTOR‡
Tube 1			
Tube 2			
Tube 3			
Tube 4			

*Dilution for this tube only, dilution from previous tube to this tube

†Total or cumulative dilution at this point in the dilution series

‡Total or cumulative dilution factor at this point in the dilution series

Colony counts:

If over 300 on a plate, report as "TNTC" ("Too numerous to count").

INOCULATED PLATES	PLATE A	PLATE B	MEAN
Plates inoculated from tube 1			
Plates inoculated from tube 2			
Plates inoculated from tube 3			
Plates inoculated from tube 4			

Calculation of CFU/ml in stock solution (show complete equation):

Name: _____   Section: _____

Course: _____   Date: _____

DILUTION PROBLEMS

1. Let's say a stock culture of *E. coli* was diluted by serial dilution as indicated below. Then, 0.1 ml of suspension was removed from each tube and spread on a plate, and the colonies were counted, producing the following results:

	TUBE CONTENTS	NUMBER OF COLONIES	INDIVIDUAL DILUTION	TOTAL DILUTION	DILUTION FACTOR
Tube 1	1 ml stock + 99 ml water	TNTC			
Tube 2	1 ml tube 1 + 99 ml water	276			
Tube 3	1 ml tube 2 + 9 ml water	153			
Tube 4	1 ml tube 3 + 9 ml water	17			
Tube 5	1 ml tube 4 + 9 ml water	1			

What is the CFU/ml count for the undiluted stock?

2. A stock culture of *E. coli* was diluted by serial dilution as indicated below. Then, 0.1 ml of suspension was removed from each tube and spread on a plate, and the colonies were counted, producing the following results:

	TUBE CONTENTS	NUMBER OF COLONIES	INDIVIDUAL DILUTION	TOTAL DILUTION	DILUTION FACTOR
Tube 1	1 ml stock + 99 ml water	TNTC			
Tube 2	1 ml tube 1 + 9 ml water	TNTC			
Tube 3	1 ml tube 2 + 9 ml water	954			
Tube 4	1 ml tube 3 + 9 ml water	143			
Tube 5	1 ml tube 4 + 9 ml water	15			

What is the CFU/ml count for the undiluted stock?

3. A stock culture of *E. coli* was diluted by serial dilution as indicated below. Then 0.1 ml of suspension was removed from each tube and spread on a plate, and the colonies were counted, producing the following results:

	TUBE CONTENTS	NUMBER OF COLONIES	INDIVIDUAL DILUTION	TOTAL DILUTION	DILUTION FACTOR
Tube 1	1 ml stock + 9 ml water	TNTC			
Tube 2	1 ml tube 1 + 9 ml water	836			
Tube 3	1 ml tube 2 + 9 ml water	72			

What is the CFU/ml count for the undiluted stock?

4. A stock culture of *E. coli* was diluted by serial dilution as indicated below. Either 1.0 ml or 0.1 ml of cells was removed from each tube, mixed with melted agar, and poured into a plate, and the colonies were counted.

Use the first table to calculate the dilution factor for each of the four tubes, then enter these values in the appropriate cells in the second table

	TUBE CONTENTS	INDIVIDUAL DILUTION	TOTAL DILUTION	DILUTION FACTOR
Tube 1	1 ml stock + 99 ml water			
Tube 2	1 ml tube 1 + 99 ml water			
Tube 3	1 ml tube 2 + 99 ml water			
Tube 4	1 ml tube 3 + 99 ml water			

	DILUTION FACTOR	AMOUNT ADDED FROM EACH DILUTION TO THE MELTED POURED AGAR	NUMBER OF COLONIES
Tube 1		0.1 ml from tube 1	TNTC
Tube 2		1.0 ml from tube 2	TNTC
Tube 2		0.1 ml from tube 2	1,243
Tube 3		1.0 ml from tube 3	231
Tube 3		0.1 ml from tube 3	27
Tube 4		1.0 ml from tube 4	1

What is the CFU/ml count for the undiluted stock?

ADDITIONAL QUESTIONS

1. a. What is the advantage of a spread plate count over a simple hemocytometer count?

    b. Given your answer to (a), why would you do a hemocytometer count; that is, what is the advantage of this method over spread plating?

2. Why are counts expressed in CFU per milliliter instead of cells per milliliter?

3. If you want a final cell density in CFU per 1.0 ml, why do you apply only 0.1 ml to the spread plate?

4. In this lab, you sterilized the spreader before spreading drops of different cell densities. What if you could not sterilize the spreader between the sets of plates? Do you think that it would be better or produce a more accurate count if

    a. you spread the higher-density drops first and then, without sterilization, spread the lower-density drops

or

    b. you spread the lower-density drops first and then, without sterilization, spread the higher-density drops?

That is, if you had to choose, which would you chose, method A or method B? Explain your choice.

5.   a. How did the CFU/ml count that you generated by doing this lab compare with the actual CFU/ml count (provided by your instructor)? Was it higher or lower than the actual count?

    b. What types of mistakes in procedures or calculations could account for a CFU/ml count that is lower than the actual count? List several possibilities.

    c. What types of mistakes in procedures or calculations could account for a CFU/ml count that is higher than the actual count? List several possibilities.

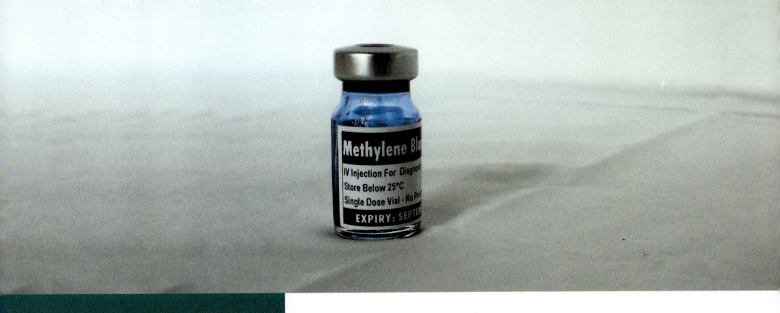

Simple Staining and Cell Morphology

Methylene blue is a remarkable chemical that has been used in medicine, biology, and chemistry for more than one hundred years. It has antifungal and virucidal properties, serves as an antidote in cyanide poisonings, and is used as a redox indicator in analytical chemistry and biochemistry research. In this lab, we use methylene blue as a water-soluble dye that stains heat-killed bacterial cells affixed to glass slides.

Learning Objectives

- Describe the reasons for staining bacterial cells.

- Understand basic staining concepts such as stain composition and the role of heat fixing.

- Compare and contrast direct simple staining and indirect simple staining.

- Perform direct and indirect simple staining and understand the scientific terms used to describe cell morphology.

WHY WE STAIN

Microbiologists have been staining bacterial cells since at least the second half of the nineteenth century. German microbiologists led the way in the development and use of new stains, in part because of the German chemical industry's advances in the then-new science of organic chemistry. Its methods led to the synthesis of many novel stains and dyes, especially aniline dyes, from coal tars (turns out organic chemistry has its uses, after all). Not surprisingly, many of the microbiologists who developed these stains were also associates of, or influenced by the work of the German doctor and microbiologist Robert Koch.

Broadly speaking, there are at least three reasons why we stain bacterial cells:

1. **To improve contrast between cells and slide.** Staining significantly improves the contrast between the usually tiny, nearly transparent bacterial cells and the white background of the glass slide used for microscopy. In fact, without staining, bacterial cells are very difficult to see with the standard microscopes that we use in this lab course (although there are other specialized types

of light microscopes, such as phase contrast microscopes, that do a better job of visualizing unstained bacterial cells). The **simple stain**, a method that uses only one type of stain, is all that is needed when the goal is primarily to improve contrast.

2. **To help identify bacteria.** Stains that reveal the distinguishing traits of different types of bacteria, called **differential stains**, are very useful in the identification of bacteria. By far the most important example of a differential stain is the Gram stain, a stain that divides the bacterial world into two large groups, the Gram-positive species and the Gram-negative species (see Lab 6). A more specialized differential stain is the acid-fast stain; this stain is used to identify certain types of bacteria that have waxy cell walls (see Lab 7). These bacteria include *Mycobacterium tuberculosis*, the cause of the lung disease tuberculosis.

3. **To improve the ability to see bacterial cell structures.** Most bacterial cell structures are too small to see with the light microscope, but some structures can be observed and studied at 1,000× magnification with the application of **structural stains**. Structural stains such as the spore stain are discussed in Lab 8.

SIMPLE STAINS

Simple staining is any staining procedure that uses a single type of dye or stain to provide contrast between the cells and the background (the slide). The stains or dyes themselves are water- or alcohol-soluble salts in which one of the charged ions is colored. The colored ion is referred to as the **chromogen**, and it is usually one of those charged organic compounds originally derived from coal tars by nineteenth-century German organic chemists. Chromogens are often synthesized by adding a **chromophore** to a colorless organic molecule such as benzene. The chromophore is the part of the molecule that absorbs and reflects particular wavelengths of light, and thus it's the chemical group that generates the particular color of the stain. Depending on whether or not the colored ion is positively or negatively charged, simple staining can be done by either a direct or an indirect process.

Direct (positive) staining

Direct stains use dye salts in which the chromogen is a **cation**, and thus is positively charged. Stains with colored cations are also called **basic stains**, and obviously, the positively charged chromogens will stick to any molecule or cell structure that has a net negative charge. Such negatively charged structures are said to be **basophilic** or, literally, "base-loving" because of their affinity for basic dyes.

As it happens, bacterial cells tend to have a net negative charge due to the presence of negatively charged molecules, such as phospholipids and nucleic acids, as well as negatively charged structures, such as cell membranes and ribosomes. So the colored cations of basic stains will stick directly to the cells themselves, dyeing the cells and increasing the contrast between the cells and the white background of the slide. In other words, in the case of bacterial cells, the phrases "direct staining," "direct stain," and "basic stain" go together.

Since the direct staining process depends on the binding of positively charged dyes to negatively charged cell bits, anything that blocks or covers up the negative charges will reduce the effectiveness of the staining process. For example, when the environment is acidic, the excess positively charged protons (H^+) may compete for binding sites with

CH₃
|
N—CH₃

FIGURE 5.1 Chemical structures of two basic stains.
Methylene blue and crystal violet use basic, or cationic, dye salts. The positively charged group in each dye (shaded yellow) reacts with negatively charged parts of bacteria, such as cell membranes. Chloride (Cl⁻) is the negatively charged ion in both of the dye salts.

Methylene blue   **Crystal violet**

the colored cations and interfere with your efforts to stain the cells (bad protons, no treat for you!). This is worth remembering because the medium in old cultures tends to be acidic due to the production of organic acids by fermenting cells. Conclusion? Staining is best done with fresh, 24-hour-old cultures.

In this lab course, almost all the stains we use are basic stains. For example, methylene blue **(Figure 5.1)** is a salt made of the positively charged methylene blue chromogen and a colorless, negatively charged chloride ion ($MB–Cl \rightarrow MB^+ + Cl^-$). Other direct or basic stains that you will use in this course include crystal violet (Fig. 5.1), safranin, carbolfuchsin, and malachite green.

Indirect (negative) staining

Indirect stains use dye salts in which the chromogen is an **anion**, and thus is negatively charged. So, if you take everything that you know about direct stains and reverse it, you've got everything you need to know about indirect stains. For instance, stains with colored anions are also called **acidic stains**, and their negatively charged chromogens will be repelled by any molecule or cell structure that has a net negative charge, including the numerous negatively charged components of bacterial cells. So, in the case of bacterial cells, "indirect staining," "indirect stain," and "acidic stain" go together.

When acidic or indirect stains are applied to bacterial cells on a slide **(Figure 5.2)**, the repelled, **negatively charged chromogen molecules are deposited on the slide itself**. This colors the background while leaving the cells colorless (white). Viewing such a slide is like looking up on a starry night when the stars are circular, rod-shaped, or spiraled, depending on the type of cells in question. Nigrosin (black) and eosin (red) are two acidic stains that can be used to create a bacterial "night sky." In this lab, you will be working with nigrosin.

Indirect or negative staining has some advantages over direct staining:

1. Direct staining usually requires heat fixing (described below), but this process is not needed for indirect staining. Since the cells are not heated in indirect staining, there is little or no distortion of cell size and shape. So negative staining is particularly useful if your goal is accurate measurement of the length and width of the cell.

2. Cells with thick, waxy cell walls resist staining with direct or basic stains, so indirect stains are useful for visualizing these cells. Waxy-walled species include those in the genus *Mycobacterium*, including the deadly tuberculosis bacterium.

3. Around pH7, the outer capsule that surrounds the cells of certain species of bacteria lacks a net charge; thus, the colored cations of basic stains will not stick to this structure. But if an indirect stain is used, the uncolored capsules become visible against the dark background of the negatively charged chromogen.

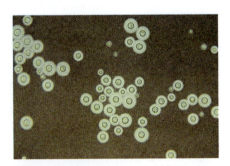

FIGURE 5.2 Capsule-covered bacteria stained indirectly.

SUMMARY

	Direct Stains (Basic stains)	Indirect Stains (Acidic stains)
Chromogen	Positively charged cation (e.g., methylene blue)	Negatively charged anion (e.g., eosin)
Colorless ion of the salt	Negatively charged anion (e.g., Cl^-)	Positively charged cation (e.g., Na^+)
Binds to	Negatively charged or basophilic cell parts	Positively charged or acidophilic cell parts
Works less well at	Low pH (acidic)	High pH (alkaline)
Effect on bacterial cells	Cells are directly colored or stained because they have a net negative charge	Colored anions are repelled by negatively charged cells, so cells remain colorless while the background is stained
Examples	Crystal violet Carbolfuchsin Safranin Methylene blue Malachite green	Eosin Nigrosin

POINTS TO REMEMBER FOR SUCCESSFUL STAINING

1. **Use a minimal amount of water.** In most cases, a single loopful of water will do. The process of staining includes an "air drying" step, and the more water you put on the slide to begin with, the longer it will take for the slide to dry. Less water means faster drying … and a shorter lab.

2. **Spread the cell-water mix into a very thin layer.** The rate at which a slide dries depends on the **surface-to-volume ratio** of the water droplet. That is it depends on the amount of droplet surface area exposed to the air per unit volume of the droplet. Thinner films of water have higher surface-to-volume ratios and will dry much faster than rounded droplets.

3. **Avoid thick smears of cells.** You don't need to cover the slide with a thick paste of cells; in fact, it will be more difficult to see the cells clearly if you do.

4. **Get all the required slides to the "air drying" stage before doing anything else.** Then, as all of your slides dry together, you can be doing other parts of the lab. It's called multitasking.

5. **Heat-fix properly.** Direct or basic staining requires that the cells be **heat-fixed** after the slides have dried. Heat fixing will (a) kill the cells by denaturing and coagulating cell proteins and (b) significantly improve the adherence of cells to the slide because adherence is partly a function of protein coagulation. If the cells are "under-fixed," or not heated enough, the cells will not be stuck firmly to the slide, and they will wash off the slide during staining. If they are "over-fixed," or heated too much, the heat will distort the shapes of the cells, producing "popcorn" cells. So watch closely when your instructor demonstrates the heat-fixing procedure, a technique that involves three passages of the slide through the Bunsen burner flame. And don't burn yourself.

6. **Stains stain.** Everything. Hands, clothes, everything. So, on days when staining is on the schedule, you may want to use latex or polyvinyl chloride (PVC) gloves for your hands and leave your finest toggery at home.

CELL MORPHOLOGY

Once cells are stained, you can observe **cell morphology (Figure 5.3)**, a term that refers to traits such as size, shape, color, and the arrangement of connected cells.

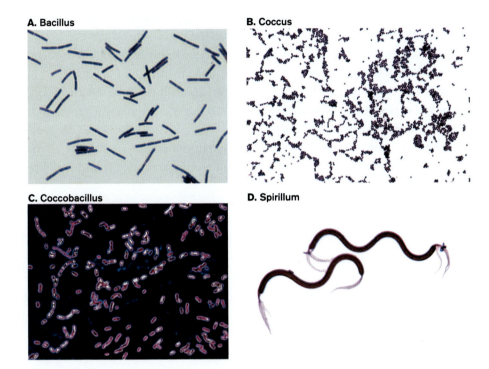

A. Bacillus

B. Coccus

C. Coccobacillus

D. Spirillum

FIGURE 5.3 Cell morphologies.

Cell shape

Cell shape can be described using a variety of terms:

Bacillus (plural, bacilli) The term **bacillus**, derived from a Latin word meaning "staff," "walking stick," or "little rod," is used to denote rod-shaped cells. (The Latin word *bacillus* is probably derived from the Greek word *bacterion*, a word that also means "staff," "stick," or "rod.") Somewhat confusingly, there is also a specific genus of bacteria called *Bacillus* (with a capital B). So don't forget, when the word "bacillus" is spelled with a lowercase b, it's a generic term that simply refers to any cells that are rod-shaped, but when the word is spelled with a capital B, it refers to a specific taxonomic collection of species in the genus *Bacillus*.

Coccus (plural, cocci) The term **coccus** comes from a Latin word that is derived from the Greek word *kokkos*, meaning "berry" or "grain." In the context of microbiology, "coccus" is used to denote spherical cells. When viewed with the standard light microscope, stained cocci appear as colored circles.

There are no bacterial genera called *Coccus* (although some genus names, such as *Staphylococcus*, end in the suffix "–coccus").

Coccobacillus Coccobacillus is a term used to describe short rods—for example, rods that are only about twice as long as they are wide. Depending on the angle at which they are resting on the slide, these cells, at first glance, may appear to be circular or spherical in shape. But careful observation of the cells will distinguish coccobacilli from true cocci.

Spirillum or spirochete The terms **spirillum** and **spirochete** refer to bacterial cells that have a spiraled or coiled form. "Spirillum," derived from *spira* (Latin) and *speira* (Greek), meaning "coiling" or "winding," is usually applied to thicker cells with two or three twists or turns. "Spirochete," a word with the suffix "-chete," derived from the Latin *chaeta*, meaning "hair" or "bristle," usually denotes very long, thin, threadlike or hairlike cells with multiple (four or more) twists along the length of the cell.

Among human-associated bacteria, the spirochete form is not as common as the bacillus and coccus forms. However, there are two medically important genera of spirochete bacteria, *Treponema* and *Borrelia*. The genus *Treponema* brings us the species *T. pallidum*, the causative agent of the sexually transmitted disease syphilis. *Borrelia* species include *B. burgdorferi*, the spirochete responsible for the tick-transmitted Lyme disease. Neither disease is a happy thought.

Cell arrangement

Cell arrangement refers to the position of one cell with respect to another after cells divide. These arrangements are described with terms typically used as prefixes.

Diplo- The term **diplo-**, from the Greek word *diplous*, meaning "double," is used to describe cells that remain connected as pairs after cell division. This arrangement is a diagnostic trait for several types of cocci that are major human pathogens, such as:

Streptococcus pneumoniae: The most common cause of pneumonia
Neisseria meningitidis: The most common cause of meningitis among young adults
Neisseria gonorrhoeae: The cause of the sexually transmitted disease gonorrhea

Tetrads The word **tetrad**, from the Greek *tetra*, meaning "four," is applied to an arrangement in which four cells are joined together in a symmetrical packet of cells. This arrangement can be seen with species in the genus *Micrococcus*.

Strepto- When cells are arranged in chains, the term **strepto-** is used to describe this pattern; the word is derived from the Greek term *streptos*, meaning "twisted" or "twisted chain." It can be used as a prefix in descriptive terms such as **streptobacilli**, an arrangement often seen with *Bacillus* species. It also appears as a prefix in genus names such as *Streptococcus*, which means, literally, "twisted chains of berries" (sounds more impressive in Latinized Greek, doesn't it?). Again, a lowercase s indicates use as a general descriptor, while a capital letter denotes a genus name.

Staphylo- The term **staphylo-**, from the Greek word *staphyle*, meaning "bunches of grapes," is used to describe cells that form clusters after cell division. While the term can be applied to any cells that grow in clusters, by far the most commonly encountered example is the clustered cells of species in the genus *Staphylococcus*. Cells in this genus can be found in abundance on the skin, and one species, *Staphylococcus aureus*, is a major human pathogen, capable of causing food poisoning and skin, lung, and blood infections.

PROTOCOL

Wear disposable gloves throughout this lab.

Note: In any lab where you are required to make slides, begin your activities for the day by getting all slides to the air-drying step (step 5 below) before going on to do other work.

Materials

- Disposable gloves
- Slide holder or clothespins
- Clean glass microscope slides (9 per student; 3 for direct, 6 for indirect)
- Staining tray with staining platform (1 per four students)
- Squirt bottle with water (wash bottle)
- Bibulous paper or paper towels
- Compound microscope with 100× oil immersion lens
- Crystal violet
- Methylene blue
- Nigrosin
- Safranin

Cultures (overnight slant cultures):

- *Escherichia coli*
- *Micrococcus luteus*
- *Bacillus megaterium*

Figure 5.4 illustrates the steps for performing a direct stain with methylene blue.

DIRECT STAINING

1. Select three unused slides and label as follows:
   a. *E. coli*
   b. *M. luteus*
   c. *B. megaterium*

   Slides can be labeled directly or by sticking a little masking tape on one end of the slide.

2. Transfer one or two loopfuls of water to the center of each slide using an inoculating loop.

   This is one of the rare occasions when the loop does not have to be flame-sterilized before use; don't get accustomed to not flaming.

When adding water to the slide, you are balancing two conflicting goals. You need enough water to spread out the cells to be stained, but any water added will have to be removed by air-drying, and air-drying takes time. So you don't want to add any more water than you really need. You will learn how much water to add to the slide by trial and error, but usually one to two loopfuls is a good starting point.

3. Flame your loop, allow the loop to cool, and then transfer cells from a given species to the drop of water on the slide with the matching species name.

   Dragging a loop over the layer of cells on the slant for a distance of about 1.0 cm should be enough to pick up an ample number of cells for staining. There is no need to overload the loop with cells by dragging the loop over the cells several times or until the loop is dripping in slime.

4. After transferring cells to the water droplet, use the loop to mix the cells into the water and spread out the cells and water.

   Drying time (next step) will be a function of the ratio of water droplet volume to surface area, so a thin, well-spread drop will dry much faster than a thick, compact drop of water.

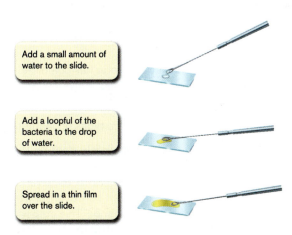

Add a small amount of water to the slide.

Add a loopful of the bacteria to the drop of water.

Spread in a thin film over the slide.

FIGURE 5.4 Procedure for direct staining with methylene blue. (*continued*)

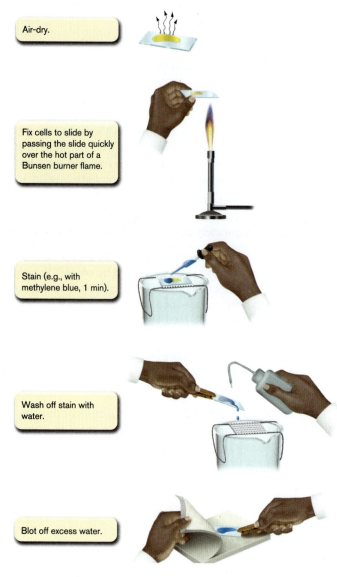

Air-dry.

Fix cells to slide by passing the slide quickly over the hot part of a Bunsen burner flame.

Stain (e.g., with methylene blue, 1 min).

Wash off stain with water.

Blot off excess water.

FIGURE 5.4 (*continued*)

Species	Stain Type	Contact Time
E. coli	Methylene blue	60 seconds
M. luteus	Safranin	60 seconds
B. megaterium	Crystal violet	30 seconds

During the staining process, the cells on the slides should be completely flooded with the appropriate stain for at least the amount of time given above as "contact time."

9. Once the stain has been in contact with the cells for the listed amount of time, use a wash bottle to rinse the stain off the slide and into the bottom of the staining tray.

10. Pick up the slides with the slide holder or clothespins (or your gloved hand) to keep the stain off your fingers, and dry the slides by gently blotting the water from them with paper towels or bibulous paper.

11. Once the slides are completely dry, observe the stained bacteria under a microscope, as described in Lab 1. Record your observations of the cells in the Results section.

INDIRECT STAINING

1. Select three unused slides and label as follows:
   a. *E. coli*
   b. *M. luteus*
   c. *B. megaterium*

   Slides can be labeled directly or by sticking a little masking tape on one end of the slide.

2. Place the three slides on a staining platform over a staining tray, and add one small drop of nigrosin to one end of each slide. The drop should be about 1 cm in diameter and should be located about 1 cm from one end of the slide.

3. Flame your loop, allow the loop to cool, and then transfer a loopful of cells from the appropriate species to the drop of nigrosin on the slide. Use the loop to mix the cells into the drop of nigrosin. Repeat with each of the other two bacterial species.

4. Use a second clean slide to smear the cell-filled nigrosin drop across the slide **(Figure 5.5)**. Hold the second slide at about a 45° angle with the edge of one end in contact with the surface of the first slide. Slowly pull the slide through the nigrosin drop and down the length of the first slide. This will spread the nigrosin drop into a thin film of variable thickness.

   Use a separate clean slide to smear each of the three samples.

5. Set the slide aside until all the water has completely evaporated. You can do other lab activities while you're waiting for the slides to dry.

6. Repeat steps 3–5 for the other two types of bacteria.

7. Once the slides have dried completely, heat-fix the cells to the slide to kill the cells and to improve their adherence to the slide. To heat-fix bacteria to a slide, pass the slide over the hot part of a Bunsen burner flame three times in 3 seconds. This step will be demonstrated by your lab instructor; **pay close attention during the demo.**

8. After heat-fixing, place the slides on the staining platform in the staining tray, and apply the different stains to the different slides as follows:

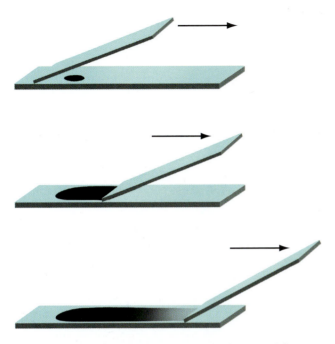

FIGURE 5.5 Making a smear slide for indirect staining.

5. Set the slides aside to air-dry. Do not blot the slides with paper towels or rinse them with water. Do not proceed until all of the nigrosin stain has dried, leaving a dull dark-colored smear across the slide.

6. After the slides have dried, observe the stained preparations under a microscope as described in Lab 1. Record your observations of the cells in the Results section. Examine several parts of the slide. In some locations, the nigrosin layer may be too thick, so cells may be completely covered, and all you will see is a dark layer of stain. In other spots, the layer may be too thin, and the cells will appear as white spheres and rods surrounded by halos of dark stain set against a light background. Like Goldilocks, you are looking for areas that are just right. These are places on the slide where the background is uniformly darkly colored and where cells appear to be circular or rod-shaped stars in a night sky (**Figure 5.6**). It may take a little time to find the right place, but in most cases, a given slide will have such areas.

FIGURE 5.6 Cells stained by indirect staining.
Cells repel stain and so appear colorless against a dark background.

Name: _____   Section: _____

Course: _____   Date: _____

DIRECT STAINING

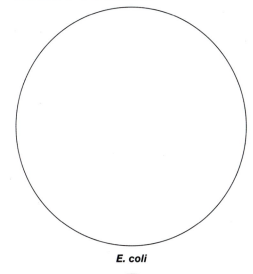

Written description: _____

E. coli

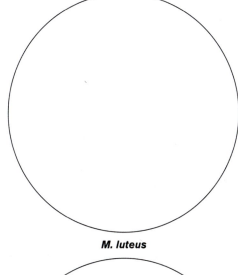

Written description: _____

M. luteus

Written description: _____

B. megaterium

INDIRECT STAINING

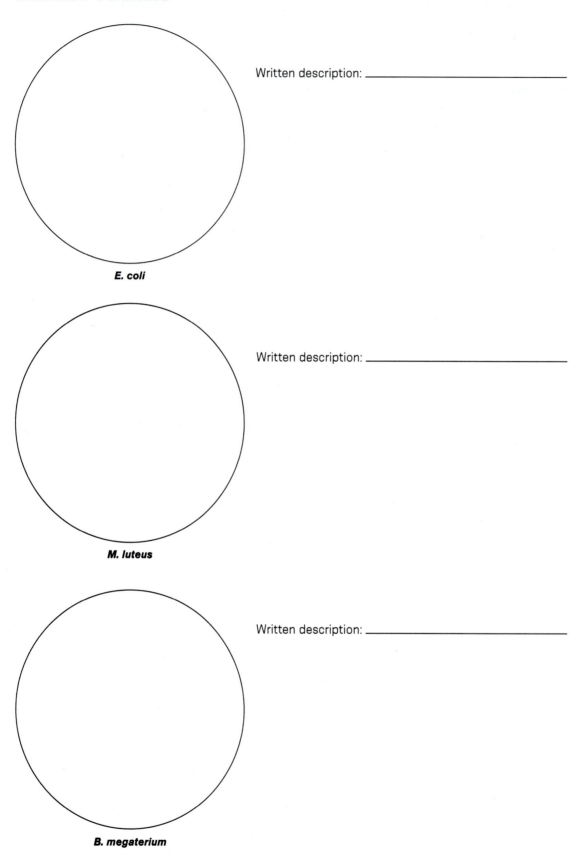

Written description: _____

E. coli

Written description: _____

M. luteus

Written description: _____

B. megaterium

QUESTIONS

Name: _____   Section: _____

Course: _____   Date: _____

1. What is the natural color of cytoplasm? How is this related to our need to stain bacterial cells?

2. Which of the following was the main goal of the staining done in this lab?

    a. Improve contrast.

    b. Identify bacteria.

    c. Visualize bacterial structures.

3. Define and sketch the following:

    a. bacillus

    b. coccus

c. strepto-

d. staphylo-

4. What is the difference between *"Bacillus"* and *"bacillus"*?

5. What happens to the cells when you heat-fix them, and what are two goals or purposes of heat-fixing cells prior to staining them?

6. How do direct stains work; that is, how or why do direct stains stain?

7. Why do cells appear to be white or colorless when you use indirect stains?

8. If direct stains work so well, why would we ever use indirect stains? Give two reasons why you might choose to use an indirect stain instead of a direct stain.

9. Assume that two dyes (dye A and dye B) were applied to a bacterial smear. The cells were stained by dye A while the slide background was stained with dye B. Which stain had an anionic chromogen? Which stain had a cationic chromogen?

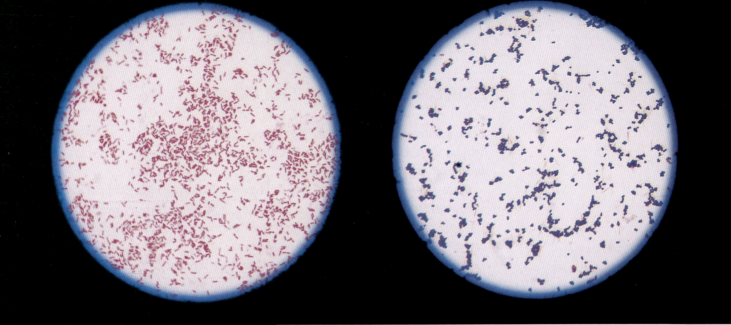

LAB **6**

The Gram Stain

Learning Objectives

- Understand the structural differences between Gram-positive and Gram-negative cell walls and consider why these differences lead to differences in staining properties.

- Know the four steps of the Gram stain process and understand what is happening to Gram-positive and Gram-negative cells at each step.

- Apply your understanding of the Gram stain to explain why Gram-positive cells might appear to be Gram-negative, and vice versa, paying particular attention to errors that can occur during decolorizing.

- Perform the Gram stain on pure and mixed cultures of Gram-positive and Gram-negative species.

INTRODUCTION TO THE GRAM STAIN

The **Gram stain** is unquestionably the single most important stain in bacteriology. It is a differential stain (see Lab 5) that allows us to divide the bacterial world into Gram-positive and Gram-negative organisms, a division that has many consequences and applications. As we'll see, the distinction between Gram-positive and Gram-negative cells is actually due to differences in cell wall structure between the two groups. By contrast, a simple stain will not distinguish between Gram-positive and Gram-negative organisms; the simple stain will tint all cell types the same color.

Determining **Gram morphology**, which is whether an organism is Gram positive or Gram negative, is often the first step in the process of identifying bacteria, especially in a clinical environment. Making this distinction not only aids in the diagnosis of infectious disease, but also guides antibiotic therapy, since the effectiveness of a given antibiotic may depend on whether the infection to be treated is caused by a Gram-positive or Gram-negative species. That is, the sensitivity of an infectious agent to a given antibiotic often correlates with the cell wall structure of the bacteria.

PRINCIPLES BEHIND THE GRAM STAIN

Gram-positive and Gram-negative species stain differently with the Gram stain because there are significant underlying structural differences between Gram-positive and Gram-negative cell walls (**Figure 6.1**).

Gram-positive cell walls have a thick layer of a latticelike biological polymer called **peptidoglycan**. The peptidoglycan accounts for about 90% of the dry weight of the wall. The pores in a Gram-positive peptidoglycan layer are very small. In addition, there is no outer phospholipid envelope or membrane.

Gram-negative cell walls have a much thinner layer of peptidoglycan. In Gram-negative organisms, the peptidoglycan accounts for only about 10% to 15% of the dry weight of the wall. Most of the cell wall is composed of an **outer envelope** or **outer membrane** of alcohol-soluble phospholipids, which is similar in structure to the inner cell membrane. The key point that is relevant to the Gram stain is that the peptidoglycan

FIGURE 6.1 Comparison of Gram-positive and Gram-negative cell wall structure.

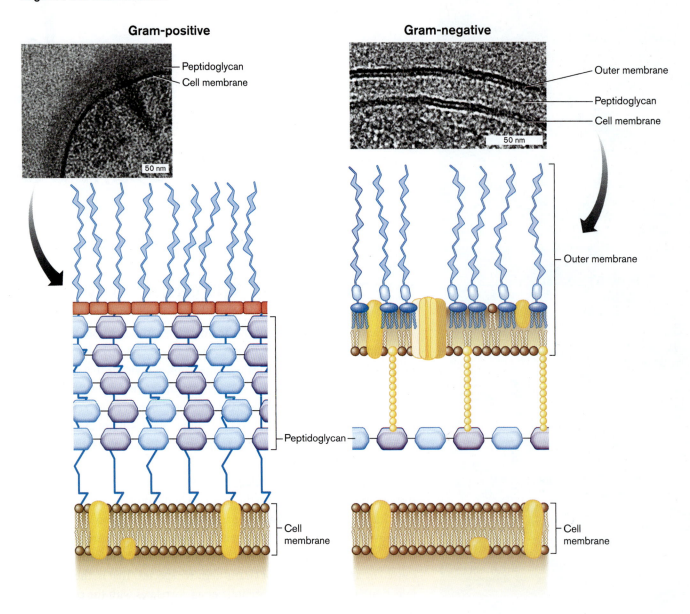

Gram-positive

Peptidoglycan
Cell membrane
50 nm

Peptidoglycan
Cell membrane

Gram-negative

Outer membrane
Peptidoglycan
Cell membrane
50 nm

Outer membrane
Peptidoglycan
Cell membrane

Dr. Gram and His Stain

The Gram stain is named for its inventor, Danish scientist Hans Christian Gram. In 1884, while working in a morgue in Berlin, Germany, Gram tried to develop a staining protocol that would differentiate *Streptococcus pneumoniae* from human lung tissue in patients that had died from this common bacterial cause of pneumonia.

He knew that Robert Koch and Koch's associate Paul Ehrlich had recently developed a differential staining technique to detect the tuberculosis bacterium (*Mycobacterium tuberculosis*) in lung tissue. Taking his cue from their research, Gram stained lung tissue infected with *S. pneumoniae* with the crystal violet stain used by Ehrlich. He then placed the slides in a solution of iodine-potassium iodine to precipitate the crystal violet. Gram then soaked the slides in alcohol to decolorize the tissue. The alcohol removed the crystal violet from the lung cells, but the dye was retained by the *S. pneumoniae* cells, creating a stark contrast between the pale yellow lung cells and the dark-purple cocci. Gram had achieved the differentiation between lung and bacterial cells that he was seeking, and he went on to use the procedure to stain other bacteria such as *Streptococcus pyogenes*, the causative agent of strep throat and scarlet fever, and the anthrax bacterium, *Bacillus anthracis*.

Had all bacteria stained in a similar manner as *Streptococcus*, Gram's invention would have been useful, but not monumental, in the history of microbiology. A hint that this stain was going to be something really big can be found in Gram's 1884 paper introducing the stain. In this report, Gram noted an odd "failure" of the new stain: When he tried the procedure on "typhoid bacilli," the bacterial cells failed to retain the crystal violet during decolorizing, even when he stained the cells for 24 hours. While it wasn't clear at the time, Gram had stumbled onto the fact that there is considerable variation in the structure of bacterial cell walls—what we now refer to as Gram-positive and Gram-negative cell walls. A Gram-negative cell wall is easily decolorized with alcohol, unlike a Gram-positive cell wall. The "typhoid bacilli" failed to retain the crystal violet because they were Gram-negative *Salmonella cells*.

How do we know that, in 1884, Gram didn't fully grasp what he'd discovered? Gram's first paper on the new stain, published in 1884, ends with the following (translated) passage:

> I am aware that as yet [my method] is very defective and imperfect; but it is hoped that in the hands of other investigators it will turn out to be useful.

This may qualify as the greatest understatement in the history of microbiology.

The Gram stain used today differs from the 1884 protocol of Hans Christian Gram. Gram didn't use a red-colored dye called safranin as a counterstain to improve the contrast of cells that had lost the crystal violet when rinsed in alcohol. The final safranin step was added a few years later by the German pathologist Carl Weigert as a way of visualizing the Gram-negative bacterial cells that were left colorless by the alcohol wash. This made it much easier to see cells from genera such as *Salmonella* at the end of the staining procedure.

And it's one of those odd historical facts that Carl Weigert was the older cousin of Paul Ehrlich. Ehrlich learned how to stain cells from his cousin Carl when the younger Ehrlich worked for Weigert before he moved on to work with Robert Koch. So, Weigert taught Ehrlich how to stain bacteria; Ehrlich's methods for staining tuberculosis bacteria guided and inspired Gram; Gram discovered that some types of bacteria retained crystal violet after an alcohol wash while other types lost the stain; and finally, it was Weigert who added the safranin step to "Gram's stain." It's the microbiologist's equivalent of the "circle of life."

Table 6.1 | COMPARISON OF GRAM-POSITIVE AND GRAM-NEGATIVE CELL WALL STRUCTURES

Characteristics	Gram-positive	Gram-negative
Percentage peptidoglycan	About 90%	About 10%–15%
Relative pore size	Smaller	Larger
Phospholipid envelope or membrane	No	Yes

layer is much thinner in Gram-negative cell walls. In addition, the pores in the Gram-negative peptidoglycan layer are larger than those in Gram-positive cell walls. These differences between the two type of walls are summarized in **Table 6.1**.

STAINING BACTERIA

The Gram stain is a four-step process. The steps and results are summarized in **Figure 6.2**. The steps are explained in detail below.

Step 1: Primary stain

The first stain used is **crystal violet**, a direct, basic stain that colors all cells of all types a blue-purple color (see Lab 5 for a discussion on direct, basic stains). If you looked at the cells at this point, there would be no visible difference between the Gram-positive and Gram-negative species.

Step 2: Mordant

The second step uses **Gram's iodine**, an aqueous solution of iodine and potassium iodide (KI), to fix the crystal violet stain to the cells. Gram's iodine is an example of a **mordant**, a word derived from the French word *mordre*, meaning "to bite." Mordants increase the affinity of a stain for the object to be dyed. The cloth-dyeing industry has used mordants such as tannins for centuries to improve the binding of dyes to textiles.

In the case of the Gram stain, the objects to be "bitten" by the dye are assorted structures in the bacterial cells. This iodine solution may help the crystel violet penetrate and adhere to the peptidolycan in the cell walls and to other cell components. The stain and mordant also form insoluble complexes. These complexes are more difficult to remove from cell walls and cytoplsam than are uncomplexed crystal violent molecules **(Figure 6.3)**. This is especially true when Gram-positive organisms are stained.

At this point in the staining process, all cells—Gram-positive and Gram-negative—will be stained the same color. While the crystal violet may be more tightly bound to the Gram-positive cell walls, the Gram-negative cells are also stained with crystal violet and iodine. At this stage, there is no way to visually distinguish between the two types of cells.

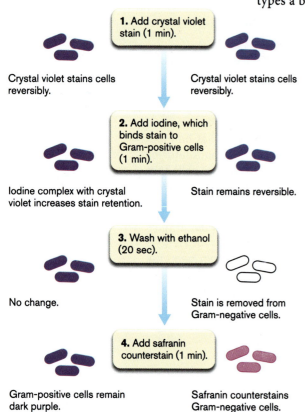

1. Add crystal violet stain (1 min).

Crystal violet stains cells reversibly.

Crystal violet stains cells reversibly.

2. Add iodine, which binds stain to Gram-positive cells (1 min).

Iodine complex with crystal violet increases stain retention.

Stain remains reversible.

3. Wash with ethanol (20 sec).

No change.

Stain is removed from Gram-negative cells.

4. Add safranin counterstain (1 min).

Gram-positive cells remain dark purple.

Safranin counterstains Gram-negative cells.

FIGURE 6.2 The Gram stain.

Step 3: Decolorizer

Now we come to the most critical step in the staining process. This is the step that will visually separate Gram-positive cells from Gram-negative cells. Either alcohol (usually ethanol) or acetone may be used as the decolorizing agent. Either will dissolve the outer phospholipid envelope of Gram-negative cells. In this lab, we'll use alcohol.

Gram-positive cells When Gram-positive cells are rinsed with alcohol, the **decolorizer** may actually decrease the permeability of the cell wall because it acts as a drying agent, and dehydration shrinks the small pores in the bacterial cell walls. Unlike phospholipid envelopes, the peptidoglycan is not dissolved by alcohol, and the dye complex that is trapped within the cell resists removal (see Figure 6.3). The net result is that the removal of the stain from Gram-positive cells is a very slow process.

While much of the crystal violet may eventually be washed out of the cells by the alcohol, as long as you show proper restraint and limit the decolorizing rinse to the prescribed amount of time, Gram-positive cells will retain the dye and remain purple in color.

Gram-negative cells When Gram-negative cells are rinsed with alcohol, the decolorizer removes the cell wall's alcohol-soluble phospholipid outer envelope. This removes any cationic stain molecules bound to negatively charged phosphate groups within the phospholipids. It also uncovers the larger pores in the thin peptidoglycan layer, and these pores cannot be completely closed by the dehydrating effects of the alcohol. Thus, the crystal violet is rapidly removed from the cells (see Figure 6.3), and within about 10–20 seconds from the start of the rinse, the cells are colorless. So Gram-negative cells are no longer stained purple.

Importance of the decolorized step Decolorizing is the first point in the Gram stain procedure where Gram-positive cells are visibly different from Gram-negative cells (and if you were Hans Christian Gram in 1884, you'd be done at this point). Since this decolorizing step is the step that differentiates positive from negative, it's the critical step in the Gram stain procedure. You'd better do it right.

If you don't decolorize enough, it is possible that Gram-negative cells will retain the crystal violet and appear falsely positive. This can result from ending decolorizing before alcohol running off the slide is colorless or when the smear on the slide is too thick. However, these

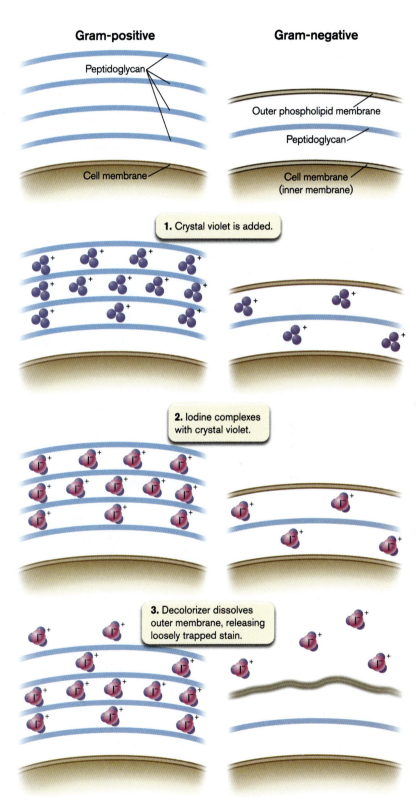

FIGURE 6.3 The fate of the crystal violet dye during Gram staining.

false positives are uncommon, because the alcohol rapidly removes the stain from the phospholipid-rich cell walls of Gram-negative cells. On the other hand, it is common for errors to result in Gram-positive cells that look red instead of purple. The key to successfully performing the Gram stain is to understand that Gram-positive cells may fail to hold the crystal violet under certain conditions, so that Gram-positive cells may appear falsely negative.

What can lead to the removal of crystal violet from Gram-positive cells?

1. **Over-decolorization.** Timing is everything. If you continue to rinse Gram-positive cells for a long time, they will eventually lose much of the crystal violet stain. So it's important to limit rinse time to 20–30 seconds, or until that instant when the alcohol draining off the slide changes from purple to colorless (see Figure 6.2). The exact time required can depend on the thickness of the smear, so it helps to practice until you can recognize the instant when the running alcohol color changes from purple to colorless.

2. **Age of the cells.** The peptidoglycan of older cells may not have the same structural integrity or ability to retain the crystal violet as do younger cells, so Gram staining should be done with fresh cultures that are less than 24 hours old.

3. **Acidity of the medium.** As we noted in Lab 5, when the pH of the environment is low, the excess hydrogen ions can interfere with the ability of basic stains, such as crystal violet, to bind to the negatively charged cell components. Since many species produce acidic waste products, medium pH tends to drop with time, and this is another reason why Gram staining should be done only with fresh cultures less than 24 hours old.

Step 4: Counterstain

We're almost finished. If the Gram stain ended with the decolorization step, it would leave the Gram-negative cells colorless and difficult to see. That's why Carl Weigert added a fourth step, a counterstain with safranin. Safranin, another direct, basic stain, binds to the colorless Gram-negative cells, and so the Gram-negative cells are stained bright red. Now we have the contrast that we need to see cells that lost the crystal violet. The purple color of the Gram-positive cells remains essentially unchanged because most of the binding sites for positively charged chromogens are already occupied by crystal violet, and adding a little red dye to a purple cell still leaves you with an essentially purple cell. Examples of Gram-stained cells are shown in **(Figure 6.4)**.

Final notes for lab

While there are Gram-negative cocci in the world, most notably in the genus *Neisseria*, we do not work with Gram-negative cocci in this lab. The cocci used in this lab are in the genera *Streptococcus*, *Enterococcus*, *Staphylococcus*, and *Micrococcus*, and all of these are Gram-positive genera. So, if you do a Gram stain, and you think you see Gram-negative cocci … something went wrong, horribly wrong. Either your Gram-positive cocci lost their crystal violet stain due to over-decolorizing or old age, or you've mistaken Gram-negative coccobacilli for true cocci. Stain again or look more closely.

A.   **B.**

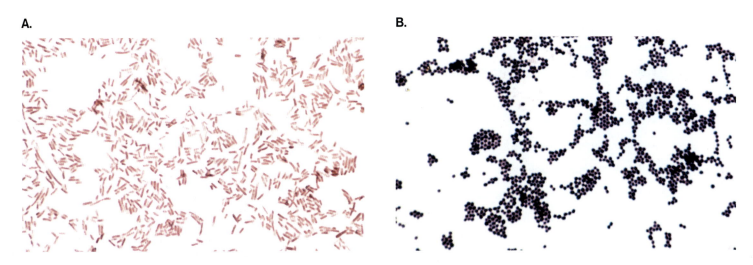

FIGURE 6.4 Gram-negative paired bacilli (A) and Gram-positive cocci (B) after Gram staining.

Cells in the Gram-positive genus *Bacillus* produce spores that are resistant to staining with the Gram stain procedure. They resist staining with crystal violet. They resist staining with safranin. So, when the Gram stain is done, the spores in the middle of the *Bacillus* cells will remain colorless. You may notice a spore as a white oval inside of a purple Gram-positive *Bacillus* cell. In Lab 8, we'll try our hand at staining these structures.

PROTOCOL

Wear disposable gloves throughout this lab.

Materials

- Disposable gloves
- Clean glass microscope slides (4 per student)
- Masking tape or label tape
- Pens or pencils for labeling tape on slides
- Slide holder or clothespins
- Bunsen burner
- Striker
- Inoculating loop
- Squirt bottle with water (wash bottle)
- Staining tray with staining platform (1 per two students)
- Bibulous paper or paper towels
- Compound microscope with 100× oil immersion lens
- Crystal violet
- Gram's iodine
- 95% ethanol
- Safranin

Cultures (overnight slant cultures):

- *Bacillus megaterium*
- *Enterococcus faecalis*
- *Escherichia coli*
- *Micrococcus luteus*

Note: In any lab in which you are required to make slides, begin your activities for the day by getting all slides to the air-drying step before going on to do other work.

1. Label four unused slides as follows:

   a. *Escherichia coli*
   b. *Enterococcus faecalis*
   c. *Bacillus megaterium*
   d. Mix of *E. coli* and *Micrococcus luteus*

   Slides can be labeled by sticking a little masking tape on one end of the slide. See Figure 6.2 for an illustration of the steps in this protocol.

2. Using an inoculating loop, transfer one or two loopfuls of water to the center of each slide. You want enough water to mix, suspend, and spread the cells, but not so much that it takes hours to air-dry the slide.

3. For each of the first three slides, flame-sterilize the loop, allow the loop to cool, transfer cells from a given species to the drop of water on the slide with the matching species name, and use the loop to mix and spread the cells and water. Remember that dragging the loop over the cell layer on the slant for a distance of 1.0 cm should be enough to pick up an ample number of cells for staining, and don't forget to flame-sterilize the loop between each species.

4. For the fourth slide (slide d, mixture of bacteria), flame the loop, allow the loop to cool, and transfer cells from the *E. coli* culture to the drop of water on the slide. Then repeat these steps to transfer cells from the *M. luteus* culture to the drop of water with the *E. coli* cells. Use your loop to mix the *E. coli* and *M. luteus* cells together in the water droplet and to spread water across the slide. When finished, be sure to flame-sterilize your loop before setting it down on the bench.

5. Set the slides aside to air-dry. Do not proceed until all the water has completely evaporated, leaving a dry, thin, whitish film of cells on the slide. While you are waiting for the slides to air-dry, keep busy with any other required lab activities.

6. Once the slides have completely and totally dried, heat-fix the slides by passing each slide over the hot part of the Bunsen burner flame three times in about 3 seconds.

7. After heat-fixing, place the slides on the staining platform inside of the staining tray.

8. Flood the first slide with crystal violet stain for a total of 60 seconds. Do not wipe the water off of the slide as this can wipe off the cells as well.

9. Pick up the slide with the slide holder or clothespins (or your gloved hand) to keep the stain off your fingers, and rinse the stain off the slide into the bottom of the staining tray using the wash bottle. Dry the slide by gently blotting the water from it with paper towels or bibulous paper.

10. Again, place the slide on the staining platform, and flood the slide with Gram's iodine (mordant) for a total of 60 seconds.

11. Rinse the iodine off the slide into the bottom of the staining tray using the wash bottle. Dry the slide by

gently blotting the water from it with paper towels or bibulous paper.

12. **This is the critical step.** Hold the slide at about a 45° angle above the slide platform. Dribble ethanol over the stained area of the slide in an unbroken, continuous stream that runs over the entire surface of the smear, and pay close attention to the color of the ethanol running off the slide. **When the ethanol at the bottom of the slide becomes colorless, STOP! You should not decolorize for more than 15–20 seconds.**

13. Use the wash bottle to <u>immediately</u> rinse the excess alcohol off the slide into the bottom of the staining

tray. Dry the slide by gently blotting the water from it with paper towels or bibulous paper.

14. Again, place the slide on the staining platform, and flood the slide with <u>safranin</u> for <u>60 seconds</u>.

15. Use the wash bottle to rinse the stain off the slide into the bottom of the staining tray. Dry the slide by <u>gently</u> blotting the water from it with paper towels or bibulous paper.

16. Repeat steps 8 through 15 for the remaining slides.

17. Once the slides are completely dry, observe them under a microscope. Record your observations in the Results section.

Name: _____   Section: _____

Course: _____   Date: _____

Draw some representative cells as seen under the microscope and write a description of the cells to the right of your drawings, including cell morphology (shape and arrangement), cell color, and whether they are Gram-positive or Gram-negative.
If colored pens or pencils are available, use the appropriate colors when drawing these cells.

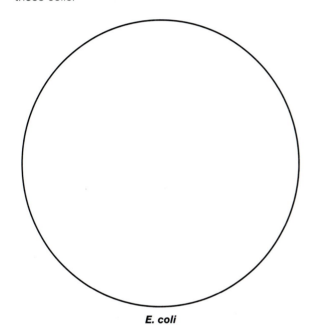

E. coli

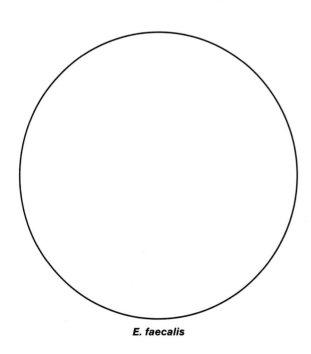

E. faecalis

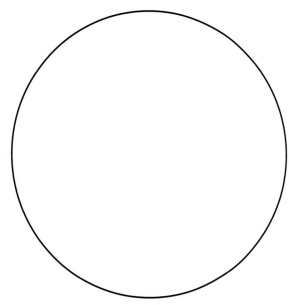

B. megaterium

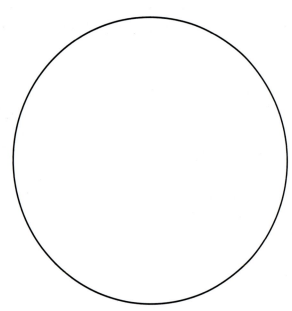

E. coli and *M. luteus*
mixture

Name: _____   Section: _____

Course: _____   Date: _____

1. Name three underlying structural differences between Gram-positive and Gram-negative cell walls that explain why these different cell types can be distinguished from each other using the Gram stain?

2. When *B. megaterium* cells are Gram stained, the cells may have white (colorless) ovals inside the cells. What are these ovals and why are they colorless? (Hint: Have a look at Lab 8.)

3. You do a Gram stain with a pure culture of *M. luteus*. When you look at the cells, half of the cells are purple and half of the cells are red. Assuming that your pure culture really was pure, explain why you see this result.

4. You do a Gram stain with a pure culture of *E. coli*. When you look at the cells, half of the cells are purple and half of the cells are red. Assuming that your pure culture really was pure, explain why you see this result.

5. You are working in a clinical laboratory engaged in disease diagnosis, and you receive a bacteria-rich sample from an infected patient. Why would you start by doing a Gram stain instead of a simple stain (Lab 5)? After all, the Gram stain takes more steps than a simple stain. In this context, why is Gram staining preferable to simple staining?

6. If you performed the Gram stain, but forgot to use the ethanol, what color would Gram-positive bacteria be at the end of the procedure? Assuming that you repeat the same mistake, what color would the Gram-negative bacteria be at the end of the procedure? In both cases, be sure to explain why you drew the conclusions that you did.

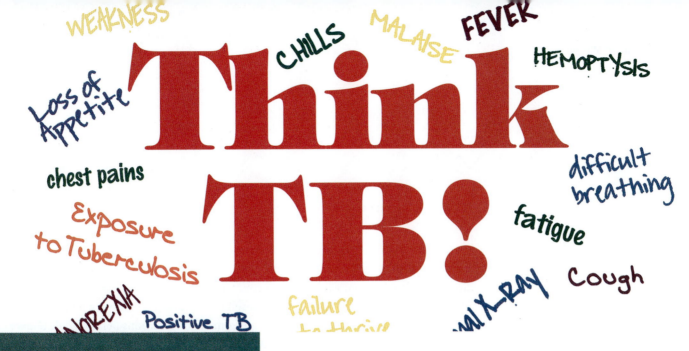

Like the Gram stain, the acid-fast stain identifies bacteria based on properties of the bacterial cell wall. Among the few bacteria that are acid-fast positive are the *Mycobacterium*, including *M. tuberculosis*, the cause of TB or tuberculosis. While we think of TB as a disease of our grandparents' generation, according to the World Health Organisation, 1.5 million people died of TB in 2014 and 9 million new cases of TB were reported. Of those 9 million people, more than 1 million of them were HIV-positive.

Learning Objectives

- Explain how the acid-fast stain acts as a differential stain and why this stain is valuable.

- Describe the structure of the acid-fast cell wall and understand the role of this waxy wall in the growth rate of acid-fast cells and in diseases cause by acid-fast bacteria.

- Understand what is happening to acid-fast cells at each step of the acid-fast stain and why this method is able to stain *Mycobacterium* and other acid-fast cells when other stains fail to do so.

- Perform the acid-fast stain on acid-fast and non-acid-fast bacterial species with an awareness of why acid-fast species might appear to be non-acid-fast.

LAB 7

The Acid-Fast Stain

The **acid-fast stain** is another example of a differential stain (see Lab 5), designed to distinguish one type of microbe from another. The acid-fast staining method differentiates bacteria with high concentrations of waxy lipids in their cell walls from microbes that lack these waxy lipids. Unlike the differential Gram stain, this stain distinguishes only a handful of genera that are "positive," or acid-fast, which limits its applications. However, these acid-fast groups include species in the medically important *Mycobacterium* and *Nocardia* genera. The genus *Mycobacterium* has particular clinical significance because it includes *M. tuberculosis*, the cause of **tuberculosis** (or TB), a disease that usually attacks the lungs but which can also involve infections of the bones, joints, and lymphatic system. Another medically important *Mycobacterium* species is *M. leprae*, the cause of **Hansen's disease** (also known as **leprosy**), a chronic disease of the nerves, skin, eyes, and respiratory tract which leads to the formation of small nodules, numbness at the sites of infection and loss of parts of the extremities. The acid-fast stain is still widely used in the diagnosis of these diseases.

The cells walls of bacteria in the *Mycobacterium* genus differ significantly from the walls found in the vast majority of Gram-positive and Gram-negative

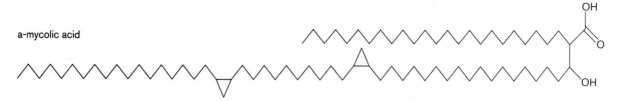

a-mycolic acid

FIGURE 7.1 Mycolic acids.

species. The *Mycobacterium* cell wall contains some peptidoglycan, but most of the wall is composed of complex and often waxy lipids such as **mycolic acids**. Mycolic acids are strongly hydrophobic molecules that account for a substantial portion of the dry weight of the acid-fast cell wall. They are composed of two unequal and very long chains of hydrogen and carbon, containing a total of from 60 to 90 carbons per molecule **(Figure 7.1)**. As a general rule, increasing the number of carbons in a fatty acid chain increases the waxiness and water repellency of a molecule. Compared with other common fatty acids, if the carbon chains in mycolic acids were like strands of hair, they'd be considered Rapunzel-esque in length. So these molecules are really, really waxy.

Mycolic acids are arranged in layers deposited just outside of the peptidoglycan, with the fatty acid chains arranged approximately perpendicular to the peptidoglycan layer **(Figure 7.2)**. This arrangement creates a thick lipid shell around the cell that significantly slows the movement of nutrients into the bacteria, leading to a much lower rate of growth and cell division compared with most other human-associated bacteria. Under ideal conditions, *E. coli* cells divide every 15–20 minutes, but cells of *Mycobacterium* species usually take several hours to reach the point at which they can divide. On the bright side (from the *Mycobacterium*'s point of view), the waxy walls

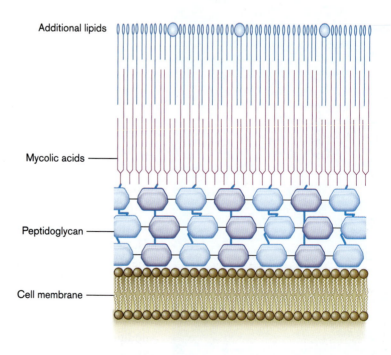

Additional lipids

Mycolic acids

Peptidoglycan

Cell membrane

FIGURE 7.2 Alpha mycolic acid.
The most common mycolic acid in the cell wall of *M. tuberculosis*.

significantly slow the rate at which water molecules leave the cytoplasm and enter the environment. Since drying can mean dying, these water-retaining walls allow the bacteria to survive outside the body in dry environments for long periods—for instance, if someone coughs up infected sputum and spits onto the floor of a railroad car (Figure 7.3).

While the sluggish movement of molecules into the cell might present a problem in nutrient acquisition, it can be a very good thing (again, from *Mycobacterium*'s point of view) if there are toxic molecules around, because keeping toxins out is one way to prevent these compounds from harming the bacterial cells. As a direct consequence of the acid-fast cell wall, *Mycobacterium* cells are among the most resistant to common disinfectants and antiseptics. It's certainly possible to kill TB bacteria with chemical agents, but compared with what is needed to control other types of microbes, it often takes more toxic chemicals, higher concentrations, and longer contact times to do the deed. Waxy walls also slow the movement of antibiotics into the cells, and not surprisingly, *Mycobacterium* species are resistant to many antibiotics. Even when the bacteria are sensitive to a drug, it's necessary to administer the treatment over many months, or even a couple of years, compared with the few days to a couple of weeks it takes to treat most other bacterial infections. To further complicate matters, many patients suffer from multiple drug resistant tuberculosis (MDRTB), a condition caused by strains of *M. tuberculosis* that have become resistant to the drugs routinely used to treat TB; these types of drug resistance can be due to mechanisms unrelated to the waxy cell wall.

In addition, the *Mycobacterium* waxy cell wall protects the bacteria against many mechanisms used by host immune systems to kill bacteria. For example, TB bacteria are resistant to lysozymes, the enzymes that protect us by degrading bacterial cell walls. *Mycobacterium* can also survive inside macrophages for a much longer time than most other types of bacteria. Macrophages are phagocytic white blood cells that engulf bacteria and other invading microbes in a process that normally kills them. So mycolic acids appear to play a major role in determining the **pathogenicity** or the ability to cause disease, of various types of *Mycobacterium* because bacterial cells that are harder to kill have a greater potential to invade, multiply, and spread through the body.

Finally, the presence of mycolic acids and related lipids in the cell wall greatly affects the staining properties of *Mycobacterium* and other acid-fast cells. Just as the waxy compounds restrict the passage of nutrients, disinfectants, and antibiotics, these same molecules make it difficult for dye molecules to move through the *Mycobacterium* cell wall, especially if the stains are dissolved in water. Acid-fast bacteria stain poorly with the Gram stain, although these species may be considered "Gram-positive," because they are more closely related genetically to other Gram-positive species than to Gram-negative species. However, Gram morphology is determined mainly by the structure of the cell wall and by staining properties, and since the cells stain poorly, these species might be better thought of as neither Gram positive nor Gram negative. Instead, they may be thought of as belonging to a third staining group, the **acid-fast bacteria**. The challenges involved in staining *Mycobacterium* cells were noted as early as the 1880s, and after much research and many modifications, what emerged were two acid-fast staining techniques: the **Ziehl-Neelsen method** and the **Kinyoun method**.

FIGURE 7.3 No Spitting.
Sign from a British railway car of the late 1800s or early 1900s. "Consumption" is another name for tuberculosis that was widely used in the nineteenth and early twentieth centuries.

Development of the Acid-Fast Stain

In 1881, Herr Doktor Robert Koch was determined to use the newly developed staining and pure-culture methods to discover the specific microbe that causes tuberculosis. He often observed long, thin rods in tubercles (nodules in the lungs), but these cells turned out to be quite difficult to stain due to their unusual waxy cell walls. Koch discovered that the cells could be stained by using an alkaline solution of methylene blue and staining times of up to 24 hours. He later heated the cells to about 40°C during the staining process, which reduced the staining time to about an hour. (Consider how these times compare to the times used in the Gram stain.)

While Koch's method did stain the cells of the species later named *Mycobacterium tuberculosis*, it also stained all other cell types blue as well. To increase the contrast between the bacteria and human cells, Koch applied a stain called vesuvin (Bismarck brown). This stain quickly colored all the human cells and other background materials brown, even if they had been previously stained blue by methylene blue. However, the *M. tuberculosis* cells held on so tenaciously to the methylene blue dye that they appeared as bright blue rods against a dark brown background of human cells.

Koch's stained *M. tuberculosis*. Figure published by Koch in 1884, in his paper entitled *Die Aetiologie der Tuberkulose* (The Cause of Tuberculosis). In these images, the bright blue *M. tuberculosis* cells stand out against brown-stained animal tissue.

Even better, while the *M. tuberculosis* cells retained the methylene blue dye when counterstained with vesuvin, almost no other bacteria did. Instead, following counterstaining, cells of most other bacterial species turned brown, just like human cells.[1] Thus, Koch had created one of the first differential stains, a stain that distinguished *M. tuberculosis* from almost all other bacterial species that might be present in a given animal. Now he could easily identify the presence or absence of *M. tuberculosis* in the lungs and other organs, and he could link the presence of a specific bacterial species to a specific disease. Koch announced his conclusion that *M. tuberculosis* caused tuberculosis in March 1882, less than a year after he set out to find the cause of TB. In 1905, Koch won the Nobel Prize in Medicine for his work on *Mycobacterium* and TB.

Building on Koch's discoveries, Dr. Paul Ehrlich found that staining times for *Mycobacterium* could be significantly reduced when he used water-saturated aniline oil to transport the stain into or through the waxy cell wall. Further, he changed the primary stain from methylene blue to a red-purple dye called carbolfuchsin or basic fuchsin, which is soluble in aniline oil. Significantly, Ehrlich discovered that aniline oil–delivered stains were difficult to remove from *Mycobacterium* cells, even when those cells were washed with strong decolorizing agents. In his words, he used "strong, even heroic concentrations of (nitric) acid" until "yellow clouds are given off," and yet the TB bacteria remained "intensely colored." Paul Ehrlich had discovered a property of *Mycobacterium* cells known as "acid-fastness," or resistance to decolorization by acids. This trait could be used to differentiate these cells from other types of bacteria because almost all other bacterial cells (and human cells, too) very quickly lose the carbolfuchsin stain when washed with very strong acid. He then used methylene blue (instead of vesuvin) as a counterstain, so in the case of tissues infected with *Mycobacterium*, this method produced slides showing red-dyed *Mycobacterium* rods against a background of blue-stained animal cells.

By the middle of 1882, almost all the basic concepts of the acid-fast stain had been discovered or established by Ehrlich and Koch. Today, however, the acid-fast stain is called the Ziehl-Neelsen stain or the Kinyoun stain (depending on the protocol) in recognition of modifications to the method made by other scientists. In August 1882, the bacteriologist Franz Ziehl published a paper in which he described the use of phenol (carbolic acid) in place of Ehrlich's aniline oil. In 1883, the pathologist

Friedrich Neelsen described a method for staining *Mycobacterium* that used carbolfuchsin in a phenol-alcohol-water solution, methylene blue as a counterstain, and a step in which the staining solution was heated to 100°C to partially liquefy the wax in the cell wall and improve the stain's ability to penetrate the cell wall. By the early 1890s, the practice of staining acid-fast bacteria using (1) carbolfuchsin in phenol (in alcohol and water), (2) heating to boiling during staining, and (3) methylene blue as a counterstain had become known as the Ziehl-Neelsen method.[2]

Finally, in 1915, the American bacteriologist Joseph Kinyoun published a method for staining *Mycobacterium* that did not require heating during staining. Kinyoun's method used higher concentrations of both phenol and carbolfuchsin to partially solubilize and penetrate waxy cell walls in the absence of heat. Though not quite as effective as the Ziehl-Neelsen method at getting the stain into the cell wall, it does have the advantages of being simpler to do and less likely to leave boiled stain all over the lab.

---

[1]Koch noted that bacterial cells found in leprosy patients were an exception to the rule that non-TB bacteria turned brown with counterstaining. It was later established that leprosy is caused by another species in the genus *Mycobacterium*, *M. leprae*.

---

[2]Unfortunately, Friedrich Neelsen had little time to revel in the glory, because he died of tuberculosis in 1898, at the age of just 44 years. The TB that killed him may have reached his lungs from material in his laboratory, or he may have been infected by the more mundane route of exposure to another individual with tuberculosis.

OUTLINE OF THE ACID-FAST STAINING PROCEDURE

Two traits of *Mycobacterium tuberculosis* and other acid-fast species are relevant to the methods used for staining these cells:

1. These cells have waxy cell walls that make it difficult to move dye molecules into and through the cell wall, so we'll need to use heat or high concentrations of phenol and dye molecules, or both, in order to stain these cells.

2. If you can get the stain into and through the cell walls, these cells are difficult to decolorize, even when washed in strong agents such as acid alcohol (alcohol that has been acidified with hydrochloric acid).

Step 1: Primary stain

In both the Ziehl-Neelsen and Kinyoun methods, the primary stain is **carbolfuchsin** (also called basic fuchsin), a stain that is soluble in phenol. The Ziehl-Neelsen method uses heat to partially liquefy the waxy layers of the cell wall, while the Kinyoun method relies on higher concentrations of phenol and dye to partially dissolve and stain the waxes. The Kinyoun method is a "cold stain," and in the absence of liquefying heat, longer staining times are required.

Step 2: Decolorizer

As with the Gram stain, the decolorization step is the key to differentiating types of bacteria. Acid alcohol will quickly remove carbolfuchsin from non-acid-fast species, rendering these cells colorless. In contrast, once the dye molecules have penetrated the waxy cell walls of acid-fast species, the layers of mycolic acids make it difficult to

A.

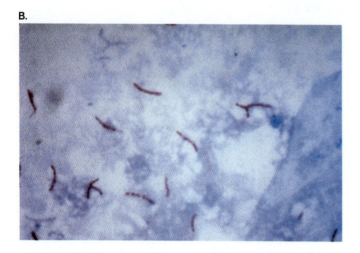

Step	Acid-fast	Non-acid-fast
Carbolfuchsin (primary stain)		
Acid alcohol (Decolorizer)		
Methylene blue (Counterstain)		

B.

FIGURE 7.4 Kinyoun staining.
A. Summary of the Kinyoun acid-fast staining method and expected results. **B.** Acid-fast positive cells (red-purple) against counter-stained background (blue).

remove the stain. Acid-fast cells will retain the stain longer, in part because (1) carbolfuchsin is more soluble in phenol and in the cell-wall lipids than in the acid-alcohol decolorizes and (2) the waxy mycolic acids are not very soluble in alcohol.

Always remember that acid alcohol is a very powerful decolorizer. This is not a big problem when staining pathogenic *M. tuberculosis* cells taken directly from the lungs of a patient because such cells have thick layers of mycolic acids and can hold onto the carbolfuchsin stain for up to a minute when washed with acid alcohol. Mycolic acid concentrations are probably correlated with both pathogenicity and the ability to resist destaining by acid alcohol, so the most virulent TB bacteria should be very good at holding onto carbolfuchsin.

In the teaching lab, however, it would be a bit beyond the pale if we exposed you to *M. tuberculosis*. Instead, we use nonpathogenic species of *Mycobacterium* such as *M. smegmatis*. As well as being much less likely to cause disease, these cells have become adapted to environments in which mycolic acids are not really needed for survival. So, while much safer to work with, these strains may have thinner mycolic acid layers and may not be as strongly acid-fast as *M. tuberculosis*. (Again, remember the likely correlation between mycolic acids, acid-fastness, and pathogenicity.) As a result, we usually use shorter decolorization times than might be used in a clinical lab, especially when using the Kinyoun method, which is not quite as effective at staining acid-fast cells.

Step 3: Counterstain

As noted above, after the cells are rinsed in acid alcohol, acid-fast (positive) cells will be purple-red in color, but non-acid-fast (negative) cells will be colorless. So **methylene blue** is used as a counterstain to visualize non-acid-fast bacterial cells and any human cells that might be present in a clinical sample. Thus, at the end of the staining procedure, acid-fast cells will be purple-red and non-acid-fast cells will be blue. **Figure 7.4** shows the expected results.

The Ziehl-Neelsen method is quite effective, but since it uses an open flame, it is more hazardous and often quite messy. In this Lab, we'll use the cold-stain Kinyoun method.

PROTOCOL

Wear disposable gloves and protective safety glasses throughout this lab.

Materials

- Safety glasses
- Disposable gloves
- Clean glass microscope slides (3 per student)
- Slide holder or clothespins
- Bibulous paper or paper towels
- Inoculating loop
- Staining tray with staining platform
- Compound microscope with 100× oil immersion lens
- Acid alcohol
- Kinyoun carbolfuchsin
- Methylene blue

Cultures:

- *Mycobacterium smegmatis* (48–72-hour culture)
- *Staphylococcus epidermidis*

THE KINYOUN METHOD

The acid alcohol decolorizer and the phenol in the Kinyoun stain can cause chemical burns, so you must wear disposable gloves and protective safety glasses throughout the staining procedure. **Since phenol is a toxic and hazardous chemical,** you should follow your institution's guidelines when disposing of this stain. Ask your instructior if you are unsure.

1. Select three unused slides and label as follows:

   a. *Mycobacterium smegmatis*
   b. *Staphylococcus epidermidis*
   c. Mixture

	Slide a	Slide b	Slide c
M. smegmatis	✓		✓
S. epidermidis		✓	✓

2. Transfer one or two loopfuls of water to the center of each slide using an inoculating loop.

3. Flame-sterilize the loop and allow it to cool. Transfer cells as shown in the table below to the drop of water on the slide with the matching label, and use the loop to mix and spread the cells and water.

4. Set the slides aside to air-dry.

   Do *not* proceed until all the water has completely evaporated, leaving a dry film of cells on the slide. The waxy cells of *M. smegmatis* tend to form clumps on the slide instead of a thin white film.

5. Once the slides have completely dried, heat-fix each slide by passing it over the hot part of the Bunsen burner flame three times in about 3 seconds.

   Due to the lipids they contain, *M. smegmatis* cells do not stick to the slide quite as well as other types of cells, so a little extra heat-fixing may be required.

6. After heat-fixing, place slides on a staining platform placed over a staining tray and flood the slides with Kinyoun's carbolfuchsin stain. Incubate for 10 minutes.

   A longer staining time is required with the Kinyoun method because we are not using heat to drive the stain into the waxy cells. **Monitor the slides during the 10-minute incubation period,** and add a little stain solution if the slides begin to dry out.

7. Pick up a slide with the slide holder or a clothespin (or your gloved hand) and use water to rinse the stain off the slide and into the bottom of the staining tray.

   Since phenol is not as water-soluble as alcohol, it might take a little more rinsing than usual to remove all the stain.

8. Dry the slides by gently blotting water from them with paper towels or bibulous paper.

9. **This is the critical step.** Hold the slide at about a 45° angle above the staining platform. Dribble acid alcohol over the stained area of the slide in a continuous stream for just 4 or 5 seconds.

   Acid alcohol is a very powerful decolorizing agent, and this short decolorizing time should be sufficient to remove the stain from non-acid-fast species while retaining at least some carbolfuchsin in the moderately acid-fast *M. smegmatis* cells.

   To prevent chemical burns, gloves should be worn as you are carefully disposing of waste acid alcohol, and the waste reagents should be diluted thoroughly with water when discarded.

10. Use water to immediately rinse the excess acid alcohol off the slide and into the bottom of the staining tray. Dry the slides by gently blotting the water from them with paper towels or bibulous paper.

11. Again, place the slides on the staining platform, and flood the slides with methylene blue. Incubate for 45 seconds.

12. Use water to rinse the stain off the slides and into the bottom of the staining tray. Dry the slides by gently blotting the water from them with paper towels or bibulous paper.

13. Once the slides are completely dry, observe the stained bacteria under a microscope. Record your observations of the cells in the Results section.

Name: _____   Section: _____

Course: _____   Date: _____

Draw some representative cells as seen under the microscope, and write a description of the cells to the right of your drawings, including cell morphology (shape and arrangement), cell color, and whether they are acid-fast-positive or acid-fast-negative. If colored pens or pencils are available, use the appropriate colors when drawing these cells.

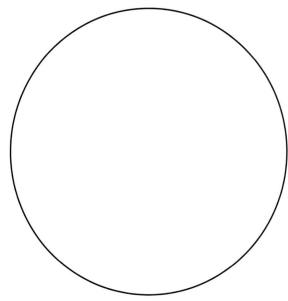

M. smegmatis cells only, stained by acid-fast staining

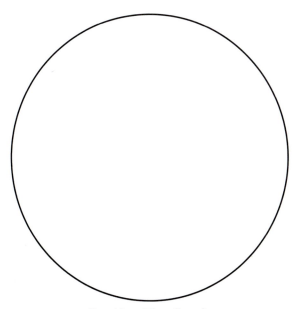

S. epidermidis cells only, stained by acid-fast staining

Mixture of *S. epidermidis*
and *M. smegmatis* cells,
stained by acid-fast staining

Name: _____   Section: _____

Course: _____   Date: _____

1. How is the acid-fast stain able to stain the waxy cell walls of *Mycobacterium* cells when other staining procedures fail to stain these cells?

2. How is the capacity of the *Mycobacterium* cell wall to retain carbolfuchsin in the presence of acid alcohol related to our ability to use the acid-fast stain as a differential stain?

3. What color or colors were observed with the *S. epidermidis* cells? Explain what you saw; that is, why did you see the colors that you saw?

4. What color or colors were observed with the *M. smegmatis* cells? What colors did you expect to see, and why was this your expectation?

5. Give two reasons why you might see some blue *M. smegmatis* cells.

6. The acid fast stain is an example of a differential stain. Name another differential stain.

7. While working with tuberculosis, Robert Koch used methylene blue to stain the bacteria *Mycobacterium*. Yet when working with the same disease, Paul Ehrlich used the same dye, methylene blue, to stain the tissue not the bacteria. How could the one dye stain two different things and yet both methods demonstrate the presence of acid-fast bacteria?

8. When Robert Koch set out to discover the cause of tuberculosis, he tried to isolate bacterial colonies from samples taken from TB-infected lungs. However, after a day of incubation, there were no visible colonies on the solid media used in Koch's lab. After 2 days, still nothing. In fact, it often took up to a week before anything appeared on the surface of the media. This was quite different from what Koch observed when growing other types of bacteria, such as *Bacillus anthracis*, the cause of anthrax. Fortunately for our understanding of TB, Koch was a patient man.

    a. How is the *Mycobacterium* cell wall structure connected to what was happening in Koch's lab?

    b. How are these observations by Koch related to difficulties in treating tuberculosis?

9. Once the tuberculosis bacterium and its transmission modes were identified in the late 1800's, many municipalities passed laws making it illegal to spit on the street. Why was this a concern? Would it still be a concern today for tuberculosis or any other diseases?

10. If you were a doctor, describe two cases or situations in which you might order a lab to perform an acid-fast stain. That is, what types of disease symptoms would prompt the use of this stain in the examination of a medical specimen in a clinical laboratory? (You may need to do some research to learn more about tuberculosis.)

LAB

8

The Endospore Stain

Learning Objectives

- Explain how the structure of endospores makes them resistant to heat, chemical toxins, and other environmental stressors.

- Know some of the diseases associated with specific endospore-forming bacteria.

- Understand how the endospore stain is able to stain endospores while conventional stain procedures such as the Gram stain fail to stain these structures.

- Know the steps of the endospore stain and understand what is happening to endospore-containing cells at each step.

The **endospore stain** is an example of a **structural stain**, which is designed to visualize particular structures in or on a bacterial cell—in this case, **endospores**. Other structural stains include the capsule stain and the flagellum stain. Endospores are extremely tough, durable structures produced by just a few genera of bacteria. At least two of those genera, *Bacillus* and *Clostridium*, both common in soils, are of interest to us because species in these genera are sources of human disease, and paradoxically, because *Bacillus* species are a source of some antibacterial drugs.

ENDOSPORE FORMATION AND STRUCTURE

Endospores are formed by bacteria when environmental conditions become unfavorable; for example, when certain amino acids are no longer available or when water becomes less available. Given the choice between dying or **sporulating** (forming endospores), the cells of *Bacillus* and *Clostridium* will typically choose sporulation. Wouldn't you?

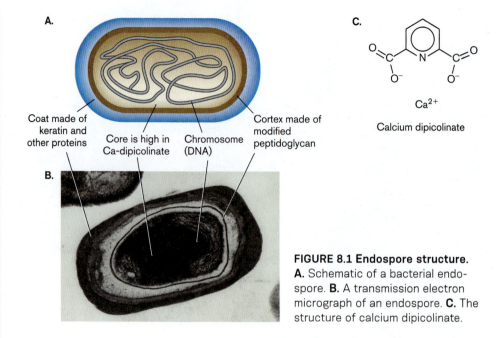

FIGURE 8.1 Endospore structure.
A. Schematic of a bacterial endo-spore. **B.** A transmission electron micrograph of an endospore. **C.** The structure of calcium dipicolinate.

Once formed, the metabolically inactive endospores are much more resistant than active vegetative cells to the damaging effects of drying, heat, radiation, and a variety of toxic chemicals, including chemicals often used by humans to kill bacteria. They are not indestructible, but broadly speaking, bacterial endospores are probably the most difficult of all living things to kill. They are mighty tough little critters.

The durability of the endospores is a product of their structure **(Figure 8.1)**. Endospores are composed of multiple, complex layers of resilient materials such as keratin and peptidoglycan. **Keratin** and other proteins are abundant in the thick outer endospore coat. Keratin is the same protein that is used by your body to make skin, hair, and fingernails. Within the protein shell, a modified form of peptidoglycan forms the middle layers or cortex of the endospore. In the interior, or core, of the endospore, there is a high concentration of **calcium dipicolinate** (calcium salt of dipicolinic acid), a compound that is unique to endospores. It is thought that the calcium dipicolinate plays a major role in the endospore's resistance to heat and radiation. It should also be noted that the water content of the endospore is very low. The lack of water makes metabolic activity almost impossible and contributes to the endospore's resistance to chemical toxins, as most toxic chemical reactions in cells either require water or must occur in an aqueous environment.

ENDOSPORES AND DISEASE

Endospore formation is a concern to us because some *Bacillus* and *Clostridium* species can cause human disease. Prevention of these diseases often relies on techniques that manipulate the environment to either kill or prevent the germination of endospores.

Bacillus species

Bacillus cereus Vegetative or metabolically active *B. cereus* cells can produce **enterotoxins** ("gut poisons"), which can cause two forms of **food poisoning**. Cooking the food doesn't always prevent this type of food poisoning, because endospores can survive certain cooking times and temperature and then later germinate into toxin-producing cells. The **emetic form** of food poisoning, characterized by cramps and vomiting (emesis), is the result of the consumption of a heat-stable toxin that is produced in food by *B. cereus* cells before the food is consumed. The **diarrheal form**, characterized by cramps and diarrhea, occurs when a large number of ingested cells and endospores produce a different set of *B. cereus* toxins within the digestive tract. These toxins can be produced in the gut either by existing cells that survive the passage through the very acidic stomach or by new cells produced when endospores germinate in the intestines. Fortunately, both forms are rarely fatal, and most sufferers recover within a day or two.

Bacillus anthracis *B. anthracis* is the causative agent of **anthrax**, a deadly disease that usually involves either a skin, digestive tract, or a lung infection; these infections are called cutaneous, gastrointestinal, and pulmonary anthrax, respectively. *B. anthracis* has recieved attention as a possible biological weapon because the durable endospores can be loaded into artillery shells for delivery. In 2001, someone mailed anthrax endospores in envelopes addressed to the U.S. Senate and a few media outlets. Five people died, including two employees at a post office that processed some of the letters.

Clostridium species

Most *Clostridium* species **(Figure 8.2)** are soil dwellers, and all are strict anaerobes. *Clostridium* spores can survive in the presence of oxygen, and they remain viable in the soil and other environments for many years. However, the cells produced from germinating spores are inhibited, and are often killed, by oxygen. Thus, the development of diseases caused by *Clostridium* requires an oxygen-free environment at some point in the disease process. Several diseases are caused by *Clostridium* species, often resulting from exposure to the organism through wounds or ingestion of contaminated food.

Clostridium botulinum *C. botulinum* causes **botulism**, a deadly type of food poisoning that occurs when endospores germinate under anaerobic conditions, such as in sealed cans of food. The cells produce a protein neurotoxin that blocks the transmission of signals from nerve cells to skeletal muscles. The resulting **flaccid paralysis** causes death when muscles essential to breathing fail to contract. If the toxin is injected around the eyes in the form of Botox, it paralyzes tiny muscles that cause wrinkles, producing a persistently surprised expression.

Clostridium tetani *C. tetani* is responsible for the disease **tetanus**, an illness that develops after the bacteria multiply in oxygen-free tissue. For example, a puncture wound can inject these soil bacteria deep into the body, and this trauma may also disrupt the blood supply to the area. If the injury results in extensive cell death and a spreading zone of anaerobic tissue, conditions will be ideal for *C. tetani*. The bacteria produce a neurotoxin that causes muscles to contract uncontrollably, resulting in a horrifying death from a **spastic paralysis** of chest muscles. So, get your tetanus shot.

Clostridium perfringens *C. perfringens* is the most common cause of **gas gangrene**, a type of wound infection in which the bacteria multiply in dead, anaerobic tissue. *C. perfringens* cells produce a toxic phospholipase, an enzyme that degrades the phospholipids in cell membranes, as well as a variety of other exotoxins (secreted toxins).

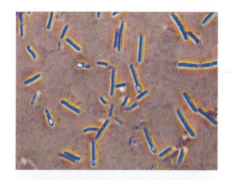
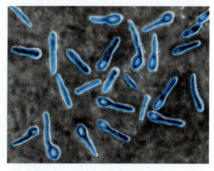
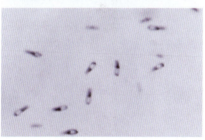

FIGURE 8.2 Endospores in *Clostridium* species.
Endospores vary in size and location within the bacteria. The size and location of the endospore can be a useful marker for identifying the bacterial species. (top) *C. botulinum*, (middle) *C. tetanus*, and (bottom) *C. perfringens*.

John Roebling and the Brooklyn Bridge

John Augustus Roebling, the designer of the Brooklyn Bridge, died from tetanus in 1869 after his foot was crushed in an accident at the bridge site. His son, Washington, also an engineer, took over the bridge project after his father's death and saw it through to its completion in 1883.

John Augustus Roebling.

The completed bridge is shown as it appeared in 1900.

The net result is a spreading infection that can cost the victim a limb or a life. During the Civil War, thousands of arms and legs were hit by what soldiers called "minnie balls," conical lead bullets designed by a French captain, Claude-Étienne Minié. Bones were shattered by the minnie balls, and infection followed. In these infected wounds, *C. perfringens* grew, causing gangrene and death. Amputation was the only effective treatment, although it often failed to stop the infection or save the life.

C. perfringens can also contaminate food products. Consumption of the food can result in the bacteria producing enterotoxins in the intestines followed by typical food poisoning symptoms of cramps, diarrhea, and vomiting.

Clostridium difficile *C. difficile* is adapted to the relatively anaerobic large intestines, and usually does not cause problems in healthy people whose digestive tracts are already well-colonized by a mix of beneficial and protective microbial species, referred to as the **normal flora organisms** (see Lab 29). However, *C. difficile* cells are more resistant to many types of antibiotics than many of the normal flora species, and of course, the *C. difficile* endospores are very resistant to these drugs. So, when certain antibiotics are used, many of the normal flora species are wiped out while the *C. difficile* cells and endospores survive. In the relative absence of competing species, a *C. difficile* population can expand rapidly into the intestines, where it can become a dominant species in the gut environment. This is a problem, because *C. difficile* cells produce toxins that can lead to **chronic diarrhea** and a condition called **pseudomembranous colitis**, an inflammation of the colon characterized by the formation of a membrane on the intestinal lining composed of fibrin, white blood cells, dying intestinal cells, and bacteria. Remember, these cells and endospores are resistant to many antibiotics, and so this condition can be very difficult to treat.

STAINING ENDOSPORES (THE SCHAEFFER-FULTON METHOD)

Part of the secret to the endospore's success is that its coat of many layers (see Figures 8.1 and 8.2) is almost completely impermeable to external chemicals; that is, outside chemicals cannot get inside the endospore. These chemicals include all the routine stains, such as the Gram stain, that we might otherwise use to visualize the endospores. These normal staining procedures leave the endospores unbowed, unstained, and colorless.

Given the resistance of the endospores to chemicals, how can we storm the Bastille of the endospore coat walls? In brief, **we will use heat**. Heat can be used to drive the primary stain through the protective layers and into the endospore. We can also take advantage of the fact that once a chemical, including a stain, has penetrated the endospore, the same layers that once kept the chemical out of the endospore make the stain very difficult to remove. **Figure 8.3** summarizes the steps for staining endospores.

Step 1: Primary stain

In the Schaeffer-Fulton version of the endospore stain, the primary stain is **malachite green**, a stain that normally binds relatively weakly to cell structures. Despite the name, there is no copper-containing malachite $[Cu_2CO_3(OH)_2]$ in malachite green **(Figure 8.4)**. The name of the dye refers to the similarity of the stain's green color to the color of the mineral malachite.

The key to staining endospores with malachite green is the application of heat during the staining process. If the stain is applied to spores or to endospore-containing cells for a short time at room temperature, the dye will be unable to penetrate the endospore coat, the endospores will not be stained, and nothing will be gained. However, if heat is applied during the staining process—and it will be—then the heat will drive the dye into the spores, staining them bright green. At temperatures around 100°C (steam heat), it takes about 5 minutes to do the trick. During this heating process, stain must be applied constantly, so that the cells and stain don't dry out. Lower temperatures (55°C–60°C) can be used if staining times are increased. In addition to staining endospores, this procedure will color all other cell parts green, so at this point, there will be no difference in color between endospores and the rest of the cell.

FIGURE 8.4 Malachite green.

FIGURE 8.3 Staining endospores.

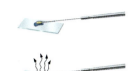

1. Place a loopful of water onto a slide. Add a loopful of bacteria and mix them together.

2. Spread into a thin film. Air dry.

3. Heat fix the cells to the slide.

4. Place the slide in a dish and cover the smear with a paper towel cut to fit. Flood the slide with malachite green. Incubate at 55°C–60°C for 15–30 minutes.

5. Discard the paper towel. Wash off the stain with water.

6. Blot away excess water.

7. Counterstain with safranin for 1 minute.

8. Wash off the stain with water.

9. View under microscope.

Step 2: Decolorizer

As stated above, malachite green binds weakly to many cell components, including non-spore structures. Since the binding is weak, the stain can be removed from the cell walls of both Gram-positive and Gram-negative cells and from cell parts other than endospores simply by rinsing the cells with water. There is no need to use typical decolorizers such as alcohol or acetone.

By contrast, once the stain has penetrated the multiple layers of the spore coat, it is difficult to remove from endospores. A quick rinse with water will certainly not be sufficient to wash out the stain; the malachite green will remain in the endospore. Thus, decolorizing leaves the green endospore inside an otherwise unstained cell. For the malachite green, it's as the Eagles put it way back in 1976, in the song "Hotel California," "You can check out any time you like, but you can never leave."

Step 3: Counterstain

While it's easy enough to see the green endospores at 1,000× magnification, everything else will appear colorless. Cell walls, cytoplasm, and all cell parts that were metabolically active before the staining will lose the malachite green when rinsed with water. It would be a shame if we missed these other bits; therefore, a counterstain is called for to improve the contrast and visibility of the decolorized structures.

The Schaeffer-Fulton method uses **safranin** as the counterstain. The stain is applied without heat, so no safranin will enter the endospores, but everything else will be stained bright red, including cells without endospores as well as cells that have formed endospores but that have not yet disintegrated. In the latter case, the staining produces a pattern of green endospores wrapped inside red cells (**Figure 8.5**). It's a happy combination of greens and reds that brings a touch of the holiday season to bleak microbiology labs all over the world. **Figure 8.6** summarizes the results of the endospore staining procedure.

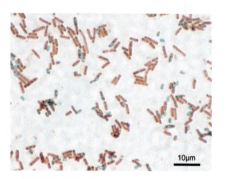

FIGURE 8.5 Stained endospores in *Bacillus*.

ENDOSPORE STAINING OPTIONS

There are several ways to do the endospore stain, and most methods use a Bunsen burner to provide the heat required to drive the green stain into the endospores. In some protocols, the flame is applied directly to the underside of the slide during the staining process. In other cases, the flame is used to boil water in a metal cup, and the slide to be stained is placed across the top of the cup so that steam from the boiling water heats the slide from below.

These methods are effective, but often quite messy because the malachite green stain tends to drain off the slide during heating and spread all over the lab. Fortunately, endospores can usually be stained by heating the slides to temperatures of only 50°C–60°C for 15–30 minutes, so it is not necessary to use burners. Descriptions of the various Bunsen burner methods may be found in numerous other lab manuals, but the protocol presented here does not use an open flame.

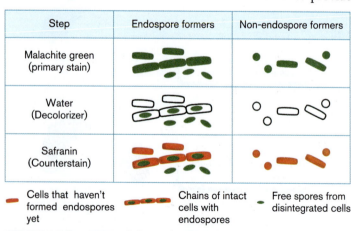

Step	Endospore formers	Non-endospore formers
Malachite green (primary stain)		
Water (Decolorizer)		
Safranin (Counterstain)		

— Cells that haven't formed endospores yet

⬤— Chains of intact cells with endospores

– Free spores from disintegrated cells

FIGURE 8.6 Summary of the endospore stain.

PROTOCOL

Wear disposable gloves throughout this lab.

Materials

- Disposable gloves
- Slide holder or clothespins
- Inoculating loop
- Clean glass microscope slides
- Kimwipes or paper towels
- Bibulous paper or paper towels
- Empty plastic petri dishes
- Staining tray with staining platform
- Compound microscope with 100× oil immersion lens
- Malachite green
- Safranin

Cultures:

- *Staphylococcus epidermidis*
- *Bacillus megaterium*[1]

PROCEDURES

1. Select two unused slides and label as follows:

   a. *Staphylococcus epidermidis*
   b. *Bacillus megaterium*

2. Transfer one or two loopfuls of water to the center of each slide using an inoculating loop.

3. Flame-sterilize the loop and allow it to cool. Transfer cells from a given species to the drop of water on the slide with the matching species label, and use the loop to mix and spread the cells and water.

4. Set the slides aside to air-dry.

   Do not proceed further until all the water has completely evaporated, leaving a dry, thin, whitish film of cells on the slide.

5. Once the slides have completely dried, heat-fix each slide by passing it over the hot part of the Bunsen burner flame three times in about 3 seconds.

6. After heat-fixing, place each slide in the bottom of an empty plastic petri dish. Cover the smear on each slide with a small section of paper towel or Kimwipe cut to a size that matches the area of the smear. The paper covering should not extend beyond the edges of the slide.

7. Flood the slides with malachite green stain such that the paper cover remains directly on top of the smear.

8. Place the petri dish in an incubator set to 55°C–60°C.

   When carrying the petri dish to the incubator, be certain to hold the dish level so that the pool of stain remains on top of the smear to be stained.

9. Incubate the slides at 55°C–60°C for at least 15 minutes.

   Endospores will be more intensely stained after a total incubation time of 30 minutes, but 15 minutes at 55°C–60°C is usually a sufficient time to stain them. If staining for more than 15 minutes, you may need to add stain to the slide during the incubation.

10. After incubation for 15 minutes or more, discard the paper towel, pick up each slide with the slide holder or clothespins (or your gloved hand), and wash it with water for about 30 seconds to remove the malachite green from the non-spore or cytoplasm parts of the cells. Dry the slides by gently blotting the water with paper towels or bibulous paper.

11. Place the slides on the staining platform inside of the staining tray, and flood each slide with safranin. Incubate for 60 seconds.

12. Pick up each slide with the slide holder or clothespins (or your gloved hand), and rinse the stain off with water. Dry the slides by gently blotting the water from each slide with paper towels or bibulous paper.

13. Once the slides are completely dry, observe the stained bacteria under a microscope. Record your observations in the Results section.

---

[1]Good endospore production can be achieved by incubating *B. megaterium* cultures on brain-heart infusion (BHI) medium slants for 24 hours at 37°C followed by incubation at 30°C for another 24 hours.

Name: _____   Section: _____

Course: _____   Date: _____

Draw and color some representative cells as seen under the microscope and write a description of the cells.

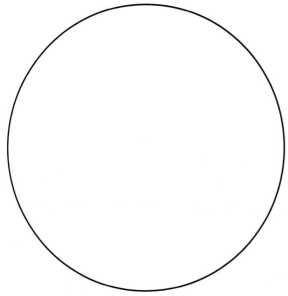

Written description of cells:

B. megaterium cells stained by endospore staining

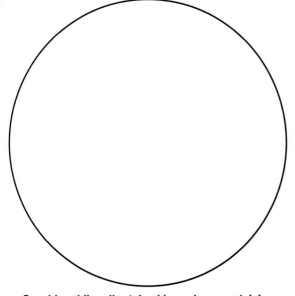

Written description of cells:

S. epidermidis cells stained by endospore staining

Name: _____   Section: _____

Course: _____   Date: _____

1. The endospores within *Bacillus* cells are colorless when the cells are stained with the Gram stain. How is the absence of color after Gram staining related to specific components of the endospore's structure?

2. How does the endospore's structure and metabolic rate provide resistance to heat, toxins, and other environmental stressors?

3. What color or colors were observed after using the endospore stain to stain the *S. epidermidis* cells? Explain what you saw when you observed the *S. epidermidis* cells.

4. If you used an endospore stain with *B. megaterium* cells and you didn't see green-colored structures in these cells, what are two possible explanations for this observation? What might you do to improve your chances of seeing something green?

5. The endospore stain is generally described as a "structural stain," but in what sense could this stain also be considered a "differential stain"?

6. Fill in the table below:

	Diseases Caused	Disease Description or Symptoms
B. cereus		
B. anthracis		
C. botulinum		
C. tetani		
C. perfringens		

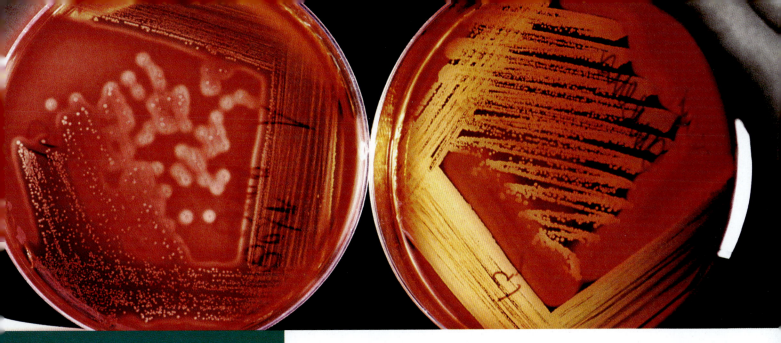

Look in a microbiology supply catalog, and you'll find hundreds of types of microbiological growth media. Given that bacteria and fungi grow practically everywhere, many of these media contain everyday nutrients, like egg yolk, oatmeal, liver, and blood. Blood agar plates can grow a wide variety of bacteria, because blood is a rich medium containing many nutrients that cells need to thrive. Here *Staphylococcus* colonies are growing on the left and *Streptococcus* is growing on the right.

LAB 9

//

Media Components, Bacterial Nutrition, and Sterilization

Learning Objectives

- Understand the basic metabolic needs of microorganisms and how microbiological media meet these needs.

- Consider the differences between complex and defined media and explain the advantages and disadvantages of each.

- Describe how autoclaves are used to sterilize a variety of materials, and when and why alternative methods should be used in place of autoclaving.

- Be aware of the rules and principles that will allow you to operate an autoclave safely.

When bacterial or fungal cells are transferred to growth media such as broths or agars, these cells must acquire everything they need from the environment that we have provided for them. The broths and agars must meet all their nutritional needs because, for the microbes, the tubes or plates represent the entire "world" in which the growing cells exist. And since "needs" vary with species, knowing what a given species needs and desires, and understanding how to meet those requirements, is a key to successful culturing of microorganisms. That is, we must answer the question "What do bacteria want?"

In addition, growth media must be sterilized before we inoculate them. Sterilization is essential for maintaining pure cultures and for correct interpretation of results. If there are microbes growing in a medium before we use it, then it is essentially worthless for our purposes. So, in this lab, we will also consider the various methods available to sterilize media.

BASIC NEEDS

Water

The first [cell], like the last and the one after that is, here, there and everywhere, for its vehicle, its medium, its essence is water. Water—the ace of elements. Water dives from the clouds without a parachute, wings or safety net ... water walks on fire and fire gets the blisters. Stylishly composed in any situation—solid, gas or liquid—speaking in penetrating dialects understood by all things—animal, vegetable or mineral—water travels intrepidly through four dimensions . . . (it has even been said that human beings were invented by water as a device for transporting itself from one place to another, but that's another story). Always in motion ... , the ongoing odyssey of water is virtually irresistible. And wherever water goes, [life goes] along for the ride.

—adapted from Tom Robbins, preface from *Even Cowgirls Get the Blues* (1976)

Water is obviously fundamental to life, and bacteria and fungi need water as much as we do. Fortunately, almost all media automatically provide plenty of water. Broths, of course, are almost all water with a few solutes added, but even "solid" media are usually at least 95% water. Nutrient agar, for example, is 97.7% water. If this seems surprising, given the firm nature of agars, remember that a typical agar-based medium is solidified at a concentration of just 1.5% (15 grams per liter) agar. As a rule, water availability becomes an issue only when the medium contains a high concentration of dissolved solutes; that is, when the environment is very hypertonic (see Lab 12).

Energy and carbon sources

In addition to water, all life must have a source of energy and carbon. **Carbon** is needed because cells are built of organic molecules, and the backbones or **carbon-skeletons** of organic molecules are built of rings and chains of carbon atoms. In short, life on Earth is carbon-based. **Energy** is required to build organic molecules, to transport atoms and molecules across the cell membrane, to move the cell, and for many other tasks as well. Without energy, there is no life.

When all microbial species are considered, we can see that single-celled organisms have evolved the ability to acquire energy and carbon from several possible sources. However, in this lab, life is simpler because we work almost exclusively with bacteria that, like us, are **chemoheterotrophs**. That means that in meeting the needs of our lab bacteria, we must provide organic molecules as a source of carbon and reduced organic molecules as a source of energy. Many reduced organic molecules can meet both the need for carbon to build molecules and the need for energy to operate the cell.

Organic compounds that can provide both carbon and energy for chemoheterotrophic bacteria are, on the whole, the same types of nutrient molecules that meet our needs, including sugars, other carbohydrates, fats, amino acids, and proteins. The specific subset of organic compounds required by a given microbial species will depend on which metabolic enzymes it can produce. As we'll see, in some cases, a bacterial species may have the metabolic capacity to derive all of the carbon and energy it needs from just one specific, organic molecule, such as citrate.

In some cases, these organic molecules are added to our media as pure chemicals; many media contain highly purified sugars such as glucose, lactose, and sucrose. Alternatively, carbon and energy may be supplied by crude preparations, including beef extracts and digests of casein, gelatin, meat proteins, and soy meal (see "Nitrogen Sources" below). These extracts and digests provide carbon and energy almost entirely in the form of amino acids and short-chain peptides, although soy meal digests also add significant amounts of carbohydrates.

As an example, nutrient agar (NA) contains only protein digests and beef extract, so most of the carbon and energy for the chemoheterotrophs growing on this agar is supplied by amino acids. But the medium can be supplemented with the high-energy carbohydrate glucose to create glucose nutrient agar (GNA). Well, why not add a little glucose? There's nothing like a spoonful of sugar to get your flagella spinning.

Nitrogen sources

Nitrogen is required because it is one of the five or six elements in amino acids, proteins, nucleotides, DNA, and RNA; all of these biological chemicals have nitrogen atoms in their molecules. We humans acquire almost all of our nitrogen from the amino acids in our diet. To get our daily nitrogen fix, we consume plant and animal protein and then absorb the amino acids derived from their enzymatic digestion in our small intestines.

Some of the 20 different types of amino acids that we need to build proteins and nucleotides can be made from other biological molecules, often from other amino acids. However, there are eight to ten **essential amino acids** that must be supplied in our food, either because we cannot synthesize these molecules from carbohydrates or other amino acids, or because we make too little of these amino acids. As a consequence, you can't keep a human body going by supplying it with inorganic nitrogen salts, such as $NH_4H_2PO_4$ or $(NH_4)_2SO_4$, as the sole source of nitrogen; this is noteworthy because, as we'll see, bacteria are different.

Amino acids as a nitrogen source: When bacteria act like humans

Most bacteria can acquire nitrogen by absorbing amino acids from their environments, just as we absorb amino acids in our small intestine. Many types of media provide nitrogen in the form of amino acids, either as pure chemicals (pure L-phenylalanine, pure L-lysine, etc.) or, more often, in the form of crude preparations. These preparations include beef or yeast extracts; infusions created by boiling animal organs, such as brain-heart infusions; and digests of animal and plant proteins. Many of these extracts and digests do double duty as nitrogen sources and as energy and carbon sources.

Protein digests include **peptones** and **tryptones**, terms used to describe a variety of amino acid–rich preparations. The protein sources used for these digests include casein, gelatin, meat proteins, and soy meal. However, most products sold specifically as peptones or tryptones use meat proteins or casein as the protein source (as opposed to, say, soy meal). The long peptide chains of the proteins are degraded by exposure to hot acidic solutions or by treatment with proteolytic enzymes. Enzymes that can be used include (1) **papain**, a papaya juice protease; (2) **pepsin**, a protease found in the stomach; and (3) the pancreatic proteases **trypsin** and **chymotrypsin**. The pancreatic enzymes are the ones most widely used in generic peptone production. The resulting peptone and tryptone products of these processes are crude mixtures of individual amino acids and short peptide (amino acid) chains in which the exact concentration of any given amino acid is unknown.

Inorganic nitrogen as the sole nitrogen source: When bacteria do something that exceeds our grasp No doubt bacteria are quite happy to acquire nitrogen in the form of organic nitrogen; that is, in the form of amino acids. But unlike humans, some species can acquire all the nitrogen they require from any of a variety of **inorganic nitrogen salts**. This is something that is physiologically impossible for us humans to do, so it would be appropriate at this time to applaud in wonder and amazement. Are you clapping?

The process of using inorganic nitrogen as a nitrogen source begins with **ammonium (NH_4^+) ions**. In aqueous (watery) solutions, this type of inorganic nitrogen may also be present in the form of uncharged **ammonia (NH_3)**. The exact ratio of NH_4^+ to NH_3 at equilibrium depends on pH; at pH 7, it's the ammonium form that is, by far, the most common species, accounting for 99% of the molecules in solution.

In a microbiological medium, the ammonium ions may be supplied by a variety of **inorganic ammonium salts**, including $(NH_4)_2SO_4$ (ammonium sulfate) and $NH_4H_2PO_4$ (ammonium dihydrogen phosphate). **Nitrate salts** such as $NaNO_3$ (sodium nitrate) can also be used by bacteria as a source of nitrogen, but the nitrates (NO_3^-) must be reduced to ammonium or amino nitrogen before they can be incorporated into amino acids, and this reduction is energetically expensive.

Once inorganic NH_4^+ ions are available in cells, they can be used to create the amino groups ($-NH_2$) of certain types of amino acids, such as **glutamate** and **glutamine**. That is, cells can take an inorganic nitrogen compound, add it to a carbon-skeleton organic molecule such as 2-oxoglutarate (alpha-ketoglutarate), and produce glutamate, and then glutamine **(Figure 9.1)**. Newly synthesized glutamates and glutamines can then donate their amino groups to other organic molecules, such as pyruvate and phosphoglycerate, to create yet more types of amino acids.

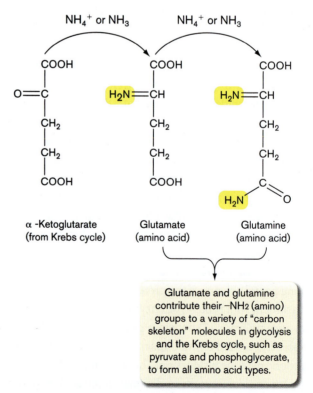

FIGURE 9.1 Synthesis of amino acids from inorganic nitrogen.

Cells can live on inorganic nitrogen as their sole source of nitrogen if the following activities can collectively produce all the required types of amino acids from other organic molecules:

1. Cells acquire or manufacture all the required precursor or intermediate non-amino-acid carbon-skeleton molecules from other **metabolites** (products of biological metabolism).

2. Cells transfer $-NH_2$ originally derived from NH_4^+ to those carbon-skeleton molecules to create new and different types of required amino acids.

3. Cells use one type of internally synthesized amino acid to create other required types of amino acids (for example, use aspartic acid to synthesize methionine, threonine, and lysine).

Here's where we humans fail. Our cells can add NH_4^+ to 2-oxoglutarate to create glutamate. However, we can't make all the required amino-accepting carbon-skeleton precursors, and we can't use certain types of internally synthesized amino acids to create other required types of amino acids. Hence, we need certain specific essential amino acids in our diet. In contrast, some species of bacteria can make all the organic precursors for all the amino acids, and with just a pinch of inorganic ammonium for nitrogen atoms, they can make all the required amino acids. If such bacteria are chemoheterotrophic, then it may take little more than glucose and ammonium salts to persuade them to grow and divide. Impressed? You should be.

Sulfur and phosphorus

In addition to carbon and nitrogen, cells also need **sulfur** and **phosphorus** atoms for a variety of biological molecules, including proteins and nucleic acids.

Sulfur atoms are found in the amino acids **cysteine** and **methionine**. They can be supplied by adding protein digests such as peptones, which contain cysteine and methionine in the mix of amino acids, to media. Alternatively, sulfur may be provided in the form of inorganic sulfur-containing salts such as $(NH_4)_2SO_4$ (ammonium sulfate) or $MgSO_4$ (magnesium sulfate). Such salts have the obvious additional benefit of providing useful cations such as NH_4^+ and Mg_2^+.

Phosphorus atoms are needed for the phosphate (PO_4) groups of ATP, for cell membrane phospholipids, and for the nucleotides used to synthesize the nucleic acids RNA and DNA. Crude meat extracts, such as beef extract, usually supply plenty of phosphates to a medium, and protein digests usually have some phosphates mixed in with the amino acids. Alternatively, phosphates may be added as inorganic salts such as HK_2PO_4 (dipotassium phosphate) or $NH_4H_2PO_4$ (ammonium dihydrogen phosphate); the latter salt also adds inorganic nitrogen to the mix.

Vitamins

Vitamins are needed as cofactors for many metabolic enzymes; these enzymes cannot function without their vitamin "assistants." For most bacteria cultured in this lab, it is not necessary to add vitamins to the medium as a separate component because, unlike humans, most of the bacteria that we work with are able to synthesize the required cofactors from other metabolites. In addition, trace amounts of vitamins are often present in widely used media components such as meat extracts and protein digests. However, if you're feeling generous, the vitamin content of the medium can be enhanced by adding vitamin-rich **yeast extract**, an excellent source of the many different types of B vitamins.

Trace elements or trace minerals

Finally, bacteria need tiny, trace—but not zero—amounts of elements such as **iron (Fe)**, **manganese (Mn)**, and **copper (Cu)**. Since the amounts required are very small, it's usually not necessary to add these minerals to media as separate, additional components because they will be present as contaminants of other media components. But magnanimous media makers may choose to mingle in a little mineral in the form of metal-containing salts such as $MgSO_4$ (magnesium sulfate) or $Fe(NH_4)_2(SO_4)_2$ (ferrous ammonium sulfate).

COMPLEX VERSUS DEFINED MEDIA

You may have noticed that we have been speaking of growth media components in two broad categories, (1) pure chemicals and (2) "crude," undefined extracts and digests. In this case, "undefined" means that we do not know the exact chemical composition of the extract or digest. That is, we don't have a complete list of which biological molecules are present; there may be hundreds of different types of organic molecules in an extract. In addition, we usually don't know the exact quantities for each of the chemicals in the mix, even when we do know for certain that a particular organic molecule is present.

Differences in media components lead to differences in media types: media may be either complex or defined.

Complex media

If any of the components of a given medium is undefined, then we call it a **complex medium**. That is, complex media have at least one component in "crude" form, meaning that you don't have complete knowledge of which specific biological molecules are present or you don't know the exact quantities or concentrations for each and every chemical in the mix. Therefore, the addition of any meat extract, any yeast extract, or any protein digest (peptone, tryptone, soytone) produces a complex medium. (**Figure 9.2** shows examples of complex media and components that might be added to make some complex media.)

Complex media are useful as general-purpose media; they meet the needs of most types of bacteria, at least most types of bacteria found in introductory microbiology labs. Nutrient agar and broth, tryptic soy (T-Soy) agar and broth, and brain-heart infusion agar and broth all contain undefined components, such as infusions, extracts, and digests, and all are used for routine culturing of common bacterial species. As a general rule, they are more likely than defined media to support the growth of **fastidious species**—that is, species that have complex nutritional requirements—because the extracts and digests provide many different types of biological molecules.

Fastidious species have strict requirements for particular nutrient molecules, such as a specific amino acid, vitamin, or mineral. If they don't get exactly what they need, they show their contempt for the microbiologist by not growing, or even dying. They are not unlike Morris the Cat from the old cat food commercials, whose philosophy was, "The cat who doesn't act finicky soon loses control of his owner"; Morris would rather waste away than eat anything other than 9 Lives cat food when called for "din-din." So, you'd better give Morris—and fastidious bacteria—exactly what they want.

FIGURE 9.2 Some complex media and media components.

Defined media

Defined media attempt to meet all the nutritional needs of the bacterial species to be cultured by providing known quantities of specific pure chemicals. Such media are good for "less fastidious" species, which can manufacture all the metabolites or carbohydrates, fats, and proteins they need from just a few types of organic molecules plus some inorganic salts for nitrogen, sulfur, phosphorus, and trace metals.

An example of a defined medium would be minimal agar, which contains just two organic compounds, glucose and citrate, and several different inorganic salts.

Minimal agar (per liter)

Glucose	1.0 g
Dipotassium phosphate	7.0 g
Monopotassium phosphate	2.0 g
Sodium citrate	0.5 g
Magnesium sulfate	0.1 g
Ammonium sulfate	1.0 g
Agar	15.0 g

Fastidious species are not likely to find the specific types of biological molecules they need on a list of ingredients such as the one for minimal agar. So, unless you know exactly what your fastidious microbe requires, and unless those specific types of molecules can be provided as pure chemicals in the defined medium, you'd better stick to complex media for culturing these organisms.

Since complex media seem to be better at meeting the nutritional needs of a wider range of bacterial species, why bother with defined media at all? As we'll see later in the course, there are times when your goal is to be very selective about what you culture in a given broth or plate. Microbiologists often start with samples that are mixtures of many, many different species, and we may want to favor the growth of just one or two of those species. If we can create a defined medium that promotes the growth of the species of interest while preventing the growth of other species by denying them essential nutrients, then we will have a mighty useful tool in the isolation and identification of specific bacteria.

Media used in this lab

In this lab, you will work with several different types of complex and defined media to determine the growth requirements for several species of bacteria. One of these media, citrate agar, will also be used later in the semester to identify species in the family Enterobacteriaceae. In that identification lab, a pH indicator will be added to the medium, but here, no pH indicator is used, as it might inhibit some of the non-Enterobacteriaceae species used in this lab.

Formulae for the media used in this lab are given below. Amounts are per liter of media.

Nutrient broth and agar (per liter)		Brain-heart infusion broth and agar (per liter) (EMD Millipore formula)	
Beef extract	3.0g	Nutrient substrate[1]	27.5g
Peptone	5.0g	Glucose	2.0g
		Sodium chloride	5.0g
Agar (if for plates)	15.0g	Disodium hydrogen phosphate	2.5g
		Agar (if for plates)	15.0g

[1]Nutrient substrate contains calf brain extract, beef heart extract, and peptones.

Citrate agar (per liter)

Ammonium dihydrogen phosphate	1.0g
Dipotassium phosphate	1.0g
Sodium chloride	5.0g
Sodium citrate	2.0g
Magnesium sulfate	0.2g
Agar	15.0g

Citrate + glucose agar (per liter)

Ammonium dihydrogen phosphate	1.0g
Dipotassium phosphate	1.0g
Sodium chloride	5.0g
Sodium citrate	2.0g
Magnesium sulfate	0.2g
Glucose	5.0g
Agar	15.0g

Citrate + peptone agar (per liter)

Ammonium dihydrogen phosphate	1.0g
Dipotassium phosphate	1.0g
Sodium chloride	5.0g
Sodium citrate	2.0g
Magnesium sulfate	0.2g
Peptone	5.0g
Agar	15.0g

Citrate + glucose + peptone agar (per liter)

Ammonium dihydrogen phosphate	1.0g
Dipotassium phosphate	1.0g
Sodium chloride	5.0g
Sodium citrate	2.0g
Magnesium sulfate	0.2g
Glucose	5.0g
Peptone	5.0g
Agar	15.0g

SUMMARY OF CITRATE AGAG MEDIA

CITRATE AGAR TYPE	CARBON SOURCE			NITROGEN SOURCE	
	Citrate	**Glucose**	**Peptones**	**Inorganic ammonium**	**Peptones**
Citrate	Yes			Yes	
Citrate + glucose	Yes	Yes		Yes	
Citrate + peptone	Yes		Yes	Yes	Yes
Citrate + glucose + peptone	Yes	Yes	Yes	Yes	Yes

STERILIZATION METHODS

Success in the microbiology lab, and in many other clinical and research environments as well, is dependent on our ability to sterilize gases, liquids, and solids. As noted above, unsterile growth media are of little use to us. Several sterilization methods are available, but we will focus here on autoclaving because this method is very widely used, it's almost always the way in which we sterilize the various media that we use in this course, and autoclaves are close at hand in the lab for demonstration purposes.

Autoclaves

Autoclaves are essentially large steam pressure cookers **(Figure 9.3)**. Steam, or high-temperature water vapor, is a key ingredient in the sterilization process because steam heat is moist heat. At any given temperature, moist air is much more effective at killing microbes than dry air because moist air contains more water molecules per unit volume than dry air. Compared to the oxygen and nitrogen molecules in the air, water vapor molecules are more efficient at transferring heat energy to the objects they bump into; and it's the heat that kills the cells in the autoclave. This efficiency is due in part to the heat of vaporization, which is released as the water vapor condenses to a liquid on the surface of an object, and in part to the transfer of heat energy when tiny, tiny droplets of very hot water impact cells, test tubes, flasks, plates, and other objects in the autoclave. The transferred heat energy then kills by denaturing proteins and liquefying cell membranes.

You can feel the effectiveness of moist heat when you compare the sensation of the heat of an oven (dry heat) as you reach in to remove cookies baking at 175°C (350°F) with the immediate pain caused by putting your hand into the 100°C (212°F) steam produced at the surface of a pot of boiling water. Do **NOT** deliberately do this experiment! A thermometer will tell you that the steam is cooler than the oven air, but the speed with which the skin is damaged by the steam will tell you that moist heat has a lot more killing power. Dry heat, such as the heat of an oven, can also sterilize, but sterilization usually requires higher temperatures and longer times than in the steam or moist heat of an autoclave.

Pressure is an important component of autoclave sterilization as well because when water vapor is under increased pressure, the temperature of the steam can also be increased. At atmospheric pressure (outside of an autoclave), the maximum temperature of steam at sea level is 100°C (212°F). But inside of an autoclave, where the pressure is raised an additional 15 to 20 pounds per square inch (psi), steam temperatures can be raised to 120°C (250°F), considerably increasing the killing power of the steam.

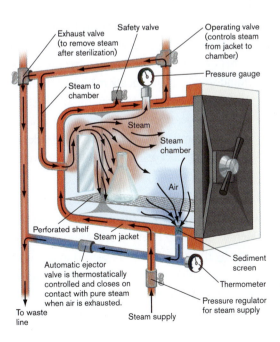

FIGURE 9.3 A steam autoclave.
An autoclave acts like a steam pressure cooker. When the autoclave door is closed, high temperature steam enters the chamber from the surrounding jacket, forcing the cooler air out through an opening in the bottom of the chamber.

At these temperatures, even bacterial endospores can be killed relatively quickly, and almost all materials can be sterilized in 15–20 minutes, although more time may be needed in the case of high-volume liquids.

Autoclaves are widely used and enormously useful, but their use can result in serious burns unless certain basic precautions are taken.

1. **Always use the "slow exhaust" setting when autoclaving liquids.** Autoclaves heat liquids, including any water-containing media, to temperatures of 120°C, or 20°C above the boiling point of water at atmospheric pressure. The liquids will not boil during the sterilization process because they are kept under 15–20 PSI of pressure. But if the pressure is rapidly removed by a "fast exhaust" and the liquids do not have time to cool to below 100°C before the pressure returns to atmospheric levels, then the liquids will bubble up and boil out of flasks, beakers, and test tubes, creating a rather unpleasant and dangerous mess. "Slow exhaust" releases the pressure slowly, keeping it high enough to prevent boiling as the liquids cool to below 100°C.

2. **Always check to be certain that the gauge showing the chamber pressure reads "zero PSI" before attempting to open the door of the autoclave.** One does not want to be hit by a blast of high-pressure steam when the autoclave is opened. Fortunately, most autoclaves are designed to make it difficult to open the door unless the pressure has dropped to zero (atmospheric pressure), but check the gauge anyway.

3. **Always put on autoclave gloves and safety glasses** before opening the door of the autoclave, and <u>always</u> wear autoclave gloves and safety glasses when removing material from the autoclave.

4. **Stand back a bit while opening the autoclave door.** When opening the door, even when the chamber pressure is zero, you can still be hit with residual hot water vapor if you are too close to the opening. This will reduce the odds of suffering from "lobster face" (first-degree facial burns).

5. **Do not swirl flasks, beakers, or test tubes that have just been removed from the autoclave.** Though the slow exhaust cycle should give liquids an opportunity to cool to below 100°C before the chamber pressure reaches "zero" (atmospheric pressure), liquid solutions and media may still be at temperatures close to or slightly above 100°C. At these temperatures, the slightest input of energy can push the liquids over the edge, so unless you enjoy the sensation of very hot liquids boiling out of containers and over your hands. Swirl a flask as you remove it from the autoclave and you will have a "lobster claw" to go with your "lobster face."

Other sterilization methods

Autoclaving works very effectively to kill all microbial life, but it cannot be used for all materials that must be sterilized. Many types of plastic objects are melted and rendered useless by the heat if they are placed in an autoclave. Some types of molecules, such as vitamins and pharmaceuticals, are **heat-labile**; that is, they are chemically altered or denatured and become biologically nonfunctional if exposed to autoclave temperatures. Other materials, including linens and certain types of paper, may be damaged or made less usable by the very high humidity inside the autoclave's chamber.

So, alternative methods of sterilization are needed. These methods are described in **Table 9.1.**

Table 9.1	STERILIZATION METHODS	
Method	**Mode of Action**	**Applications and Value**
Autoclaving	Uses pressure (15–20 psi) to raise steam temperature to 120°C; high-energy water vapor transfers heat to cells and kills by denaturing proteins and disrupting cell membranes	Very widely used to sterilize heat- and moisture-tolerant liquids and solids, including most media and glassware in microbiology labs Most materials can be sterilized in 15–20 minutes at 120°C
Ionizing radiation	High-energy gamma and X-ray radiation can strip electrons from atoms, creating very reactive free radicals that react destructively with proteins and DNA; inactivates proteins and mutates DNA	Used for some food products Widely used to sterilize heat-sensitive solids such as plastic petri dishes, plastic pipettes, and plastic syringes
Filtration	Physically removes or retains pathogens from air or liquids; typically removes all bacteria and may sterilize if pore size is small enough to restrict virus passage	Used to remove bacteria and viruses from the air to create sterile work areas and to prevent the escape of pathogens from research facilities Used in place of autoclaving to remove microbes from heat-sensitive or heat-labile liquids such as beverages and pharmaceutical solutions
Bunsen burner flames and large-scale incineration	Combusts, burns, and completely and totally oxidizes organic materials to CO_2, NO_2, SO_2, other gases, and solid ash	Used to sterilize loops and other high heat–tolerant metal or ceramic instruments for repeated use Also used to sterilize biohazard and medical waste, converting the waste to mostly gases and ash in the process
Ethylene oxide (C_2H_4O)	Used in gas form, alkylates (adds hydrocarbon chains to) proteins and DNA, and thus inactivates proteins and mutates DNA	Used as a "chemical sterilant" in "gas chambers" in place of autoclave to sterilize heat- and moisture-sensitive equipment, instruments, and linens Sterilizes materials after about 4–6 hours contact time

PROTOCOL

Day 1

Materials

Per instructor:

- 2 one-liter flasks, clean but not sterile
- Balance
- Spatulas
- Weigh boats
- Nutrient broth (NB) powder
- Brain-heart infusion (BHI) powder
- Distilled water (not sterile)
- 250-ml flasks, clean but not sterile (2 per student)
- Test tube racks

Per pair of students:

- 2 brain-heart infusion agar plates
- 2 nutrient agar plates
- 2 citrate agar plates
- 2 citrate + glucose agar plates
- 2 citrate + peptone agar plates
- 2 citrate + glucose + peptone agar plates
- Wax pencils or Sharpies for labeling plates
- Inoculating loop
- 250-ml flask with 50 ml of NB medium
- 250-ml flask with 50 ml of BHI medium
- 16 test tubes, clean but not sterile
- 16 test tube caps
- 10-ml pipettes
- 10-ml (green-barrel) Pipette Pump

Bacteria (broth cultures):

- *Citrobacter freundii*
- *Enterococcus faecalis*
- *Escherichia coli*
- *Micrococcus luteus*
- *Serratia marcescens*
- *Staphylococcus epidermidis*

GROWTH ON COMPLEX AND DEFINED MEDIA

1. Pick up two plates of each medium type (twelve plates total).

2. All the plates you will be working with look alike, regardless of medium composition, so be sure to label the plates immediately or take any and all other steps needed to keep track of the medium in each plate.

3. Divide one plate of each pair (that is, one plate of each medium type) into thirds by marking the bottom of the plate with a wax pencil or Sharpie, and label each plate as follows:

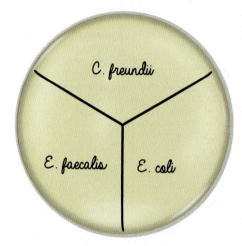

4. Divide the other plate of each pair into thirds by marking the bottom of the plate with a wax pencil or Sharpie, and label each plate as follows:

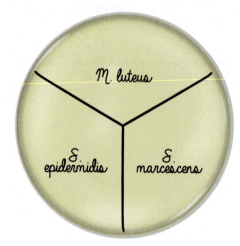

5. Dip a flame-sterilized inoculating loop into one of the pure-culture broths. Draw the loop out of the broth, and while the loop is still in the tube, gently tap the loop on the inside wall of the tube to remove excess culture material.

6. Use the loop to make a <u>short</u>, <u>single-line</u> streak in the appropriate third in each of the six plates labeled with the species in that broth.

Do not streak the entire third of the plate, and do not streak in such as way as to produce isolated colonies.

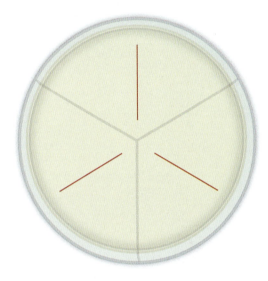

On day 2, we'll be looking for evidence that a given species can grow on a given medium, so do <u>not</u> transfer so much material that it looks as if cells have already grown on the plate and produced visible slime lines.

7. Repeat steps 5 and 6 until all twelve plates have been inoculated with three species per plate.

8. Incubate the plates at 37°C for 24 hours.

MEDIA PREPARATION AND AUTOCLAVE USE

1. Observe how your instructor prepares 1 liter of nutrient broth and 1 liter of brain-heart infusion broth, so that you understand how broth media are prepared. Broths will be dispensed into 250-ml flasks (about 50 ml per flask), and you will use one flask of nutrient broth and one flask of BHI broth for the next steps in the lab.

2. Label a total of 16 clean, but not sterile, test tubes with your names and the date.

3. Divide your test tubes into four sets of four tubes each and label the sets as follows:

a. Nutrient broth, autoclaved
b. Nutrient broth, not autoclaved
c. BHI broth, autoclaved
d. BHI broth, not autoclaved

4. Use a 10-ml pipette to dispense 5 ml of nutrient broth to each of the eight tubes labeled "Nutrient broth" (dispense to both the "Autoclaved" and "Not autoclaved" tubes).

5. Use the same 10-ml pipette to dispense 5 ml of BHI broth to each of the eight tubes labeled "BHI broth" (dispense to both the "Autoclaved" and "Not autoclaved" tubes).

6. Cap all tubes, and place the nutrient and BHI broth tubes that are to be autoclaved in the test-tube rack marked "Autoclaved." Place the remaining tubes in the rack marked "Not autoclaved."

7. Instructors should autoclave the tubes that are to be autoclaved immediately following the end of lab and then incubate them at 37°C for 48 hours.* The "Not autoclaved" tubes should also be incubated at 37°C for 48 hours.

*Instructors may wish to give students a tour of the autoclave area and provide them with further instruction about the safe use of the autoclave.

Day 2

Materials
- Plates and tubes from day 1

GROWTH ON COMPLEX AND DEFINED MEDIA

1. Examine your plates, and for each species and medium type, grade the amount of growth on and around the streak as 0, 1, 2, 3, or 4, where 0 means "no growth" and a 4 indicates very strong and thick growth. (Hint: For most species, growth on the BHI plate would be an example of a 4.)

2. Record your results in the upper table in the Results section.

USE OF THE AUTOCLAVE IN MEDIA PREPARATION

1. Examine all the nutrient broth and BHI broth tubes for turbidity and other indicators of microbial growth.

2. For each tube, record in the lower table in the Results section, the presence or absence of growth and estimate the relative amount of growth (more turbid versus less turbid, more sediment versus less sediment) when growth is present. Growth can also be described using terms from Lab 2.

Name: _____   Section: _____

Course: _____   Date: _____

Complete the table by grading the amount of growth on and around the streak as 0, 1, 2, 3, or 4.

GROWTH ON COMPLEX AND DEFINED MEDIA						
Species	Nutrient Agar	BHI Agar	Citrate Agar	Citrate + Glucose Agar	Citrate + Peptone Agar	Citrate + Glucose + Peptone Agar
Citrobacter freundii						
Enterococcus faecalis						
Escherichia coli						
Micrococcus luteus						
Serratia marcescens						
Staphylococcus epidermidis						

Complete the table by describing the turbidity and other characteristics of the growth in tubes.

USE OF THE AUTOCLAVE IN MEDIA PREPARATION				
Tube Number	Nutrient Broth, Autoclaved	Nutrient Broth, not Autoclaved	BHI Broth, Autoclaved	BHI Broth, not Autoclaved
Tube 1				
Tube 2				
Tube 3				
Tube 4				

Name: _____   Section: _____

Course: _____   Date: _____

1. Describe two different things that chemoheterotrophic cells might do with glucose.

2. Do all bacterial cells require glucose? Explain your answer; you can cite lab data in your explanation.

3. Unlike humans, some types of bacteria can grow in environments that lack amino acids; that is, they don't need amino acids in their "diet." How are bacterial cells able to do this? How can they do what we can't do?

4. What types of components would you add to a medium to meet a cell's need for each of the following?

   a. Sulfur

   b. Phosphorus

   c. Vitamins

   d. Iron

5. Which of the following media used in this lab are complex media and which are defined media? Explain your answer in each case.

a. Brain-heart infusion agar

b. Nutrient agar

c. Citrate agar

d. Citrate + glucose agar

e. Citrate + peptone agar

f. Citrate + glucose + peptone agar

6. Did you see any differences in terms of whether or not a given species grew on nutrient agar compared with BHI agar? That is, did any of these species fail to grow on one or both of the media? Explain your observations.

7. Did you see any differences in the amount of growth of various species on nutrient agar compared with BHI agar? If so, explain why there were differences in the amount of growth.

8. List all the species used in this lab that could grow using inorganic ammonium salts as their sole source of nitrogen. Cite data from this lab to support your answer.

9. List all the species used in this lab that require both glucose and preformed amino acids in order to grow. Cite data from this lab to support your answer.

10. List a species that you would consider to be "less fastidious." Explain your answer.

11. List a species that you would consider to be "more fastidious." Explain your answer.

12. Let's say that you were trying to isolate *Citrobacter freundii* from a sample that contained a mixture of species. Which medium would you use and why?

13. A patient specimen (body fluid) arrives in the lab. You plate the specimen on citrate agar and incubate. Nothing grows. Can you therefore assume that the body fluid was free of bacteria and no infection is present? Explain your answer.

14. When the pressure inside an autoclave is raised to 15–20 psi, does that increased pressure alone, by itself, kill the cells in the chamber? What is the purpose or value of raising the pressure inside the autoclave?

15. Autoclaves are very good at sterilizing liquids and solids of all types, so why don't we use autoclaves to sterilize everything?

   a. List <u>two</u> reasons why you would not use an autoclave to sterilize certain objects.

   b. Describe one thing that can't be autoclaved.

   c. Explain how the item you named could be sterilized.

16. Why are higher temperatures required when using dry heat to sterilize materials? That is, if you were using an oven to sterilize materials, why would you need a higher temperature than if you were autoclaving?

17. List three things to keep in mind for the safe use of autoclaves. (Yes, these things are already in the manual, but these are really important things, so write them down here, too.)

18. Did the results from the "Use of the autoclave in media preparation" part of this lab demonstrate that autoclaving is an essential part of media preparation? How did the results show this?

19. Did the appearance of the growth in each of the eight unautoclaved or "not autoclaved" broth types differ from tube to tube? If so, explain why there were differences.

20. Did the amount of growth in the four unautoclaved nutrient broth tubes differ from that in the four unautoclaved BHI broth tubes? If so, explain why there were differences.

21. Would you describe the cells that you observed in the unautoclaved tubes as belonging to a "pure culture?" That is, when there was growth in the unautoclaved tubes, did the unautoclaved tubes contain pure cultures? Explain your answer.

Starch is a polymer of glucose that is used as a storage molecule by most green plants. The polymer is much too large to be transported into bacterial cells, but the glucose would be a bounty of energy for bacteria that could get their hands on it (so to speak). Thus, many bacteria secrete amylases, enzymes that digest starch into simple sugars that the bacteria can take up. A sample of soil from this cornfield would contain many species of bacteria that produce and secrete these "exoenzymes."

LAB 10

Bacterial Enzymes and Use of Nutrients

Learning Objectives

- Distinguish and characterize extracellular and intracellular enzymes, and understand the critical roles these enzymes play in microbial metabolism.

- Understand and perform assays used to detect starch hydrolysis, carbohydrate fermentation, casein hydrolysis, and lysine decarboxylation.

- For each assay, know the critical components of the media used in the assay, the biochemical reactions involved in a positive reaction, the appearance of positive and negative results, sources of error, and the use and value of the assay.

In Lab 9, we learned about the energy, carbon, and nitrogen requirements of microorganisms. Like humans, many types of chemoheterotrophic bacteria meet these needs by using carbohydrates as sources for energy and carbon and by using amino acids and proteins as sources for energy, carbon, and nitrogen. Cellular metabolism of these molecules involves biochemical reactions catalyzed by specific **enzymes**, which convert specific substrates to specific end products. To utilize these molecules, microbes must have the genes for these enzymes.

We can detect the presence of these enzymes by measuring a decrease in the concentration of specific substrates or an increase in specific end products. Since the presence or absence of specific enzymes is a direct result of the presence or absence of specific genes, bacteria can be identified, in part, by the biochemical reactions they can perform. Biochemically based identification of bacteria is the focus of Labs 19–28. For now, our concern is with the different classes of bacterial enzymes and some examples of specific, detectable biochemical reactions.

CLASSES OF ENZYMES

Extracellular enzymes

Extracellular enzymes, or **exoenzymes**, are released by microbes into the surrounding environment, where they act on substrates outside the cells. Typically, these enzymes are secreted in order to break down high-molecular-weight substrates, such as **starch, proteins**, and **triglycerides**, that are too large to pass through the cell membrane. This is similar to what happens in the human small intestine, where digestive enzymes are secreted in response to the presence of food. In either case, microbial or human, once the large molecules have been split into lower-molecular-weight sugars, amino acids, and fatty acids, those smaller molecules can be absorbed to provide energy, carbon, and nitrogen.

Synthesizing enzymes is an expensive business for cells. Making these proteins consumes energy and ties up amino acids, both of which are precious commodities. So bacteria that live where lower-molecular-weight nutrients are available are generally less likely to produce and secrete exoenzymes. As an example, bacteria living in your large intestine may have less of a need for these enzymes, as your body has already done the heavy lifting involved in breaking down most types of large organic molecules.

In contrast, for microbes that live in a world where much of the energy, carbon, and nitrogen is sequestered in large, high-molecular-weight molecules, it is advantageous to produce enzymes that hydrolyze starch, lipids, and proteins. Microorganisms that are likely to produce high concentrations of these enzymes can be found in soils, where organic material and nutrients come mostly from decaying plant and animal material. In this environment, there is a significant advantage to being able to break down the starches, proteins, and lipids released by decomposing plants and animals. For example, *Bacillus* species are common soil bacteria, and most are champion exoenzyme producers (eukaryotic fungi also produce a very wide range of exoenzymes).

Intracellular enzymes

Most enzymes are intracellular enzymes, or **endoenzymes**. These enzymes act within the cell on nutritional substrates that have been transported into the cell and on other molecules synthesized internally by various metabolic processes. Endoenzymes may be involved in energy-releasing catabolism or in macromolecule-building anabolic pathways, such as protein synthesis. In the microbiology lab, most of the assays that we use are designed to detect catabolic enzymes involved in catabolic reactions, including respiration, fermentation of carbohydrates, and the use of amino acids.

STARCH HYDROLYSIS

Starch and amylases

Starch is a term used to describe very large, plant-produced polysaccharides composed of hundreds or even thousands of glucose molecules. There is some variation in the way the glucose molecules are linked together: **amylose** starch molecules are composed of glucose molecules linked in long, tightly coiled unbranched chains, while **amylopectin** starch molecules are composed of loosely coiled and branching chains of glucose (**Figure 10.1**). Starch molecules are too large to be transported across the cell membrane. But if these polymers can be broken down (hydrolized) into smaller molecules, the end products of hydrolysis can be taken up by cells and used as a source of energy and carbon.

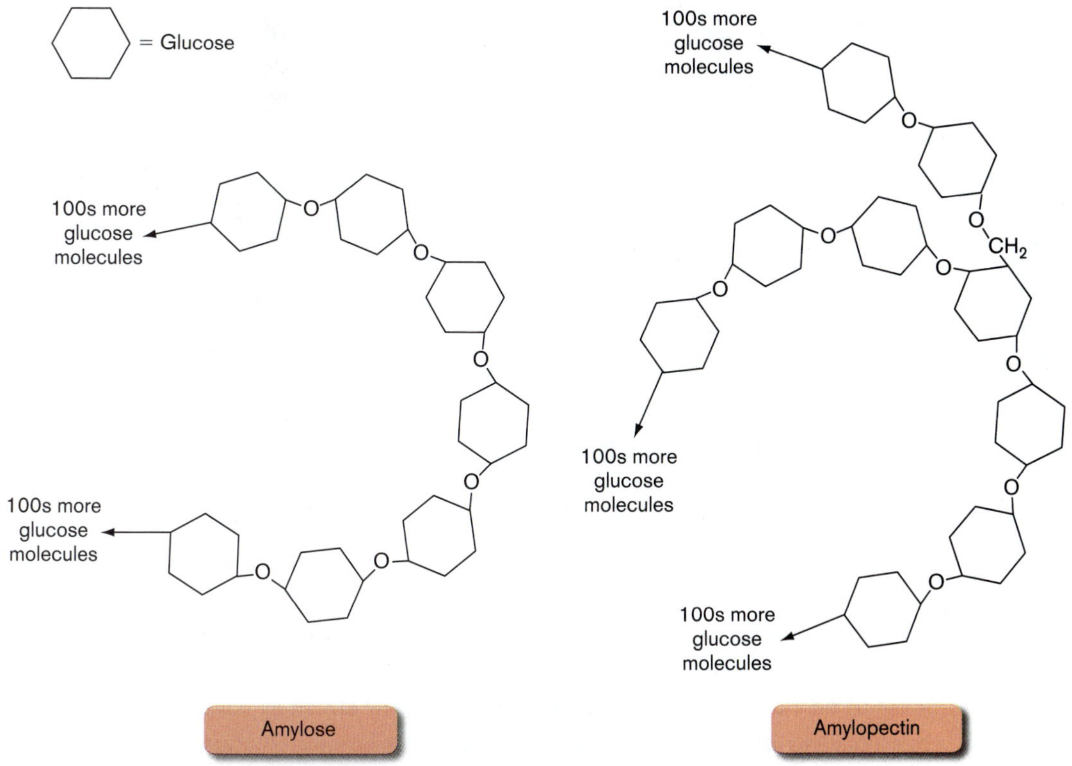

FIGURE 10.1 Structures of amylose and amylopectin starch molecules.

Many, but not all, bacterial species produce exoenzymes called **amylases**. Secreted amylases can convert a large, coiled starch molecule into a collection of glucose monomers or molecules containing two to about ten glucose subunits apiece. After the starch is broken down outside the cells, the bacteria transport the glucose molecules across the cell membrane and into the cytoplasm of the cell.

By the way, human saliva contains amylases. If you wish to experience the "amylase effect" firsthand, chew on a piece of bread for several minutes. The amylases in your saliva should begin to release glucose from the bread starches, and the chewed food should taste sweeter. This may be the only way to make cafeteria rolls taste good, but it takes patience and a lot of free time. And don't you have studying to do?

Detection of amylases: The starch hydrolysis assay

The **starch hydrolysis assay** is designed to detect amylases. The assay uses a medium called **starch agar**, which contains beef extract to provide energy, carbon, nitrogen, and vitamins for microbial growth and soluble starch as a substrate to monitor amylase activity. To perform the amylase assay, bacterial or fungal cells are streaked onto a starch agar plate. If the cells produce and secrete starch-hydrolyzing amylases, then there will be a zone extending out from the margins of the colonial growth from which the starch has been removed. To see if such a zone exists, we need to be able to detect the starch itself.

The detection of starch is based on the observation that iodine reacts with starch to produce colored complexes. Iodine binds strongly to amylose to produce a blue-violet iodine-amylose complex, and it interacts weakly with amylopectin to produce a red-brown or red-violet iodine-amylopectin complex. But iodine does not form colored complexes with glucose, so if the starch is gone as the result of amylase activity, there will be no blue-violet or red-violet to brown coloration of a starch-containing medium.

FIGURE 10.2 Positive starch hydrolysis assay.

Positive starch hydrolysis reaction If the cells growing on the starch agar medium do produce amylases, then there will be a starch-free zone extending out from the edges of the colonial growth. This zone will contain a mix of low-molecular-weight carbohydrates such as glucose, which do not interact with iodine, so no colored complexes are produced. Eventually, most of these nutrient molecules will be absorbed by the cells. Thus, when iodine is added to the plates, there will be a colorless zone out from the edges of the streak or a kind of colorless "halo" around the growing cells where the starch has been removed (**Figure 10.2**). Beyond the colorless halo, the medium will be colored red-violet to brown in those areas that the secreted amylases couldn't reach by diffusion.

So, if there is a colorless zone or halo extending out from the microbial growth surrounded by areas stained blue-violet or red-violet to brown, this is a <u>positive</u> reaction; <u>the cells produced amylase</u>.

Negative starch hydrolysis reaction If the cells growing on the starch agar medium do <u>not</u> produce amylases, then the starch in the agar will remain intact. There will <u>not</u> be a starch-free zone extending out from the margins of the colonial growth. When iodine is added to the plates, the medium will be colored blue-violet or red-violet to brown right up to the edge of the streak or up to the edges of any isolated colonies, and there will **not be a colorless zone** extending out from the streak or colonies. Note that the microbial growth in the streak itself will usually not be strongly stained.

So, if all areas of the plate up to the edge of the microbial growth are stained blue-violet or red-violet to brown, it's a negative reaction; the cells did not produce amylase.

PHENOL RED CARBOHYDRATE FERMENTATION ASSAYS

Phenol red carbohydrate fermentation assays are used to assess a microorganism's ability to produce the endoenzymes needed to ferment monosaccharides and disaccharides, including sugars such as glucose, lactose, and sucrose. Many bacteria use sugars as energy and carbon sources, but species differ in terms of (1) which types of sugars they can use and (2) whether they break down sugars by respiration or by fermentation. In Labs 24–27, we'll use these differences to help characterize and identify bacterial species.

Use of different sugars

Most chemoheterotrophic species can use glucose as the starting material for energy-releasing, ATP-producing glycolysis. But to use disaccharides such as lactose or sucrose, cells must produce additional enzymes specific for each substrate. Some of these additional enzymes split disaccharides into monosaccharides; for example, beta-galactosidase splits the disaccharide lactose into the monosaccharides glucose and galactose, and beta-fructosidase breaks down the disaccharide sucrose into glucose and fructose. Other enzymes are needed to convert galactose and fructose into glucose or into some other intermediate in glycolysis. So the lesson is this: depending on which enzymes they can produce, bacterial species vary with regard to which sugars they can use.

Respiration versus fermentation

The second possible difference between species is whether they metabolize sugars by aerobic respiration or by fermentation. We can determine whether the cells are aerobically respiring or fermenting by designing assays that detect the end products of these metabolic pathways (**Figure 10.3**).

Aerobic respiration In **aerobic respiration**, bacterial cells (like human cells) use oxygen as the electron acceptor in a complex, multienzyme, multistep process that oxidizes glucose, releases large amounts of energy for the generation of lots of ATP per glucose, and produces carbon dioxide (CO_2) and water (H_2O) as end products. In the carbohydrate fermentation assays described below, you will see that it's the end products that are detected, so in the case of aerobic respiration, it's the carbon dioxide and the water that are the important molecules to remember. You should note that since aerobic respiration requires oxygen, the CO_2 and H_2O are produced by cells growing near the top of the liquid medium, where oxygen is diffusing into the medium across the air-water interface. That observation will be important when it comes to interpreting the results of these assays.

Fermentation What do microorganisms do if oxygen is unavailable? One option is **fermentation**, which does not require oxygen or even genes for an electron transport system. Another option is **anaerobic respiration**, which like aerobic respiration, generates energy using electron transport systems, but these systems don't require oxygen as an electron acceptor (see Labs 14 and 33). But it is fermentation that we will focus on here. There are many different fermentative pathways in the biological world, but they all use organic molecules in place of oxygen as electron acceptors in energy-releasing metabolic pathways. Using organic molecules instead of oxygen as electron acceptors will allow cells to grow in the absence of oxygen. However, fermentation produces far less ATP per glucose than respiration does, so in the case of facultative anaerobes—species that can carry out both respiration and fermentation—growth and cell

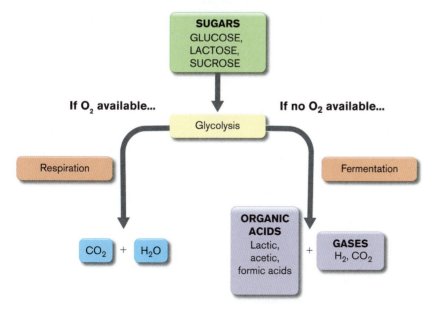

FIGURE 10.3 Respiration and fermentation.

division rates will be lower when oxygen is absent and the cells are forced to rely on fermentation.

Many different fermentation pathways are utilized in the microbial world, and these pathways produce a mix of many different products. Fortunately, your life will be made a little simpler by the fact that these products fall into only a few major categories, such as **organic acids** and **gases**. Depending on the microbial species, fermentation may produce any number of different organic acids, including lactic, acetic, formic, and butyric acids, as well as many different types of gases, including carbon dioxide (CO_2) and hydrogen (H_2) gas.

Detection of sugar fermentation

Phenol red carbohydrate fermentation assays are designed to detect the fermentation of sugars into a variety of organic acid and gas end products. The procedure can't tell you which specific organic acids or which specific acids plus gases are produced, but it will show that a given type of sugar was fermented by indicating the presence of some type of acid or some type of acid plus some type of gas.

These assays use nutritionally rich phenol red broths, which contain peptones and beef extract to meet the basic metabolic needs of a wide range of species. A single, specific type of sugar or other carbohydrate is added at concentrations of 0.5%–1.0% (5–10 grams per liter) as a substrate for fermentation. Note that a given phenol red fermentation broth contains one and only one type of carbohydrate. Common sugars used are glucose, lactose, and sucrose.

Phenol red: Detecting a change in pH Phenol red is a pH indicator molecule that is yellow in color when it is uncharged and red in color when it loses a proton (or hydrogen ion, H^+) and becomes negatively charged **(Figure 10.4)**. The color of any given solution containing phenol red depends on the ratio of yellow-colored molecules to red-colored molecules.

At pH 6.3, there is an excess of protons in the solution (that's what makes a solution acidic), so the phenol red molecules tend to pick up protons. Uncharged phenol red molecules outnumber negatively charged molecules by ten to one, so there are ten times more yellow-colored molecules than red-colored molecules, and the broth is clearly yellow in color.

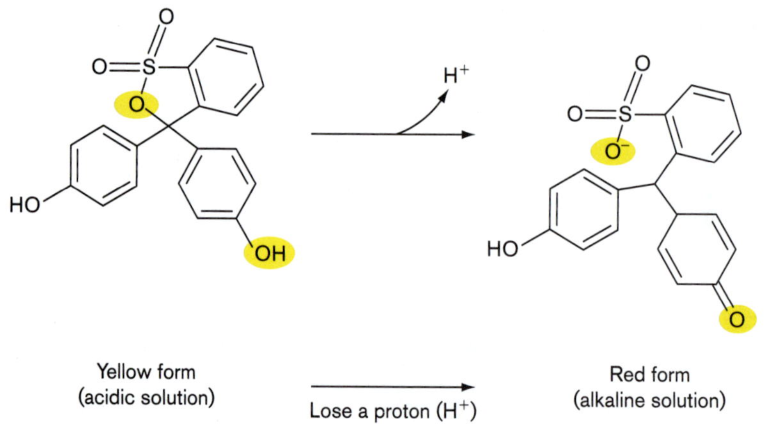

Yellow form
(acidic solution)

Lose a proton (H^+)

Red form
(alkaline solution)

FIGURE 10.4 Phenol red.

Between pH 6.8 and pH 7.8, the solution will go through a transition from yellowish orange to orange to reddish orange. At pH 7.3, there are equal numbers of yellow-colored and red-colored molecules, and the broth appears orange in color. Remember first grade? You learned that red and yellow make orange. Now orange you glad you learned that?

At pH 8.3, there's a relative shortage of free protons. So it's the negatively charged red form that is ten times more common than the yellow form, and the broth is red.

SUMMARY

SUMMARY OF PHENOL RED COLORS AT DIFFERENT pH VALUES		
pH	Ratio of Yellow to Red Phenol Red Molecules	Color of Broth
5.3	100 to 1	Yellow
6.3	10 to 1	Yellow
6.8	3 to 1	Yellowish orange
7.3	1 to 1	Orange
7.8	1 to 3	Reddish orange
8.3	1 to 10	Red
9.3	1 to 100	Red

Durham tube: Detecting gases To capture any gases that might be produced by fermentation at the anaerobic bottom of a broth, a small glass tube called a **Durham tube** is dropped into the test tube containing the broth prior to autoclaving. The opening of the Durham tube is at the bottom of the broth; that is, it is inverted relative to the test tube itself. When the broth is then sterilized by autoclaving, the heat and pressure generated by the autoclave dissolve the gases in the Durham tube into the liquid medium and drive the broth up into the upside-down tube. After autoclaving, there should be no bubble visible in the Durham tube prior to inoculation.

Positive and negative reactions

Fermentation of a given sugar can produce (1) organic acids only or (2) organic acids plus gases. Production of acids and production of gases are assessed independently of each other.

Organic acids: Positive reaction If the cells in the broth use the carbohydrate in a fermentative pathway to produce organic acids, then the pH of the broth will usually be below pH 6.0, and will almost always be below pH 6.5. At these pH levels, a very high percentage of the phenol red molecules are uncharged and yellow in color. Since it takes only a slight acidification of the medium (pH 6.0–6.5) to convert a high percentage of

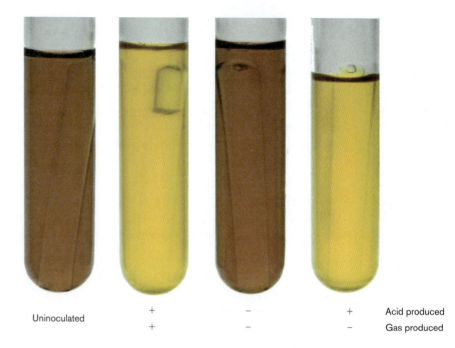

| Uninoculated | + | − | + | Acid produced |
| | + | − | − | Gas produced |

FIGURE 10.5 Positive and negative phenol red fermentation results.

phenol red molecules to the yellow form, the assay is considered positive for fermentation to organic acids only if the color is a bright, Tweety-Bird yellow (**Figure 10.5**).

So, if the broth has a bright yellow color, this is considered a positive reaction for the fermentation of a carbohydrate to organic acids.

Remember, different enzymes are needed to ferment different sugars, so a given species may show a positive reaction for some sugars and negative for others.

Organic acids: Negative reaction If the cells in the broth do not use the carbohydrate in a fermentative pathway to produce organic acids, then the pH of the broth will usually be above pH 6.8. At pH levels above 6.8, a high percentage of the phenol red molecules will have lost a proton and will be negatively charged, so the broth will be orange to red in color (see Figure 10.5).

So, any orange to red color is considered a negative reaction for the fermentation of a carbohydrate to organic acids. The cells may be using the sugars via an aerobic respiratory pathway, but if they do so, this will produce CO_2 and H_2O and there won't be sufficient quantities of organic acids to change the broth color from orange to yellow. If the cells are using sugars by respiration, the broth will usually be turbid or cloudy due to the growth and reproduction of the cells, but don't forget, this is a fermentation assay, so that is still considered a negative result for this test.

Furthermore, if the respiring cells used the amino acids derived from the peptones as an additional energy source, the amino group ($-NH_2$) will be removed as an ammonia (NH_3) molecule. This yields a carbon skeleton that can be converted to an intermediate found in glycolysis and the Krebs cycle, and the energy in the carbon skeleton will be extracted in these pathways. When amino acids are used as an energy source, they release ammonia, and the alkaline ammonia raises the broth pH, changing the indicator color to a deeper red than the orange-red of the uninoculated medium.

Caution: False Negatives If the pH of an incubated phenol red carbohydrate fermentation broth is above 6.8 and the broth is red in color, the conclusion is that fermentation did not occur. But under certain conditions, this could be a "false negative"; that is, the broth could be red in color even though carbohydrate fermentation really did occur. False negatives can occur if the bacteria exhaust the supply of sugars by rapid fermentation and then, rather than giving up and dying, begin to use amino acids derived from peptones as a primary source of energy and carbon. As described above, this process produces ammonia, and if enough NH_3 is generated, it will neutralize the organic acids, raise the pH back up to 6.8 or above, and make the broth red in color, despite the fact that carbohydrates were fermented. To prevent this, the assay is read at 24 to 48 hours, or before any ammonia production can mask the effects of organic acids on the pH of the broth. This is why some prefer to use carbohydrate concentrations of 1.0%, instead of 0.5%: if more sugar is present at the start, it will take longer to exhaust the supply, and more organic acids will have accumulated at the point at which amino acids begin to be used as the primary energy source.

Gases: Positive reaction If the carbohydrate is fermented and the end products include gases such as carbon dioxide and hydrogen, then the bubbles of gas produced at the bottom of the test tube will enter the opening of the Durham tube and gradually displace the broth in the tube. When enough gas has accumulated, there will be a visible bubble of gas trapped at the top of the inverted tube (see Figure 10.5). Gases will also be produced during aerobic respiration, but they will be produced at the top of the medium at the air-water interface and will not be trapped by the Durham tube.

So, the presence of a bubble at the top of the Durham tube is a positive reaction for fermentation of carbohydrates to gases.

Gases: Negative reaction If the carbohydrate is not fermented, or if the fermentation does not produce gases as end products, then no gases will be produced at the bottom of the test tube, and there will be no visible bubble of gas trapped at the top of the inverted Durham tube.

So, the absence of a bubble at the top of the Durham tube is a negative reaction for fermentation of carbohydrates to gases.

Many species ferment carbohydrates and produce organic acids as end products, but do not produce gases. So a given microbe can be positive for fermentation of carbohydrates to organic acids, but negative for fermentation to gases. In other words, a "negative" for gases does not automatically mean that the carbohydrate was not fermented. Before you can conclude that the species in question does not ferment the carbohydrate in the broth in any way, shape, or form, you must observe a negative result for both organic acid and gas production.

CASEIN HYDROLYSIS ASSAY

Hydrolysis of proteins

The **casein hydrolysis assay** is an example of an assay that tests for the ability of cells to produce exoenzymes called proteases. **Proteases** degrade proteins, producing a mix of short peptide chains and free amino acids as end products. These much smaller molecules are usually water-soluble and can be absorbed or transported into the cells to be used as sources of nitrogen and amino acids. Proteases are usually not very specific in terms of which proteins they will degrade. Instead of breaking down just one particular type of protein, many proteases break peptide bonds located next to specific amino acid residues within the peptide chain in many different protein types. Other proteases break the peptide bond that holds the last or terminal amino acid at the end of the peptide chain, again, in many different proteins. That is, the specificity of the enzyme is usually based on where it cuts within a protein, instead of which specific protein it cuts.

Detection of casein hydrolysis

The casein hydrolysis assay enables us to differentiate between species that can hydrolyze the milk protein called casein and species that cannot. This type of biochemical differentiation between species will become increasingly important later in the course when we attempt to identify bacteria. For now, a positive result for the ability to hydrolyze casein shows that the microorganism in question can produce one or more types of proteases and peptidases that can specifically degrade casein to small, soluble peptides and amino acids. Such enzymes are collectively called **caseinases**, and they are often produced by soil bacteria, such as *Bacillus* species.

This hydrolysis assay uses **skim milk agar** or **casein agar**, terms that can be used to describe a variety of media developed to test a given pure culture's ability to produce and secrete caseinases. Beef extract and peptones are included to meet the cells' fundamental nutritional needs for energy, carbon, and nitrogen. Then, skim milk is added to provide a casein substrate. Casein is present in milk in clusters of protein called micelles, and in skim milk, almost all of the cloudiness or opacity of the milk is due to these micelles. If cells growing on a casein or skim milk agar plate secrete caseinase enzymes into the surrounding environment, the enzymes will hydrolyze the casein protein, removing the micelles. The lower-molecular-weight end products, a mix of short peptides and amino acids, will either be transported into the cells or dissolved into the medium, which in either case forms a clear region in the agar.

Positive casein hydrolysis reaction In the regions where the proteases are active and the casein micelles disintegrate, the medium will become clear. There will be a transparent zone extending out from the edges of the streak or a kind of transparent halo around the growing cells where the casein has been degraded (**Figure 10.6**). Beyond the transparent halo, in areas the secreted proteases couldn't reach by diffusion, the medium will remain opaque and milky white.

So, if there is a transparent zone or transparent halo extending out from the microbial growth surrounded by opaque, milky white areas, this is a positive reaction; the cells produced casein-hydrolyzing proteases.

Negative casein hydrolysis reaction If the cells growing on the skim milk agar do not produce casein-hydrolyzing proteases, then large casein molecules and micelles will remain intact. There will not be a casein-free zone or transparent halo extending out from the margins of the colonial growth, and the medium will be opaque and milky white right up to the edge of the streak or up to the edges of any isolated colonies.

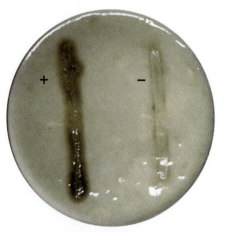

FIGURE 10.6 Positive and negative results for casein hydrolysis.

So, if all areas of the plate, including areas surrounding the microbial growth, are still opaque and milky white, this is a negative reaction; the cells did not produce casein-hydrolyzing proteases.

Casein in the presence of high concentrations of acids

In some cases, cells growing on the casein or skim milk produce a high concentration of organic acids by fermenting sugars in the medium. These acids diffuse out of the cells and lower the pH of the surrounding medium. If the acid concentration of the medium is high enough, then the casein in the medium will be denatured, and the casein micelles will coagulate and produce a bright white precipitate in places where there is intact, unhydrolyzed casein. (As an aside, this precipitation of the casein is similar to what happens in a container of spoiled milk when fermenting bacteria lower the pH of the milk and chunky curds of casein tell you that this food item may be, uh … slightly past its expiration date.) So, species that are able to both ferment sugars and hydrolyze casein will produce a dense white outer halo or outer white ring surrounding the inner transparent ring of hydrolysis, and fermenting species unable to hydrolyze casein will produce a white ring immediately adjacent to the colonial growth.

LYSINE DECARBOXYLASE ASSAY

Decarboxylation of amino acids

Decarboxylases are enzymes that remove the carboxyl groups (—COOH) from amino acids. Since each amino acid has a unique structure, different decarboxylases are needed to remove the carboxyl groups of different amino acids. The end products of decarboxylation may have a variety of uses in cell metabolism, depending on the microbial species and the particular product in question. A given bacterial species produces its own particular set of decarboxylases, so we can begin to identify bacteria by determining whether a specific decarboxylase is present or absent in that species. For example, in this lab you may notice that species such as *E. coli* and *P. vulgaris*, both members of the Family Enterobacteriaceae (Lab 24), can be distinguished from each other on the basis of their ability or inability to decarboxylase lysine.

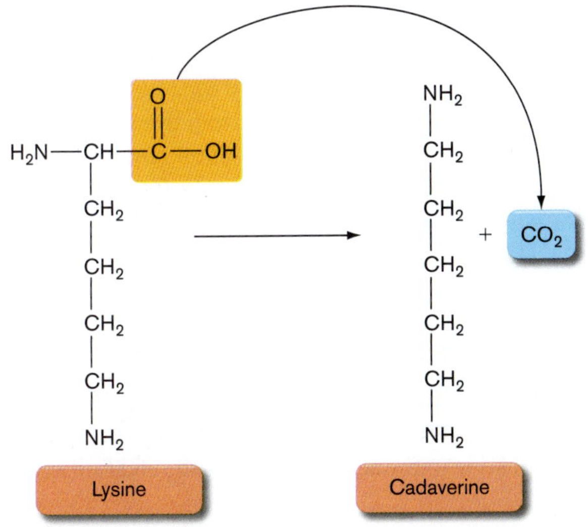

FIGURE 10.7 Conversion of lysine to cadaverine.

Lysine decarboxylase, under acidic and anaerobic conditions, removes the carboxyl (—COOH) group from **lysine** (no surprise there), producing carbon dioxide (CO_2) and an amine called **cadaverine** as end products **(Figure 10.7)**.

Cadaverine: A Lysine Decarboxylase Product

Man is a museum of diseases, a home of impurities; he comes today and is gone tomorrow; he begins as dirt and departs as stench.

—Mark Twain, *The Mysterious Stranger*, published posthumously (1916)

In some bacterial species, cadaverine plays a useful role in the cells. For example, in a few species of anaerobic Gram-negative bacteria, cadaverine is an essential component in peptidoglycan, and thus in cell wall synthesis. In many other species, including *E. coli* and other Gram-negative intestinal bacteria, cadaverine interacts with protein channels found in cell membranes and in the outer phospholipid envelopes of the Gram-negative cell walls. This interaction alters membrane permeability and tends to reduce the rate at which molecules move through and across the membrane. This, in turn, can protect the cell from chemicals such as antibiotics and prevent an excess inflow of protons (H^+) in acidic environments.

Probably not coincidentally, lysine decarboxylase genes are often expressed only at low pH levels; that is, cadaverine production is likely an adaptive response to acidic conditions. Lysine decarboxylase activity may have much less value at neutral pH levels, so there may be little advantage in "wasting" energy and amino acids in the production of the enzyme unless the cell is in an acidic environment. And under high-pH or alkaline conditions, cadaverine may actually have a negative or toxic effect on the cells; this may be another reason why synthesis of lysine decarboxylase occurs only in acidic environments. Keep all of this in mind when we examine how the lysine decarboxylase assay works.

One last point about cadaverine: like many other amines produced by decarboxylation of amino acids, cadaverine stinks. Think … dead fish. When an animal dies, bacteria begin the process of breaking down or decomposing its body tissues, including the protein-rich muscle tissue. This process yields the lysine that can be decarboxylated to the odoriferous cadaverine. This is something that you can experience at home if you forget to take out the trash. Remember that "meat" is muscle tissue, so when meat begins to rot or spoil in a garbage can, cadaverine will be produced. Beef, pork, chicken, fish … all will quickly become offensive unless steps are taken to reduce or prevent microbial growth. As Ben Franklin once noted, "Fish and visitors stink after three days." And perhaps it's a good thing that cadaverine is so easily detected by our noses when meat does spoil, despite our best efforts to prevent it. Cadaverine and other amines produced by decarboxylation of amino acids can be toxic and can cause food poisoning if present in meat and fish in high concentrations. As it is, the odor of amines is usually enough to dissuade most would-be carnivores from consuming the rotten muscle tissue.

And, yes, as you have probably already guessed, the name "cadaverine" is derived from the word "cadaver." "Cadaver" comes from the Latin term *cadere*, meaning "to fall"; this was a Latin metaphor or euphemism for "to die." So, while it's not a pleasant thought, this amine's name reflects the fact that the decomposition of human muscle tissue or the decaying of cadavers also produces cadaverine.

Detection of lysine decarboxylase

The **lysine decarboxylase assay** is designed to detect the production of cadaverine and carbon dioxide that results from the removal of the carboxyl group from lysine. Specifically, the assay reveals a rise in pH due to the accumulation of the alkaline amine in the medium.

The nutritionally rich decarboxylase broth contains peptones and beef extract to meet the basic metabolic needs of a wide range of species. Glucose is added as a substrate for fermentation; fermentation produces organic acids, and the subsequent drop in medium pH stimulates the synthesis of decarboxylases. Lysine is added as a substrate for decarboxylation when the goal is to detect lysine decarboxylase. However, other amino acids can be substituted for lysine if one wishes to detect other types of decarboxylases.

Changes in pH are detected using **bromcresol purple** as a pH indicator. This indicator has a structure similar to phenol red, but bromcresol purple is yellow in color when uncharged at lower pH levels and purple when it loses protons at higher pH levels. The color of the indicator depends on the ratio of uncharged to charged bromcresol purple molecules, and its color changes from **yellow at pH 5.8 or lower**, to dark yellow to yellow-purple between pH 5.8 and 6.8, and **to purple at pH 6.8 or higher**.

To perform the lysine decarboxylase assay, a tube of broth is inoculated and then sealed with mineral oil to promote the formation of an anaerobic environment. Under these conditions, the cells will ferment the glucose in the broth, producing organic acids as end products and creating the low-pH environment that is needed for decarboxylase synthesis. Acid production will drop the pH to around pH 6.0 or lower, and the medium will initially turn yellow, as the bromcresol purple indicator is yellow-orange at pH 6.0 and clearly yellow below pH 5.5 or so.

If the species in question is able to decarboxylate lysine, then over time, cadaverine will begin to accumulate in the medium. The pH will begin to rise, and the color of the medium will begin to change from yellow to dark purple as the broth becomes increasingly alkaline; once the pH has risen to 7.0, the solution is usually purple. While in some species amine production may be rapid enough to turn the medium purple within 24 hours of inoculation, other species decarboxylate lysine at a slower rate. Therefore, in the case of apparent negative reactions, test tubes should be checked for up to 96 hours before concluding that lysine was not decarboxylated.

Positive lysine decarboxylase reaction If the cells in the broth can decarboxylate lysine to produce cadaverine and carbon dioxide, then eventually, the pH of the broth will usually be well above 7.0. At these pH levels, a large majority of the bromcresol purple molecules are negatively charged and purple in color **(Figure 10.8)**.

So, if the lysine decarboxylase broth has a light purple to dark purple color, this is considered a positive reaction for the decarboxylation of lysine. The cells produced lysine decarboxylase and the end product, cadaverine.

There is one catch. The initial color of the lysine decarboxylase broth is closer to purple than to yellow. So, if we inoculate a lysine decarboxylase broth and absolutely nothing happens (no growth or fermentation or decarboxylation), then the broth color will remain purple. This purple color could be incorrectly interpreted as a positive reaction—that is, as evidence of lysine decarboxylation. To guard against such false positives, we need to be certain that the species being tested is capable of fermenting glucose. A purple color is considered a positive reaction for lysine decarboxylation only if it is preceded by a fermentation-caused change to acidic conditions and a temporary change to a yellow color in the broth.

One way to be certain that our lysine decarboxylase–positive species can ferment glucose to organic acids would be to watch the tube continuously for 24 hours as the broth turns from purple to yellow and then back to purple. But we assume that you

Lysine   +   −   +   −

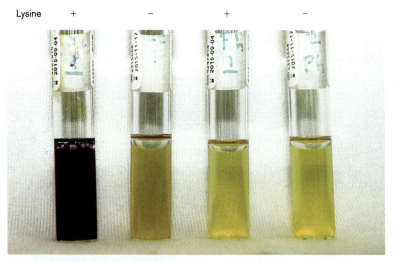

FIGURE 10.8 Lysine decarboxylase results.
Two types of bacteria were grown in decarboxylase medium with (+) or without (−) lysine. All four tubes show growth and fermentation of glucose. Only one tube (purple) also shows decarboxylation of lysine.

have more important things to do. So, instead, for each species tested, we will inoculate a control tube containing lysine-free decarboxylase broth. In this broth, if the test species can ferment the glucose in the broth, the fermentation of the sugar will produce organic acids, the pH of the broth will drop, and the tube will turn yellow. This tells us that fermentation-caused change to acidic conditions will occur when we subsequently do an assay with lysine in the medium.

It should be emphasized that a yellow color in a broth without added lysine establishes only that the species can ferment glucose to create acidic conditions. If there's no lysine in the medium, then obviously cadaverine cannot be produced by decarboxylation, and so the yellow color should not be interpreted as a negative reaction for the lysine decarboxylase assay. In the absence of lysine, you cannot draw any conclusions at all about the ability to decarboxylate lysine.

Negative lysine decarboxylase reaction If the cells in the lysine decarboxylase broth do not decarboxylate lysine to produce cadaverine and carbon dioxide, then the pH of the broth will be at or below the starting pH of 6.8, and if fermentation of glucose occurred, the pH will usually be well below 5.5–6.0. At these lower pH levels, almost all of the bromcresol purple molecules are uncharged. Uncharged molecules appear yellow in color, and the broth may be various shades of yellow, depending on the exact pH (see Figure 10.8).

So, if the lysine decarboxylase broth has a yellow to dark yellow color, this is considered a negative reaction for the decarboxylation of lysine. The cells did not produce lysine decarboxylase and cadaverine.

PROTOCOL

Day 1

Materials

Per pair of students:

- 2 decarboxylase broths without lysine
- 2 lysine decarboxylase broths
- 4 phenol red glucose broths with Durham tubes
- 4 phenol red lactose broths with Durham tubes
- 4 phenol red sucrose broths with Durham tubes
- 2 skim milk agar plates
- 2 starch agar plates
- Gram's iodine
- Sterile mineral oil

Cultures:

- *Bacillus cereus*
- *Enterococcus faecalis*
- *Escherichia coli*
- *Proteus vulgaris*
- *Staphylococcus epidermidis*

STARCH HYDROLYSIS PLATES

1. Using a flame-sterilized inoculating loop, streak one starch agar plate with *E. coli* and the other with *B. cereus*. Streaks should be single lines across the plate's diameter.

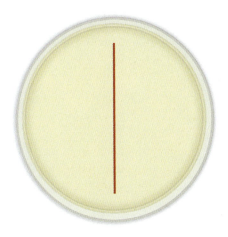

2. Incubate the plates at 37°C for 48 hours.

PHENOL RED CARBOHYDRATE FERMENTATION BROTHS

1. For each of the three types of phenol red broths (glucose, lactose, and sucrose broths), label four tubes as follows:

   a. *Escherichia coli*
   b. *Enterococcus faecalis*
   c. *Proteus vulgaris*
   d. Uninoculated control

   Note that all of these different media look alike, so be sure to label the tubes immediately or take any and all other steps needed to keep track of the sugar type in each tube of broth.

2. Inoculate each labeled tube with the appropriate species using a flame-sterilized loop and the techniques described earlier in the course for transferring cells to broths. Do not add cells to the uninoculated control tubes.

3. Incubate all the tubes at 37°C for 48 hours.

CASEIN HYDROLYSIS PLATES

1. Using a flame-sterilized loop, streak a skim milk (casein) agar plate with *S. epidermidis*. Flame the loop again, and streak another skim milk agar plate with *B. cereus*. In both cases, the streak should be a single line across the plate's diameter, as with the starch agar plates.

2. Incubate the plates at 37°C for 48 hours.

LYSINE DECARBOXYLASE BROTH

1. Inoculate one lysine decarboxylase broth with *Proteus vulgaris* and one broth of the same type with *E. coli* using a flame-sterilized loop and the techniques described earlier in the course for transferring cells to broths.

2. Inoculate one decarboxylase broth without lysine (control) with *Proteus vulgaris* and one broth of the same type with *E. coli*.

3. Layer about 1 cm (about ¼ inch) of sterile mineral oil on top of the broths in all the tubes.

4. Incubate the tubes at 37°C for 48 hours.

Day 2
ASSAYS FOR ENZYMES ACTING ON CARBOHYDRATES

1. Add iodine to the starch agar plates, so that the surface is covered; but do not flood it with iodine. Record the presence or absence of colorless zones around the streak in the "Starch Hydrolysis" section on the Results page. The colorless zones may be easier to see if you place the plate on top of a blank sheet of white paper.

2. Record colors (red or yellow) and the presence or absence of bubbles in the Durham tubes in the phenol red fermentation broths in the "Sugar Fermentation" section on the Results page.

ASSAYS FOR ENZYMES ACTING ON AMINO ACIDS AND PROTEINS

1. Record the presence or absence of zones of clearing on the skim milk agar plates in the "Casein Hydrolysis" section on the Results page.

2. Record the colors (yellow or purple) of the lysine decarboxylase broths and the colors of the decarboxylase broths without lysine in the "Lysine Decarboxylase" section on the Results page.

Name: _____   Section: _____

Course: _____   Date: _____

STARCH HYDROLYSIS (STARCH AGAR PLATES)

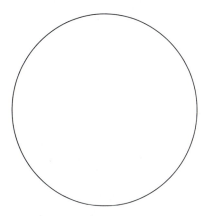

In the circle to the left, draw in color, or provide photographs of the *B. cereus* growth. In the space below, provide a written description of the growth and be sure to describe any zones of hydrolysis.

B. cereus plate

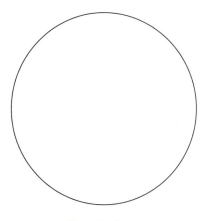

In the circle to the left, draw in color, or provide photographs, of the *E. coli* growth. In the space below, provide a written description of the growth, and be sure to describe any zones of hydrolysis.

E. coli plate

SUGAR FERMENTATION (PHENOL RED CARBOHYDRATE BROTHS)

Glucose	Broth Cloudy?	Broth Color	Broth ph*	Were Acids Produced by Fermentation?	Bubble (Yes/No)	Were Gases Produced by Fermentation?
E. coli						
E. faecalis						
P. vulgaris						
Control						

Lactose	Broth Cloudy?	Broth Color	Broth ph*	Were Acids Produced by Fermentation?	Bubble (Yes/No)	Were Gases Produced by Fermentation
E. coli						
E. faecalis						
P. vulgaris						
Control						

Sucrose	Broth Cloudy?	Broth Color	Broth Ph*	Were Acids Produced by Fermentation?	Bubble (Yes/No)	Were Gases Produced by Fermentation?
E. coli						
E. faecalis						
P. vulgaris						
Control						

*Broth pH: Choose from (a) pH < 6.8, (b) pH 6.8–7.8, or (c) pH > 7.8.

CASEIN HYDROLYSIS (SKIM MILK AGAR PLATES)

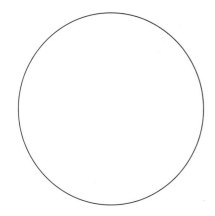

In the circle to the left, draw in color, or provide photographs of the *B. cereus* growth. In the space below, provide a written description of the growth and be sure to describe any zones of hydrolysis.

B. cereus plate

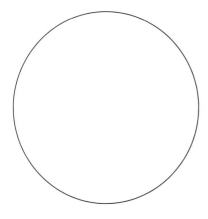

In the circle to the left, draw in color, or provide photographs of the *S. epidermidis* growth. In the space below, provide a written description of the growth and be sure to describe any zones of hydrolysis.

S. epidermidis plate

LYSINE DECARBOXYLASE

Species and Broth Type		Broth Color	Broth ph*	Does This Species Ferment Glucose?	Positive or Negative Reaction for Decarboxylase Assay	Was Cadaverine Produced by Decarboxylation?
E. coli	Broth w/o lysine					
	Lysine decarboxylase broth					
P. vulgaris	Broth w/o lysine					
	Lysine decarboxylase broth					

*Broth pH: Choose from (a) pH < 5.8, (b) pH 5.8–6.8, or (c) pH > 6.8

Name: _____   Section: _____

Course: _____   Date: _____

1. Would amylase production be more advantageous in soil or in the large intestine? What does your answer suggest about which environment is most likely to support a wider range of exoenzyme-producing bacteria? Explain your answer.

2. *Bacillus* cells are common in the soil. Does this surprise you? Why or why not? In answering, cite relevant lab results.

3. Did all of the lab species capable of fermenting glucose have the same set of fermentation pathway enzymes? How could you tell that they did or did not have the same enzyme set?

4. If you lost the labels on your *E. coli* and *P. vulgaris* stock culture tubes, how could you figure out which tube contained the *E. coli* cells?

5. If a phenol red carbohydrate fermentation broth was inoculated with a species that is unable to ferment the sugar, but the cells were able to use amino acids as carbon and energy sources, what color would the medium have been? Explain your answer.

6. If *E. coli* cells were streaked onto skim milk agar plates and incubated for 24 hours, would you expect to find a white precipitate in the medium around the cells? Cite evidence from this lab to support your position.

7. Why is it important to encourage fermentation in the lysine decarboxylase assay?

8. If an organism is "lysine decarboxylase positive," why might the lysine decarboxylase broth be bright yellow at some point during the incubation period?

9. Fill in the table below to review and summarize these assays.

Assay	Main Substrate	pH Indicator	Added Reagents	Positive Reactions (Description)	Negative Reactions (Description)
Starch hydrolysis assay		NONE			
Phenol red carbohydrate fermentation assays			NONE		
Casein hydrolysis assay		NONE	NONE		
Lysine decarboxylase assay			NONE		

For centuries it was assumed that the bottom of the oceans were sterile wastelands. But in the 1970s, biologists discovered complex ecosystems surrounding hydrothermal vents—cracks in the Earth's crust where water and gases spew out, sometimes in excess of 400ºC. At the base of these food webs are bacteria and archaea that thrive under these extreme temperatures, crushing pressure, and ocean salinity. These microbes inhabit the tissues of tube worms. The bacteria within the tube worms use chemosynthesis (analogous to photosynthesis) to make organic matter from the inorganic minerals rushing out of the vents. These minerals often include sulfides, which are responsible for the dark clouds emanating from these, so called, black smokers.

Learning Objectives

- Describe the concept of temperature growth range, including the difference between optimum and maximum growth temperatures.

- Understand the significance of temperature growth range in the control of microbes.

- Know the terms used to describe different temperature classes and the types of microbes in each class.

- Use experimental data to ascertain the appropriate temperature class for selected bacteria.

LAB **11**

Temperature and Bacterial Growth

All cellular life is affected by environmental temperature, and bacterial cells are no exception. Every bacterial species has a specific temperature range in which it can grow. This temperature range is defined by the organism's minimum, maximum, and optimum growth temperatures.

TEMPERATURE GROWTH RANGE

The **temperature growth range** is the range of temperatures within which a given species can carry out metabolic tasks, grow, and divide. Remember, for single-celled organisms, cell division essentially equals whole-organism reproduction. So the temperature growth range might also be described as the range of temperatures within which populations can increase.

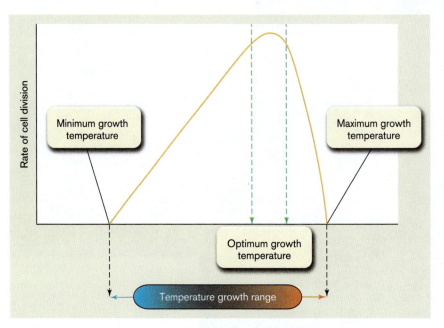

FIGURE 11.1 Bacterial growth as a function of temperature.

Minimum growth temperature

The **minimum growth temperature** is the lowest temperature at which growth and cell division occur **(Figure 11.1)**. At low temperatures, the phospholipid-based **cell membrane** becomes less fluid or more solid, which slows the movement of nutrients across this barrier, and thus slows metabolism. Low temperatures also slow chemical reactions within the cell, including enzyme-dependent chemical reactions involved in metabolism. Below the minimum growth temperature, metabolism occurs too slowly to sustain growth.

Maximum growth temperature

The **maximum growth temperature** is the highest temperature at which growth and cell division occur. Above the maximum growth temperature, metabolic reactions cannot be carried out at a rate that will sustain growth because the heat **denatures** proteins; that is, heat causes proteins to unfold. Protein shape, or three-dimensional structure, is essential for proper protein function. Denatured enzymes cannot properly interact with their metabolic substrates, so denatured enzymes cannot increase the rate of biochemical reactions. Thus, metabolism slows, and eventually, stops as temperatures approach and then exceed the maximum growth temperature.

In addition to affecting proteins, high temperatures also alter the stability and function of cell membranes. As it gets hotter, the phospholipids in a membrane become more mobile. Beyond a certain temperature, the membrane becomes too fluid to act as a selectively permeable barrier between the cytoplasm and the surrounding environment, and the barrier may fail completely.

Further, high temperatures can affect DNA and chromosome stability. If it's hot enough, the hydrogen bonds between the two strands of chromosomal DNA will break and the strands will separate. Heat can also disrupt the process of microtubule polymerization, a process needed for cell division as well as numerous other activities of the cell.

Of course, if the temperature is high enough, as it is in a Bunsen burner flame, the cells are converted into crispy critters, but that's not really what we're talking about here.

Optimum growth temperature

The **optimum growth temperature** is the relatively narrow range of temperatures at which cell division, or the rate of reproduction, for a given species is at a maximum. Since the "optimum temperature" is the temperature at which cell division rates are at a maximum, it's easy to confuse it with "maximum growth temperature," but these are two different things. This optimum temperature is not necessarily the optimum temperature for the activity of every single type of enzyme in the cell, but when the activity of all enzymes is considered collectively, growth and reproduction occur at a higher rate at this temperature than at any other temperature.

Metabolism is fundamentally a set of complex, interconnected chemical reactions. Since the rate of chemical reactions roughly doubles with every 10°C increase in environmental temperature, the rate of metabolic activity and cell division also increases with temperature until enzymes begin to denature. So the optimum growth temperature is usually close to, but still less than, the maximum growth temperature. That is, it is the point at which the input of heat energy has greatly increased the rate of chemical reactions, but it is still below the temperature at which significant denaturation of enzymes begins. Protein denaturation typically begins a few degrees below the maximum growth temperature.

The optimum growth temperature for a given species, not coincidentally, is usually close to the typical temperature of the environment in which the species is usually found. This is especially so if the bacteria are adapted to environments with relatively constant temperatures, such as the environment found in the bodies of warm-blooded (homeothermic) animals. Species, such as *E. coli*, that are adapted to the digestive tracts of birds and mammals can count on living in a tropical paradise where temperatures are almost always between 35°C and 39°C, depending on the vertebrate species. The internal temperature of the human body is usually around 37°C, and after living for millions and millions of generations in this warm sea of tranquility, most human intestinal bacteria have evolved enzyme systems that work very, very well at 37°C.

Effects of other environmental factors on temperature growth range

While each species has a temperature growth range, it should be noted that the range for a particular species is also dependent on other environmental factors. That is, the minimum, maximum, and optimum growth temperatures are not values that are fixed for all conditions and for all times. Generally speaking, the temperature growth range is greatest when other environmental conditions, such as environmental pH, are ideal. But should the environment become more physiologically stressful—should it become more acidic, for example—then the temperature growth range usually shrinks: maximum growth temperatures are lower, minimum growth temperatures are higher, and the peak rate of cell division at the optimum growth temperature may also decline.

TEMPERATURE CLASSES

One way to characterize microorganisms is by the range of temperatures in which they will grow **(Figure 11.2)**.

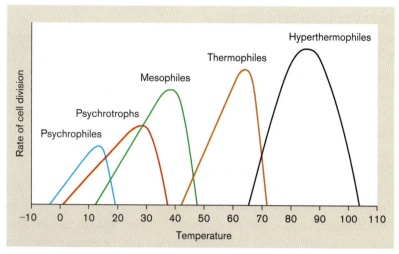

FIGURE 11.2 Bacterial temperature classes.

FIGURE 11.3 Psychrophilic bacteria.
Rhodo globus vestalii, a cold-loving
species with a temperature growth
range of –2°C–21°C, was isolated from
a lake in Antarctica.

Psychrophiles

The term "psychrophile" is derived from the Greek words *psukhros* and *philos,* meaning "cold" and "love," respectively. **Psychrophiles** are "cold-loving" microbes with temperature growth ranges between –5°C and 15°C–20°C, and all true psychrophiles can grow at 0°C–5°C. Optimum growth temperatures are no higher than 15°C, and for most species, optimum growth occurs at temperatures between 5°C and 15°C. Their enzyme systems may be so well adapted to cold temperatures that temperatures of only 20°C can denature their enzymes and stop growth and cell division. On the cold side, psychrophiles can grow at temperatures as low as –5°C if internal or environmental solute concentrations are high enough to keep the water in liquid form. Like all organisms, psychrophiles need liquid water for growth and metabolism.

Psychrophiles can be found in alpine and arctic soils, in glaciers and snowfields, and in high-latitude and deep ocean waters; 90% of the ocean's waters are at temperatures below 5°C **(Figure 11.3)**. Given that these environments account for a large fraction of Earth's surface, there are plenty of locations for hunting psychrophiles if you don't mind freezing your tuchis.

Psychrotrophs

Psychrotrophs, or psychrotolerant microbes, fall between and overlap psychrophiles and mesophiles on the temperature growth range scale. These species typically have temperature growth ranges between 0°C–5°C at the low end and 30°C–37°C at the high end, with optimum growth temperatures at around 20°C–30°C. These broad temperature growth ranges mean that some of these species can grow at both refrigerator (2°C–4°C) and human body (37°C) temperatures. Since we rely on cold temperatures for food preservation, psychrotrophs create major problems for those concerned with food spoilage and food-borne diseases. These species will grow slowly in refrigerators, and they will tolerate the cold and will remain alive for long periods. For example, the psychrotrophic bacterium *Listeria monocytogenes* can grow slowly on refrigerated foods such as cheese and precooked meats. If the live cells are consumed, *Listeria* can cause a rare, but frequently fatal, food-borne infection called listeriosis.

Mesophiles

The term "mesophile" is derived from the Greek word *mesos,* meaning "middle"; in this case, "middle" refers to the middle of the temperature spectrum. **Mesophilic** microbes have temperature growth ranges between 10°C–15°C and 45°C–50°C. Mesophiles associated with soils and plants are adapted to the cool end of this range and have optimum growth temperatures at 25°C–30°C. Those mesophilic species that are associated with warm-blooded animals, including humans, are adapted to grow best and divide most rapidly at around 35°C–40°C.

Human bodies provide microbes with a relatively stable temperature environment of around 32°C–37°C, depending on the body part. So it is not surprising that bacteria adapted to growing in or on humans have evolved enzyme systems that work best at these temperatures. And almost all human-associated bacteria, including almost all pathogenic species, are mesophilic species with maximum rates of cell division around 35°C–37°C.

Thermophiles

"Thermophile" is a term derived from the Greek words *thermos* and *therme,* meaning "hot" or "heat." **Thermophilic** species have temperature growth ranges between 40°C–45°C and 65°C–70°C, with optimum growth temperatures at around 55°C–65°C.

Thermophilic microbes find human bodies too cold for growth; instead, these species thrive in environments such as compost piles and hot springs. To survive and grow, they must have enzymes that maintain their shape at temperatures that would denature our enzymes. There are several mechanisms for making proteins heat-resistant, including (1) tighter coiling of the amino acid chain, (2) more covalent disulfide bonds between cysteine residues in the chain, and (3) a reduction in the number of glycine residues, which are points of flexibility; glycine is the smallest of the amino acids, so proteins can bend in more directions at points where glycine residues are found. All of these mechanisms or changes in proteins could slow or inhibit the unfolding of the proteins at higher temperatures.

In addition, heat-tolerant bacterial species adapt to higher temperatures by increasing the percentage of saturated fatty acids in their membrane phospholipids because saturated fatty acids can pack in more tightly, and have a higher melting point, than unsaturated fats. Thus, membranes with a higher percentage of saturated fats will be less fluid at a given temperature than membranes with a lower percentage of these fats. Cells adapted to high temperatures may also have a higher frequency of *trans* double bonds in their unsaturated fatty acids because *trans* unsaturated fats can bunch together more closely than unsaturated fats with *cis* double bonds.

Hyperthermophiles

Some textbooks make a distinction between thermophiles that grow well at "cooler" temperatures, such as 55°C–65°C, and **hyperthermophiles**, which are species that thrive at temperatures in excess of 70°C. Other manuals and texts condense all species in both groups into a single class labeled "thermophiles." When the designation "hyperthermophile" is used, it refers to species with temperature growth ranges between 65°C–70°C and 90°C–105°C, or even higher, with optimum growth temperatures at around 75°C–90°C.

If some, like Jack Lemmon and Tony Curtis, like it hot, then hyperthermophilic microbes like it really, really hot. They are found in places such as high-temperature

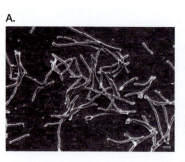

FIGURE 11.4 Hyperthermophiles.
A. *Thermus aquaticus*: what microbiologists go to see on vacation.
B. Old Faithful: what normal people go to see on vacation.

hot springs and deep-ocean hydrothermal vents. At hydrothermal vents, cells can exist and carry out metabolic activities at temperatures above 100°C because water heated to above 100°C remains in liquid form as a result of the immense pressure of the overlying seawater. By contrast, at sea level and at atmospheric pressure, water boils at 100°C, and cells cannot survive once they are turned into tiny teakettles.

Carrying out metabolic processes at temperatures as high as 80°C requires enzymes that can hold their shape in the face of a very high input of heat energy. Such enzymes can be enormously useful to us, especially in the biotechnology industry. For example, the widely used technique called the polymerase chain reaction (PCR) depends on DNA polymerase enzymes that retain their shape and remain functional at temperatures as high as 95°C (Lab 20). PCR uses high temperatures to repeatedly separate the two strands in double-stranded DNA during a process that allows us to use DNA polymerase to copy DNA outside of cells. In most PCR applications, the polymerase of choice is an enzyme called Taq, purified from the hyperthermophilic bacterium *Thermus aquaticus*. This bacterium was first isolated in the 1960s from a hot spring in Yellowstone National Park **(Figure 11.4)**. "Old Faithful" may get all the gawkers and the glory, but it's the humble *T. aquaticus* that has changed the world.

Thermodurics

The term "thermoduric" uses the suffix "-duric," derived from the Latin word *durare,* meaning "to last" or "to make hard"; this is also the root of our word "endure." **Thermoduric** microbes can outlast or endure heating; specifically, they can survive temperatures of up to 100°C for up to 10 minutes. Often, their endurance is due to the presence of endospores (Labs 8 and 19). These microbes don't like such high temperatures, they don't grow at high temperatures, and if it weren't for the honor of the thing, they'd just as soon walk. But they do survive temperatures that would kill most other species.

Thermoduric species don't fit neatly onto a temperature spectrum because their notable trait is their ability to endure, but not grow at, high temperatures. So this term doesn't tell us much about the temperatures at which they metabolize, grow and divide. As it turns out, most of these species have temperature growth ranges that would categorize them as either mesophiles or psychrotrophs.

Endospore formers, such as some *Bacillus* and *Clostridium* species, and other thermoduric microbes can create significant problems when we are trying to use heat to control food spoilage and food-borne illness. For example, pasteurization typically uses temperatures around 60°C–75°C, depending on the type of food being pasteurized. These temperatures will kill non-thermoduric species but are not hot enough to completely eliminate endospores or cells of certain other non-spore-forming thermoduric species in genera such as *Lactobacillus*.

The various temperature classes of bacteria are summarized in the following table.

SUMMARY

Temperature Class	Temperature Growth Range	Optimum Growth Temperature	Notes
Psychrophiles	−5°C to 15°C–20°C	5°C–15°C	All will grow at 0°C–5°C, all have optimum growth temperatures below 15°C.
Psychrotrophs	0°C–5°C to 30°C–37°C	20°C–30°C	Some food-borne psychrotrophs can grow at both refrigerator <u>and</u> human body temperatures.
Mesophiles	10°C–15°C to 45°C–50°C	25°C–40°C	Soil and plant bacteria have optima around 25°C–30°C; human-associated bacteria have optima around 35°C–40°C.
Thermophiles	40°C–45°C to 65°C–70°C	55°C–65°C	Thermophiles grow at temperatures found in compost piles and hot springs.
Hyperthermophiles	65°C–70°C to 90°C–105°C	75°C–90°C	Growth at 100°C or higher requires that water be under pressure (deep sea, etc.).
Thermodurics	Most have optimum growth at 25°C–35°C, so most considered mesophiles for growth.		Can survive temperatures of 100°C for up to 10 minutes.

PROTOCOL

Materials

- Tryptic soy agar (TSA) or brain-heart infusion agar (BHI) slants (5 per student)
- Test-tube racks labeled 5°C, 20°C, 37°C, 45°C, 55°C

Cultures:

- *Arthrobacter* species
- *Bacillus stearothermophilus*
- *Escherichia coli*
- *Micrococcus luteus*

Day 1

1. Each student will be assigned one and only one species for this lab, although eventually you'll record results for all the species. Once given your assignment, be sure to write it down immediately so that you don't forget.

2. Label five TSA or BHI slants with your name, the date, and a specific incubation temperature. The first tube should be labeled 5°C; the second 20°C; the third 37°C; the fourth 45°C, and the fifth 55°C.

3. Inoculate all five of the labeled slants with your assigned species, using the techniques described in Lab 2 for transferring pure-culture material to sterile slants.

   After incubating these slants at the appropriate temperatures, we'll examine them for evidence of growth, and we'll compare tubes incubated at different temperatures. Make sure when you inoculate these slants (a) that you do not transfer so much material that you start with visible slime on the slants, as this may be mistaken for "growth," and (b) that you transfer a similar amount of material to each slant.

4. Place the tubes in the appropriate test-tube racks. There will be one rack for each of the five temperatures. Tubes will be incubated for 48 hours.

Day 2

1. Examine each of your slants incubated at the five different temperatures and grade the amount of growth on each slant as 0, 1, 2, 3, or 4, where 0 means "no growth" and 4 means "hard to imagine there could be more slime than this on a slant."

   Obviously, scoring these slants will involve making judgment calls, but do your best, and try to score them in such a way as to differentiate among tubes incubated at different temperatures.

2. Record your results for your particular species in the Results section. Post your results with those of the entire class and record the pooled class results.

3. There will be several scores for each combination of species and temperature. Use these scores to calculate mean scores for growth for each species at each temperature.

4. Return all of your tubes to the various marked racks for additional incubation.

Day 3

1. Repeat the procedures from day 2 and record both individual and pooled results as before.

2. Place your slants in the test-tube racks designated for tubes to be autoclaved.

Name: _____   Section: _____

Course: _____   Date: _____

FIRST INCUBATION TIME (LAB DAY 2: FIRST LAB PERIOD IN WHICH GROWTH IS CHECKED)

Assigned species:

INDIVIDUAL RESULTS

Growth Temperature	Grade (0 through 4) Indicating Amount of Growth
5°C	
20°C	
37°C	
45°C	
55°C	

POOLED RESULTS

Species	5°C	20°C	37°C	45°C	55°C
E. coli					
	Mean:	Mean:	Mean:	Mean:	Mean:
M. luteus					
	Mean:	Mean:	Mean:	Mean:	Mean:
B. stearothermo-philus					
	Mean:	Mean:	Mean:	Mean:	Mean:
Arthrobacter					
	Mean:	Mean:	Mean:	Mean:	Mean:

SECOND INCUBATION TIME (LAB DAY 3: SECOND LAB PERIOD IN WHICH GROWTH IS CHECKED)

Assigned species:

INDIVIDUAL RESULTS

Growth Temperature	Grade (0 through 4) Indicating Amount of Growth
5°C	
20°C	
37°C	
45°C	
55°C	

POOLED RESULTS

Species	5°C	20°C	37°C	45°C	55°C
E. coli					
	Mean:	Mean:	Mean:	Mean:	Mean:
M. luteus					
	Mean:	Mean:	Mean:	Mean:	Mean:
B. stearothermo-philus					
	Mean:	Mean:	Mean:	Mean:	Mean:
Arthrobacter					
	Mean:	Mean:	Mean:	Mean:	Mean:

Name: _____   Section: _____

Course: _____   Date: _____

1. Assign each species to a temperature class (e.g., "mesophile") and explain the reasons for your choices.

    a. *E. coli*

    b. *M. luteus*

    c. *B. stearothermophilus*

    d. *Arthrobacter*

2. What is the difference between maximum and optimum growth temperatures?

3. Compared with mesophiles, would you expect psychrophiles to have a higher or a lower percentage of unsaturated fats in their membrane phospholipids? Explain your reasoning.

4. Assume that *Arthrobacter* cells do <u>not</u> secrete toxins into food, and consider what you've learned from the lab results. Given that, would you be worried about the possibility of *Arthrobacter* multiplying in your body if your latest meal had been full of *Arthrobacter*? Why or why not?

5. If you were trying to kill bacteria with heat in order to preserve food or prevent food spoilage, of the species we used in this lab, which would be most difficult to control or eliminate? Explain your answer.

Every living organism needs the sodium and chloride found in salt or NaCl. But like most things, too much salt can be bad for you. For most bacteria, living in a world surrounded by high concentrations of salt can drain them of their precious water supply. But other bacteria, like the ones living in your armpits, can tolerate higher concentrations of NaCl. And some archaea require an extremely salty environment to thrive. Shown here is the Salar de Uyuni in Bolivia. It is the largest salt pan in the world, a flat expanse of land where evaporation of water has left behind valuable salts, like NaCl and huge reserves of lithium salts.

LAB 12

Osmotic Pressure

Learning Objectives

- Define and describe osmosis.

- Predict the direction of water movement across cell membranes and the effects of this movement on cells in isotonic, hypotonic, and hypertonic environments.

- Describe how we can manipulate the osmolarity of the environment to control microbial growth.

- Examine growth of bacteria and fungi in environments with variable salt and sugar concentrations to determine the microbes' responses to environments of different osmolarities.

OSMOSIS

Single-celled microorganisms face the challenge of living in a world in which the concentration of water-soluble molecules outside of the cell is usually different from the concentration of water-soluble molecules inside the cell. In other words, the **cell cytoplasm solute concentration is usually different from the environmental solute concentration**. That means that osmosis and the effects of osmotic pressure, or the tendency of a solution to take in water across a semi-permeable membrane, can have a significant effect on the cell's ability to survive and grow. (Keep in mind that in a biological context, **solutes** are usually dissolved salts, sugars, and amino acids.)

Osmosis, a term derived from the Greek word *osmos*, meaning "push" or "thrust," is the movement ("thrusting") of water from one side of a semipermeable cell membrane to the other side. Like every other molecule, water moves according to a concentration gradient, so the net direction of movement is from the side with the higher concentration of water molecules to the side with the lower concentration of water molecules. In effect, the osmotic pressure "pushes" water molecules from the high water side to the low water side. But who talks about water concentration? Nobody.

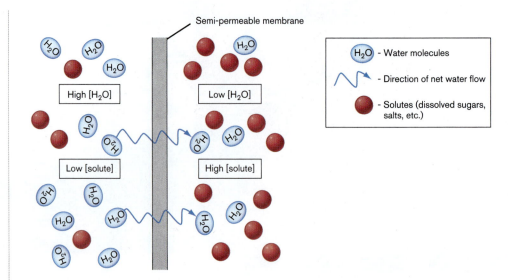

FIGURE 12.1 Osmotic pressure.
When two solutions are separated by a semi-permeable membrane, there will be a net movement of water molecules from the high water concentration side (low solute side) to the low water concentration side (high solute) side. In this figure, water movement is from left to right.

Instead, the direction of net water flow is usually described as being from the side with the lower solute concentration to the side with the higher solute concentration **(Figure 12.1)**. This language can be confusing, because osmosis is a type of diffusion, and by definition, the movement of the molecules during diffusion is always "from high to low," not "from low to high." But remember, despite the emphasis on solute concentration in discussing osmosis, we're ultimately concerned about the direction in which the water is going.

To make sense of all this, remember that low solute concentration is correlated with high water concentration, because fewer solute molecules mean that you can stuff more water molecules into a given volume. So the "low solute concentration side" and the "high water concentration side" are the same side. And water diffusing from the high-water side to the low-water side is simultaneously moving from the low-solute side to the high-solute side.

OSMOTIC ENVIRONMENTS AND EFFECTS ON CELLS

Since the solute concentration outside cells can be equal to, lower than, or higher than the concentration of solutes in the cytoplasm, there are three ways of characterizing environments with respect to osmosis.

Isotonic environments

The prefix "iso-" is derived from the Greek word *isos*, meaning "equal." **Isotonic** environments are those in which the solute concentration is equal to the solute concentration of the cytoplasm. Under these conditions, while water molecules do cross the cell membrane, there are about as many molecules entering the cell as there are leaving the cell **(Figure 12.2)**. As a result, there is no net movement of water. Being in an isotonic environment is usually a positive thing for cells, but since single-celled organisms can't control their environments, this may be an uncommon situation for most microbes. By contrast, most human cells are bathed in isotonic body fluids; such are the advantages of being part of a multicellular animal with an ability to maintain internal homeostasis, or near-constant internal conditions.

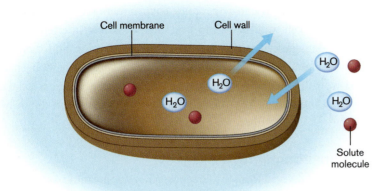

FIGURE 12.2 A bacterial cell in an isotonic environment.
There are an equal number of water molecules on both sides of the cell membrane. There are also an equal number of solute molecules on both sides of the cell membrane.

Hypotonic environments

The prefix "hypo-" is derived from the Greek word *hypo*, meaning "under." **Hypotonic** environments are those in which the environment's solute concentration is lower than ("under") the solute concentration of the cytoplasm. Under these conditions, **water flows in** from the high-water, low-solute environment to the low-water, high-solute cytoplasm side of the membrane (**Figure 12.3**). As water flows in due to higher osmotic pressure outside the cell, cytoplasmic volume increases, and this pushes the cell membrane in an outward direction. The cell membrane expands like the rubber lining of a balloon filling with water.

When human cells, such as red blood cells, are placed in a hypotonic environment, water flows in, the cell membrane expands … and the balloon explodes. That is, the red blood cells lyse. This is why we use saline solutions instead of distilled water in clinical intravenous fluid therapy (IV drip lines).

However, the outcome for single-celled organisms with strong, intact cell walls is different. Cell walls are made of polymers, or "biological plastics," such as the peptidoglycan of bacteria or the chitin and glycans of fungal cells. The polymers form rigid boxes around the far more flexible cell membranes, and these boxes can contain the swelling "water balloon" of the cell membrane. As a result, bacterial and fungal cells can usually survive in hypotonic environments; in fact, they may do better in a slightly hypotonic environment than in an isotonic environment because the slight net inflow of water may help to move nutrients into the cells.

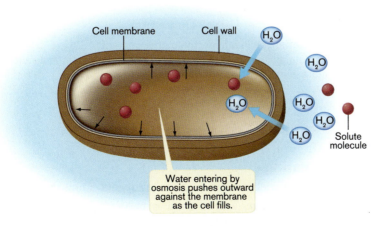

Water entering by osmosis pushes outward against the membrane as the cell fills.

FIGURE 12.3 A bacterial cell in a hypotonic environment.
There are more solute molecules inside the cell, and more water molecules outside the cell. Thus, more water flows into the cell than out.

But never forget that survival in a hypotonic environment is dependent on the structural integrity of the cell wall. If we can find chemicals that interfere with cell wall synthesis or that damage cell walls, then when these microbes find themselves in a hypotonic environment, they will pop just like red blood cells. Have we been able to do this? Does the word "penicillin" ring a bell? Penicillin damages the bacterial cell wall, and once cell wall integrity is lost, the cells subsequently burst and die in hypotonic environments.

Hypertonic environments

The prefix "hyper-" is derived from the Greek word *hyper*, meaning "over" or "beyond." **Hypertonic** environments are those in which the environment's solute concentration is higher than ("beyond") the solute concentration of the cytoplasm. Under these conditions, **water flows out** to the low-water, high-solute environment from the high-water, low-solute cytoplasm side of the membrane **(Figure 12.4)**. As water flows out due to higher osmotic pressure inside the cell, cytoplasmic volume decreases, and the cell membrane collapses in an inward direction. Just as a water-filled balloon collapses when the water is drawn off, the cell membrane shrinks away from the cell wall in a process called **plasmolysis**.

The loss of water and the collapse of the cell membrane may be enough to stop metabolism and cell division. So, if your goal is to halt microbial growth, then creating a hypertonic environment is a really good way to do it. For centuries, humans have been controlling microbial spoilage of food by adding salts and sugars to create hypertonic environments. Until the nineteenth century, no one understood why salting and sugaring worked … it just did. And that was good enough for hungry people who needed to preserve their vittles.

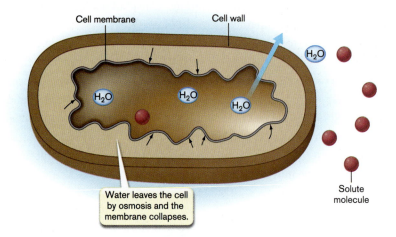

FIGURE 12.4 A bacterial cell in a hypertonic environment.
There are more solute molecules outside the cell, and fewer water molecules outside the cell. Thus, more water flows out of the cell than flows in.

CREATING HYPERTONIC ENVIRONMENTS

Sugars and sugaring

Sugar can be added to a variety of food products in concentrations that are high enough to reduce or prevent spoilage. This method is particularly effective in controlling

Condensed Milk

Condensed milk was first produced on a large scale in the 1850s by an inventor named Gail Borden. Borden's first patented attempt at creating food products was called the "meat biscuit" or "portable soup bread"; he created this item around 1850 by boiling meat to produce a concentrated, paste-like extract, which he then combined with flour. Apparently, the meat biscuit was so repulsive that even the U.S. Army wouldn't buy it. The biscuit business failed, and Borden was bankrupt by 1852.

But by 1853, Borden was working on another idea. He developed a cooker that used a vacuum to evaporate large quantities of water from milk at temperatures that wouldn't ruin the milk itself. Removing about 50%–65% of the water raised the natural sugar concentration of the milk from about 5% to between 10% and 15% for the "evaporated milk." The milk could be canned and used at that point, but excellent long-term preservation required the addition of sucrose (cane sugar) to bring the final concentration up to 55% to 65% sugar.

By 1856, Borden had secured a patent on his vacuum cooking method and apparatus, and he marketed his new condensed milk under the name "Eagle Brand." His product was not an immediate success, but when the Civil War began in 1861, the need for portable, preserved milk went through the roof. This time, the U.S. Army couldn't get enough of Borden's product, and Gail Borden became a very rich man. Borden did not understand that his method worked because it created a hypertonic environment, drawing water from bacterial cells and stopping the fermentation of the milk sugars into casein-curdling lactic acid. But he could see that his method worked, and at the time, that was enough to acquire a fortune.

A final digression: of course, with all that sugar, condensed milk is also very, very sweet ... sweet enough to attract the attention of a "bear of very little brain":

> Rabbit: Help yourself, Pooh. Would you like condensed milk, or honey on your bread?
>
> Pooh: Both. But, never mind the bread.
>
> —Dialogue from "Winnie the Pooh and the Honey Tree"

bacterial spoilage. Many fungal species, however, are **osmotolerant**; that is, they can tolerate very hypertonic environments. Some fungi are actually **osmophilic**, meaning that they thrive in high-solute-concentration environments. Therefore, spoilage of sugary foods is usually due to fungi. Just think of what you found growing on that old strawberry jam; it was probably a fuzzy fungus, right?

A few types of food are naturally preserved by their inherently high sugar concentrations. Maple syrup, for example, is approximately 60% sugar by weight, and thus is so hypertonic that additional preservatives are usually unnecessary. In other cases, sugars must be added in order to reach concentrations sufficient to slow or stop microbial growth. This is often done when preserving fruits as jams and jellies: fruits begin with natural sugar concentrations of 5% to 20%, whereas the typical finished product is 50% to 60% sugar. Finally, condensed milk offers a classic example of a food preserved by a combination of water removal and the addition of sugar.

Salts and salting

Like sugar, salts have been used for centuries to preserve food. At a time when there were very few effective ways to prevent spoilage and food-borne disease, salt's value as a preservative was immeasurable, and its production and trade influenced the fate and fortunes of empires. At the time of the Roman Empire, salt was almost literally money; our word "salary" is derived from the Latin phrase *salarium argentum*, a phrase that refers to a soldier's allowance for the purchase of salt. *Sal* is the Latin word for "salt," and *salarium argentum* can be translated as "salt money" or "money (*argentum*) of salt (*salarium*)." *Argentum* more literally translated means "silver," hence the use of "Ag" as the atomic symbol for silver, but the word was also used more generically to refer to "cash" or "money" because Roman coins were often made of silver.

While salt may enhance the flavor of food, if the inhibition of microbial growth required the same concentrations that are needed with sugar—that is, if it took concentrations of 40%–60%—then salt-preserved food would be almost inedible. (Imagine biting into a meal that is 50% salt.) Fortunately, salt is an effective inhibitor at concentrations that are much lower than those required to inhibit microbes with sugar. For example, a concentration of 10% salt has about the same preservative effect as a concentration of 50% sugar. The antimicrobial potency of salt is probably due to the fact that it does more than just create a hypertonic environment; salt also inhibits microorganisms by denaturing proteins and disrupting the transport of nutrients across cell membranes. And, of course, unlike sugar, it can't be used as a carbon and energy source.

Salt is used as a preservative for many foods, although the list was much longer in the past. Beef, pork, fish, and other meats have been salted for centuries. During the Civil War, salt pork was a staple of the army diet, although it is clear from first-person accounts that salt pork was not high on the list of the soldier's favorite foods, especially if had to be eaten without first soaking it in water to reduce the salt content. To a lesser extent, salting is still used today to slow or prevent the spoilage of meat.

Salt has also been used for centuries to favor the growth of selected microbes in fermented foods while inhibiting the growth of other species that would spoil the product. The process of fermenting milk to make butter, cheese, and yogurt almost always includes the addition of salt at some point to create a favorable environment for the bacteria that will convert milk sugars into lactic acid, and more salt may be added at the end for long-term storage. Fermented vegetable products, such as pickles, also use added salt to prevent spoilage.

Many foods may be preserved with salt, but it should be remembered that some bacterial species are able to survive or even thrive in environments with salt concentrations that inhibit other microbes. **Halotolerant** species can grow at low to moderate salt concentrations, and **halophilic** species thrive in and require high-salt conditions. The prefix "halo-" is derived from the Greek *hals*, meaning "salt," and the Latin word for salt, *sal*, is likely derived from the Greek word.

Some places where halotolerant and halophilic species are found are marine environments, salt lakes, and on skin. For example, halotolerant *Staphylococcus* species **(Figure 12.5)** thrive on sweaty, salty human skin, and being adapted to salty environments, *Staphylococcus* species can grow in salty foods as well. Since most other bacterial species are inhibited by salt, such foods are actually ideal environments for *Staphylococcus* species. One species, *S. aureus*, secretes an **enterotoxin** ("gut toxin") into its environment; consumption of the contaminated food results in a type of food poisoning characterized by vomiting and diarrhea. Some halophilic microbes actually require an environment that is three to five times saltier than the ocean. For example, the halophilic archaeon *Halobacterium* lives in salt flats **(Figure 12.6)**.

Another extremely salty habitat is the Dead Sea. It has a saline level of somewhere between 35% and 38%. The world's saltiest oceans are only 3% to 6% saline. No animals or plants can live in the Dead Sea due to its high salt content – hence its name. But some halophilic bacteria do live in the harsh conditions of the Dead Sea. So technically it isn't really a dead sea at all.

 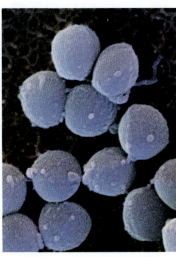

FIGURE 12.5 *Staphylococcus* cells are salt tolerant.

 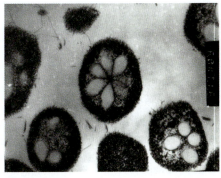

FIGURE 12.6 Halophilic salt flats and halophilic bacteria.

SUMMARY OF OSMOTIC CONDITIONS

Description of Environment	Environment Solute Concentration	Water Direction	Consequences for Cells
Isotonic	Equal to cytoplasm solute concentration	No net movement	Little or none
Hypotonic	Lower than cytoplasm solute concentration	Water flows in	Expanding membrane contained by cell wall
Hypertonic	Higher than cytoplasm solute concentration	Water flows out	Collapse of membrane; slows, stops metabolism

PROTOCOL

Materials

Per pair of students:

- 0.85% NaCl plate
- 3.5% NaCl plate
- 7.5% NaCl plate
- 15% NaCl plate
- 0.5% sucrose plate
- 15% sucrose plate
- 30% sucrose plate
- 60% sucrose plate

Bacterial and fungal cultures:

- *Escherichia coli*
- *Penicillium* species[1]
- *Saccharomyces cerevisiae*
- *Staphylococcus epidermidis*

Day 1

1. Note that all of the plates you will be working with look alike, regardless of salt or sugar concentration, so be sure to label the plates immediately or take any and all other steps needed to keep track of the salt or sugar concentration of each plate.

2. Divide each of the eight plates (one plate of each salt and sugar concentration) into quarters by marking the bottom of the plate with a wax pencil or Sharpie, and label each plate as follows:

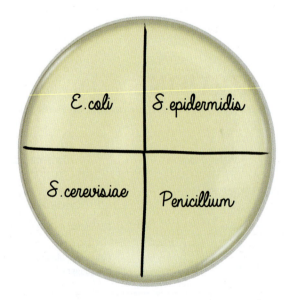

3. For each species, use a flame-sterilized inoculating loop to transfer a small amount of pure-culture material to each plate by making a short, single-line streak in the appropriate quarter of the plate.

   Do not streak the entire quarter of the plate, and do not streak in such as way as to produce isolated colonies.

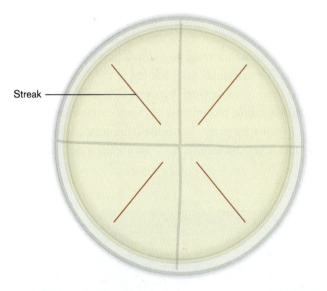

On day 2, we'll be looking for evidence that a given species can grow at a given concentration of salt or sugar, so do not transfer so much material that it looks as if cells have already grown on the plate and produced visible slime lines. In the case of *Penicillium*, even if you're careful, you'll probably be able to see mold on the plate after you transfer the cells, but note the amount of mold at the start, and look for changes by day 2.

4. Incubate the plates at 28°C–30°C for 48 hours.

Day 2

1. Examine each plate and grade the amount of growth on and around each streak as 0, 1, 2, 3, or 4, where 0 means "no growth" and 4 means "filled the quarter and threatening to crawl out of the plate."

2. Record your results.

---

[1] Use of *Penicillium* species is optional. The lab can be done using only *E. coli*, *S. cerevisiae*, and *S. epidermidis*. In this case, plates should be divided into thirds, not quarters.

Name: _____   Section: _____

Course: _____   Date: _____

SALT PLATES

Species	0.85% NaCl	3.5% NaCl	7.5% NaCl	15% NaCl
E. coli				
S. epidermidis				
S. cerevisiae				
Penicillium				

SUGAR PLATES

Species	0.5% sucrose	15% sucrose	30% sucrose	60% sucrose
E. coli				
S. epidermidis				
S. cerevisiae				
Penicillium				

Name: _____   Section: _____

Course: _____   Date: _____

1.

    a. Which was more effective at inhibiting the growth of microbes, 15% NaCl or 15% sucrose? Cite data to support your answer.

    b. Explain why one was more effective than the other, despite the fact that the concentrations were the same in both cases.

2. Which species used in this lab is (are) likely to spoil foods preserved with high concentrations of salt (you can name more than one species)? What is the evidence for your conclusion?

3. Grape juice has a concentration of up to 25% sucrose. Assume, for the moment, that *E. coli*, *S. epidermidis*, and *S. cerevisiae* all fermented sucrose to produce ethanol and carbon dioxide. If all of these cell types produced the same fermentation products, would there be any reason to use *S. cerevisiae* to make wine instead of the other two species? Cite evidence from lab to back up your answer.

4. Based solely on data from this lab, describe two very specific ways to prevent the growth of *E. coli* in food. We want to see some concentrations here.

5. Penicillin damages cell walls (as do many other antibiotics).
   a. Would penicillin kill bacterial cells in a hypotonic environment? If so, how or why would exposure to this antibiotic kill bacterial cells?

   b. Would penicillin kill bacterial cells in an isotonic environment? If so, how or why would exposure to this antibiotic kill bacterial cells?

We encounter pH every day of our lives, such as in the foods we eat. We sense pH by taste and touch, and sometimes by smell. The bracing acidity of a lemon is a familiar example. Many bacteria live in a world where the pH is in flux. So, they have developed ways to stabilize their internal pH despite the changing pH around them. In addition, bacterial metabolism often produces products that alter the environmental pH. This may affect bacterial growth, but changing pH also serves as a useful diagnostic indicator in many tests we run to identify bacteria in a sample.

Learning Objectives

- Define pH and understand the effect of pH on cell activity and cell growth.

- Understand that microbial species vary in their ability to grow at a given pH value.

- Explain how microbial activity can change environmental pH and how this change can be modulated by the presence of buffers.

- Examine the growth of specific bacteria and fungi in environments with variable pH levels to determine the tolerance of these microbes to acidic and alkaline environments.

LAB 13

Environmental pH, Fermentation, and Buffers

Unlike most human cells, most bacterial cells are not constantly bathed in a fluid of near-constant pH. Instead, bacteria often have to cope with a wide range of environmental pH values.

WHAT IS pH?

The term **pH** is used in chemistry to describe the hydrogen ion (H^+) concentration or hydrogen ion activity of a solution. Specifically, pH is the inverse or negative log of the hydrogen ion concentration of the solution. A hydrogen ion is a proton, so pH can also be defined as the negative log of the proton concentration of the solution. Note that the pH scale is an inverse scale, so if the hydrogen ion or proton concentration is high, then the pH value is low, and the environment is said to be **acidic**. If the hydrogen ion proton concentration is low, then the pH value is high, and the environment is said to be **alkaline** or **basic**. It is also

Concentration of hydrogen ions compared to distilled water

Example of solutions at this pH

Concentration	pH	Example
10,000,000	pH = 0	Battery acid (strong), hydrofluoric acid
1,000,000	pH = 1	Hydrochloric acid secreted by stomach lining
100,000	pH = 2	Lemon juice, gastric acid, vinegar
10,000	pH = 3	Grapefruit, orange juice, soda
1,000	pH = 4	Acid rain, tomato juice
100	pH = 5	Soft drinking water, black coffee
10	pH = 6	Urine, saliva
1	pH = 7	"Pure" water
1/10	pH = 8	Seawater
1/100	pH = 9	Baking soda solution
1/1,000	pH = 10	Great Salt Lake, milk of magnesia
1/10,000	pH = 11	Ammonia solution
1/100,000	pH = 12	Soapy water
1/1,000,000	pH = 13	Bleaches, oven cleaner
1/10,000,000	pH = 14	Liquid drain cleaner

FIGURE 13.1 pH values of common items.

important to remember that the pH scale is a logarithmic or exponential scale, so a change of one pH unit means a tenfold change in H^+ concentration. So, for every decrease of one pH unit, the hydrogen ion concentration increases tenfold. For example, in an environment with a pH value of 6, the proton concentration is 10 times higher than in an environment with a pH value of 7. An environment with a pH of 5 has a H^+ concentration that is 100 times higher than that found in a pH 7 environment. **Figure 13.1** shows examples of pH values for everyday items.

HOW pH AFFECTS MICROBIAL GROWTH

A given bacterial species is adapted to live within a certain range of pH values or H^+ concentrations. Each species has a minimum and maximum environmental pH beyond which cells cannot grow and divide. When environmental pH values are extremely acidic or extremely basic, there are many negative effects on cells. Even if the environment is only moderately acidic or alkaline compared with the pH range tolerated by a species, it may be difficult for cells to maintain the proper internal proton concentration. If this concentration is too high or too low, it can denature proteins. Denaturation changes protein shape, and since protein function depends on protein structure, denatured proteins cannot do their jobs. Among other things, the shapes of enzymes will change to a point at which metabolism slows and eventually stops. Recall that temperature has a similar effect on protein structure and function (see Lab 11).

A.

B.
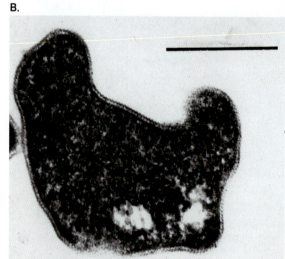

FIGURE 13.2 Acidophiles live in acidic environments.
A. An acidic hot spring. **B.** *Sulfolobus acidocaldarius.*

While most microbes are adapted to do well around neutral pH levels, **acidophiles** prefer or are adapted to more acidic environments, and **alkaliphiles** do better in more alkaline environments. For example, the *Acidithiobacillus* species found in acidic mine drainage are acidophiles adapted to pH levels as low as pH 2 or 3 (refer to Lab 34). *Sulfolobus acidocaldarius* lives in acidic hot springs that are rich in sulfur (**Figure 13.2**), and, in general, most fungal species are somewhat acidophilic and are more tolerant of high-acid/low-pH environments than are most bacterial species. This can be a useful difference if one is trying to isolate fungi from a mixed microbial community. By using low-pH media, we can manipulate the environment so that we favor the growth of most types of fungi while inhibiting the growth of many bacteria. Acid-tolerant fungi are the main culprits in the spoilage of acidic foods such as strawberries, citrus fruits, and tomatoes. Check out a jar of spaghetti sauce after a month or two in the refrigerator, and odds are that you will find it fuzzy with mold as opposed to slimy with bacteria.

Alkaliphiles grow best at pH 8.5 to pH 11. Many of these species carry cell membrane antiporters that pump out sodium (Na^+) ions while simultaneously pumping in hydrogen (H^+) ions to maintain a higher H^+ concentration (lower pH) inside the cell compared with the alkaline environment. The habitats where alkaliphiles are found include saline soda lakes such as Lake Magadi in Kenya. These lakes are rich in carbonates, which accounts for their high pH (**Figure 13.3**). They are ideal habitats for alkaliphiles such as the cyanobacterium *Spirulina* and the salt-loving *Natronobacterium*. Lake Magadi is home to numerous wading birds such as the flamingo. However, due to the extreme alkalinity and the high salinity of the water, it supports just a single species of fish.

In this lab, we'll use glucose nutrient agar (GNA) plates adjusted to different pH values (ranging from pH 3 to pH 9) to assess the ability of selected bacterial and fungal species to grow at various hydrogen ion concentrations and to determine if any of these microbes might be classified as acidophiles or alkaliphiles.

In many cases, the changes in environmental pH experienced by a microorganism are the product of the metabolism of the microbe itself (see Lab 10). If a bacterium is busy decomposing proteins, peptones, or tryptones, then the ammonia produced by the removal of amino groups ($-NH_2$) from amino acids can raise the pH of the environment. By contrast, the fermentation of sugars usually leads to an accumulation of organic acids, such as lactic acid, which lower the pH of the environment. Eventually, microbes may change the environment to the extent that the pH is outside the range within which the cells can grow, and growth is halted by the cells' own metabolic waste.

PHOSPHATE BUFFER MECHANISM AROUND NEUTRAL pH

Changes in pH may stop cell growth, but the rate of these changes can be slowed by the presence of a **buffer**. Buffer molecules work by binding H^+ or OH^- ions and preventing them from contributing to the pH of the solution. Thus, the presence of a buffer slows the rate of change in the pH value as more acid or more base is added to the solution. For example, phosphate buffers are widely used in biological laboratories because monobasic phosphate ($H_2PO_4^-$) and dibasic phosphate (HPO_4^{2-}) can buffer a solution around pH 6.8 to 7.6 when these ionized forms of phosphate are about equal in concentration.

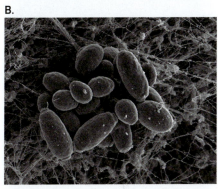

FIGURE 13.3 An alkaline lake and some of the organisms that live in it.
A. Lake Magadi. **B.** *Natronobacterium*. **C.** Pink flamingo, which gets its color from eating large amounts of *Spirulina*.

The phosphate buffering reaction is

$$H_2PO_4^- \rightleftharpoons H^+ + HPO_4^{2-}$$

But metabolic reactions such as fermentation add protons (H^+) to the medium. This alters the balance of the reaction. The H^+ are consumed in a reaction with HPO_4^{2-}, which slows the increase in proton concentration, slows the decline in pH, and pushes the equilibrium toward the left side of the equation, increasing the $H_2PO_4^-$ concentration:

$$H_2PO_4^- \rightleftharpoons H^+ + HPO_4^{2-}$$

When a buffer is added to a medium in which bacteria are respiring by fermentation, more organic acid molecules have to be generated by fermentation before the pH of the medium will drop to a level that will inhibit enzyme systems and cell growth. As a result, the decline in pH is slowed, more sugar molecules are fermented, more ATP is made, and more growth and cell division occur before the accumulation of organic acids and a decline in pH finally stops the cells. But always remember, while buffers slow the rate of pH change caused by the addition of acid or base, they cannot entirely prevent pH changes, especially if something like microbial metabolism continues to add H^+ or OH^- to the solution. Eventually, all the buffer molecules will be tied up with added protons or hydroxide ions, much as a sponge becomes saturated with water, and then the pH will change rapidly with the addition of still more acid or base.

To test the effect of adding a buffer to bacteriological media, we will use a species called *Enterococcus faecalis*, which uses sugar by fermentation alone; that is, these cells do not use oxygen and are not capable of respiration. *E. faecalis* is one of a large group of Gram-positive species known as the "lactic acid bacteria." As these bacteria grow, they break down glucose via glycolysis, then convert pyruvate (the end product of glycolysis) to lactic acid. The process of lactic acid fermentation is of value to the cell because it oxidizes NADH to NAD^+, and the NAD^+ is an essential substrate in glycolysis for the continued production of ATP. However, the lactic acid product will accumulate as waste, increasing the H^+ concentration outside the cells and lowering the pH of their environment. As a result of the buildup of lactic acid, if no buffer is present, the pH of the medium drops relatively rapidly, inhibiting the growth and reproduction of the *E. faecalis* cells.

We can measure how well different media support the growth of *E. faecalis* cells by measuring the **turbidity**, or cloudiness, of *E. faecalis* broth cultures. Turbidity is directly related to the number of cells in a broth; that is, the more cells there are, the cloudier the broth will be. To measure turbidity, we'll use a **spectrophotometer**, a device that sends a beam of light at 650 nm (red light) through a tube of broth. When the beam passes through the tube, some of the light is scattered or absorbed by cells in the broth. The absorbance value displayed by the spectrophotometer is directly related to turbidity, and thus to the number of *E. faecalis* cells in the broth. So the absorbance value provides a measure of how well the cells grew under the conditions present in the broth. Higher absorbance values are correlated with greater turbidity and more cells, and thus more cell growth and cell division.

PROTOCOL

Day 1

Materials
Per pair of students:

- pH 3 glucose nutrient agar (GNA) plates
- pH 5 GNA plates
- pH 7 GNA plates
- pH 9 GNA plates
- Broth A (no sugar, no buffer)
- Broth B (1.0% glucose, no buffer)
- Broth C (1.0% glucose, buffer)
- Broth D (0.1% glucose, buffer)
- pH paper, range pH 4 to 7
- Spectrophotometer
- Tubes of uninoculated broths (spectrophotometer blanks)

Bacterial and fungal cultures:

- *Escherichia coli*
- *Enterococcus faecalis*
- *Penicillium* (filamentous fungi)
- *Saccharomyces cerevisiae* (yeast)
- *Staphylococcus epidermidis*

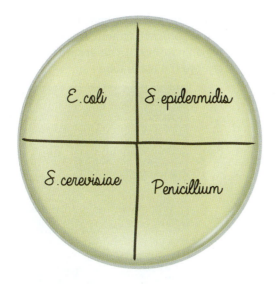

VARIABLE pH GLUCOSE NUTRIENT AGAR PLATES

1. Note that all of the GNA plates look alike, regardless of pH, so be sure to label the plates immediately or take any and all other steps needed to keep track of the pH of each plate.

2. Divide each of the four plates into quarters by marking the bottom of the plate with a wax pencil or Sharpie, and label each plate as follows:

3. For each species, use a flame-sterilized inoculating loop and transfer a small amount of pure-culture material to each plate by making a short, single-line streak in the appropriate quarter of the plate.

   Do not streak the entire quarter of the plate, and do not streak in such as way as to produce isolated colonies.

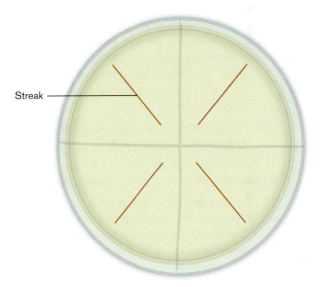

Streak

On day 2, we'll be looking for evidence that a given species can grow at a given pH, so do **not** transfer so much material that it looks as if cells have already grown on the plate and produced visible slime lines. In the case of Penicillium, even if you're careful, you'll probably be able to see mold on the plate after you transfer the cells, but note the amount of mold at the start, and look for changes by day 2.

4. Incubate the plates at 30°C for 48 hours.

TYPE A, B, C, AND D BROTHS

1. Suspend the cells in the *E. faecalis* stock culture broth by holding the test tube in the middle with one hand while gently flicking the bottom of the test tube with the other hand to create a swirl inside of the tube. Alternatively, if there is a vortex mixer on your lab bench, use that to mix the cells.

   It's critical that the cells be evenly distributed within the tube before you proceed.

2. Use a flame-sterilized inoculating loop to transfer one (and only one) loopful of cells to each of the four tubes of broth (A, B, C, and D).

   It's very important to add a similar number of cells to each tube.

3. Incubate the tubes at 37°C for 24 hours.

Day 2

VARIABLE pH GLUCOSE NUTRIENT AGAR PLATES

1. Examine each plate and grade the amount of growth on and around each streak as 0, 1, 2, 3, or 4, where 0 means "no growth" and 4 means "filled the quarter and threatening to crawl out of the plate." Record your results in the table in the Results section.

2. Dispose of the plates in the receptacle designated for trash to be autoclaved.

MEASURING pH OF THE A, B, C, AND D BROTHS

1. Dip one end of a strip of pH paper into the tube of type A broth. Compare the color of the pH paper with the chart on the pH paper dispenser. Record your observations in the table in the Results section. Dispose of the used pH paper strip in the receptacle designated for trash to be autoclaved.

2. Repeat step 1 with the tubes of type B, C, and D broth.

3. Post your results with those of the entire class, record the pooled class results, and calculate the mean values for all four broth types.

MEASURING ABSORBANCE OF TYPE A, B, C, AND D BROTHS

1. Determine the turbidity (cloudiness) of each of the four broths by measuring their absorbance in a spectrophotometer set to a wavelength of 650 nm. Follow the instructions below for the Spectronic 20 spectrophotometer (**Figure 13.4**). For a different model, follow the instructions provided by your lab instructor.

FIGURE 13.4 The Spec20 Spectrophotometer.

a. Take the cap off the tube of broth labeled "A, B, and D Blank" (provided next to the Spec 20), insert the blank tube into the spectrophotometer, and close the lid over the tube.

b. Using only the right-hand knob, adjust the absorbance to zero. Do not touch the left-hand knob. Ever. Once absorbance reads 0.0, remove the blank tube.

c. Pick up your A broth, and while holding the tube in the middle with one hand, resuspend the cells by gently flicking the bottom of the tube with the index finger of your other hand. Accurate results depend on getting all the cells suspended in the broth.

Do not hold the tube by the cap when doing this because the caps are loose and the test tube may slip out and smash on the ground.

d. Remove the test-tube cap, insert the type A broth into the Spec 20, close the lid, and read the absorbance. Record the absorbance in the table in the Results section. Remove the tube, replace the cap, and put the tube back in the test-tube rack.

e. Repeat steps c and d for the B and D broths. Since the colors or tints of uninoculated broths A, B, and D are almost exactly the same, you do not have to reset the Spec 20 for each of these broths.

f. After measuring the absorbance of the A, B, and D broths, repeat steps a and b using the tube labeled "C Blank." Remember, turn only the right-hand knob to adjust absorbance. Once the absorbance of the C blank reads 0.0, you're ready to measure the absorbance of your C broth.

g. Repeat steps c and d for your C broth.

2. Place the culture tubes in the location designated for materials to be autoclaved. Leave the blanks next to the Spec 20 for the next group.

3. Post your results with those of the entire class, record the pooled class results, and calculate the mean values for all four broth types.

Name: _____   Section: _____

Course: _____   Date: _____

VARIABLE pH PLATES

Species	pH 3.0	pH 5.0	pH 7.0	pH 9.0
E. coli				
S. epidermidis				
S. cerevisiae				
Penicillium				

Graph the medium pH versus the amount of growth for each species. Put lines for all four species on the one graph and use different colors or different symbols for each species so that the lines for the different species can be easily differentiated from each other.

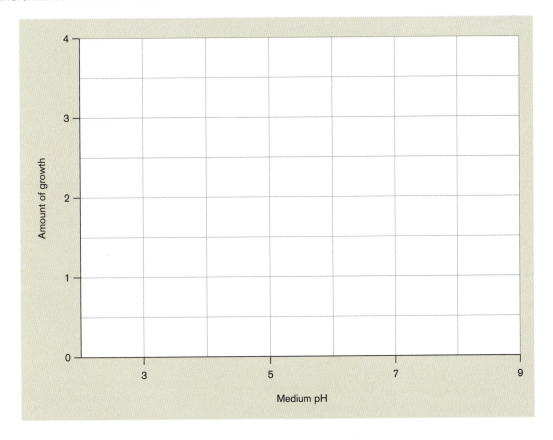

FERMENTATION AND BUFFERS (BROTHS A, B, C AND D)

INDIVIDUAL RESULTS FOR BROTH pH AND BROTH ABSORBANCE		
Broth Type	Broth pH	Broth Absorbance
A		
B		
C		
D		

POOLED RESULTS FOR BROTH pH AND BROTH ABSORBANCE				
Broth Type	Broth pH (pooled values)	Broth pH (mean)	Broth Absorbance (pooled values)	Broth Absorbance (mean)
A				
B				
C				
D				

QUESTIONS

Name: _____   Section: _____

Course: _____   Date: _____

1. What is the difference in hydrogen ion concentration between a pH 3 solution and a pH 6 solution?

2. Which of the species used in this lab appeared to be most tolerant of alkaline conditions (it could be more than one species)? Cite data from this lab to support your answer.

3. Which of the species used in this lab appeared to be most acidophilic or acid tolerant (it could be more than one species)? Cite data from this lab to support your answer.

4. Assuming that these lab species are representative of bacteria and fungi as groups, which group (bacteria or fungi) is more acid tolerant? Cite data from this lab in your answer.

Below are the recipes for each of the broth types used in this lab. Note that the glucose and potassium phosphate buffer (K_2HPO_4) concentrations are different in each broth. Also note that the addition of the buffer raises the initial pH. Use the information provided by this table to answer Questions 5 through 9.

Broth Type	Initial pH	Yeast Extract	Tryptone	Glucose	K_2HPO_4
A	6.5 – 6.8	0.5%	0.5%	0.0%	0.0%
B	6.5 – 6.8	0.5%	0.5%	1.0%	0.0%
C	7.5 – 7.8	0.5%	0.5%	1.0%	0.5%
D	7.5 – 7.8	0.5%	0.5%	0.1%	0.5%

5. For each broth type, explain how microbial activity changed the initial pH of the broth. For example, what was the role of fermentation in the pH changes?

6. Which broth type (A, B, C, or D) produced the least amount of growth? Explain why this broth type produced the least amount of growth.

7. Compare broth types B and C, and explain why you measured different pH values and different amounts of growth. Specifically, what was the effect of the buffer on pH change and the amount of *E. faecalis* growth?

8. Compare broth types C and D, and explain why you measured different pH values and different amounts of growth.

9. Find three food products at home or in a grocery store in which an ingredient has apparently been added to alter pH for the purpose of food preservation. List these three items along with the chemical that you think has been added to manipulate environmental pH.

Gaseous oxygen, or O₂, is in the air around us and in the water that covers much of the Earth. For us, O₂ is a vital substance; deprived of O₂ for more than ten minutes and we begin to slip into the arms of death. But for bacteria, it is more complicated. Some require O₂, as we do. Others prefer it, but can do without it for a time. Some prefer less O₂ than we require, and for some bacteria, O₂ is a killer.

Oxygen

Learning Objectives

- Consider the variable responses of microbes to oxygen, and understand that oxygen is highly reactive and even deadly to some microbial species.

- Describe the categories of microorganisms based on their ability to use or tolerate oxygen.

- Explain the characteristics of oxidizing and reducing environments, and know which types of microbes are favored in each environment.

- Understand and apply a variety of methods used to culture anaerobic bacteria.

We tend to think of **atmospheric oxygen (O_2)** as a universal good because without oxygen, humans turn blue and die. We do not do well under **anaerobic** conditions (from the Greek, *an-*, "without" + *-aer-*, "air" + *-bios-*, "life"); that is, we do not do well in environments with no oxygen. But the response of microorganisms to oxygen is highly variable, and not all microbes greet O_2 with irrational exuberance. Some species must have oxygen to grow, while other microbes find oxygen toxic and sometimes fatal.

The problem with oxygen is that it can occur in two highly reactive and toxic forms in cells: superoxide (O_2^-) and hydrogen peroxide (H_2O_2). Both of these molecules react with many cell molecules and structures, thoroughly trashing these components in the process. To cope with the potential destruction caused by reactive types of oxygen, some microbial species produce oxygen-detoxifying enzymes. These enzymes include **superoxide dismutase**, which catalyzes the reduction of superoxide to either molecular oxygen (O_2) or hydrogen peroxide, and **catalase**, which converts hydrogen peroxide to molecular oxygen and water. Cells that lack these enzymes cannot detoxify the reactive products of oxygen metabolism, so they cannot grow—and will likely be killed—if oxygen is present.

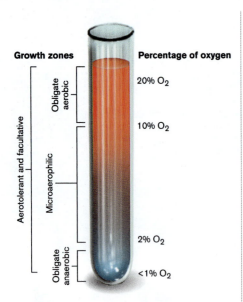

FIGURE 14.1 Oxygen tolerance among bacteria.

Thus, we can divide microbes, especially bacteria, into several different categories based on their ability to either use or tolerate oxygen. **Figure 14.1** illustrates some of those categories.

CATEGORIZING BACTERIA BASED ON RESPONSES TO OXYGEN

Obligate aerobes

Obligate aerobes are microorganisms that must have oxygen to grow and multiply. These microbes depend completely on aerobic respiration (the Krebs cycle and electron transport systems) for the production of ATP. *Micrococcus* species are good examples of strict aerobes.

Microaerophiles

Microaerophiles require oxygen, but grow best at oxygen concentrations below 20%; that is, at levels below the concentration of oxygen in the atmosphere. More specifically, typical microaerophiles are adapted to grow at oxygen concentrations of 2%–10%. Many of these species also prefer carbon dioxide levels that are higher than the usual atmospheric CO_2 concentrations. One example of a microaerophile is *Helicobacter pylori*, a species that lives in the stomach and causes peptic ulcers. There is usually some oxygen in the stomach due to the swallowing of air, but the oxygen concentration there is typically below the 20% O_2 found in the atmosphere.

Facultative anaerobes

Facultative anaerobes (often called simply **facultative** oganisms without the "anaerobes" part added) are microorganisms that can use oxygen if it's available, but don't require it to grow and multiply. As we saw in Lab 10, aerobic respiration is more efficient than either fermentation or anaerobic respiration at producing ATP from the energy released from molecules such as sugars. If oxygen is present, these microbes will preferentially use aerobic metabolic pathways, which allow the cells to grow much faster, and the microbial populations to increase more rapidly, than they would in the absence of oxygen. However, if there is little or no oxygen in the microorganisms' environment, the cells will not give up and curl into a fetal position. Instead, they will switch to fermentation or anaerobic respiration, both of which are capable of generating ATP in the absence of O_2. Growth will be slower than if oxygen was present, but facultative anaerobes are able to grow, divide, and expand their populations under anaerobic conditions. Many of the species that we use in this course are facultative anaerobes, including *E. coli* and *Staphylococcus* species.

Aerotolerant microorganisms

Aerotolerant microorganisms are incapable of using oxygen, and they rely on anaerobic metabolic pathways such as fermentation to produce ATP. While they do not use oxygen, they can tolerate its presence, although they grow best under low oxygen concentrations. One large group of aerotolerant bacteria is the lactic acid bacteria, which ferment sugars and produce large quantities of lactic acid as a result. This group includes *Streptococcus* and *Enterococcus* species.

Obligate anaerobes

Obligate anaerobes lack oxygen-detoxifying enzymes and other mechanisms for coping with oxygen, and so are unable to tolerate or grow in the presence of oxygen. Some obligate anaerobes, called **oxyduric** obligate anaerobes, can endure or survive exposure to oxygen and can grow later if oxygen is removed. But other species, the **oxylabile** obligate anaer-obes, are quickly killed by contact with oxygen. Perhaps the best-known examples of obli-gate anaerobes are the *Clostridium* species (see Lab 8). One species, *C. perfringens*, causes gas gangrene, a very dangerous, life-threatening type of wound infection, in tissue where injury has led to oxygen-free conditions. Gas gangrene may be treated by hyperbaric (high pressure) oxygen therapy which increases the amount of oxygen dissolved in the blood plasma. This, in turn, can lead to more *Clostridium*-killing oxygen at the site of infection.

ENVIRONMENTAL OXIDATION/REDUCTION POTENTIAL (ORP)

It is helpful to be able to describe the relative amount of oxygen in a microbial environment. One way of doing so is to consider whether the environment is "more oxidizing" or "more reducing." Recall that a molecule that is being oxidized will lose electrons, while a molecule that is being reduced gains electrons. The character of the environment may be quantified or measured as the **oxidation/reduction potential (ORP)** of the habitat. Environmental ORP is a measure of the habitat's overall capacity or tendency to accept or donate electrons.

In highly oxidizing environments (high ORP), there are a large number of molecules with a high affinity for electrons, including high-electron-affinity oxygen molecules. Such molecules readily accept electrons from other molecules, and so many of these other molecules will be in an oxidized state. A high ORP environment favors microbial species that can operate with many of their biological molecules in an oxidized state, including obligate aerobes, microaerophiles, and facultative species. In contrast, aero-tolerant species often do not grow well in high ORP habitats, and obligate anaerobes will not grow at all.

In low ORP or highly reducing environments, there are a large number of molecules with a low affinity for electrons and little or no O_2 is present. Low-electron-affinity mol-ecules readily donate electrons to other molecules, thus many of these other molecules will be in a reduced state. A low ORP habitat favors aerotolerant and obligate anaerobe microbes; these species can operate with many of their biological molecules in a re-duced state. In contrast, obligate aerobes will not grow at all and facultative species will grow more slowly than they do in oxidizing environments.

ANAEROBIC CULTURE METHODS

Bacterial species that cannot abide the presence of oxygen are usually more difficult to culture than those that can use or tolerate oxygen because the growth of obligate anaerobes requires the removal and exclusion of oxygen from the environment. Several methods have been developed for the cultivation of anaerobes:

Boiling and sealing

Hot liquids have a lower capacity to hold dissolved oxygen than cold liquids, so one way to remove dissolved oxygen is to boil the medium. Depending on the length of boiling time and the volume of the liquid, this will drive almost all (if not all) of the oxygen out of the me-dium. If the medium is then quickly cooled and inoculated and the air-medium interface sealed with oil or wax, then most obligate anaerobes will be able to grow in the medium.

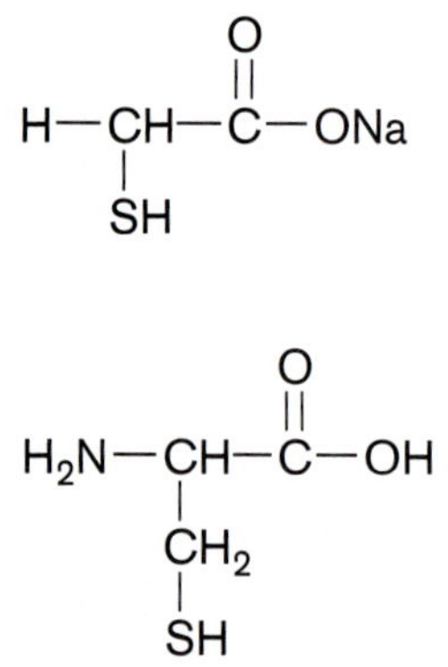

FIGURE 14.2 Cysteine (top) and sodium thioglycollate (bottom).

Use of reducing agents

Oxygen can also be removed, or the effects of oxygen minimized, by the addition of reducing agents. **Reducing agents** are molecules that (1) bind free oxygen or (2) donate electrons and H^+ (protons, hydrogen ions) to keep other molecules in a reduced state. Molecules with **sulfhydryl groups** are often effective reducing agents because the $-SH$ group readily donates electrons and protons to other molecules, thus reducing them. In some cases, the sulfhydryl-bearing compounds react directly with certain reactive forms of oxygen, removing the reactive oxygen species by reducing them to water.

Two reducing agents that are often added to media used in the cultivation of strict anaerobes are (1) the amino acid **cysteine** and (2) **thioglycollate**, usually in the form of sodium thioglycollate **(Figure 14.2)**. Thioglycollate is particularly effective because it reacts directly with oxygen to remove it from the environment. In addition, thioglycollate broths are usually relatively viscous, which reduces the rate at which oxygen diffuses into the medium from the air. This viscosity also creates oxygen concentration gradients in broth-containing tubes, with some oxygen dissolved in the top layer of the broth, lower oxygen concentrations in the middle, and essentially no oxygen at the bottom of the tube.

How do we know that reducing agents are doing their job? Anaerobic media often include **redox indicators**, or molecules whose color reflects the ORP of the media. For example, in this lab, we will use resazurin as a redox indicator. Resazurin is a dye that, in certain biological assays, is blue in color when fully oxidized. However, in this particular lab, when used in the thioglycollate medium and in the anaerobic chamber indicator strips, resazurin will appear essentially colorless under anaerobic or reducing conditions, and will turn pink under aerobic or oxidizing conditions. So when a resazurin-containing medium turns pink in color, this is a sign that there is probably too much free oxygen dissolved in the medium for the growth of strict anaerobes.

Removal of oxygen from a sealed chamber

Strict anaerobes can also be cultured in an oxygen-free chamber. If there is no oxygen in the surrounding air, then there is no oxygen to diffuse into the medium in the chamber, so the medium remains oxygen-free, too. One approach to creating oxygen-free chambers is to remove, or "purge," the oxygen from a hood or glove box **(Figure 14.3A)** by replacing it with a flow of nitrogen (N_2) or carbon dioxide (CO_2) gas.

Another method uses a small container called an **anaerobe jar** or **Brewer jar** **(Figure 14.3B)**, from which the free oxygen can be removed by chemical reactions after the jar has been sealed. Inoculated media are placed inside the jar, and a GasPak packet is added to the jar. The GasPak contains sodium bicarbonate, citric acid, sodium borohydride, and a palladium metal catalyst. When water is added to the packet and the jar is sealed, several chemical reactions create an atmosphere inside the sealed container that is ideal for the growth of strict anaerobes.

Initially, when the water reaches the sodium bicarbonate and citric acid tablet in the GasPak, the tablet releases carbon dioxide (CO_2) gas in a reaction similar to the one produced in everybody's favorite science project, the baking soda and vinegar volcano. In this case, the organic acid triggering the release of the CO_2 from the bicarbonate (baking soda) is citric acid, not acetic acid (vinegar). Carbon dioxide is important because most anaerobes grow better under high CO_2 concentrations.

A.

B.

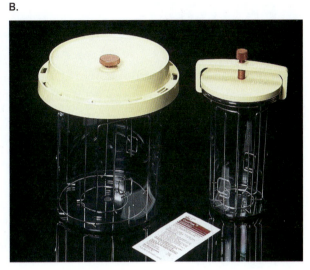

FIGURE 14.3 Sealed chambers from which oxygen can be removed.
A. Glove box for cultivations of anaerobes. **B.** Brewer jars with GasPak.

However, it's the second set of reactions that is essential for cultivating obligate anaerobes. When water comes in contact with the sodium borohydride, hydrogen gas (H_2) is released from the hydride. The hydrogen gas combines with the free oxygen in the air in a reaction catalyzed by the palladium metal to produce water. The reduction of free oxygen to water removes the oxygen from the atmosphere inside the jar, creating an anaerobic environment.

It is critical to the success of the operation that the jar be well sealed during the incubation of the anaerobes. There is a limit to the amount of CO_2 and H_2 that the GasPak can produce, and if the jar leaks, eventually the oxygen in the surrounding air will reach and stop the growing anaerobes. One way to monitor conditions inside the jar is through the use of a oxygen indicator strips.

Indicator strips typically contain either resazurin or methylene blue. As noted above, when used in indicator strips, resazurin will turn dark pink in the presence of oxygen, but once the oxygen is removed from the air in the jar, the color will fade to a light pink to white color. Methylene blue, in addition to being a direct stain, is a redox indicator **(Figure 14.4)**. Like resazurin, the methylene blue molecule changes color depending on whether it is in an oxidized or a reduced state. If oxygen is present and methylene blue is oxidized, it will be blue in color, but in the absence of oxygen, the methylene blue will be reduced and will be colorless. So if we place a methylene blue strip inside an anaerobe jar, we will be able to see from the color of the strip if oxygen is entering the jar after we have attempted to seal it. A light blue to white color means that the oxygen has been removed, but a dark blue strip means that there is probably too much free oxygen entering the jar for the successful cultivation of strict anaerobes.

In this lab, you will study the effects of oxygen on a variety of bacterial species by (1) creating oxygen gradients in yeast-tryptone agar tubes prior to inoculating them; (2) inoculating broths to which thioglycollate has been added to create a reducing environment; and (3) incubating inoculated nutrient agar plates in Brewer jars.

FIGURE 14.4 Methylene blue.
When the reduced and colorless form of methylene blue loses electrons in an oxidizing environment, the methylene blue is oxidized and turns blue in color.

PROTOCOL

Materials

Per pair of students:

- 3 melted yeast-tryptone agar (YTA) tubes
- 2 thioglycollate broths
- 2 nutrient agar plates
- 47°C water bath
- Brewer jar with GasPak

Cultures:

- *Clostridium sporogenes*
- *Escherichia coli*
- *Micrococcus luteus*

Day 1

OXYGEN GRADIENT USING MELTED YTA

1. Prepare three masking tape labels with your name, the date, and the following species names:

   Escherichia coli
   Micrococcus luteus
   Clostridium sporogenes

2. Remove three melted YTA tubes from the 47°C water bath and attach the labels.

   As soon as the tubes are removed from the water bath, the agar will begin to cool and solidify. This is why you should prepare your labels ahead of time.

3. For each bacterial species, use a flame-sterilized inoculating loop to transfer one loopful of cells to one of the melted YTA tubes.

4. Gently move the loop up and down in the column of agar to distribute the cells throughout the agar. Do not agitate the agar too much, because you don't want to introduce air, and thus oxygen, into the agar column.

5. After removing the loop, gently shake or flick the tube for a minute or so to continue mixing cells throughout the tube, but do not shake or flick the tube too vigorously, because you don't want to introduce oxygen (air) into the agar column.

6. Let the agar solidify and incubate the tubes at 37°C for 48 hours.

THIOGLYCOLLATE BROTHS

1. Label two freshly boiled thioglycollate broths with your name, the date, and the following species names:

   Micrococcus luteus
   Clostridium sporogenes

   The thioglycollate broths will probably have a pinkish-colored zone at the top of the tube extending down about one-half inch. The pink color should not extend beyond the top half of the tube because pink shows where oxygen has diffused into the medium. We want at least 50% of the broth to be oxygen-free, so do not use the broth if there is any pink color in the bottom half of the tube.

2. For each thioglycollate broth tube and bacterial species, use a flame-sterilized inoculating loop and transfer one loopful of cells to the bottom of the tube.

   As you transfer the cells, be careful not to agitate the broth too much, as this will introduce oxygen into the broth. In the case of *Clostridium sporogenes*, it is particularly important that the loop touches the bottom of the tube because you want to introduce the cells into a part of the broth where there is no oxygen.

3. Incubate the tubes at 37°C for 48 hours.

PLATES IN AND OUTSIDE BREWER JARS

1. Divide each nutrient agar plate into thirds by marking the bottom of the plate with a wax pencil or Sharpie, and label each plate as follows:

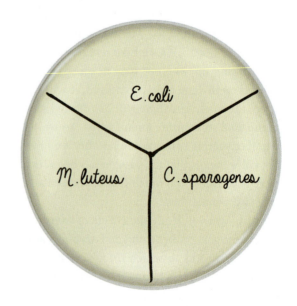

2. Each plate will be inoculated with three species. For each species, use a flame-sterilized inoculating loop and transfer a small amount of pure-culture material to the plate by making a short, single-line streak in the appropriate sector of the plate.

Do not streak the entire sector of the plate, and do not streak in such a way as to produce isolated colonies.

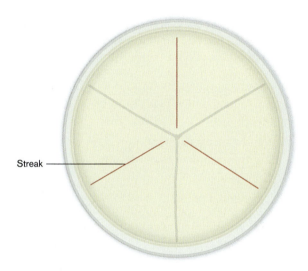

Streak

On day 2, we'll be looking for evidence that a given species can grow in the presence or absence of oxygen, so **do not transfer so much material that it looks as if cells have already grown on the plate** and produced visible slime lines.

3. After both plates have been inoculated, place one plate in a Brewer jar to be grown under anaerobic

conditions. The other plate will be grown in the presence of oxygen.

4. When the jar is full, your instructor will activate the GasPak, and then incubate the anaerobic plate in the Brewer jar, and the aerobic plate in the open air, at 37°C for 48 hours.

Day 2

OXYGEN GRADIENT USING MELTED YTA

1. Examine the YTA tubes to determine if and where cells were able to grow. Growth may take the form of turbidity or cloudiness within the agar, slime on top of the agar, or both. If gas was produced under the anaerobic conditions within the agar, then the agar may be cracked. Record your observations in the Results section.

2. Remove the labels from the tubes and place them in the racks designated for tubes to be autoclaved.

THIOGLYCOLLATE BROTHS

1. Examine the thioglycollate broth tubes to determine whether and where cells were able to grow within or on top of the broth. Take note of any pink zones at various points in the tube. Record your observations.

2. Remove the labels from the tubes and place them in the racks designated for tubes to be autoclaved.

PLATES IN AND OUTSIDE BREWER JARS

1. Examine the plates to determine whether species were able to grow on the surface of the nutrient agar in the presence or absence of oxygen (outside and inside the Brewer jar, respectively). Record your observations.

2. Dispose of the plates in the receptacle designated for trash to be autoclaved.

RESULTS

Name: _____   Section: _____

Course: _____   Date: _____

OXYGEN GRADIENT USING MELTED YTA

For each tube, fill in areas of turbidity.

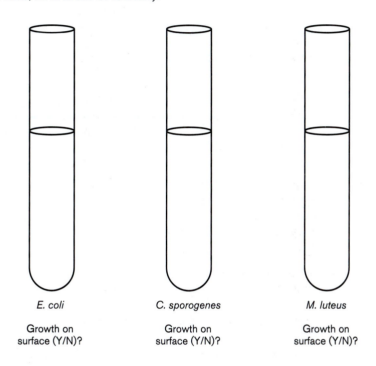

E. coli	*C. sporogenes*	*M. luteus*
Growth on surface (Y/N)?	Growth on surface (Y/N)?	Growth on surface (Y/N)?

THIOGLYCOLLATE BROTHS

For each tube, fill in areas of turbidity and note any pink zones.

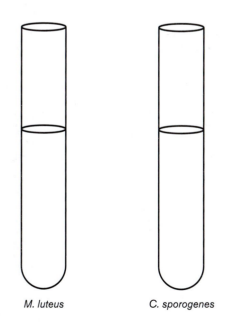

M. luteus   *C. sporogenes*

PLATES IN AND OUTSIDE BREWER JARS

Species	Amount of Growth Outside Brewer Jar	Amount of Growth Inside Brewer Jar
E. coli		
M. luteus		
C. sporogenes		

QUESTIONS

Name: _____   Section: _____

Course: _____   Date: _____

1. Based on the location of growth in the YTA tubes, classify each species with respect to the cell's ability to use or tolerate oxygen ("strict aerobe," etc.). In each
   case, cite evidence from the YTA tubes to support your conclusions.

   a. *E. coli*

   b. *M. luteus*

   c. *C. sporogenes*

2. Which part of the YTA tube has a higher oxidation/reduction potential or region
   favoring oxidation, the top or the bottom? Based on your answer, where would you
   expect *E. coli* cells to grow more rapidly (top or bottom)? Explain your answer.

3. Based on the location of growth in the thioglycollate broths, how would you classify
   each species with respect to the cell's ability to use or tolerate oxygen? In each
   case, cite evidence from the thioglycollate broths to support your conclusions.

   a. *M. luteus*

   b. *C. sporogenes*

4. Why do we care about pink zones in the thioglycollate broths?

5. Based on what you actually observed on the nutrient agar plates, how would you classify each species with respect to the cell's ability to use or tolerate oxygen? In each case, cite evidence from the nutrient agar plates to support your conclusions.

   a. *E. coli*

   b. *M. luteus*

   c. *C. sporogenes*

6. Looking at the nutrient agar plate data for *M. luteus*, would you conclude that the Brewer jar worked? Explain your answer.

7. Looking at the nutrient agar plate data for *C. sporogenes*, would you conclude that the Brewer jar worked? Explain your answer.

8. Did *E. coli* grow better inside of the Brewer jar or outside of the Brewer jar? Explain your answer.

Disinfectants, such as bleach, alcohol, and iodine kill "germs"—disease-causing bacteria, fungi, and viruses. By 1910, when this advertisement was produced, people understood that killing germs meant reducing the spread of infectious diseases ("Charbon" is another word for anthrax).

Learning Objectives

- Define and understand the significance of terms containing the suffixes "-static" or "-cidal" and the terms "disinfectant" and "antiseptic."

- Understand how antimicrobial chemical agents are evaluated; in particular, consider how the phenol coefficient is determined.

- Know the modes of action of disinfectants and antiseptics used in this lab.

- Assess the effectiveness of various chemicals against Gram-positive and Gram-negative species.

LAB **15**

Disinfectants and Antiseptics

The Antiseptic Baby and the Prophylactic Pup,
Were playing in the garden when the Bunny gamboled up;
They looked upon the Creature with a loathing undisguised;—
It wasn't Disinfected and it wasn't Sterilized.
—Arthur Guiterman, Strictly Germ-proof (1906)

CLASSES OF ANTIMICROBIAL AGENTS

The use of chemicals to control microbial growth and reduce microbial populations began thousands of years ago. For centuries, people didn't understand why certain substances prevented food spoilage or helped to heal wounds, but they could observe and make use of the positive effects of these substances. The development of germ theory in the mid-1800s provided an underlying explanation for how and why certain chemicals might act to preserve food and prevent or treat disease. This led to a more systematic effort to discover and apply antimicrobial agents, and most of the disinfectants and antiseptics in use today have been used since at least the late 1800s, including alcohols, phenolics, iodine, and hypochlorite (bleach). Antimicrobial agents can be categorized by their effects as well as by their uses.

Definitions based on effects

Static agents The suffix **-static** is added to prefixes such as "germi-," "bacteri-," or "fungi-," as in "germistatic," "bacteristatic," and "fungistatic." Static agents inhibit or stop microbial growth, but may not kill the microbes. If the agent is removed from the environment, the microorganisms that it held in check may resume growth and cell division.

Cidal agents The suffix **-cidal** is used in words such as "germicidal," "bactericidal," and "fungicidal." **Cidal agents** kill microorganisms. Obviously, if the cells are killed, there will be no further growth and cell division, even if the chemical agent is removed from the environment.

A cidal agent may be able to sterilize an object or environment depending on factors such as (1) how the cidal chemical is used, (2) the length of time the chemical is in contact with the microorganisms, and (3) the nature of the microbes that are present. Bacterial endospores are usually the last living things to be killed by a potential sterilizing chemical. Chemicals capable of sterilization, sometimes called **high-level disinfectants**, include agents such as formaldehyde.

Definitions based on use

Disinfectants The term **disinfectant** is used to describe chemicals that are applied to **nonliving or inanimate surfaces and objects** such as instruments, benchtops, and floors. These agents are not intended for use on living tissue and tend to be more toxic than antiseptics. Almost all of these chemicals are cidal agents. In most applications, disinfectants greatly reduce the microbial population, but only highly toxic disinfectants that are in contact with the treated surfaces or objects for an extended period of time are able to sterilize them.

Antiseptics The term **antiseptic** is used to describe chemicals applied to **living tissue**. Specifically, antiseptics are applied to skin and mucous membranes, such as the membranes lining the inside of the mouth, to reduce the number of microbes in these locations. Since antiseptic chemicals come in direct contact with cells we don't want to kill—that is, with our cells—they are usually less toxic than disinfectants.

In some cases, a disinfectant can be used as an antiseptic if the chemical is used at a lower concentration, because lowering the concentration reduces the toxic effects. That is, by varying the concentration, the same chemical can be used as both a disinfectant and an antiseptic. For example, a 5%–10% phenol solution can be used as a strongly bactericidal disinfectant, but at these concentrations, phenol is also very irritating and corrosive to living tissue. If the concentration is reduced to around 0.5%–1.0%, then phenol can be used as an antiseptic (and as a local anesthetic in sore throat lozenges). In other cases, the chemical of choice is an agent that isn't toxic enough to be widely used as an effective disinfectant, but is still potent enough to be useful as an antiseptic.

While generally not as toxic as disinfectants, most of the antiseptics in use still have a killing effect on microorganisms that we are trying to control. So, it is not accurate to say that disinfectant = cidal agent and antiseptic = static agent. Antiseptics may be "safer" than disinfectants, but they can still do damage to our cells. Therefore, they are rarely applied at a concentration that would kill all the microbes on living

Table 15.1 | DISINFECTANTS AND ANTISEPTICS

Name	Forms	Mode of Action	Disinfectant Applications	Antiseptic Applications
Phenol and phenolics	Phenol	Denatures proteins, disrupts membranes	May be used on inanimate hard surfaces, but not widely used today; has strong odor	Lister's old surgical antiseptic "carbolic acid"; rarely used as an antiseptic today, too toxic
	Phenylphenol (Lysol)	Denatures proteins, disrupts membranes	Disinfects surfaces, including floors, institutional showers, etc.	Too toxic for topical use, not used as antiseptic
Alcohols	70%–75% ethanol (grain alcohol) 70%–75% isopropanol (rubbing alcohol)	Denatures proteins, disrupts membranes	Disinfects lab benches and lab or medical instruments	Used externally on unbroken skin to reduce pathogen population before hypodermic injections, insertion of IV lines, or other procedures that break the skin
Chlorine	Hypochlorite (ClO^-) solutions, such as Clorox bleach (NaOCl)	Chemically oxidizes many types of cell components and biochemicals	Disinfects drinking water, pools, and treated sewage; used on many surfaces in lab and clinical settings (floors, etc.)	Rarely used on skin, too corrosive
Quaternary ammonium compounds	Many examples, including benzalkonium chloride and cetylpyridinium chloride	Amphipathic molecules interact with phospholipids in cell membranes, act as detergents to disrupt lipid membranes	May be used to disinfect instruments and hard surfaces such as floors	Main use is in mouthwashes, toothpastes, throat lozenges, and in sore throat, breath, and nasal sprays

tissue. Furthermore, antiseptics are too toxic to be taken internally or injected into the body. Their main value is in reducing, though not totally eliminating, microbial populations on skin and mucous membranes.

Some examples of antiseptics and disinfectants are shown in **Figure 15.1** and described in **Table 15.1**.

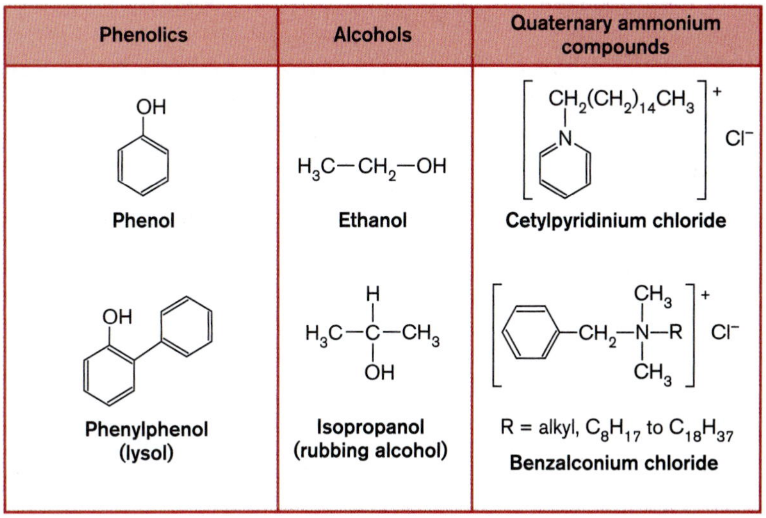

EVALUATING ANTIMICROBIAL ACTIVITY

The antimicrobial activity of chemical agents is influenced and altered by many factors, including, but not limited to, the following:

1. pH
2. Temperature
3. Solubility of the agent

FIGURE 15.1 Structures of some common antiseptics and disinfectants.

4. Interactions with other chemicals in the environment

5. Initial size of the microbial population

6. Types of microbes present

So how can we tell whether one agent is more potent than another? In short, we have to compare or test these chemicals against a standard disinfectant under standard conditions against a standard microorganism. In this context, the word "standard" means that the disinfectant, microorganisms, and test conditions have been agreed upon ahead of time by those involved in comparing various chemical control agents. This is the principle behind the phenol coefficient. The **phenol coefficient** is a value that can be determined for a test chemical by comparing its effectiveness against that of the disinfectant phenol. The comparison must be done under standard conditions against a Gram-positive species, *Staphylococcus aureus*, and a Gram-negative species, *Salmonella typhi*.

In the phenol coefficient assay, the test chemical is **repeatedly diluted** until the dilution is so great that the agent is no longer able to kill *S. aureus* and/or *S. typhi*—that is, until it is no longer effective. The greatest dilution that kills the test bacteria is recorded as the **maximum effective dilution**. (This information is also valuable for determining the concentration at which a chemical agent should be used when it is applied in microbial control efforts.) Then the same dilution and testing procedure is performed using phenol to determine the maximum effective dilution for that standard disinfectant.

At the end of the assay, the phenol coefficient is determined by dividing the reciprocal of the maximum effective dilution for the test chemical by the reciprocal of the maximum effective dilution for phenol to determine the ratio of the toxicity of the test chemical to the toxicity of phenol. If the value is less than 1.0, then the chemical is less effective than phenol, and if the value is greater than 1.0, then the agent is more effective than phenol.

For example, if the maximum effective dilution for a hypothetical disinfectant called Killemdead is 1/10,000 (10^{-4} or 0.0001), and the maximum effective dilution for phenol is 1/100 (10^{-2} or 0.01), then the phenol coefficient for Killemdead is

$$\text{Phenol coefficient} = \frac{\text{reciprocal of 0.0001 (max effective dilution of Killemdead)}}{\text{reciprocal of 0.01 (max effective dilution of phenol)}}$$

$$10{,}000/100 = 100$$

Most disinfectants in use today are more toxic than phenol and have phenol coefficient values ranging from about 10 to 200.

In this lab, we will assess the activity of antimicrobial chemicals using a method that is simpler than calculating the phenol coefficient and it does not require using pathogens. However, it is still important for you to understand how the phenol coefficient is determined.

PROTOCOL

Day 1

Materials

Per pair of students:

- 2 nutrient agar plates
- 2 empty sterile test tubes
- 2 sterile 1-ml pipettes
- 2 sterile 10-ml pipettes
- 2-ml (blue-barrel) Pipette Pump
- 10-ml (green-barrel) Pipette Pump
- 10% bleach
- 70% ethanol
- 0.5% Lysol
- 5.0% Lysol
- Mouthwash

Cultures:

- *Bacillus megaterium*
- *Escherichia coli*

1. Each pair of students will be assigned one chemical to test. Once given your assignment, immediately make a note of which disinfectant or antiseptic you are to test so that you don't forget.

   One member of the pair should follow the steps below using *E. coli* and the other member should follow the steps using *B. megaterium*.

2. Transfer 5 ml of your assigned chemical from the stock solution to one of your empty sterile test tubes using a sterile 10-ml pipette and a 10-ml (green-barrel) Pipette Pump.

3. Label a nutrient agar plate either *E. coli* or *B. megaterium*, depending on which species you are working with. Divide the plate into quarters by marking the bottom of the plate with a wax pencil or Sharpie, and label each plate as follows:

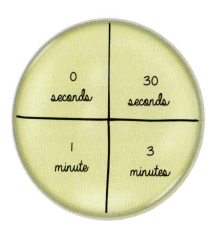

4. Use a flame-sterilized inoculating loop to transfer a loopful of *E. coli* or *B. megaterium* cells from the pure-culture broth to the plate by making a short, single-line streak in the "0 (zero) seconds" quarter of the plate.

   Do not streak the entire quarter of the plate, and do not streak in such a way as to produce isolated colonies. Instead, follow the pattern below.

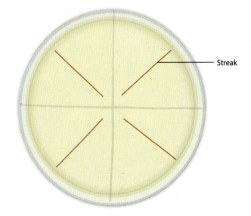

5. Using a sterile 1-ml pipette and a 2-ml (blue-barrel) Pipette Pump, transfer 1 ml of *E. coli* or *B. megaterium* cells to the test tube containing the 5 ml of your assigned disinfectant or antiseptic.

6. **As soon as cells are transferred to the chemical solution**, begin tracking the amount of time the cells are in contact with the disinfectant or antiseptic. Immediately dispose of the pipette in the receptacle designated for trash to be autoclaved.

   Do not put contaminated pipettes down on the benchtop.

7. Mix cells into the chemical solution by holding the test tube in the middle with one hand while gently flicking the bottom of the test tube with the other hand to create a swirl inside the tube.

8. After 30 seconds, 1 minute, and 3 minutes of contact time, repeat step 4 above. That is, use a flame-sterilized inoculating loop to transfer a loopful of cells from the chemical solution test tube to the plate by making a short, single-line streak in the appropriate quarter of the plate.

   Once you and your partner have finished, you should now have one *E. coli* plate and one *B. megaterium* plate with cells streaked after four different contact times.

9. Incubate plates at 37°C for 24 hours.

Day 2

1. Examine each plate and grade the amount of growth in each quarter as 0, 1, 2, 3, or 4, where 0 means "no growth" and 4 means an amount of growth about equal to the growth in the "0 seconds" quarter.

   Obviously, scoring these streaks will involve making judgment calls, but do your best, and try to score the streaks in such a way as to differentiate among the streaks from different contact times.

2. Record your results for your particular chemical solution in the Results section. Post your results with those of the entire class and record the pooled class results.

3. There will be several scores for each combination of species, contact time, and disinfectant or antiseptic. Use these scores to calculate mean scores for each species, contact time, and chemical solution.

4. Dispose of the plates in the receptacle designated for trash to be autoclaved.

Name: _____   Section: _____

Course: _____   Date: _____

INDIVIDUAL RESULTS

Name of assigned disinfectant or antiseptic: _____

Species	0 seconds	30 seconds	1 minute	3 minutes
E. coli				
B. megaterium				

POOLED RESULTS

10% BLEACH

Species	0 seconds	30 seconds	1 minute	3 minutes
E. coli				
	Mean:	Mean:	Mean:	Mean:
B. megaterium				
	Mean:	Mean:	Mean:	Mean:

70% ETHANOL

Species	0 seconds	30 seconds	1 minute	3 minutes
E. coli				
	Mean:	Mean:	Mean:	Mean:
B. megaterium				
	Mean:	Mean:	Mean:	Mean:

0.5% LYSOL

Species	0 seconds	30 seconds	1 minute	3 minutes
E. coli				
	Mean:	Mean:	Mean:	Mean:
B. megaterium				
	Mean:	Mean:	Mean:	Mean:

5.0% LYSOL

Species	0 seconds	30 seconds	1 minute	3 minutes
E. coli				
	Mean:	Mean:	Mean:	Mean:
B. megaterium				
	Mean:	Mean:	Mean:	Mean:

MOUTHWASH

Species	0 seconds	30 seconds	1 minute	3 minutes
E. coli				
	Mean:	Mean:	Mean:	Mean:
B. megaterium				
	Mean:	Mean:	Mean:	Mean:

For each species, graph the amount of growth versus time of exposure for all five chemicals, putting the lines for all chemicals on the same graph.

Effect of chemicals on *E. coli* growth

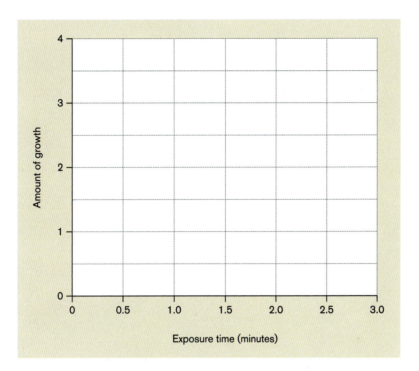

Effect of chemicals on *B. megaterium* growth

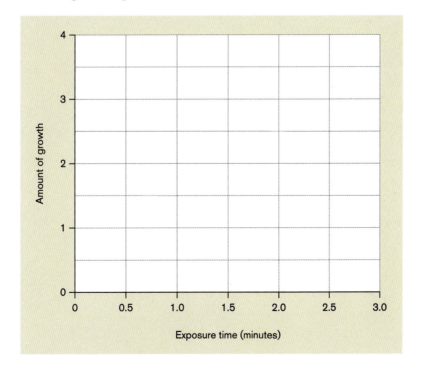

Name: _____   Section: _____

Course: _____   Date: _____

1. Which chemical (or chemicals) was (or were) most effective at killing *E. coli*? How long did it take the most effective chemical(s) to kill all the cells? Cite lab data, including evidence from the graphs, to support your conclusion.

2. Which chemical (or chemicals) was (or were) most effective at killing *B. megaterium*? How long did it take the most effective chemical(s) to kill all the cells? Cite lab data, including evidence from the graphs, to support your conclusion.

3. Which chemical (or chemicals) was (or were) least effective killing *E. coli*? Cite lab data, including evidence from the graphs, to support your conclusion.

4. Which chemical (or chemicals) was (or were) least effective at killing *B. megaterium*? Cite lab data, including evidence from the graphs, to support your conclusion.

5. Which species, *E. coli* or *B. megaterium*, was apparently more difficult to kill? Cite lab data, including evidence from the graphs, to support your conclusion.

6. Was the answer to Question 5 (which species was more difficult to kill) what you expected? Why or why not? If the results did not match your expectations, explain what might have gone wrong.

7. When you looked at Table 15.1, did you notice that the modes of action of phenol, phenolics, and alcohols are all quite similar? Look at the structures of these different molecules in Figure 15.1 and explain why these molecules have a similar mode of action; that is, how do the structures of the molecules explain their mode of action?

8. A newly developed disinfectant was tested under the standard conditions used to determine the phenol coefficient for the new chemical. Testing produced the following results:

Dilution	Growth of Cells after Exposure to Chemical
0.005 (0.5% solution)	No growth
0.004 (0.4% solution)	No growth
0.003 (0.3% solution)	No growth
0.002 (0.2% solution)	No growth
0.001 (0.1% solution)	Growth

Assuming that the maximum effective dilution for phenol is 0.01, what is the phenol coefficient of the new chemical?

9. In 2009, a vascular surgery ward in a hospital in Scotland experienced an outbreak of intestinal illnesses caused by *Clostridium difficile*. During the outbreak, dispensers of alcohol-based hand gels were actually <u>removed</u> from the ward. Why was this done? (Think about what *Bacillus* and *Clostridium* cells have in common.)

Ultraviolet light can damage the DNA inside cells. For that reason, it is used as a physical means of killing microorganisms on solid surfaces, like countertops, floors, and walls. Ultraviolet light also allows us to see things or features of objects that are invisible to the visible light rays that our eyes detect. Here is a collection of minerals that absorb ultraviolet light, and fluoresce longer wavelengths of beautiful colored light.

LAB 16

Ultraviolet Light

Learning Objectives

- Describe the physical properties of ultraviolet (UV) light.

- Consider the effects of UV light on DNA structure and mutation rates, and understand why these effects allow us to use UV light to control microorganisms.

- Understand the variables that influence and limit the effectiveness of UV light as a control agent.

- Assess the capacity of UV light to control specific bacterial species by exposing the bacteria to UV light under different conditions.

In laboratory and clinical settings, microorganisms may be controlled by a combination of chemicals used as disinfectants and antiseptics (Lab 15) and physical control methods, such as the use of ultraviolet light.

ULTRAVIOLET LIGHT

The term **ultraviolet (UV) light** refers to the segment of the electromagnetic spectrum between about 100 and 400 nanometers (nm) **(Figure 16.1)**. These wavelengths of light have more energy than the longer wavelengths of visible light, but they have much less energy than the much shorter gamma and X-ray radiation wavelengths. In addition to having more energy than visible light, UV light has significant effects on cells because the DNA within cells absorbs the energy in UV light. The maximum absorption of UV light energy by DNA occurs at wavelengths between 260 and 265 nm.

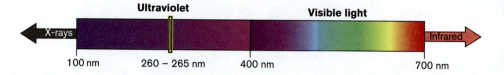

FIGURE 16.1 The ultraviolet and visible portions of the electromagnetic spectrum.

When DNA is exposed to UV light, especially UV light with wavelengths of 260–265 nm, the DNA may be damaged and mutated. Absorption of UV light energy leads to the formation of **pyrimidine dimers**, or covalent links between adjacent pyrimidine bases on the same strand. (Recall that pyrimidines include cytosine, uracil, and thymine.) In most cases, the links are formed between two thymines, creating a **thymine dimer** **(Figure 16.2)**.

The formation of thymine dimers distorts the shape of the DNA molecule and prevents the accurate copying or replication of DNA. Thus, when the cells copy their chromosomes prior to cell division, mutations appear in the new strands. If too many mutations occur, the microbes can't make the proteins that are essential for staying alive, and so the cells die.

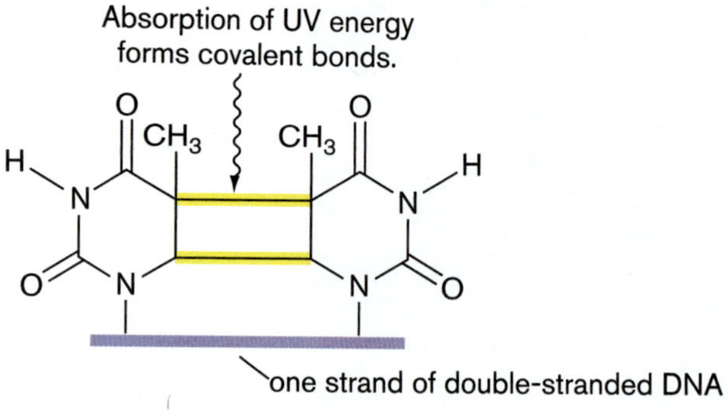

FIGURE 16.2 A thymine dimer.

VARIABLES AFFECTING THE IMPACT OF ULTRAVIOLET LIGHT ON MICROBES

Several variables affect the impact of UV light on microorganisms. These variables include the following:

1. **The wavelength used.** Since the maximum absorption of UV light energy by DNA occurs at wavelengths between 260 and 265 nm, these wavelengths are usually the most effective; that is, they are the most germicidal.

2. **The time of exposure.** Longer exposure times usually result in a greater germicidal effect resulting in a greater reduction of microbial populations.

3. **The distance between the light source and the cell.** The intensity of light decreases by the reciprocal of the square of the distance; for example, if the distance between the light and the cell is doubled, then the intensity is reduced four-fold.

4. **The type of organism and its life stage when exposed to the UV light.** Bacterial spores are more resistant to UV light than vegetative cells of the same species, so UV light will be less effective against *Bacillus* and *Clostridium* populations if there are large numbers of spores present.

5. **The type of cell wall.** As a broad generalization, Gram-negative species are more sensitive to, and more easily killed by, UV light than Gram-positive species. However, these are two very large and diverse groups of microbes, and there are plenty of exceptions to this observation.

6. **The presence of UV-blocking or UV-absorbing pigments.** Many bacterial species are routinely exposed to the UV rays of sunlight, including airborne bacteria and those living on the surfaces of plants or human skin. Some of these species have adapted to such high-UV environments by producing UV-blocking or UV-absorbing pigments that act as microbial "sunscreens." These pigments reflect or absorb the energy of the UV light, and so reduce damage to DNA.

7. **The presence of mutation repair systems.** Bacterial cells typically possess one or more enzyme systems that can locate and repair damaged sections of DNA. Some of these systems are **photoactivated repair systems**; that is, they require light to be activated. These systems cleave the covalent bonds that create thymine dimers, restoring the DNA to its original state without physically replacing the thymines. There are also **dark repair systems**, which do not require light for activation; dark repair systems cut out the thymine dimers and replace them with new, separated thymines.

USES AND LIMITATIONS OF ULTRAVIOLET LIGHT

UV light is often used to disinfect or reduce microbial populations on lab benches and in lab hoods, and it can also be used to disinfect drinking water. But its usefulness is limited because UV light lacks much of the surface-penetrating power of high-energy gamma radiation. Even a thin layer of glass or plastic usually blocks UV light, so therefore, UV light is most effective at disinfecting surfaces that are in the direct path of the beam. In addition, UV radiation can cause thymine dimer formation and mutations in the DNA of exposed skin cells, so **it can increase the risk of skin cancer**. While UV light can reduce microbial populations on the skin, there are far safer ways to disinfect the surface of the body prior to, say, medical procedures in which the skin is going to be broken by injections or surgery (see Lab 15).

In this lab you will examine the capacity of UV light to control microbes by exposing (a) a pigmented Gram-positive species (*Micrococcus luteus*), (b) a Gram-positive spore former (*Bacillus megaterium*), and (c) a Gram-negative species (*E. coli*) to UV light for varying lengths of time. You will also observe the limits of UV light by testing its ability to pass through plastic that is transparent to visible light. Finally, since the presence of spores has a significant effect on *B. megaterium*'s ability to survive exposure to UV light, you will do a spore stain of the *B. megaterium* culture to check for spore production.

PROTOCOL

Day 1

Materials

Per pair of students:

- 2 nutrient agar plates
- Disinfectant in beakers
- Sterile swabs
- UV lamps in light boxes
- Clean glass microscope slides
- Empty plastic petri dishes
- Incubator set to 50°C–60°C
- Disposable gloves
- UV-blocking goggles or face shields
- Compound microscope with 100× oil immersion lens
- Staining tray with staining platform
- Malachite green
- Safranin

Cultures:

- *Bacillus megaterium*
- *Escherichia coli*
- *Micrococcus luteus*

> **WARNING:** This lab uses UV light at a wavelength that can harm your eyes and skin. UV lamps should be operated in boxes that can be completely closed when the lamps are turned on, or students should use UV-blocking goggles or face shields when operating UV lamps.

EXPOSING BACTERIA TO ULTRAVIOLET LIGHT

1. Each pair of students will be assigned one species of bacteria to test. Once given your assignment, immediately make a note of which species you are to test so that you don't forget.

2. Label a nutrient agar plate with your name, the date, and the name of the species you were assigned, and mark this plate "30 seconds."

3. Insert a sterile swab into a broth culture of the assigned species and gently swirl the broth to completely resuspend the bacterial cells.

4. Use the swab to "paint" the surface of the nutrient agar plate. The goal is to touch every point on the plate and to generate a continuous sheet, or **lawn**, of bacterial cells across the surface of the plate.

5. Dispose of the contaminated swab in a beaker of disinfectant.

6. Using the assigned species, repeat steps 2–5 with a second nutrient agar plate. Label this second plate with your name, the date, and the name of the species you were assigned, and mark this plate "180 seconds."

7. Take both plates to a UV lamp set to emit light at 254 to 265 nm, and place the plates side by side under the lamp. The plate marked "30 seconds" should be to the left of the center line under the lamp, and the plate marked "180 seconds" should be to the right of the line **(Figure 16.3A)**.

8. Move the lid of each petri dish so that about half of the surface of the agar is exposed and about half of the surface is still covered or shaded by the lid **(Figure 16.3B)**.

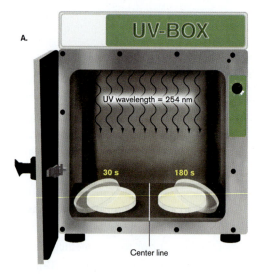

FIGURE 16.3 Experimental setup.
Two plates inoculated with a given species should be placed side-by-side in the UV light box *(A)* with the lids halfway off of the plates *(B)*. UV lamps should not be turned on until the door of the box is closed.

9. When the lids have been shifted so that the agar surfaces are partly exposed, close the box, turn on the lamp for 30 seconds.

10. After 30 seconds, turn off the lamp, replace the lid on the plate marked "30 seconds," and remove that plate.

11. Turn the lamp on for another 150 seconds (2½ minutes). At the end of this exposure, the second plate will have been exposed to UV light for a total of 180 seconds.

12. After the additional 150 seconds (180 seconds total), turn off the lamp, replace the lid on the second plate, and remove that plate.

13. Incubate the two plates at 37°C for 24 hours.

SPORE-STAINING *BACILLUS MEGATERIUM*

1. Use an inoculating loop to transfer one or two loopfuls of water to the center of an unused slide, then use a flame-sterilized inoculating loop to transfer one or two loopfuls of *Bacillus megaterium* cells to the drop of water on the slide.

2. Use the loop to mix the cells into the water and to spread the cell suspension across the slide. Set the slide aside to air-dry.

Do not proceed until all the water has completely evaporated, leaving a dry, thin, whitish film of cells on the slide.

3. Complete steps 5 through 13 of Lab 8 ("The Spore Stain").

Day 2

1. Examine each plate and grade the amount of growth on each plate in both the shaded (covered by lid) and unshaded (uncovered) area. Score the amount of growth in each area as 0, 1, 2, 3, or 4, where 0 means "no growth" and 4 means a solid lawn of bacteria.

2. Record your results for your particular species in the Results section. Post your results with those of the entire class and record the pooled class results.

3. There will be several scores for each combination of species, exposure time (30 or 180 seconds) and shading. Use these scores to calculate mean scores for each species, exposure time, and shading combination.

4. Dispose of the plates in the receptacle designated for trash to be autoclaved.

Name: _____   Section: _____

Course: _____   Date: _____

Assigned species: _____

INDIVIDUAL RESULTS

Draw the growth seen on the plates.

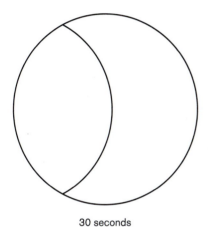

30 seconds

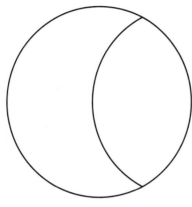

180 seconds

Scores for growth of your assigned species (scale of 0 to 4)

30 seconds:

  Shaded:

  Unshaded:

180 seconds:

  Shaded:

  Unshaded:

POOLED RESULTS

Species	30 SECONDS		180 SECONDS	
	SHADED	UNSHADED	SHADED	UNSHADED
E. coli				
	Mean:	Mean:	Mean:	Mean:
B. megaterium				
	Mean:	Mean:	Mean:	Mean:
M. luteus				
	Mean:	Mean:	Mean:	Mean:

Name: _____   Section: _____

Course: _____   Date: _____

1. How did covering or shading part of the agar surface with a petri dish lid alter the amount of bacterial growth on the plates? (Cite evidence from all three species to support your answer.) What do these results tell you about the ability of 254-nm UV light to penetrate clear plastics that are transparent to visible light?

2. How did the amount of growth on the unshaded areas of the plates differ between the "30 seconds" and "180 seconds" plates? Was either time period sufficient to sterilize the unshaded surface of the agar? Cite evidence to support your answer.

3. How might differences among the stock cultures in initial cell density affect the outcome of this experiment?

4. Which species (*E. coli*, *M. luteus*, or *B. megaterium*) produced the least growth in the unshaded areas after 180 seconds of UV exposure? Give two possible reasons or explanations for why that species (instead of the others) produced the least growth.

5. Which species (*E. coli*, *M. luteus*, or *B. megaterium*) produced the most growth in the unshaded areas after 180 seconds of UV exposure? Give two possible reasons or explanations for why that species (instead of the others) produced the most growth.

6. Was the answer to Question 5 (which species produced the most growth) what you expected? Why or why not? If your results did not match your expectations, explain what might have happened.

7. The figure below shows the absorption by double-stranded DNA of a range of UV light wavelengths. Now assume that the only UV lamp that you have available for the control of microbes emits light with a wavelength of 275 nm.

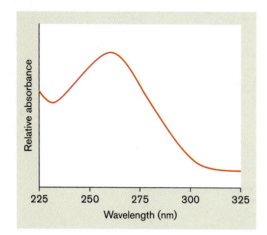

a. How would the rate of mutation generated by exposure to 275 nm light compare with that produced by exposure to 255 nm light (the approximate wavelength used in this lab)?

b. How would the rate at which the bacterial cells die be changed by the use of 275 nm light instead of 255 nm light?

c. If your goal was to kill a given number of cells, how would you alter or adjust the time of exposure at 275 nm compared with 255 nm?

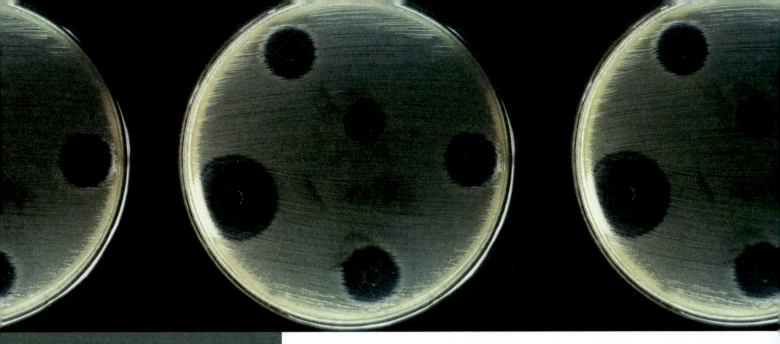

LAB **17**

Antibiotic Sensitivity Test

Learning Objectives

- Define and describe key concepts related to antibiotics, including selective toxicity, antibiotic spectra, and bacterial sensitivity and resistance.

- Understand how the Kirby-Bauer assay is used to determine the spectrum of a given antibiotic and the sensitivity or resistance of a given bacterial strain to a variety of drugs.

- Use the Kirby-Bauer assay to determine the spectrum of activity of several antibiotics and the drug sensitivity of several bacterial strains.

DESCRIPTION OF ANTIBIOTICS

Antibiotics are chemical control agents, often of microbial origin, that have the very important property of **selective toxicity**. That is, these chemicals are usually much more toxic to microbes than they are to us or to our cells. While certain antibiotics can inhibit or kill viruses, fungi, and protozoa, the vast majority of antibiotics are specifically antibacterial drugs. Therefore, some use the term "antibiotic" to refer exclusively to those selectively toxic chemicals that inhibit or kill bacteria. In this lab, we will be discussing only the antibacterial antibiotics.

Many antibiotics are selectively toxic because they target cell structures, enzymes, or pathways that are unique to bacteria, such as cell wall and folic acid synthesis pathways. Other antibiotics interfere with the activity of cell components that are sufficiently different in bacteria compared to human cells that our cells can tolerate the toxic effects of the drug. For example, many antibiotics stop protein synthesis in bacterial cells by binding to the 70S prokaryotic ribosome. These same drugs have a lower affinity for the larger 80S ribosomes found in the cytoplasm of eukaryotic cells, and hence have less of an effect on us.

235

As a result of selective toxicity, antibiotics can be used **chemotherapeutically**. That is, they can be used after an infection has been established, and unlike antiseptics and disinfectants, they can be administered internally by injection or oral ingestion. This allows us to use antibiotics to stop life-threatening internal infections, and their discovery and development in the twentieth century has saved millions of lives. However, the safe and effective use of antibiotics requires both (1) knowledge of the spectrum of each drug and (2) the infecting bacterium's response to a particular antibiotic.

ANTIBIOTIC SPECTRA

The **spectrum** of an antibiotic describes the range of bacterial species that can be inhibited or killed by that drug. **Narrow-spectrum** antibiotics inhibit or kill fewer species than broad-spectrum antibiotics. Often this shorter list of species is correlated with cell wall type. For example, the original penicillin, penicillin G, is a narrow-spectrum antibiotic because it is much more effective against Gram-positive species than against most Gram-negative species. By contrast, **broad-spectrum** antibiotics act against many more species of bacteria, inhibiting or killing Gram-positive and Gram-negative bacteria alike.

SENSITIVITY AND RESISTANCE

The terms "sensitivity" (or "susceptibility") and "resistance" refer to the response of a given bacterial strain to a given antibiotic; in other words, sensitivity and resistance are characteristics of the microbe, not the drug. A **sensitive** bacterial strain can be inhibited or killed by a concentration of a given antibiotic that is low enough that patients can usually tolerate the drug. That is, when a strain is sensitive to a given antibiotic, that drug should be effective in treating an infection caused by that strain. A **resistant** bacterial strain is not effectively inhibited or killed by a concentration of a given drug that can usually be tolerated by a patient. Note that resistance is not an absolute quality; in some cases, a higher dose of the drug would inhibit or kill the bacteria causing an infection. However, if the microbes are resistant to an antibiotic, then the concentrations required to inhibit or kill them may be too high to use safely on a patient.

Kirby-Bauer Assay

How can we determine the spectrum of an antibiotic or the response of a particular bacterial strain to a specific antibiotic? This information is obviously of great importance in guiding antibiotic therapy because we want to know which drugs work against which infectious bacterial strains. Further, since bacterial strains can acquire resistance to antibiotics over time, we can't just do this once for a given antibiotic and strain; instead, we have to continue to test antibiotic sensitivity for as long as we use these chemicals.

One of the most widely used methods for determining spectrum and response is the **Kirby-Bauer assay**. In this test, a standard number of cells of a given bacterial strain are spread across the surface of a Mueller-Hinton agar plate; Mueller-Hinton agar is a rich medium that supports the growth of most clinically important bacteria. In this lab, we will use a **McFarland turbidity standard** to regulate the number of cells applied to the

A.

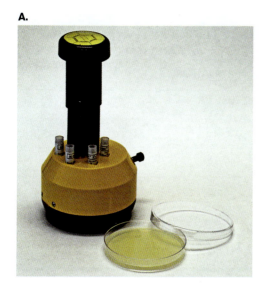

B.

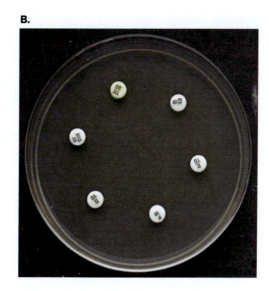

FIGURE 17.1 Performing the Kirby-Bauer assay.
A. Disk dispenser loaded with antibiotic cartridges for Kirby-Bauer analysis. **B.** Disks dispensed onto an assay plate.

plate. The turbidity, or cloudiness, of a broth culture is directly related to the density of the cells in that culture. We can adjust the density of cells in a broth by adding a sterile liquid (such as saline) until broth culture turbidity matches a standard turbidity corresponding to a particular density of cells. For example, if the turbidity of the broth matches a 0.5 McFarland turbidity standard, then there are about $1.0–1.5 \times 10^8$ cells per milliliter in the broth.

After the bacteria are spread on the plate, paper disks impregnated with standard amounts of a given antibiotic are applied to the plate with a disk dispenser **(Figure 17.1)**. The antibiotic diffuses out of the disks and into the surrounding agar as an expanding circle of the drug. Plates are incubated for 24 hours to give the bacteria an opportunity to grow and produce a visible lawn, or opaque layer of cells.

If microbial cells are sensitive to the concentration of a particular antibiotic on a disk, there will be a circular area of clearing, or **zone of inhibition**, around the disk after incubation **(Figure 17.2A)**. Such clear zones show where the concentration of a given drug was too high for the bacteria to tolerate, and where the antibiotic thus prevented any growth or cell division, or even killed the cells. But if the cells are resistant to an antibiotic, then the presence of the drug in the medium will be of much less consequence to the cells, and they will grow and divide at points much closer to the disk; they may even produce an opaque layer that extends right up to the edge of the disk.

Diffusion of the drug into the medium dilutes the drug, so even if the cells on the agar are sensitive to the antibiotic, there will be a point where the drug concentration is too low to inhibit growth, and the sensitive cells will be able to produce a visible lawn beyond the zone of inhibition. We would not want to confuse the growth of cells in the lawn far from the disk with resistance to the chemical agent. So, how big does a zone of inhibition have to be before we conclude that the drug is effective or that the strain is sensitive to the antibiotic? This is something that has been determined by extensive testing, and the results of this testing are available in the form of antibiotic disk sensitivity tables, which give standard zone of inhibition values for standard antibiotic concentrations **(Table 17.1)**.

To use these tables, we measure the diameter of the zone of inhibition around a disk impregnated with a particular antibiotic **(Figure 17.2B)** in millimeters, then

A.

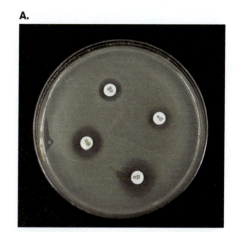

B.

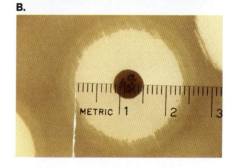

FIGURE 17.2 Results of Kirby-Bauer assay.
A. Kirby-Bauer assay plate with four different antibiotics and corresponding zones of inhibition. **B.** Measuring the diameter of the zone of inhibition in millimeters (zone is about 24 mm in diameter).

Table 17.1 | ANTIBIOTIC DISK SENSITIVITY TABLE

Antibiotic	Amount of Antibiotic on a Disk	Code	Resistant (R) (equal or **less** than)	Intermediate (I) (between)	Sensitive (S) (equal or **more** than)	Mode of Action
Bacitracin	10 units	B	8 mm	9–12 mm	13 mm	Inhibits cell wall synthesis
Erythromycin	15 µg*	E	13 mm	12–17 mm	18 mm	Inhibits protein synthesis
Penicillin G	10 units	P	20 mm	21–24 mm	25 mm	Inhibits cell wall synthesis
Streptomycin	10 µg	S	11 mm	12–14 mm	15 mm	Inhibits protein synthesis
Tetracycline	30 µg	TE	14 mm	15–18 mm	19 mm	Inhibits protein synthesis
Triple sulfa	300 µg	SSS	12 mm	12–16 mm	17 mm	Inhibits folic acid synthesis

*µg = micrograms.

compare that result to standard values for that drug. If the measured zone is **less** than the value given in the table in the "resistant" column, then we can conclude that the strain of bacteria on the plate is **resistant** to that particular antibiotic. If the measured diameter of the zone of inhibition is **greater** than the value given in the table in the "sensitive" column, then we can conclude that the bacterial strain is **sensitive** to the drug, and that this may be the antibiotic to use to treat infections caused by this bacterial strain. "Sensitive" and "resistant" are relative terms, or two ends of a continuous spectrum, so the diameter of the zone of inhibition could also be intermediate between the resistant and sensitive values. In these cases, we can conclude that the strain is "intermediate" in its response to the drug. It should also be noted that there might be a few scattered colonies within the zone of inhibition. These colonies could originate either from cells that started to grow and divide before the antibiotic diffused into this part of the medium, or from a tiny portion of cells in the original population that happened to have and express resistance genes against the antibiotic.

The Kirby-Bauer assay allows us to draw qualitative conclusions about the sensitivity or resistance of a given bacterial species to a given antibiotic. However, it may not provide a good quantitative answer to the question of what dose or concentration will be needed to control a given species during an infection. A different assay, the minimum inhibitory concentration (MIC) assay, is designed to supply information about the dose required to effectively treat an infection; that assay is described in Lab 18.

PROTOCOL

Day 1

Materials

- Mueller-Hinton agar plate
- 1 0.5 McFarland turbidity standard
- 1 empty sterile test tube
- Sterile saline (10-ml aliquots)
- Card with sharp black lines
- 9-ml sterile pipette
- 1-ml sterile pipette
- 2-ml (blue-barrel) Pipette Pump
- 10-ml (green-barrel) Pipette Pump
- Sterile swab
- Antibiotic-impregnated disks (see Table 17.1)
- Antibiotic disk dispenser or forceps
- Disinfectant in beakers

Cultures:

- *Escherichia coli*
- *Pseudomonas aeruginosa*
- *Staphylococcus epidermidis*

1. Each student will be assigned one species of bacteria to test. Once given your assignment, immediately make a note of which species you are to test so that you don't forget.

2. Label the Mueller-Hinton agar plate with your name, the date, and the name of the species you were assigned.

3. Transfer about 2 ml of your assigned broth culture to a sterile test tube of the same diameter as the 0.5 McFarland turbidity standard.

   In a spectrophotometer, the 0.5 McFarland standard has an absorbance of 0.08 to 0.10 at 600 nm.

4. Gently swirl the broth culture and the 0.5 McFarland standard until the cells and particles in both tubes are evenly suspended and turbidity is uniform throughout each tube.

5. Hold the tube with the transferred culture and the tube with the McFarland standard upright in front of a card with a sharp black line drawn on it. Be sure to hold the card so that you are viewing the black line through the liquid in the culture and the standard **(Figure 17.3)**.

FIGURE 17.3 Using the McFarland Standards.
From left to right: (1) distilled water control; (2) 0.5 McFarland standard; (3) *S. epidermidis* culture adjusted by dilution to match the turbidity of the standard; (4) *S. epidermidis* culture before dilution.

6. As a result of the turbidity or cloudiness of the liquid in the tubes, the black line will look a little fuzzy or blurry when viewed through the liquid. You are looking for a match in blurriness or turbidity between the culture and the McFarland standard; that is, you are looking to see if the black line is equally blurry when seen through the two tubes.

7. If there is difference between the two tubes in the blurriness of the black line, then add about 0.5 ml of sterile saline to the broth culture in the sterile test tube and use the card with the black line to compare again.

8. If a difference remains, continue to add 0.5-ml aliquots of sterile saline to the broth culture checking the turbidity in the two tubes after each addition. When the black line looks about the same when viewed through the two tubes, stop adding saline.

9. Insert a sterile swab into the diluted broth culture of the assigned species and gently swirl the broth to completely resuspend the bacterial cells. Before removing the swab, touch it to the inner surface of the test tube to remove excess broth.

10. Think of your swab as a paintbrush, and use the swab to "paint" the surface of a Mueller-Hinton agar plate. Your objective is to completely cover the agar surface with "paint," that is, the swab should touch every point on the agar surface. This will generate a continuous sheet or lawn of bacterial cells across the surface of the plate.

11. Dispose of the contaminated swab in a beaker of disinfectant.

    P. aeruginosa **can cause skin infections,** so exercise extra caution when using this species. Do not let the contaminated swab touch anything other than the surface of the agar, and immediately after use, put the swab into the beaker of disinfectant.

12. After swabbing the plate, add antibiotic disks to the bacterial lawn. This may be done using a commercial disk dispenser capable of placing several disks onto the lawn at once. Alternatively, forceps sterilized by dipping them in alcohol followed by a brief flaming can be used to transfer individual disks from sterile petri dishes to the bacterial lawn. In this case, it is very important to sterilize the forceps between each transfer, because the forceps will become contaminated when the disks are placed on the lawn.

13. Incubate plates at 37°C for 24 hours.

Day 2

1. Use a metric ruler to measure the diameter of the zone of inhibition in millimeters (see Figure 17.2B). **Do not open the plate to do this measurement.**

    In some cases, the zones of inhibition may be so large that they overlap with each other or extend beyond the edge of the plate. In these cases, there will not be a full circle of growth around the disk. However, as long as there is a short arc of growth at some point around the disk, then you can measure the distance from the center of the disk to the edge of the arc, that is, you can measure the radius of the circle. Then multiply by two to determine the diameter of the inhibition zone.

    If there are a few scattered colonies within the zone, ignore them; they are either from cells that were able to grow and divide before the antibiotic diffused into the medium, or they are from the tiny proportion of cells in the original population that happened to have resistance genes.

2. Use the measured diameter and the antibiotic disk sensitivity table (Table 17.1) to determine whether your species is sensitive or resistant to each type of antibiotic.

3. Record your results for your particular species in the Results section. Post your results with those of the entire class and record the pooled class results.

4. There will be several results for each combination of species and antibiotic. Use these results to draw conclusions about the resistance or sensitivity of a given species to a given antibiotic.

5. Dispose of plates in the receptacle designated for trash to be autoclaved.

Name: _____ Section: _____

Course: _____ Date: _____

INDIVIDUAL RESULTS

Assigned species: _____

MEASURED DIAMETERS

Antibiotic	Diameter of zone of inhibition (in millimeters)	Conclusions (sensitive, resistant, or intermediate)
Bacitracin (B)		
Erythromycin (E)		
Penicillin G (P)		
Streptomycin (S)		
Tetracycline (T)		
Triple sulfa (SSS)		

POOLED RESULTS

E. COLI

Antibiotic	Class data (R, S, or I)	Conclusions (sensitive, resistant, or intermediate)
Bacitracin (B)		
Erythromycin (E)		
Penicillin G (P)		
Streptomycin (S)		
Tetracycline (T)		
Triple sulfa (SSS)		

P. AERUGINOSA

Antibiotic	Class data (R, S, or I)	Conclusions (sensitive, resistant, or intermediate)
Bacitracin (B)		
Erythromycin (E)		
Penicillin G (P)		
Streptomycin (S)		
Tetracycline (T)		
Triple sulfa (SSS)		

S. EPIDERMIDIS

Antibiotic	Class data (R, S, or I)	Conclusions (sensitive, resistant, or intermediate)
Bacitracin (B)		
Erythromycin (E)		
Penicillin G (P)		
Streptomycin (S)		
Tetracycline (T)		
Triple sulfa (SSS)		

Name: _____ Section: _____

Course: _____ Date: _____

1. Which species was sensitive to the greatest number of different antibiotics? Was this species an example of a Gram-positive or a Gram-negative species?

2. Assuming that all the diseases caused by the species used in this lab are equally severe or equally dangerous (not true, but let's pretend), which species would you least like to be infected with? Explain your answer.

3. Would you describe penicillin as a narrow-spectrum or broad-spectrum antibiotic? Cite evidence to support your answer.

4. Would you describe tetracycline as a narrow-spectrum or broad-spectrum antibiotic? Cite evidence to support your answer.

5. How might your results have been different if you had applied the bacterial cells to the plate at a cell density that was 100 times greater than the density used in this lab? Consider the difference in time required for dividing cells to produce a visible lawn on two different plates on which the initial cell densities differ by a hundred-fold, and remember that it takes time for the antibiotic to diffuse into the medium.

Minimum Inhibitory Concentration Assay

Apples are one of the most popular fruits. You might be surprised, then, to learn that their seeds contain a poison. Amygdalin is metabolized by our bodies producing cyanide, an extremely toxic substance. Apples are safe to eat though, because the amount of amygdalin in apples is 0.6 grams per kilogram of seeds, a tiny amount that is far too small to cause us harm. This is an example of the adage, "The dose makes the poison." For a poison, like an antibiotic or antimicrobial agent, to be effective, it needs to be present at a concentration—a dosage—high enough to kill the intended target.

Learning Objectives

- Understand how dose or concentration affects the impact of a chemical on a bacterial cell or human body.

- Define minimum inhibitory concentration (MIC) and describe how it can be determined.

- Consider how absorption, distribution, metabolism, and excretion (ADME) influence the dose required to treat an infection with an antibiotic.

- Calculate the MIC for ampicillin when the antibiotic is used against *E. coli*.

THE DOSE MAKES THE POISON

In the previous lab, we determined the sensitivity of a variety of bacterial species to different antibiotics. This can be an important step in guiding drug therapy because, obviously, we want to treat infections with drugs that have been shown to inhibit or kill the organism that is causing a patient's disease. However, it's not enough to know that a given chemical can harm a given bacterium; we also need to know the concentration of the drug that is required to harm the bacterium. That is, we need to know the **dose**, or the amount of antibiotic that is to be consumed, applied, or injected into the body.

Dose matters. Dose always matters. The response of any organism to any chemical depends on the dose or concentration of the chemical. This fact was nicely described about 500 years ago in an axiom attributed to the German-Swiss physician-botanist-alchemist Paracelsus (1493–1541):

All things are poison and nothing is without poison; only the dose makes a thing not a poison.

This is a rather cumbersome phrase, so it has been summed up this way: **The dose makes the poison**. This adage can be expressed in the form of a couple of important rules:

1. For most drugs, including antibiotics, there is a dose or concentration below which there are no observable toxic effects on cells; in other words, an antimicrobial drug must be present at a certain minimum concentration to cause harm to microbes.

2. In most cases, the negative effects of a drug or toxin increase with increasing dose or concentration.

There are exceptions to these generalizations, but we won't worry about them in this lab because these rules usually hold for the use of antibiotics.

MINIMUM INHIBITORY CONCENTRATION

So, let's use these rules. Applying the first rule to antibiotic therapy, we'd like to be able to determine the minimum dose or concentration needed to inhibit or kill a particular bacterial pathogen that is causing a particular infection. We know that there will be doses or concentrations that are too low to cause harm to the bacteria. What we want to know is the **minimum inhibitory concentration (MIC)**, or the lowest dose that will inhibit the growth of the bacteria. In other words, where does the line fall between "no toxic effects" and "toxic effects" for the pathogen?

Why are we interested in the minimum dose? Why not simply pour in as much antibiotic as a body can hold? Remember the second rule. Toxic effects usually increase with dose, and while antibiotics may be more toxic to bacteria than to us, this is not the same as saying that they are not toxic to us. So we'd like to find and use the minimum dose that will effectively control the bacterial pathogen because lower doses will reduce the probability of poisoning the patient. The minimum inhibitory concentration is also known as the "therapeutic dose" or the "effective dose"; ideally, it will be much lower than the dose that is toxic to us.

The difference between the therapeutic or effective dose of a drug and the dose that is clearly toxic to the human body is quantified by the calculation of a ratio known as the **therapeutic index**. The therapeutic index is defined as the drug dosage that causes harmful side effects in humans divided by the dose that kills or inhibits the microbe.

If you think about it, you'll realize that we want to see high values for the therapeutic index. That is, we want a drug for which a low dose is sufficient to harm the microbe, while significant harm to us occurs only at higher doses. Then we can easily find a minimum dose that is therapeutic (because it inhibits or kill microbes), but at the same time is relatively safe for humans taking the drug. Most widely used drugs based on the penicillin molecule have a high therapeutic index. In contrast, other antibiotics, such as gentamicin, have a low therapeutic index, and they are used only if there are no alternative drugs or in the case of life-threatening infections.

DETERMINING MIC

One method used in the determination of the therapeutic dose is the **minimum inhibitory concentration (MIC) assay**. In this assay, an antibiotic is serially diluted to create solutions covering a range of concentrations. Next, an equal number of cells of a given bacterial species are added to each dilution. If the antibiotic concentration in the tube is too low to stop the growth of or kill the cells, then the cells will multiply, and the broth will become cloudy in 24–48 hours. On the other hand, **if the drug can inhibit cell growth at the concentration found in a given tube, then the broth in that tube will remain clear**. Our goal is to identify the lowest antibiotic concentration that stops growth. Therefore, we will look for the last tube in the serial dilution in which the broth is clear. This is the tube with the lowest antibiotic concentration that inhibited growth, so the concentration in this tube is the minimum inhibitory concentration.

While we've used serial dilutions in other labs, they have generally been tenfold and hundredfold dilutions. In the MIC assay we'll use twofold dilutions, so the difference in concentrations between the tubes will be relatively small. This gives us a much more precise measure of the MIC than we would obtain if we used, say, tenfold dilutions. When administering drugs, we need this level of precision because a tenfold difference in concentration can make a huge difference inside a human body. For example, a blood alcohol concentration of 0.06% may help some feel more relaxed in an awkward social situation, but a tenfold increase to a blood alcohol concentration of 0.60% is likely to leave one dead and food for worms.

ABSORPTION, DISTRIBUTION, METABOLISM, AND EXCRETION (ADME)

The MIC assay is certainly useful in determining the dose that we would use to treat a bacterial infection with an antibiotic. If you don't start somewhere, you're gonna go nowhere. However, it's important to remember that the human body is not like a laboratory test tube. Grasping what happens when a drug is used to treat disease requires an understanding of four processes; absorption, distribution, metabolism, and excretion. Collectively, these activities are known by the acronym **ADME**.

1. **Absorption.** Some diseases, such as skin infections, are treated by topical application of antibiotics, and absorption of the drug is not a significant issue. However, in other cases, the drugs must reach the bloodstream to be effective. In these cases, one must consider how well the drug is going to be absorbed into the blood in organs such as the small intestine. If a drug is stable in the stomach and well absorbed in the gut, then it can usually be given orally. If the antibiotic must reach the blood, but it is destroyed in the stomach or poorly absorbed across the intestinal lining, then even a really big pill will have little value, and the drug will probably have to be injected or delivered through an IV line. For example, some forms of penicillin-type drugs, such as Penicillin G (benzylpenicillin), are unstable in stomach acid, and so they must be injected to be effective.

2. **Distribution.** After absorption, the drug must reach the infection site at concentrations sufficient to harm the bacteria. In other words, the treatment must generate the minimum inhibitory concentration at the location where the pathogens are growing and spreading. Absorbed drugs can reach some locations, such as the lungs, relatively easily. By contrast, it can be very difficult for the body to transport chemicals from the blood across the blood-brain barrier and into the central nervous system. So, drug concentrations may be high in the blood, but very low in the brain.

3. **Metabolism.** Drugs begin to break down as soon as they enter the blood and other body tissues. To the body, an antibiotic is just another potentially toxic chemical, and it has several mechanisms adapted to modifying or detoxifying

such chemicals. Cytochrome P450 enzymes, for example, may alter the structure of the antibiotic so that the drug becomes less effective, increasing the dose that is required to control a pathogen. (Conversely, in other cases, the action of enzymes may actually activate a drug. This was the case with the original sulfa drug, prontosil.) In addition, the modifications mediated by enzymes usually make the drug more water-soluble, which leads to an increase in its excretion rate.

4. **Excretion.** Eventually, the drug, or some enzyme-created derivative of the drug, will be eliminated from the body, most often via the digestive and urinary tracts. If a drug is water-soluble, it will probably be eliminated relatively quickly, and the concentration of the drug will drop rapidly. Fat- or lipid-soluble drugs are usually retained in the body for much longer periods. Longer retention may sound like a positive thing, but it should be remembered that this means that it's easier for the drug to accumulate to levels that are toxic to the patient.

These four processes all affect the outcome of antibiotic use. So ultimately, while the MIC assay is valuable, eventually you're going to have to put the drug in a human body to determine the actual therapeutic dose. **Figure 18.1** shows a graph of ampicillin concentrations in the blood serum over time. The graph also includes the MIC for ampicillin so that it is possible to see how long the antibiotic remains useful in the body.

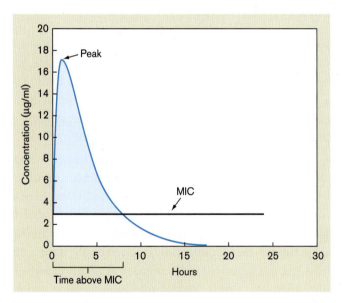

FIGURE 18.1 Antibiotic serum concentrations.
The graph illustrates changes in the concentration of ampicillin in serum over time.

PROTOCOL

Day 1

Materials

- Ampicillin solution, filter-sterilized and at a concentration of 128 µg/ml

- 10 Mueller-Hinton broths, 1.0 ml each

- 1 empty test tube

- 1-ml sterile pipettes

- 2-ml (blue-barrel) Pipette Pump

Cultures:

- Overnight broth culture of *Escherichia coli* diluted in sterile 0.5% saline to a turbidity of McFarland standard 0.5

If McFarland standards are not available, an overnight broth culture of *E. coli* should be diluted 1:1,000 (10^{-3}) in saline before student use.

PROCEDURES

For steps 1 through 13, see **Figure 18.2**.

1. Label the ten tubes containing 1 ml of Mueller-Hinton broth per tube with your names and the date.

2. Label eight of the ten tubes as follows:

   Tube 1: 64 µg/ml
   Tube 2: 32 µg/ml

Tube 3: 16 µg/ml
Tube 4: 8 µg/ml
Tube 5: 4 µg/m
Tube 6: 2 µg/ml
Tube 7: 1 µg/ml
Tube 8: 0.5 µg/ml

3. Label the ninth tube "Positive control," and label the tenth tube "Negative control."

4. Use a sterile 1-ml pipette to transfer 1.0 ml of 128 µg/ml ampicillin solution to the tube labeled 64 µg/ml. Gently draw the broth in the tube up and down to the 1.0-ml mark a couple of times to mix the ampicillin into the broth. Do not discard this pipette, but hold it in such a manner as to prevent the bottom half of the pipette from being contaminated.

If you know that you are allergic to penicillin-type antibiotics, a drug class that includes ampicillin, do not handle the ampicillin solution or tubes containing this solution. The ampicillin solution should be handled by your lab partner only.

5. Using the same sterile 1-ml pipette as in step 4, transfer 1.0 ml of 128 µg/ml ampicillin solution to the tube labeled "Negative control." Draw the liquid in the tube up and down in the pipette to mix the ampicillin into the broth. Set this tube aside until the end of the day 1 inoculation steps, and discard the first pipette.

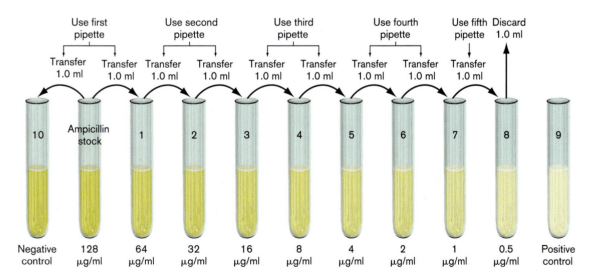

FIGURE 18.2 MIC serial dilutions.

6. Using a second pipette, transfer 1.0 ml from the 64 µg/ml tube to the 32 µg/ml tube. With the same pipette, gently draw the broth up and down to the 1.0-ml mark a couple of times to mix the ampicillin into the broth. Do not discard the pipette.

7. Transfer 1.0 ml from the 32 µg/ml tube to the 16 µg/ml tube with the second pipette. Gently draw the broth up and down to the 1.0-ml mark a couple of times to mix the ampicillin into the broth. Now discard the second pipette.

8. Using a third pipette, transfer 1.0 ml from the 16 µg/ml tube to the 8 µg/ml tube. With the same pipette, gently draw the broth up and down to the 1.0-ml mark a couple of times to mix the ampicillin into the broth. Do not discard the pipette.

9. Transfer 1.0 ml from the 8 µg/ml tube to the 4 µg/ml tube with the third pipette. Gently draw the broth up and down to the 1.0-ml mark a couple of times to mix the ampicillin into the broth. Now discard the third pipette.

10. Using a fourth pipette, transfer 1.0 ml from the 4 µg/ml tube to the 2 µg/ml tube. With the same pipette, gently draw the broth up and down to the 1.0-ml mark a couple of times to mix the ampicillin into the broth. Do not discard the pipette.

11. Transfer 1.0 ml from the 2 µg/ml tube to the 1 µg/ml tube with the fourth pipette. Gently draw the broth up and down to the 1.0-ml mark a couple of times to mix the ampicillin into the broth. Now discard the fourth pipette.

12. Using a fifth pipette, transfer 1.0 ml from the 1 µg/ml tube to the 0.5 µg/ml tube. With the same pipette, gently draw the broth up and down to the 1.0-ml mark a couple of times to mix the ampicillin into the broth. Do not discard the pipette.

13. With the fifth pipette, remove 1.0 ml from the 0.5 µg/ml tube, and discard this 1.0 ml in an empty test tube.

This 1.0 ml is removed from the 0.5 mg/ml tube so that all tubes have the same final volume before the *E. coli* are added.

For steps 14 through 16, see **Figure 18.3**.

14. Rotate your test tube rack so that the positive control (tube 9) is on the far left.

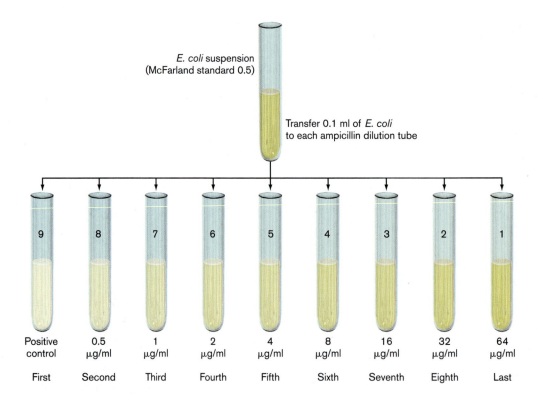

E. coli suspension (McFarland standard 0.5)

Transfer 0.1 ml of *E. coli* to each ampicillin dilution tube

9	8	7	6	5	4	3	2	1
Positive control	0.5 µg/ml	1 µg/ml	2 µg/ml	4 µg/ml	8 µg/ml	16 µg/ml	32 µg/ml	64 µg/ml
First	Second	Third	Fourth	Fifth	Sixth	Seventh	Eighth	Last

FIGURE 18.3. Addition of *E. coli* to MIC tubes.
Add the *E. coli* to the tubes in the order shown, starting with the positive control tube, then the tube with 0.5 µg/ml ampicillin, and so forth.

15. Gently swirl the *E. coli* stock suspension to evenly distribute the cells. Using a graduated 1.0-ml pipette, draw up 0.9 ml of *E. coli* cells in 0.5% saline.

16. Starting with tube 9 (positive control), add 0.1 ml of the *E. coli* suspension to each of the nine tubes in this order:

    Positive control tube, 0.5 µg/ml tube, 1 µg/ml tube, 2 µg/ml tube, 4 µg/ml tube, 8 µg/ml tube, 16 µg/ml tube, 32 µg/ml tube, 64 µg/ml tube. Do not add any *E. coli* to tube 10.

    When adding *E. coli* to each of the dilutions, do not allow the pipette tip to touch any unsterile surfaces, such as the outside of the tube. In addition, avoid touching the ampicillin solution with the pipette tip. **Do not add any *E. coli* cells to the Negative control tube.**

    If micropipetters are available, the transfer of *E. coli* cells can be done a little more accurately using a micropipetter set to 100 microliters. In this case, sterile tips should be changed between transfers.

17. When you've finished adding *E. coli* to the nine tubes, gently flick each tube with your index finger to mix the cells into the broth.

18. Incubate all ten of the broths at 37°C for 48 hours.

Day 2

Examine all eight experimental tubes and the two control tubes for evidence of growth. Record your observations in the Results section.

If a given tube is as clear as the uninoculated tube (Negative control), this indicates that the ampicillin concentration was high enough to inhibit or kill the *E. coli* cells. If the tube is cloudy or turbid, this indicates that the ampicillin concentration was too low to stop *E. coli* cells from growing and dividing.

Name: _____

Course: _____

Section: _____

Date: _____

Ampicillin concentration	Observation (cloudy or clear)
64 µg/ml	
32 µg/ml	
16 µg/ml	
8.0 µg/ml	
4.0 µg/ml	
2.0 µg/ml	
1.0 µg/ml	
0.5 µg/ml	
Positive control (No ampicillin)	
Negative control (No *E. coli*)	

1. What is the purpose of the positive and negative controls? That is, what can you conclude from what you saw in the positive and negative control tubes, or what did these tubes show you?

2. Before the start of lab, the *E. coli* suspension used in the lab was significantly diluted in 0.5% saline. Why was this done? That is, why does the initial *E. coli* cell density matter? Think about how you determined which tube contained the minimum inhibitory concentration of ampicillin.

3. In this experiment, what was the minimum inhibitory concentration for ampicillin against *E. coli*?

4. Why is this assay referred to as a minimum inhibitory concentration assay instead of a minimum lethal concentration assay? Hint: When you look at an inoculated tube that is clear, not cloudy, what can't be concluded from the tube's appearance alone?

5. Let's say that you did want to perform a minimum lethal concentration assay. How could this be done? Briefly describe a possible experimental design for a minimum lethal concentration assay.

6. Say that you plan to inject ampicillin to treat a blood infection caused by *E. coli*. An average body holds about 5 liters of blood, and to make this exercise less complicated, let's assume that the injected ampicillin mixes into the blood so that the concentration of the ampicillin in the blood will be the same in all parts of the body. Discount the possibility of metabolism and excretion of ampicillin.

Given these oversimplifications, and given the results of this lab, how much ampicillin (in milligrams taken into the body) would you need to inject to achieve the minimum inhibitory concentration of ampicillin in the blood? That is, what ampicillin dose would you use? (In practice, metabolism and excretion matter, and you'd use much higher doses of ampicillin to treat a life-threatening blood infection.)

To convert the MIC (μg/ml) to dose (mg/body), you will have to convert micrograms to milligrams and milliliters to liters. Show your work.

Bacillus is a genus of motile, Gram-positive rod-shaped bacteria that form endospores. They are ubiquitous in soils and likely found nearly everywhere. Different Bacillus species form varied colony morphologies due to environmental factors like pH, the motility of the organisms, whether they swarm, and numerous other factors.

LAB 19

Introduction to Isolating and Identifying Bacteria: Isolation of *Bacillus* Species

Learning Objectives

- Understand the need to isolate and identify bacteria and consider the roles played by selective media, differential media, and biochemical fingerprints in meeting this need.

- Describe traits of the genus *Bacillus* and understand how we can use these traits to select for and isolate *Bacillus* species cells.

- Use exposure to high temperatures (80°C), plating on skim milk agar, and Gram staining to isolate and identify *Bacillus* species cells in soil samples.

- Compare colony counts from heated and unheated soil samples to examine the heat tolerance of various types of microbes and to estimate the relative abundance of *Bacillus* spores compared with other types of microbes in soil.

There are countless reasons why it is extremely useful to be able to isolate and identify bacteria. For example, isolation and identification are valuable in:

- Understanding and treating human infections
- Revealing contamination of food and water
- Understanding the role of bacteria in ecosystems
- Discovering commercially useful strains of bacteria

Now, the task of identifying bacteria would be much simpler if the world was filled with neatly separated pure cultures of bacteria, but reality is seldom so kind

FIGURE 19.1 Mannitol salt agar is both selective and differential.
The high salt concentration favors *Staphylococcus* species, while the addition of mannitol and pH indicators allows differentiation of species within the *Staphylococcus* genus based on the ability to ferment mannitol (if the medium turns yellow, this is a positive reaction for mannitol fermentation).

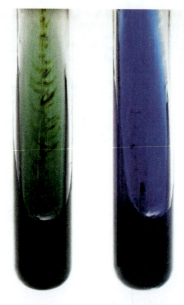

FIGURE 19.2 Citrate agar is a beautiful example of a differential medium.
When species can grow on an agar in which citrate is the sole carbon source available to the cells, the medium will turn blue in color. If a species cannot survive when citrate is the only carbon source, the medium will remain green.

or so accommodating. Almost all clinical and environmental samples are complex mixtures of many microbial species, so in almost all cases, the first step in working with them is to separate the cells and select for specific bacteria of interest. We can, of course, use streak plating and pour plating (see Lab 3), and we shall, but the task will be easier if we can streak and pour onto selective and differential media, which are carefully designed to assist in the isolation and identification of given types of bacteria.

While many of the media used in microbiology are either selective or differential, some media are both selective and differential; the two properties are not mutually exclusive.

SELECTIVE AND DIFFERENTIAL MEDIA

Selective media

Selective media and selective techniques are designed to select for and favor the growth of specific desired microbial groups or species of interest, or to prevent or reduce the growth of interfering or undesired species. A given medium or technique may be designed to favor or promote the growth of one kind of microorganism while simultaneously inhibiting the growth of certain unwanted wee beasties.

Some selective media accomplish the goal of selection by enrichment. These media are enriched with nutritional substrates that a particular microbial group of interest uses especially well. The idea is to give these selected groups a significant competitive advantage over other groups by creating an environment that fits the adaptations of the desired microbes. For example, a medium could be enriched with lactose to favor species able to metabolize lactose over species that cannot use it, or the lactic acid concentration of a medium might be boosted to select for groups, such as *Propionibacterium* species, that use lactic acid as a primary source of energy.

Other selective media work by inhibiting, or preventing the growth of microorganisms that would interfere with our goal of isolating other types of microbes. If our goal were to isolate fungal species, for example, we could use a very acidic medium, because fungi are usually more acid tolerant than bacteria. Or if we wished to select for *Staphylococcus* species, we could add a large amount of salt to the medium **(Figure 19.1)**. Most bacteria are inhibited by the elevated concentrations of salt that the skin-adapted *Staphylococcus* species are able to withstand.

Differential media

Differential media are used to aid in the identification of various microorganisms because these media differentiate, or allow separation of, different types of microbes based on their ability to perform particular biochemical reactions. In most cases, it is the presence of specific metabolic substrates, chemical dyes, and pH indicators that reveal a cell's ability to perform a particular reaction. Such reactions may result in characteristic differences in colony color, or in the color of the medium surrounding the colonies, which allow us to distinguish cells of different groups and species. Examples are shown in **Figures 19.1** and **19.2**.

GRAM STATUS AND CELLULAR MORPHOLOGY

Once pure cultures have been created by streak or pour plating onto selective or differential media, the next step in identification is usually Gram staining (introduced in Lab 6). For your convenience, the Gram status and cellular morphologies for various genera, many of which will be used in the "identification of unknown bacteria" exercises in this manual, are given bellow.

Some genera of Gram-positive bacteria:

Bacilli (rod-shaped)
Bacillus
Clostridium

Cocci (spherical)
Streptococcus
Enterococcus
Staphylococcus
Micrococcus

Some genera of Gram-negative bacteria:

Bacilli (rod-shaped)
Alcaligenes
Citrobacter
Enterobacter
Escherichia
Klebsiella
Proteus
Providencia
Pseudomonas
Salmonella
Shigella

Although there are some very important Gram-negative cocci genera, like *Neisseria gonorrhoeae*, we won't encounter any in this lab course.

BIOCHEMICAL FINGERPRINTS

Once we know the Gram status of our isolate, there are additional steps that we can take to identify it. Unfortunately, unlike plants and animals, for which we can use flower color, leaf shape, number of legs, or wing patterns as aids in identification, microbes have few distinctive morphological traits to go by. Beyond differences in Gram reaction, cell shape or morphology, and the arrangement of cells, most bacteria look pretty much alike. So how are we to tell the different species apart?

Fortunately, bacterial metabolism varies significantly from one species to another (as we saw in Lab 10), and so these differences can help us to tell them apart. Differences in metabolism are due to differences in which enzymes are present, because it's the enzymes that determine which substrates a given species can use, how a particular substrate is used, and which products a given species generates by metabolizing a particular substrate. And at the heart of the matter, it's the genes that determine which enzymes are produced. So, by looking at differences in enzyme activity, we can get a glimpse at the genetic differences that make one bacterial species different from another.

We can test for the presence or absence of specific enzymes by growing cells of a given species on many different media containing a variety of different substrates. Specific enzymes catalyze specific reactions that convert specific substrates into specific products. So, the presence of a specific enzyme is revealed by observing a decrease in the concentration of a specific substrate or an increase in the concentration of specific products. Therefore, assays used in the microbiology lab are designed to reveal either a decrease in the concentration of a metabolic substrate or an increase in the concentration of metabolic products.

Here's the payoff: Testing for a series of enzymes will produce a string or pattern of pluses and minuses. Pluses mean that an enzyme was present, a particular substrate was used, or a particular product was produced. Minuses mean that the enzyme was absent, so nothing happened—there was no observable change. This pattern of pluses and minuses is a kind of **biochemical fingerprint** for the unknown isolate, based on the presence or absence of enzymes, or the ability or inability to use certain substrates and produce certain products. Once determined, the biochemical fingerprint of an unknown species can be compared with the patterns of known species of bacteria. A very close match between an unknown bacterium's fingerprint and the pattern of a previously identified species would lead us to suspect that our unknown is the same as the known species.

OTHER METHODS

Besides biochemical fingerprints, there are many other ways to identify bacteria, including methods that look for specific DNA sequences and methods that detect specific, unique proteins or carbohydrates on the surface of the cells. Some of these methods are the basis of other labs in this manual, such as Labs 20 and 28.

ISOLATING AND IDENTIFYING THE GENUS *BACILLUS*

The genus *Bacillus* is a collection of Gram-positive, rod-shaped ("bacillus"), aerobic or facultative anaerobic bacterial species usually found in soil and water **(Figure 19.3)**. Although most *Bacillus* species do not cause human disease, there are two notable exceptions. *B. cereus* are usually relatively harmless, but they can thrive in certain foods and may cause food poisoning, when the cells secrete toxins that cause vomiting and diarrhea when the contaminated food is consumed. *B. anthracis* can cause life-threatening infections of the skin, digestive tract, and lungs called: **cutaneous anthrax, gastrointestinal anthrax**, and **pulmonary anthrax**, respectively. This species is a favorite among those who like to play with biowarfare weapons because of the endospore's ability to survive harsh conditions outside the body. However, it should be noted that spores must be "weaponized" before they can be used as true biowarfare agents. This involves turning spore preparations into very fine, easily air-borne powders.

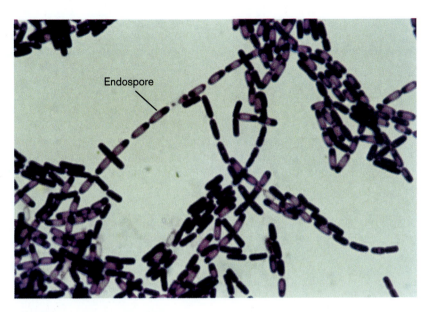

FIGURE 19.3 *Bacillus subtilis*.

Bacillus cells are widely distributed, and the genus has two traits that we can use to select for and isolate its members.

First, like the anaerobic *Clostridium* species, which are also common in soils, *Bacillus* cells produce endospores (Figure 19.3). These endospores are survival structures, and as we saw when we stained them (in Lab 8), they are very durable and very heat resistant. By contrast, fungal spores are adapted for dispersal and can be killed at lower temperatures. At temperatures of 55°C–80°C, most *Bacillus* cells will not grow, because most *Bacillus* species are mesophilic, not thermophilic. But thanks to their endospores, *Bacillus* species are thermoduric, and the spores will survive temperatures capable of killing almost everything else in the soil. So, an easy way to select for *Bacillus* species is to heat soil samples to temperatures of 70°C–80°C for about 15–20 minutes. And that's what we're going to do in this lab.

The second trait that is useful in the isolation and identification of *Bacillus* species is the fact that most of them secrete many extracellular enzymes. Most soils contain significant amounts of organic debris, mostly in the form of cellulose, starches, proteins, fats, and humic acids from decaying plant material. Bacterial species capable of secreting digestive enzymes into the surrounding soil can take advantage of these carbon and energy sources by converting large, unabsorbable macromolecules into smaller, absorbable compounds such as sugars and amino acids. *Bacillus* species that secrete amylase and proteases can be differentiated from cells that lack these enzymes by growing them on media that contain starch and proteins, respectively. In this lab, we'll be using media containing the milk protein casein as the differentiating agent (see Lab 10). *Bacillus* species and other microbes that secrete casein-digesting enzymes will yield clear zones around their colonies.

In addition, we'll dilute soil samples and plate them on glucose nutrient agar (GNA), a medium that supports the growth of a wider range of microbial species than skim milk agar. This medium will give us a more accurate count of the total number of microbes in the soil, including fungi and non-*Bacillus* bacteria. A comparison of colony counts on GNA plates before and after heating the soil may give us a rough estimate of the ratio of *Bacillus* spores to the total number of microbes in the soil.

PROTOCOL

MATERIALS

Per pair of students:

- 1 skim milk agar plate
- 2 tubes of 5-ml water blanks
- 8 tubes of 9-ml water blanks
- 3 tubes of 15-ml of melted skim milk agar
- 6 tubes of 15-ml of melted glucose nutrient agar
- 9 sterile petri dishes
- 12 1-ml pipettes
- 2-ml (blue-barrel) pipette pump
- 47°C and 80°C water baths
- Paper towels
- Vortex mixer
- Gram stain reagents
- Clean glass microscope slides
- Compound microscope with 100× oil immersion lens

Cultures:

- *Bacillus cereus*
- *Staphylococcus epidermidis*
- Sample of soil

Day 1

KNOWN SPECIES ON SKIM MILK AGAR

1. Label a skim milk agar plate with your names and the date, and divide the plate in half by drawing a line on the bottom with a wax pencil or Sharpie.

2. Label one side of the plate *B. cereus* and the other side *S. epidermidis*. Inoculate the plate with single streaks of the appropriate bacteria.

3. Incubate the skim milk agar plate at 28°C–30°C for 48 hours.

HEATED AND UNHEATED SOIL SAMPLES

1. Label two 5-ml water blank test tubes; one tube should be marked "Unheated" and the other "Heated."

2. Scoop about a cubic centimeter (about 0.40 cubic inches) of soil into each of the two water blanks.

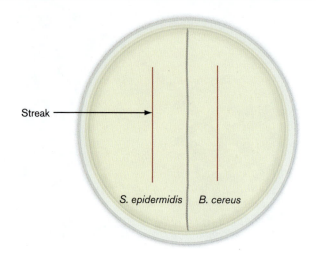

Streak →

S. epidermidis   *B. cereus*

3. Leave the Unheated tube at your bench. Take the Heated tube to the 80°C water bath and place it in a rack in the bath for 15 minutes.

4. After heating the Heated tube for 15 minutes, return the tube to your bench and follow the steps below.

SKIM MILK AGAR AND HEATED SOIL SAMPLE

1. Label the bottoms of three empty sterile petri dishes with your names, the date, and "Skim Milk—Heated Soil"; then label the first plate 10^{-2}, the second plate 10^{-3}, and the third plate 10^{-4}.

2. Label four 9-ml sterile water blanks with your names. Mark the first tube 10^{-1}, the second tube, 10^{-2}, the third tube 10^{-3}, and the last tube 10^{-4}.

Refer to **Figure 19.4** for a diagram of steps 3–14.

3. Use a vortex mixer to vortex the heated soil suspension for a few seconds, then use a sterile 1.0-ml pipette to transfer 1.0 ml of liquid from the heated soil suspension tube (the undiluted) to the 10^{-1} water blank tube.

4. Mix the transferred soil suspension into the 10^{-1} water blank by drawing the water in the blank into the pipette to a point about halfway between the 1.0-ml mark and the top of the pipette, then empty the contents of the pipette back into the tube. Repeat this step several times. Briefly vortex the tube to finish the process of mixing the cells into the water. Do <u>not</u> discard the pipette.

5. Use the same pipette as in step 4 to transfer 1.0 ml of cell suspension from the 10^{-1} water

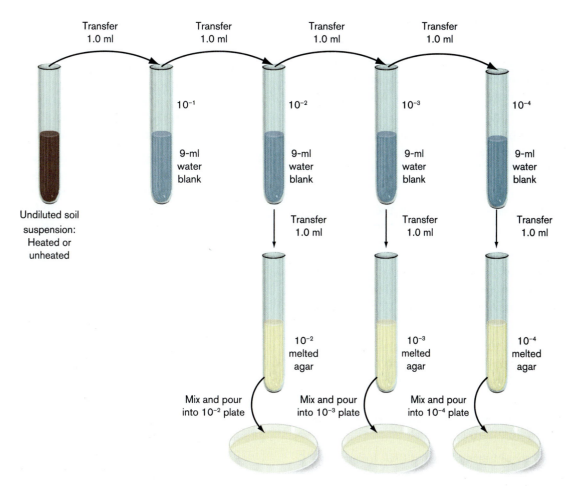

FIGURE 19.4 Serial dilutions of soil.

blank tube to the 10^{-2} water blank tube. Repeat the draw-and-empty step, briefly vortex the tube to mix the cells, and do <u>not</u> discard the pipette.

6. Using the same pipette as in step 5, transfer 1.0 ml of cells from the 10^{-2} water blank tube to the 10^{-3} water blank tube. Repeat the draw-and-empty step, briefly vortex the tube to mix the cells, and do <u>not</u> discard the pipette.

7. Using the same pipette as in step 6, transfer 1.0 ml of cells from the 10^{-3} water blank tube to the 10^{-4} water blank tube. Repeat the draw-and-empty step and briefly vortex the tube to mix the cells. Discard the pipette in the appropriate container to be autoclaved.

The serially diluted tubes of cells prepared in steps 5, 6, and 7 will be used in steps 11–14.

8. Prepare three masking tape labels as follows: 10^{-2}, 10^{-3}, and 10^{-4}. Set these labels aside.

9. Remove three tubes of melted skim milk agar from the 47°C water bath and wipe off any excess water with a paper towel. Be ready to work quickly once the agar is removed from the water bath.

10. Label the three melted agar tubes with your masking tape labels: 10^{-2}, 10^{-3}, and 10^{-4}.

11. Use a new sterile 1.0-ml pipette to transfer 1 ml of soil solution from the 10^{-2} test tube to the 10^{-2} melted agar tube. Gently stir the agar with the pipette to mix the cells into the agar, but don't stir for more than about 10–15 seconds, as the agar is cooling quickly at this point. Discard the pipette in the appropriate container to be autoclaved.

12. After stirring the agar for 10–15 seconds, the student doing the stirring should hand the tube to his or her lab partner. This partner should carefully pour the contents of the 10^{-2} melted skim milk agar

tube into the sterile petri dish labeled 10^{-2} while the "stirrer" quickly goes on to step 13.

13. Repeat steps 11 and 12, adding 1.0 ml from the 10^{-3} dilution tube to the melted agar tube labeled 10^{-3} and pouring the agar into the sterile petri dish labeled 10^{-3}.

14. Repeat steps 11 and 12, adding 1.0 ml from the 10^{-4} dilution tube to the melted agar tube labeled 10^{-4} and pouring the agar into the sterile petri dish labeled 10^{-4}.

15. After the agar has solidified in the three plates, incubate the plates at 30°C for 48 hours.

GNA AND HEATED SOIL SAMPLE

1. Label the bottoms of three empty sterile petri dishes with your names, the date, and "GNA—Heated Soil"; then label the first plate 10^{-2}, the second plate 10^{-3}, and the third plate 10^{-4}.

2. Use the 10^{-2}, 10^{-3} and 10^{-4} dilutions from the heated soil suspension you created in the "Skim Milk Agar and Heated Soil Sample" section for the subsequent steps.

3. Prepare three masking tape labels as follows: 10^{-2}, 10^{-3}, and 10^{-4}. Set these labels aside.

4. Remove three tubes of melted GNA from the 47°C water bath and wipe off any excess water with a paper towel. Be ready to work quickly once the agar is removed from the water bath.

5. Label the three melted agar tubes with your masking tape labels: 10^{-2}, 10^{-3}, and 10^{-4}.

6. Follow steps 11–15 from the previous section ("Skim Milk Agar and Heated Soil Sample") using melted GNA in place of melted skim milk agar to generate three GNA plates from the 10^{-2}, 10^{-3}, and 10^{-4} dilutions of the heated soil sample. Incubate all the plates at 30°C for 48 hours.

GNA AND UNHEATED SOIL SAMPLE

1. Label the bottoms of three empty sterile petri dishes with your names, the date, and "GNA—Unheated Soil"; then label the first plate 10^{-2}, the second plate 10^{-3}, and the third plate 10^{-4}.

2. Label four 9-ml sterile water blanks with your names. Mark the first tube 10^{-1}, the second 10^{-2}, the third 10^{-3}, and the last 10^{-4} (Figure 19.4).

3. Vortex the unheated soil suspension for a few seconds, then use a sterile 1.0-ml pipette to transfer 1.0 ml from the unheated soil suspension tube (the undiluted) to the 10^{-1} water blank tube.

4. Mix the transferred soil suspension into the 10^{-1} water blank by drawing the water in the blank into the pipette to a point about halfway between the 1.0-ml mark and the top of the pipette, then empty the contents of the pipette back into the tube. Repeat this step several times. Briefly vortex the tube to finish the process of mixing the cells into the water. Do <u>not</u> discard the pipette.

5. Use the same pipette as in step 4 to transfer 1.0 ml of cell suspension from the 10^{-1} water blank tube to the 10^{-2} water blank tube. Repeat the draw-and-empty step, briefly vortex the tube to mix the cells, and do <u>not</u> discard the pipette.

6. Using the same pipette as in step 5, transfer 1.0 ml of cell suspension from the 10^{-2} water blank tube to the 10^{-3} water blank tube. Repeat the draw-and-empty step, briefly vortex the tube to mix the cells, and do <u>not</u> discard the pipette.

7. Using the same pipette as in step 6, transfer 1.0 ml of cells from the 10^{-3} water blank tube to the 10^{-4} water blank tube. Repeat the draw-and-empty step and briefly vortex the tube to mix the cells. Discard the pipette in the appropriate container to be autoclaved.

8. Prepare three masking tape labels as follows: 10^{-2}, 10^{-3}, and 10^{-4}. Set these labels aside.

9. Remove three tubes of melted GNA from the 47°C water bath and wipe off any excess water with a paper towel. Be ready to work quickly once the agar is removed from the water bath.

10. Label the three melted GNA tubes with your masking tape labels: 10^{-2}, 10^{-3}, and 10^{-4}.

11. Follow steps 11–15 from the section "Skim Milk Agar and Heated Soil Sample" using melted GNA in place of melted skim milk agar to generate three GNA plates from the 10^{-2}, 10^{-3}, and 10^{-4} dilutions of the unheated soil sample. Incubate all the plates at 30°C for 48 hours.

Day 2

KNOWN SPECIES ON SKIM MILK AGAR

Examine and record results from the plate inoculated with *S. epidermidis* and *B. cereus*. Be sure to note any signs of casein hydrolysis (see Lab 10 for further assistance in interpreting your results).

GRAM STAIN OF HEATED SOIL SAMPLE COLONY

1. On the three skim milk agar plates that were inoculated with the heated soil sample, try to find

a colony with a zone of casein hydrolysis around it. Such a colony has a high probability of being a *Bacillus* colony, and one goal of this lab is the isolation of cells of *Bacillus* species. If there are no colonies with hydrolysis zones, look for a colony with a flat and dull or matte surface, because these characteristics are also indicative of *Bacillus* species.

2. Use a flame-sterilized inoculating loop to transfer cells from this colony to a glass slide, and Gram-stain these cells (see Lab 6 for instructions and materials).

While you're waiting for the slide to dry, go on to the rest of the day 2 steps.

POUR PLATES OF HEATED AND UNHEATED SOIL SAMPLES

1. Record the approximate number of colonies on each of the nine pour plates (you don't need a precise count, just a rough estimate). Colonies embedded in the agar may be small, so look closely for these colonies. It would not be surprising if colonies on the 10^{-2} plates are too numerous to count (TNTC).

2. For each of the nine pour plates, record a written description, including descriptions of the various colony types. For the skim milk agar plates, note roughly how many of the colonies show evidence of casein hydrolysis. For all the plates, note how many of the colonies are flat and dull, with or without evidence of hydrolysis, because many of these are likely to be *Bacillus* colonies. Also note roughly how many (if any) of the colonies look "fuzzy," as these are likely to be fungus (mold) colonies.

3. If you will be doing Lab 20, then be sure to save one or two of the plates inoculated with the heated soil sample and, ideally, containing casein-hydrolyzing colonies. The remaining plates can be placed in the receptacle designated for trash to be autoclaved. Your instructor will hold saved plates at 4°C until needed. If you will not be doing Lab 20, then dispose of all plates in the receptacle designated for trash to be autoclaved.

Name: _____   Section: _____

Course: _____   Date: _____

SKIM MILK AGAR PLATE WITH *B. CEREUS* AND *S. EPIDERMIDIS*

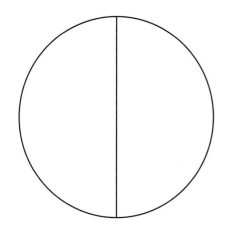

For each species, draw colonial growth and any zones of hydrolysis.

For each species, provide a written description of growth. For both species, describe the colony morphology and the width of any zones of hydrolysis.

GRAM STAIN OF COLONY FROM HEATED SOIL SAMPLE PLATE

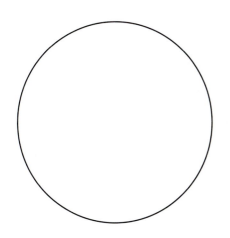

Draw cells, including any spores.

Provide a written description of cell morphology. Be sure to note the Gram reaction (positive or negative) and the presence or absence of spores.

HEATED SOIL SAMPLE ON SKIM MILK AGAR

Dilution	Colony Count	Written Description of Various Colony Types
10^{-2}		
10^{-3}		
10^{-4}		

HEATED SOIL SAMPLE ON GNA

Dilution	Colony Count	Written Description of Various Colony Types
10^{-2}		
10^{-3}		
10^{-4}		

UNHEATED SOIL SAMPLE ON GNA

Dilution	Colony Count	Written Description of Various Colony Types
10^{-2}		
10^{-3}		
10^{-4}		

1. How does skim milk agar act as a differential medium? How is this relevant in this particular lab?

2. Which species, *B. cereus* or *S. epidermidis*, showed a greater ability to hydrolyze casein? Based on the natural habitat of the species that is better at hydrolyzing casein, is there any reason to expect that this species would be better at hydrolyzing casein than the other species? Why?

3. Do you think that the cells you Gram-stained belong to the genus *Bacillus*? Describe two independent lines of evidence that support your conclusion. That is, what are the characteristics of these cells, or of the colony from which the cells originated, that lead you to your conclusion?

4. If you observed Gram-positive rods with spores in your Gram stain preparation, how do you know that the colony from which the cells were picked was not a colony of spore-forming *Clostridium* cells?

5. How did the GNA plates of the heated and unheated soil samples differ from each other in terms of the range of different types of colonies? Were your observations consistent with what you'd expect to see? Why or why not?

6. How did the GNA plates of the heated and unheated soil samples differ from each other in terms of the number or density of colonies? Were your observations consistent with what you'd expect to see? Why or why not?

7. Compare the GNA plates of the heated and unheated soil samples, and estimate the percentage of the colonies on the unheated sample plate that were the product of *Bacillus* spores. What does this suggest about the relative abundance of *Bacillus* spores in soil compared with other types of microbes?

8. Was there a difference between the heated and unheated soil plates in terms of the number of fungal colonies (fungal colonies are often dry or fuzzy in appearance)? If there was a difference, what does this suggest about the heat-tolerance of fungal cells and spores?

9. If you wanted to isolate bacteria from soil samples while simultaneously preventing the growth of fungi on your plates, describe a couple of strategies for doing this. You may need to do a little research here.

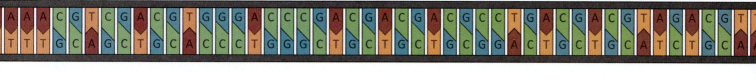

The polymerase chain reaction, or PCR, is a relatively recent invention, whose impact on biology and medicine is on par with the influence the microscope has had on these fields. Yet, PCR is remarkably simple. Short pieces of single-stranded DNA—the primers—bracket the region of DNA that will be copied by the polymerase enzyme. By repeating the replication cycle thirty times, the DNA of interest is amplified more than a billion-fold.

Learning Objectives

- Appreciate the history and applications of the polymerase chain reaction (PCR).

- Understand the roles of the key PCR components, including Taq DNA polymerase and oligonucleotide primers.

- Describe how PCR cycles produce large numbers of copies of specific sections of double-stranded DNA and how these copies can be visualized with gel electrophoresis.

- Use PCR to determine whether bacterial colonies isolated from soil are *B. cereus* colonies.

LAB 20

The Polymerase Chain Reaction and the Identification of *Bacillus cereus* from Soil Samples

INTRODUCTION

The **polymerase chain reaction (PCR)** is a powerful tool that enables us to amplify, or make multiple copies of, short stretches of DNA in vitro (in a test tube, outside of cells). The amplified sequences are usually somewhere between 100 and 10,000 base pairs in length, so this method can generate copies of entire genes, although in many cases, only a part of a gene will be copied. Furthermore, we can start with just one copy of a DNA sequence and create millions and millions of copies. In practice, however, we usually start with more than just a single copy.

So, what can we do with millions of copies of a short section of DNA? There are many, many uses for PCR and its products. Most of these applications are derived from the observation that with millions of copies of DNA in hand (as opposed to a few copies), it's (a) much easier to detect the presence of a particular DNA sequence and (b) much easier to create **recombinant DNA**, that is, DNA composed of base sequences from two or more organisms.

An incomplete list of PCR applications includes:

- Paternity testing ("On the next *Maury Show* …")
- Tissue typing and matching prior to organ transplantation
- Screening genes for mutations linked to risk of genetic diseases
- Forensic DNA fingerprinting that may link a suspect to a crime
- Creating copies of genes for insertion or splicing into DNA from other organisms (recombinant DNA technology)
- Identification of microorganisms

It's this last application, microbe identification, that will be the focus of this lab.

Invention of PCR

PCR is the invention of Kary Mullis and others working at Cetus Corporation in the early 1980s. Mullis is a chemist whose love for chemistry began at an early age with experiments he conducted using a Gilbert chemistry set. He recalled in his autobiography, *Dancing Naked in the Mind Field:*

> The first thing of any consequence that I made with my chemistry set was a substance similar to thermite. I mixed . . . and heated it over an alcohol burner. When I pulled it away from the flame, the reaction kept going. The mixture got red hot, broke the test tube, and suddenly went *Fffffsshhoooo*! Now that, I thought—being only seven years old—was cool. I didn't know what had happened then, but I decided that science was going to be fun.

By 1983, Mullis was more productively engaged doing DNA chemistry for the Cetus Corporation. His work included the creation of oligonucleotide "primers," and he was searching for uses or applications for these molecules. (*Oligos* is Greek for "few," so an oligonucleotide is a nucleic acid strand composed of a few to several nucleotides.) One evening in May 1983, he was thinking about oligonucleotides as he drove along California Highway 128 on his way to a weekend in the mountains with his girlfriend. Mullis began to ponder how he could locate specific stretches of DNA and generate a large number of copies of those sequences, such as those involved in genetic diseases.

> A short stretch of synthetic DNA could be treated in such a way that it would stick to a longer strand of DNA in a specific way if the sequences matched up somewhere on the long piece. The matching process would not be perfect. I might locate a thousand different places that were similar to the one I was searching for in addition to the correct one. A thousand out of 3 billion in the human genome would be no trivial feat, but it wouldn't be enough. I needed to find just one place.
>
> Suddenly, I knew how to do it. If I could locate a thousand sequences out of billions with one short piece of DNA, I could use another short piece to narrow the search. This one would be designed to bind to a sequence just down the (DNA) chain from the first sequence I had found. It would scan over the thousands of possibilities out of the first search to find just the one I wanted. And using the natural properties of DNA to replicate itself under certain conditions that I could provide, *I could make that sequence of DNA between the sites where the two short search strings landed reproduce the hell out of itself* [emphasis added]. In one replicative cycle, I could have two copies, and in two cycles I could have four, and in ten cycles … I thought that I remembered that two to the tenth was around a thousand. Twenty cycles would give me a million, thirty would give me a billion. … And they would always be the same size.
>
> I would be famous. I would win the Nobel Prize.

It would take two years of work by Mullis and his Cetus colleagues to fully develop his Highway 128 insight into the process that we call PCR, but by 1985, it was clear that this technique would revolutionize molecular biology. Mullis would indeed win the Nobel Prize in 1993, although he was nearly arrested on the morning of his acceptance speech when the Nobel laureate decided to entertain himself by shining a laser pointer out of his Stockholm hotel window. He aimed the laser at newspapers, onto the dashboards of cabs, and on the sidewalks in front of the Swedes strolling below. Mullis was politely asked by the police to find another way to amuse himself.

DOING PCR

Overview

PCR consists of 20–30 repeating cycles. With each cycle, a specific segment of DNA is amplified, and over time, the process dramatically increases the total number of copies of this DNA segment. Each cycle includes three basic steps or reactions: (a) denaturing, (b) annealing, and (c) extension. The individual steps in a cycle are composed of specific sets of reactions that work best at a particular temperature, so each step is done at a different temperature. The good news is that PCR is so incredibly useful that clever folks have created machines called **thermocyclers (Figure 20.1)**, which take care of all of these cyclic temperature changes for you.

(a) Denaturing In the **denaturing** step, the double-stranded template DNA that is to be amplified is heated to 90°C–95°C to denature, or separate, the two strands. As in DNA replication inside of cells, the DNA cannot be copied by PCR until the strands are separated.

(b) Annealing During the **annealing** step, **oligonucleotide primers** bind to specific complementary sites on the separated strands of the DNA at temperatures of 50°C–60°C. Primer sequences and binding sites are chosen so that they bracket the segment of DNA that is to be amplified by the PCR process.

(c) Extension In the **extension** step, which takes place at 70°C–75°C, individual nucleotides pair up with complementary nucleotides along the template DNA, and a heat-stable DNA polymerase (most often Taq polymerase) interacts with the DNA at sites where the primers are bound. The polymerase moves along the template strand, linking together the primers and the individual nucleotides to form new, complementary strands, and thus primer-specified segments of the DNA are created.

Certain components of PCR (primers and Taq polymerase) deserve a little more attention before we examine the PCR cycle itself in much more detail.

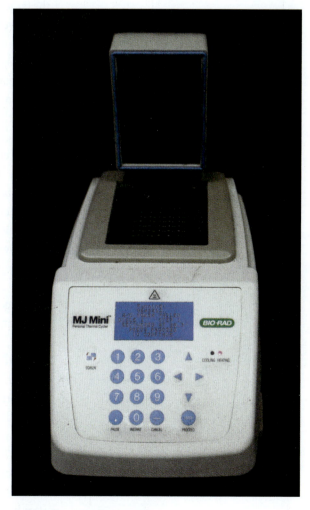

FIGURE 20.1 A thermocycler.

Primers

The polymerase chain reaction amplifies a specific region of a DNA molecule, so to do PCR, we must first know the nucleotide base sequence of the specific section of

DNA to be copied. As the salesman says in *The Music Man*, you gotta know the territory. Fortunately, such sequence information is often available in the scientific literature, and we can use that information to construct oligonucleotide primers, which are typically composed of 15–25 nucleotides. Primer sequences are chosen for their ability to complement and bind to the DNA at the beginning and at the end of the section of the DNA we wish to copy. That is, PCR reactions use two different primers, which flank the stretch of DNA to be amplified. Today, it is relatively simple to create primers using automated DNA synthesizers, and primers of a particular desired sequence can be purchased from many sources.

To use PCR for identification of a particular microorganism, we need to find primer sequences that complement base sequences that are **unique to that specific type of microbe**. That is, we want to use sequences that are found in the target organism, and only in the target organism. Then, ideally, **the primers will bind to DNA from the organism that we are looking for, but they will not stick to DNA from any other type of microbe**. This specificity lessens the risk of a false positive for amplification when the target species isn't actually present. In addition, primer sequences should be "highly conserved," or always present in closely related strains of the same species (see Lab 27 for a discussion of strain variation). This reduces the chances that strain variation within species will lead to a false negative, or the failure to detect a given species when it is actually present.

In this lab, we will be detecting and identifying *Bacillus cereus* cells in soil samples. Researchers have discovered that a nucleotide sequence found within the *B. cereus motB* gene makes a good target for PCR-based assays designed to identify *B. cereus* and its close relatives.[1] The *motB* gene codes for **motility protein B** (**MotB**), a motor protein found in the bacterial flagellum. The *motB* gene is found in many types of motile bacteria, but the specific base sequence of *motB* varies among different bacterial species that possess the gene. For example, there are sequences within the motB gene (given below) that are unique and close to identical among the many different soil-dwelling *B. cereus* strains. So, primers that complement these sequences should bind to and help amplify *B. cereus* DNA, but they should not stick tightly to and amplify DNA derived from unrelated *B. cereus* species. As a result, we can use these sequences to identify *B. cereus* strains by PCR and thus distinguish *B. cereus* strains from other bacterial species.

Taq DNA polymerase

The goal of PCR is to replicate DNA, and this process requires enzymes called **DNA polymerases**, which by using the separated DNA strands as templates, link individual nucleotides to each other to form long strands of DNA. In PCR, we use a particular DNA polymerase called **Taq polymerase**. This enzyme begins DNA synthesis at sites where primers are bound to template DNA, and the new strands include the primer sequences. Taq polymerase is used in PCR because it is very heat stable: it catalyzes DNA synthesis most efficiently at around 70°C–75°C, and it remains functional even after heating to temperatures of 90°C–95°C. This trait is critical for PCR because the denaturing step in the cycle is done at 90°C–95°C. If the polymerase were denatured and rendered permanently nonfunctional at this step, then the enzyme would have to be added at the beginning of each of the 20–30 cycles.

[1] K. Oliwa-Stasiak et al., 2010, *J. Appl. Microbiol.* **108**:266–273.

The Significance of Hot-Springs Enzymes

When PCR was first developed, Kary Mullis and his Cetus colleagues used a polymerase derived from *E. coli*. However, *E. coli* cells and their enzymes are adapted to function at temperatures around 30°C–45°C. Not surprisingly, *E. coli* DNA polymerases are permanently inactivated when temperatures are raised to 90°C–95°C in the DNA denaturing step. So, fresh polymerase had to be added at the start of every extension step, a process that became very tedious after 20–30 cycles. Continued use of *E. coli* DNA polymerase would have also eventually complicated efforts to automate PCR using thermocyclers.

Fortunately, there was another type of DNA polymerase available to the research team at Cetus. This polymerase came from a yellow-orange-pigmented bacterium called *Thermus aquaticus*. As noted above, this enzyme is known as Taq polymerase ("T-" from *Thermus* and "-aq" from *aquaticus*). Unlike *E. coli* and other human-associated bacteria, *T. aquaticus* is a thermophile that lives in hot springs and thermal pools. Survival in these environments depends on possession of enzymes capable of functioning at very high temperatures (as well as other heat-tolerance traits). This explains why Taq polymerase catalyzes DNA synthesis most rapidly at temperatures of 70°C–75°C. It also accounts for the enzyme's ability to withstand temperature of 90°C–95°C without denaturation and loss of function, even when these temperatures are reached cycle after PCR cycle. So, the Cetus group began to use Taq polymerase in PCR around 1985, eliminating the need to add polymerase with every cycle, and opening the door to automation and today's PCR thermocyclers.

Thermus aquaticus was originally isolated from a hot spring in Yellowstone National Park in the 1960s by Thomas Brock. Brock was a microbiologist with an interest in life at high temperatures, and after Taq polymerase was isolated by other scientists in the mid-1970s, his research into the little-known world of "extremophile" microbes helped to make today's PCR possible. Though *T. aquaticus* was originally found in a national park, the National Park Service received no royalties for a discovery that was ultimately worth billions of dollars. This experience led to a change in park policy. Today, "bio-prospectors" in U.S. national parks sign benefits-sharing agreements with the Park Service that return benefits to the parks when the results of cooperative research lead to the development of something that is commercially valuable.

THE PCR CYCLE (DENATURING, ANNEALING, AND EXTENSION)

PCR consists of 20–30 cycles of (a) denaturing, (b) annealing, and (c) extension.

First cycle (cycle 1)

(a) Denaturing At the start of the cycle, a PCR reaction mixture containing the DNA segments to be amplified is heated to temperatures of 90°C–95°C to denature the DNA molecules. In this context, the term "denature" refers to the disruption of the hydrogen bonds between base pairs in the adjacent strands of double-stranded DNA and the separation of the two strands. Remember, the two strands in a double helix must be parted from each other because this allows each strand to serve as a template that can be copied by a polymerase.

Each single strand of DNA acts as a template for the DNA replication that takes place during the extension phase of PCR, so the DNA used in the initial denaturing step is referred to as **template DNA**. This DNA carries or includes those segments or sequences that we wish to amplify, along with lots of additional DNA that will not be replicated. For example, in this lab, the template DNA will consist of *B. cereus* chromosomal DNA extracted from bacterial cells. The *B. cereus* chromosome contains a few thousand genes, but we'll be amplifying just part of one gene, or less than 0.1% of all the available DNA sequences in all the chromosome copies in the reaction mixture.

(b) Annealing Once the double-stranded DNA has been separated into two single-stranded molecules, the primers in the reaction mixture are able to anneal to the template DNA by hydrogen bonding. The mixture is cooled to 50°C–60°C, because temperatures much above this would tend to disrupt the hydrogen bonds between the primer and the target sequence on the template DNA. The ideal annealing temperature is typically 3 to 5 °C below the temperature that would separate the primer from the template DNA, and it's specific value depends on the primer sequence. While the primer can bind to its complementary target at 50°C-60°C, these temperatures are still hot enough to discourage nonspecific binding between the primers and the template DNA. That is, the primers won't bind unless there is a very close complementary match between the primer and template sequences. Recall that we are trying to amplify a very specific segment of DNA, and we don't want the primers binding to the DNA in other locations.

The reaction mixture contains two different primers with different sequences. One of these primers will bind to one of the template strands at the 3′ end at the start of the DNA section to be copied. The other primer will bind to the complementary or opposing template strand at the 3′ end at the end of the complementary DNA section to be copied. This will have the effect of bracketing a specific segment of DNA. This process is a little easier to follow if we use a specific example.

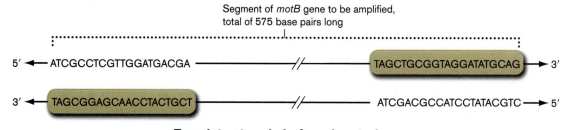

Segment of *motB* gene to be amplified, total of 575 base pairs long

5′ ◄— ATCGCCTCGTTGGATGACGA ———————//——— TAGCTGCGGTAGGATATGCAG ►— 3′

3′ ◄— TAGCGGAGCAACCTACTGCT ———————//——— ATCGACGCCATCCTATACGTC —► 5′

Template strands before denaturing.

In this lab, we are looking for and trying to amplify a section of the *B. cereus motB* gene, a segment that is a total of 575 base pairs (bp) long.

1. At the start of this part of the *motB* gene, one template strand contains a base sequence that reads 3′–TAGCGGAGCAACCTACTGCT–5′. So, a primer with a sequence reading 5′–ATCGCCTCGTTGGATGACGA–3′ will complement this template sequence exactly. After the template DNA strands are denatured, such a primer will anneal tightly to the template strand at the start of the section to be amplified.

2. At the end of the target section in the *motB* gene, in the other (complementary) template strand, the sequence is 3′–GACGTATAGGATGGCGTCGAT–5′. A primer with a sequence that reads 5′–CTGCATATCCTACCGCAGCTA–3′ will complement this part of the opposing template strand and will bind strongly to this strand.

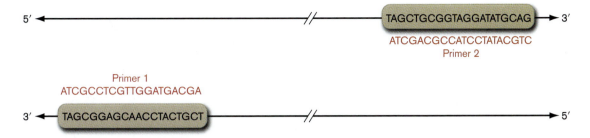

Denatured or separated template strands with primers bound during the annealing step.

(c) Extension Once primers bind to the template strands, the work of extending the DNA beyond the freshly annealed primers can begin. Now it's Taq polymerase time. Since this heat-tolerant enzyme is adapted to work best at 70°C–75°C, the extension step in the cycle is usually done at these temperatures.

As individual deoxynucleotide triphosphates, or dNTPs (provided in the reaction mixture), pair up, via hydrogen bonding, with the complementary nucleotides in the template DNA strands, Taq polymerase links these separate nucleotides together. The polymerase catalyzes the formation of a covalent bond between an –OH group on the 3′ carbon of a first nucleotide's sugar and a phosphate group on the 5′ carbon of a second nucleotide's sugar. The linking process is repeated over and over again, creating a new DNA strand that is complementary to the template strand, as the polymerase moves from the 3′ to the 5′ end of the template strand (or from the 5′ to the 3′ end of the new strand). While the Taq polymerase and dNTPs can be added to the PCR mixture separately, in practice, the Taq polymerase is usually added in a "master mix" that also contains the dNTPs as well as magnesium chloride ($MgCl_2$) and buffers.

Like other DNA polymerases, Taq polymerase cannot hop onto the DNA at any random point to begin building new DNA strands. It must have a place to start, and the bound primers provide such a location. So, the primers not only direct PCR to amplify a specific location within a DNA molecule, but are also important in marking the starting point for the polymerase.

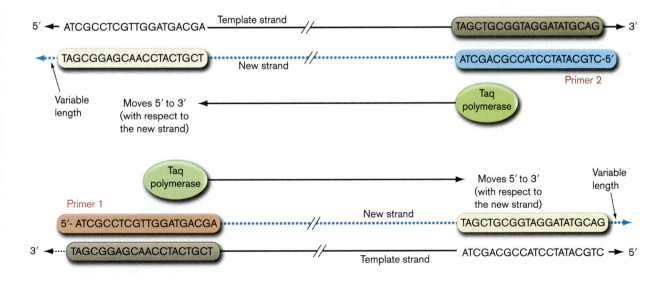

If we use the *B. cereus motB* gene primers described above, the Taq polymerase will construct a new strand of DNA starting with and including the first primer sequence, 5′-ATCGCCTCGTTGGATGACGA-3′. It will also build a strand using the second primer sequence, 5′-CTGCATATCCTACCGCAGCTA-3′.

While the bound primers tell the Taq polymerase where to start a new strand, in the <u>first</u> PCR cycle, there really isn't anything along the template DNA strand that tells the polymerase to stop adding bases when it gets to the end of the section we want to amplify. The extension process is time dependent, and it will eventually end when the temperature is again increased to 90°C–95°C at the start of the second cycle's denaturing step. The extension, however, does not stop at a particular or specific spot along the template strand, and in many cases, the polymerase will have had time to extend the new strands far beyond the end of the section to be amplified. Thus, while all the new strands made from all the available template molecules will start with one of two primer sequences, those strands will be quite variable in their final length. This variation will change with subsequent PCR cycles. Stay tuned.

Second cycle (cycle 2)

(a) Denaturing PCR is all about cycles, and the first cycle described above is just the start. At the start of the second cycle, the temperature of the reaction mixture is again raised to 90°C–95°C to separate template strands from the new strands created by the activity of the primers and Taq polymerase.

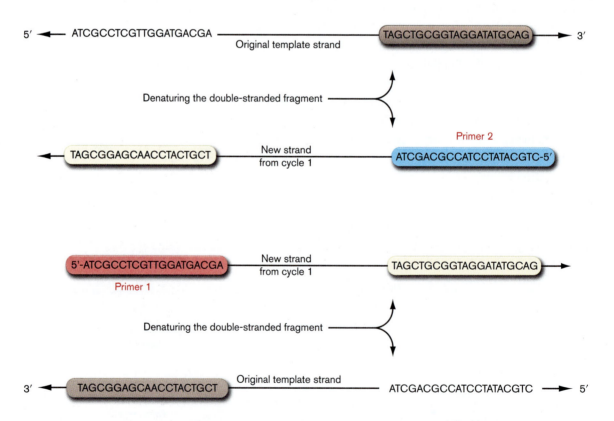

(b) Annealing After the DNA strands have been separated, the remaining free primers in the PCR reaction mixture can again anneal to template DNA by hydrogen bonding. The original template molecules are still present in the mixture, so some of the primer molecules will bind to these templates as they did in the annealing step of cycle 1. So

DNA extension using the original templates as guides can occur just as it did in cycle 1, and the DNA products will be the same as in cycle 1. However, let's set these particular template molecules aside, and instead, shift our focus to the new strands of DNA produced in the first cycle.

The first cycle produced new DNA molecules of variable length that begin with either the primer 1 or the primer 2 sequence at the 5′ end. As a result of cycle 1 extension, many single strands that start with the primer 1 sequence also have a sequence 555–575 bases downstream from the 5′ end that complements the primer 2 sequence. These strands may well be much longer than 575 bases, but their exact length is not something that we have to worry about here. Similarly, after extension, many single strands that start with the primer 2 sequence also have a sequence 555–575 bases downstream from the 5′ end that complements the primer 1 sequence.

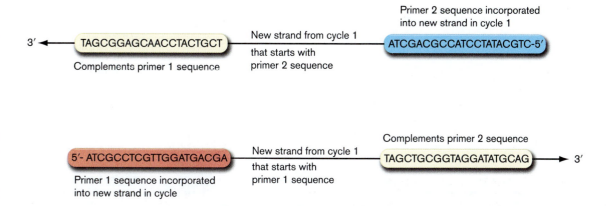

As a result, in the cycle 2 annealing step, free primer 2 will bind to complementary sites along the single-stranded DNA that begins with the primer 1 sequence, and similarly, free primer 1 will bind to complementary sites along the strands that begin with the primer 2 sequence.

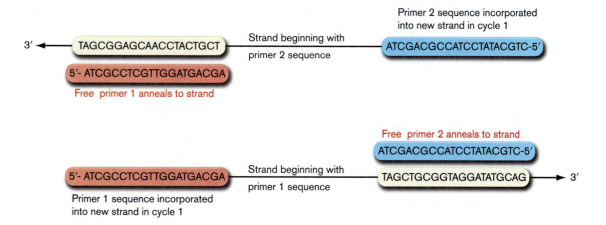

(c) Extension After the primers are bound, Taq polymerase can again link together individual dNTPs paired with complementary nucleotides in template DNA strands to create new DNA strands.

On strands created in cycle 1 to which primer 1 has annealed in cycle 2, the Taq polymerase extends a new DNA strand until the new strand reaches the point in the template strand where primer 2 was incorporated in the first cycle. This incorporated primer 2 sequence marks the end of the template strand, so extension will also stop at this point. Now we have a new strand that begins at the primer 1 site and ends at the primer 2 site, so it will be exactly 575 bases long.

Likewise, when primer 2 binds along a strand created in cycle 1, the Taq polymerase extends a new strand until it reaches the point in the template strand where primer 1 was incorporated in the first cycle. The incorporated primer 1 sequence marks the end of this template strand, and again, this creates strands that are exactly 575 bases in length.

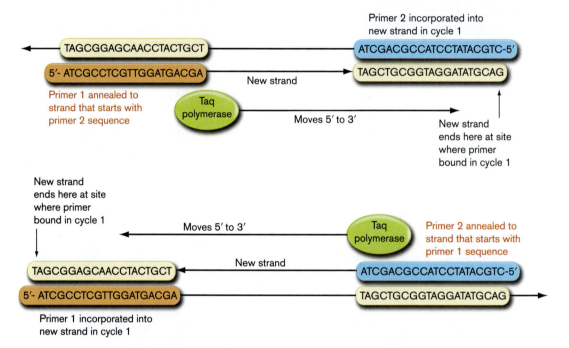

Third and subsequent cycles

At the start of the third cycle, we now have DNA strands derived from the *B. cereus* *motB* gene that are exactly 575 bases long.

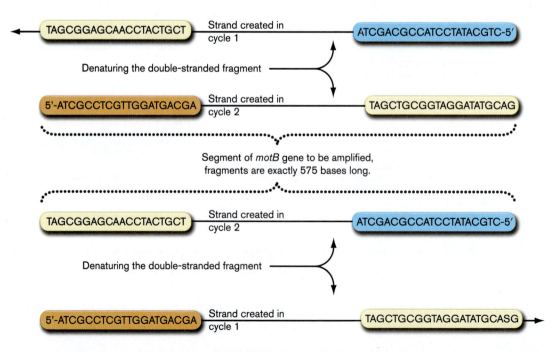

Denaturing in cycle 3.

When these 575 base long strands are used in a third cycle of denaturing, annealing, and extension, the new strands created from these strands will also be precisely 575 bases long. In fact, after the third cycle extension, we'll have double-stranded DNA molecules in which both strands are composed of exactly 575 bases.

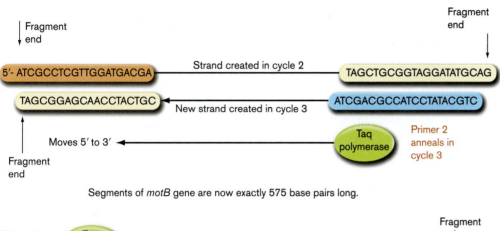

Segments of *motB* gene are now exactly 575 base pairs long.

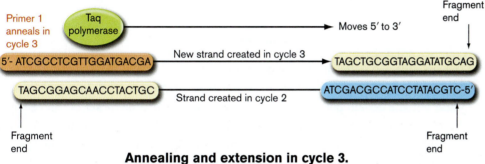

Annealing and extension in cycle 3.

Now, in subsequent PCR cycles, each 575-bp DNA molecule can yield two 575-bp double-stranded DNA molecules. So each cycle potentially doubles the number of 575-bp fragments derived from the *B. cereus motB* gene. This is the "chain reaction" stage of the polymerase chain reaction.

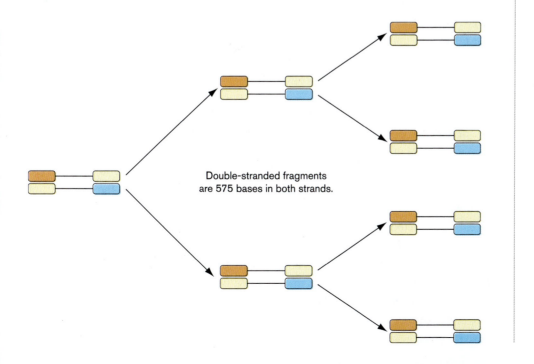

Double-stranded fragments are 575 bases in both strands.

Initially, this doubling process does not dramatically increase the total number of 575-bp DNA fragments as we go from one copy to two copies. However, as the mathematics in "The Inchworm Song" teaches us, if we double the number of copies with each cycle, then with subsequent cycles, we will go from two copies to four copies to eight copies to sixteen copies to thirty-two copies and so on.

So, extract your template DNA, add your primers, turn on your thermocycler, and follow the way of the inchworm. Soon, you and your millions of copies of DNA "will probably go far." Or at least, you'll finish this lab.

Inch worm, inch worm, measuring the marigolds,
You and your arithmetic, you'll probably go far,
Inch worm, inch worm, measuring the marigolds,
Seems to me, you'd stop and see, how beautiful they are.
—"The Inchworm Song" (Frank Loesser, 1952)

Thus, PCR is an example of an exponential progression, and the equation that predicts the number of copies after a given number of cycles is:

$$\text{Total copy number} = 2^n$$

where *n* is equal to the number of doubling cycles.

With an exponential function, the number of DNA copies may increase slowly at first, but at a certain point the count shoots upward like a skyrocket. As an example, if we start with one copy, then six cycles (2^6) can produce only 64 copies. However, another six cycles (2^{12}) could yield 4,096 copies, and six more cycles (2^{18}) could potentially produce 262,144 copies. After about twenty cycles, a single initial copy may have yielded over a million copies. In practice, the process isn't quite this efficient, but still, this is big-time DNA amplification, and it was the kind of math that Kary Mullis was doing in his head as he drove down Highway 128 to his cabin in the mountains.

So, why would we want to produce millions of copies of a short stretch of DNA? There are many answers to this question, some of which are provided in the list of PCR applications given above. If the goal is to identify specific types of bacteria, then PCR can be used to make millions of copies of a sequence that is unique to a given species or set of related strains. We want millions of copies because it's much easier to detect millions of copies of fragments of a specific size than it is to detect one copy (as we'll see below). If our primers are designed wisely, amplification will occur only if the DNA of the given species is present, and it will not occur if it is absent or if the template DNA (chromosomal DNA) originated from other unrelated bacterial species. Therefore, amplification of unique sequences can be used to identify a given species.

In this lab, we will be using primers that bind to sequences unique to the *B. cereus* group. They should bind tightly to and amplify *B. cereus* DNA, and they should not bind well to or amplify DNA from species unrelated to *B. cereus*. If we perform PCR on DNA extracted from a *B. cereus* colony, it should produce millions of copies of a specific, 575-bp DNA fragment. However, if there is no *B. cereus* DNA, or if all the DNA present is from other microbial species, then there will be no amplification of the *B. cereus*-specific DNA, and there will not be millions of copies of the fragment. If we subsequently detect millions of copies of the 575-bp fragment, then we know we have a *B. cereus* colony. In contrast, if we don't detect any 575-bp fragments, then we don't have *B. cereus*.

So, the last step in using PCR to identify a microbe is to look for millions of copies of a particular, unique DNA fragment from a specific type of bacteria. To do this, we need to use electrophoresis and agarose gels.

AGAROSE GELS AND ELECTROPHORESIS

Agarose gels and electrophoresis have been used for decades to sort and separate DNA fragments by size. **Agarose** is a component of agar, which is extracted from seaweed. The agarose molecules are very large polysaccharides composed of repeating units of a disaccharide composed of galactose and the sugar derivative 3,6-anhydro-L-galactopyranose. Gels made of purified agarose work well for separating DNA molecules by size because they are electrically neutral and thus interact very little with the migrating DNA.

Gels are made by heating agarose granules to 95°C–100°C in water or a water-based buffer to melt them. When the melted agarose is poured into a gel mold and cooled, the molecules form a polysaccharide matrix that functions as a molecular sieve or filter that will impede the movement of molecules based on their size. In **electrophoresis**, the gel is submerged in a buffer solution, suspensions of DNA are loaded into wells at one end of the gel, and an electric current is applied across the gel **(Figure 20.2)**. *Phoresis* is a Greek term meaning "being carried," and in electrophoresis, the DNA is being carried by the electric current from one end of the gel to the other. Specifically, DNA molecules have a net negative charge due to their phosphate and hydroxyl groups, so these molecules will be attracted toward the positively charged electrode, or anode. (Positively charged electrodes are usually marked in red, so DNA is said to "run to the red.")

Once the current is applied, the DNA fragments begin to migrate from the wells into the gel matrix, moving toward the positive end of the gel. This movement is hindered or restricted by the agarose meshwork because the traveling DNA molecules must slip between the polysaccharide strands. Not surprisingly, larger DNA fragments have a more difficult time moving through the gel, so they move more slowly toward the anode, while smaller or shorter fragments move at a faster rate. Over time, the smaller fragments pull ahead of the larger fragments, and thus the pieces of DNA are sorted by size (Figure 20.2).

All the DNA fragments of a given size move at about the same rate, so eventually, **all fragments of a given size will cluster together at about the same location in the gel**, forming a distinct band of DNA. For example, in this lab, all the 575-bp fragments generated by amplification of the *B. cereus motB* gene DNA will migrate to the same position in an agarose gel. The distance traveled by the DNA, and thus the final position of the band, is dependent on fragment size, so **if you know the location of the band, then you can determine the size of the fragment**. Fragment size (in base pairs) is determined by comparing the position of a given fragment's band against the positions of bands of DNA fragments of known size. A mixture of these DNA fragments of known size are called DNA markers or a "DNA ladder" (Figure 20.2).

So, when running an agarose gel of our amplified sample, if we can see specific DNA bands, then we can determine their sizes. Unfortunately, DNA is not a pigmented molecule, so it is essentially invisible as it moves through the gel. To see the bands of DNA, we must find a way to stain the DNA. In the past, almost all molecular biology labs stained DNA in gels with **ethidium bromide**, a molecule that intercalates, or inserts itself, between the stacks of bases in double-stranded DNA. In the presence of UV light, the intercalated ethidium bromide fluoresces with a bright red-orange color, so when the stained gel is exposed to UV light, the bands of DNA glow red-orange. However, ethidium bromide will also insert itself into the DNA in your cells, which is not good. In short, the same property that makes ethidium bromide a good DNA stain also makes this molecule a mutagen. So, in this lab (as in many molecular biology labs today), we will use a much safer stain, methylene blue, to visualize DNA. Our ability to see the methylene blue–stained DNA depends on the number of copies of the DNA molecules, and the stain is not as sensitive as ethidium bromide. However, since PCR will produce millions of copies of the *motB* gene fragment, methylene blue will work for our purposes. In addition, bands of DNA stained with methylene blue can be seen with visible light, eliminating the need to use mutagenic UV light.

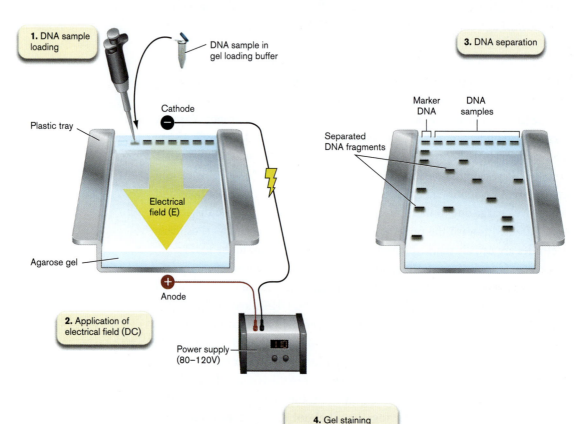

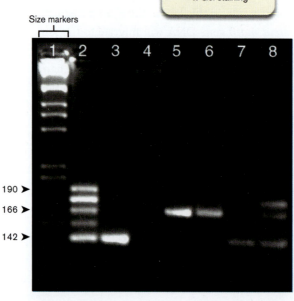

FIGURE 20.2 Agarose gel electrophoresis of DNA.

To visualize the results of the PCR reaction, the DNA samples are loaded onto an agarose gel, separated by an electric current, and stained to visualize the separated bands of DNA. The stain can be a fluorescent dye, such as ethidium bromide, which is visible to the human eye under UV light. Alternatively, as in this lab, the DNA can be made visible using methylene blue a dye that we can see under visible light.

PROTOCOL

Day 1

Materials

- Glucose nutrient agar (GNA) plates (6 per pair of students)
- Incubator set to 30°C
- Wax pencils or Sharpies for labeling plates
- Bunsen burner
- Striker
- Inoculating loop

Cultures:

- *Bacillus cereus*
- *Escherichia coli*
- Colonies from saved heated soil sample plates from Lab 19; ideally, heated soil sample plates containing casein-hydrolyzing colonies

PROCEDURES

1. Label six GNA plates with your names and the date. Then label the plates as follows:

   a. *E. coli* (Negative control)
   b. *B. cereus* (Positive control)
   c. Soil colony 1
   d. Soil colony 2
   e. Soil colony 3
   f. Soil colony 4

2. Using a sterile loop, and following the streak plating protocol used to produce isolated colonies (see Lab 3), streak the plate labeled *E. coli* using cells from the *E. coli* stock culture.

3. Using a sterile loop, and following the streak plating protocol used to produce isolated colonies (Lab 3), streak the plate labeled *B. cereus* using cells from the *B. cereus* stock culture.

4. Examine the saved plates you created by serial dilution of heated soil samples in Lab 19 for isolated colonies. You can use any and all heated soil sample plates at this point; the main goal is to find isolated colonies.

5. Find an isolated colony that has the characteristics of a *B. cereus* colony. In most soil samples, this will probably be the dominant colony type. You are looking for a colony that is flat and spreading with a dull, lusterless or matte surface. On skim milk agar plates, colonies that show casein hydrolysis are a good bet, but be aware that not all *B. cereus* strains hydrolyze casein.

6. Record the appearance of the selected colony in the Results section.

7. Using a sterile loop, and following the streak plating protocol described in Lab 3, touch the colony that you have selected and streak the soil colony 1 plate so as to produce isolated colonies.

8. Find a second colony with *B. cereus* colony traits. Record the appearance of the colony in the Results section.

9. Using a sterile loop, and following the streak plating protocol described in Lab 3, touch the colony that you have selected and streak the soil colony 2 plate so as to produce isolated colonies.

10. Find one more colony with *B. cereus* colony traits. Record the appearance of the colony in the Results section.

11. Using a sterile loop, and following the streak plating protocol described in Lab 3, touch the colony that you have selected and streak the soil colony 3 plate so as to produce isolated colonies.

12. For the soil colony 4 plate, look for a colony that does not look like a *B. cereus* colony. That is, search the plates for circular, shiny, or pigmented colonies.

13. Record the appearance of the selected colony in the Results section.

14. Using a sterile loop, and following the streak plating protocol described in Lab 3, touch the colony that you have selected and streak the soil colony 4 plate so as to produce isolated colonies.

15. Incubate plates for 24 hours at 30°C.

Day 2

MATERIALS

- Gloves
- Hot plate
- 1,000-ml beaker
- 1.5-ml microcentrifuge tubes
- Floating microcentrifuge test-tube rack
- Nuclease-free molecular biology–grade water

- Sterile toothpicks with flattened ends
- 10–100-μl micropipetters
- 10–100-μl sterile pipette tips
- 1–10-μl micropipetters
- 1–10-μl sterile pipette tips
- Sterile PCR tubes
- Thermocycler
- Ice and ice bucket
- Promega GoTAQ Green Master Mix, 2X, 200 μl aliquot in microcentrifuge tubes
- Primer 1
  - 100 μl aliquot of 10 μM primer 1
- Seq: 5′–ATCGCCTCGTTGGATGACGA–3′
- Primer 2
  - 100 μl aliquot of 10 μM primer 2
- Seq: 5′–CTGCATATCCTACCGCAGCTA–3′
- Plates inoculated on day 1

EXTRACTING TEMPLATE DNA

1. Place a 1,000-ml beaker on a hot plate and add water to a depth of about 2 inches. Turn the heat on to a setting sufficient to bring the water to a boil. While waiting for the water to heat up, go on to steps 2–9.

2. Label six 1.5-ml microcentrifuge tubes with the following abbreviations:

   a. EC (for *E. coli*)
   b. BC (for *B. cereus*)
   c. SC 1 (for soil colony 1)
   d. SC 2 (for soil colony 2)
   e. SC 3 (for soil colony 3)
   f. SC 4 (for soil colony 4)

3. Using a micropipetter and sterile pipette tips, add 100 microliters (μl) of nuclease-free molecular grade water to each of the six tubes.

4. Pick up a sterile toothpick without touching the flattened end of the toothpick. Using the flattened end, scrape some colonial material from an isolated colony on the *E. coli* plate onto the toothpick. The colony should be scraped until the flattened end holds a tiny blob of cells about the size of a pinhead (about 1 mm in diameter). Be careful to avoid scraping the surface of the GNA plate, because salts in the medium can inhibit the PCR.

5. Transfer the *E. coli* cells in the colonial scraping to the water in the microcentrifuge tube labeled <u>EC</u>. While holding the flat end of the toothpick in the water, twirl the toothpick between your fingers to remove as many *E. coli* cells as possible.

6. Close the tube and place it in a small floatable test-tube rack. Discard the toothpick in an autoclave bag.

7. Repeat steps 4–6 using a colony from the *B. cereus* plate, in this case, putting the colonial material in the tube labeled BC. *B. cereus* colonies can spread quickly, so you may not have any truly isolated colonies on the plate. If necessary, just use a colony that has a minimum of contact with other colonies.

8. Repeat steps 4–6 using a colony from the soil colony 1 plate, in this case, putting the colonial material in the tube labeled SC 1. Again, *B. cereus* colonies can spread quickly, so you may not have any truly isolated colonies on the plate, so if necessary, use a colony that has a minimum of contact with other colonies.

9. Repeat steps 4–6 using a colony from the soil colony 2 plate, in this case, putting the colonial material in the tube labeled SC 2. Follow this same pattern for soil colony plate 3 and soil colony plate 4.

10. Once the water in the beaker is boiling, place the floating test-tube rack with the six microcentrifuge tubes in the boiling water. Heat the tubes for 15 minutes to extract the DNA from the cells. Remove the rack and turn off the hot plate. You now have six tubes of DNA.

    The DNA is a mixture of the template (or target) DNA plus the remaining chromosomal DNA. The method used in step 10 provides a quick way to extract chromosomal DNA. More DNA could be obtained if we also used proteases and lysozyme in the extraction procedure.

POLYMERASE CHAIN REACTION

Wear gloves for all of the following steps.

1. Place one microcentrifuge tube of each of the following on ice:

   a. Primer 1 (10 μM)
   b. Primer 2 (10 μM)
   c. GoTAQ Green Master Mix (2X)

2. Label six PCR tubes with your initials and the following abbreviations:

   a. EC
   b. BC
   c. SC 1
   d. SC 2
   e. SC 3
   f. SC 4

3. Using a 10–100-μl micropipetter and sterile pipette tips, transfer 25 μl of GoTAQ Green Master Mix (2X) to each of the six tubes. As long as the pipette tip doesn't touch anything other than the GoTAQ solution and the inside of the six labeled tubes, you can use the same tip for all tubes.

   The GoTAQ master mix contains Taq polymerase, dNTPs, $MgCl_2$, buffers, and dyes. All of these components are present at the proper concentrations for PCR, and the use of the mix saves a great deal of time when compared with the time needed to add each component individually.

4. Using a 1–10-μl micropipetter and sterile pipette tips, transfer 7.5 μl of primer 1 to each of the six tubes. Discard pipette tips between each of the six tubes and after the last tube.

5. Using a 1–10-μl micropipetter and new sterile pipette tips, transfer 7.5 μl of primer 2 to each of the six tubes. Discard pipette tips between each of the six tubes and after the last tube.

6. Using a 10–100-μl micropipetter and a sterile pipette tip, transfer 10 μl of *E. coli* template DNA from the microcentrifuge tube marked EC to the PCR tube marked EC and immediately close the cap. Then discard the pipette tip.

7. Using a 10–100-μl micropipetter and a sterile pipette tip, transfer 10 μl of *B. cereus* template DNA from the microcentrifuge tube marked BC to the PCR tube marked BC and immediately close the cap. Discard the pipette tip.

8. Using a 10–100-μl micropipetter and a sterile pipette tip, transfer 10 μl of soil colony 1 template DNA from the microcentrifuge tube marked SC 1 to the PCR tube marked SC 1 and immediately close the cap. Discard the pipette tip.

9. Continue to follow the pattern in step 8 to transfer 10 μl of soil colony template DNA from the microcentrifuge tubes marked SC 2, SC 3, and SC 4 to the appropriate PCR tubes. Close the caps immediately after DNA transfer and discard the pipette tip between each transfer.

10. Be certain that your PCR tubes are marked with your initials (because you'll be sharing space in the thermocycler). Load the six PCR tubes into the thermocycler.

11. Program the thermocycler as follows, then start the machine:

    a. Initial denaturing: 94°C for 3 minutes
    b. PCR cycles (total 24 to 25 cycles):

       Denaturing: 94°C for 1 minute

       Annealing: 55°C for 1 minute

       Extension: 72°C for 1 minute

    c. Final extension: 72°C for 4 minutes

12. At the end of the thermocycler run, the PCR tubes should be stored at −20°C until the mixtures can be run on a gel (day 3).

Day 3
MATERIALS

- DNA standard or ladder
- PCR reaction mixes from day 2
- Balance
- 1X TBE buffer
- Agarose powder
- Deionized water
- 100-ml graduated cylinder
- 250- to 500-ml Erlenmeyer flasks
- Microwave or hot plate
- 55°C water bath (optional)
- Power supply (constant voltage of 100 V)
- Electrophoresis gel boxes, molds, and combs
- Gloves
- CarolinaBLU DNA staining kit
- 10–100-μl micropipetters
- 10–100-μl sterile pipette tips
- DNA staining boxes

PROCEDURES

Wear gloves for the following steps. While the methylene blue stain we will use is much less toxic than ethidium bromide, it is still a stain, and stains stain.

1. Use a balance and 100-ml graduated cylinder to prepare a 1.5% agarose gel by adding agarose powder to 1X TBE buffer in an Erlenmeyer flask at a ratio of 1.5 grams agarose per 100 ml of buffer.

   To prevent boilover, the flask volume should be about five times the buffer volume. The amounts of agarose and buffer required will depend on the volume of your particular agarose gel mold and electrophoresis gel box. A 5 cm × 8 cm gel works well, and molds of these dimensions can usually be filled with a little less than 50 ml of agarose. Since we are looking for a single specific 575-bp band, the gel doesn't have to be more than about 8–10 cm long.

2. Melt the agarose by heating the buffer to the boiling point in a microwave or on a hot plate. Use caution when boiling agarose suspensions and when handling hot glassware! Once the buffer begins to boil, it usually takes 20–30 seconds to completely melt all the agarose; the agarose is completely melted when there are no granules visible in the buffer.

3. Let the melted agarose cool to about 50°C–55°C before pouring. If available, use a water bath set to 55°C to cool the agarose and hold it at 55°C.

4. Before pouring the agarose into the gel mold, add concentrated CarolinaBLU stain to the melted agarose at a ratio of 80 μl of concentrated stain per 50 ml of melted agarose. Gently swirl the flask to evenly mix the stain throughout the agarose.

5. Prepare the gel mold to hold the agarose, then pour the cooled agarose into the mold and carefully insert a comb with eight teeth into the agarose to form an eight-well gel (**Figure 20.3**).

6. Wait for the gel to completely solidify; this will probably take about 20 to 30 minutes. Then submerge the gel in 1X TBE buffer in the electrophoresis gel box. Gently remove the comb. Flush the wells by drawing some of the TBE buffer into a pipette and then discharging it over the top of the wells. Do not pierce or tear the wells with the pipette tip.

7. Load each well with about 30 μl of a sample as follows:

Well 1: *E. coli* PCR mixture
Well 2: *B. cereus* PCR mixture
Well 3: DNA standard
Well 4: Soil colony 1 PCR mixture
Well 5: Soil colony 2 PCR mixture
Well 6: Soil colony 3 PCR mixture
Well 7: Soil colony 4 PCR mixture
Well 8: DNA standard

The GoTAQ master mix contains glycerol and loading dyes, so there is no need to add these components to the PCR samples before loading the gel. Glycerol increases the density of the DNA samples, which helps them sink through the buffer and settle evenly at the bottom of the wells. The dyes will tell you when the DNA is approaching the end of the gel during the electrophoresis run. If your DNA standards don't contain glycerol and loading dyes, then these should be mixed in with the standards before they are pipetted into wells 3 and 8.

8. After all the wells are loaded, place the electrode-containing lid on the gel box with the red (positive) electrodes at the far end of the gel (opposite

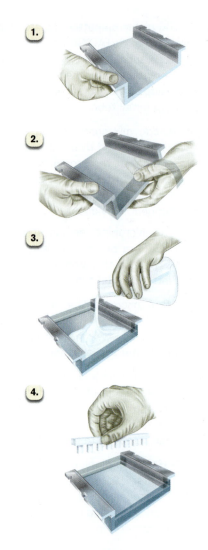

FIGURE 20.3 Pouring an agarose gel for electrophoresis.

the wells). Set the power supply for a constant voltage of 100 V, and turn on the power.

9. Once the power is turned on, observe the movement of the loading dyes toward the positive electrode. Two dyes (provided in the GoTAQ master mix) will separate as the current pulls them toward the positive electrode. The yellow dye moves at a much faster rate than the blue dye. Times can vary, but in an 8- to 10-cm gel, in most gel boxes, it will probably take about 30 minutes for the yellow dye to reach the end of the gel. At this point, any 575-bp fragments will be about halfway between the wells and the end of the gel.

Some of the CarolinaBLU DNA stain will be removed from the gel during the electrophoresis, but this will not affect the final results.

10. When the yellow dye reaches the far end of the gel, turn off the power supply, remove the gel

box lid, and move the gel to a staining box filled with the stain called Final CarolinaBLU Stain. The depth of the stain should be sufficient to completely cover the gel. To reduce the amount of stain needed, the box shouldn't be much wider or longer than the gel.

11. Stain the gel for 30–45 minutes, then move the gel to an empty staining box. Fill the box with deionized water to begin the destaining process.

12. After 15 minutes of destaining, pour out the destaining water, fill the box with deionized water again, and leave the gel for several more hours or overnight for further destaining.

13. After destaining, your instructor will wrap the gels in plastic wrap and hold them in a refrigerator or cold room until the next lab period.

Day 4

MATERIALS

- Gloves
- Metric rules
- Visible light box

PROCEDURES

Wear gloves. Stains stain.

1. After destaining overnight, pour out the destaining water, and move the gel to a visible light box. Or if your instructor has already completed the destaining process and stored the gel, then unwrap the gel and move it to a visible light box.

2. Turn on the light and measure the distance of each band from its well with a metric ruler. Record your results on the graph provided in the Results section. Take the top line of the graph as equivalent to the position of the wells, and draw a line for each band at the appropriate distance from the top.

   As an alternative, if a camera (such as your cellphone) is available, the gel can be photographed, and a printout of the photo can be taped to the Results page in the "Graph the location of all bands" section.

3. Note the size (in base pairs) of the bands using information provided by the DNA markers in wells 3 and 8. Record these data on the graph paper in the Results section just to the right of the lines for each band.

4. Rinse any stain off the metric ruler, and discard the gel in the receptacle designated for trash to be autoclaved.

Name: _____   Section: _____

Course: _____   Date: _____

APPEARANCE OF SOIL COLONIES CHOSEN FOR PCR

Describe the appearance of soil colony 1:

Describe the appearance of soil colony 2:

Describe the appearance of soil colony 3:

Describe the appearance of soil colony 4:

DNA BANDS

Record the location of all DNA bands for all lanes at the measured distances from the wells (wells are at the 0 cm line). In the case of the DNA standards, note the size of each fragment using information provided by your instructor.

	E. coli	B. cereus	DNA Standard	Soil colony 1	Soil colony 2	Soil colony 3	Soil colony 4	DNA Standard	
0 cm									0 cm
									0.5 cm
1 cm									1 cm
2 cm									2 cm
3 cm									3 cm
4 cm									4 cm
5 cm									5 cm
6 cm									6 cm
7 cm									7 cm
8 cm									8 cm

Name: _____   Section: _____

Course: _____   Date: _____

1. Assume that we start at the point in the PCR process where there is a single copy of double-stranded DNA with both strands exactly 575 base pairs in length (end of Cycle 3 above), and further assume that the PCR process works perfectly with every cycle.

   a. Given these assumptions, starting with just this single copy of double-stranded DNA, predict how many copies of this one copy could be produced after ten cycles of PCR.

   b. Again, given these assumptions, predict how many copies of this single copy of double-stranded DNA could be produced after twenty cycles of PCR.

   c. What was the effect of doubling the number of cycles from ten cycles to twenty cycles? That is, in this case, when the number cycles were increased two-fold, the number of DNA increased how many-fold? What does this tell you about exponential functions?

2. Consider the PCR results for *B. cereus* and *E. coli*.

   a. Did amplification occur in the case of *B. cereus*? What evidence supports this conclusion?

   b. Did amplification occur in the case of *E. coli*? What evidence supports this conclusion?

c. What do the results with *B. cereus* and *E. coli* tell you about the specificity of the primers, and how is this related to the possibility of using this method to identify *B. cereus*?

3. Which of your soil colonies are not likely to be *B. cereus* colonies? What PCR evidence supports your conclusion? Is the PCR evidence consistent with your observations of colony morphology?

4. Which of your soil colonies are likely to be *B. cereus* colonies? What PCR evidence supports your conclusion? Is the PCR evidence consistent with your observations of colony morphology?

5. Why might too much template DNA create a problem in this experiment? Remember that throughout all of the PCR cycles, primer molecules can continue to bind to the original template DNA—that is, the template DNA that we started with in cycle 1.

6. Say that there is a species of *Bacillus* that is distinct from *B. cereus* but still closely genetically related to *B. cereus*. If you performed PCR on a colony of this related species, what might you see on the gel? How could you adjust the conditions during the PCR run to improve your ability to distinguish between *B. cereus* and this closely related *Bacillus* species?

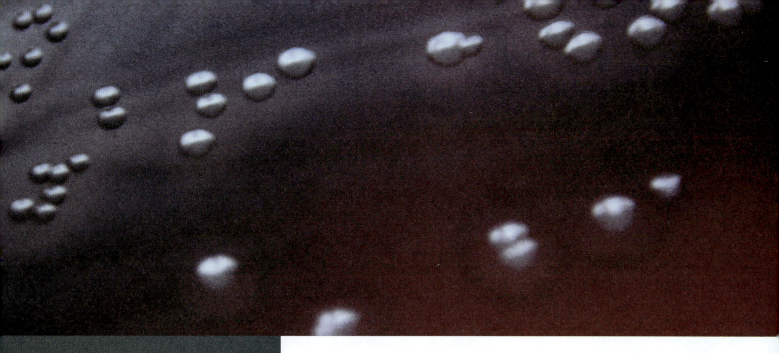

Enterococcus and Streptococcus bacteria are closely related genera that are found in many parts of the gastrointestinal and respiratory tracts. Although a number of species of these bacteria are commensal organisms in humans, notably, *E. faecalis* and *E. faecium*, they do cause opportunistic infections. Here is a close-up view of glistening *Enterococcus* colonies.

LAB 21

Streptococcus and *Enterococcus*

Learning Objectives

- Describe the traits that characterize the *Streptococcus* and *Enterococcus* genera and understand how these genera may be distinguished from *Staphylococcus* and *Micrococcus* through the use of assays such as the catalase test.

- Become familiar with and describe diseases caused by *Streptococcus* and *Enterococcus* species.

- Understand how hemolysis patterns and Lancefield group antigens are used to identify species within the *Streptococcus-Enterococcus* group.

- Use throat swab samples streaked onto blood agar plates to examine your own oral flora, including oral *Streptococcus* species.

Streptococcus and *Enterococcus* species are bacteria that are often found in the human body, especially in the oral cavity, in other parts of the upper respiratory tract, and in the large intestine **(Figure. 21.1)**. Most of these species were classified as *Streptococcus* species until close examination of their DNA revealed enough differences to justify separating them into *Enterococcus* and *Streptococcus*.

Common traits of these two genera include the following:

1. Cells are Gram-positive cocci.
2. Cells are often observed linked together in long chains (*strepto-*) because cells divide in one plane and often remain attached after cell division (Figure 21.1A); in some *Streptococcus* species, the cells are found in pairs (*diplo-*).
3. Cells do not produce endospores.
4. Cells are aerotolerant; they can grow in the presence of oxygen, but they do not use it. Instead, they rely exclusively on fermentation, producing lactic acid as the main end product. Cells grow better under reduced oxygen and elevated CO_2 concentrations (5%–10% CO_2 compared with 0.03% CO_2 in the atmosphere).

FIGURE 21.1 *Streptococcus* and *Enterococcus*.
A. *Streptococcus pyogenes* chains (stained red) in a human pus specimen.
B. *Enterococcus* sp. from a pneumonia patient. The bacteria are stained dark purple.

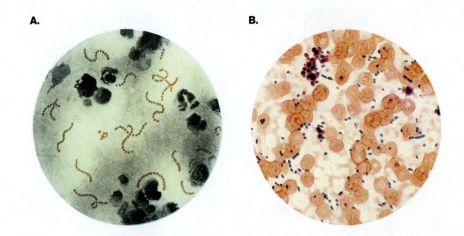

5. Cells show a negative response to the catalase assay, as is typical for cells that do not use oxygen. The catalase assay is especially useful for separating *Streptococcus* and *Enterococcus* from Gram-positive cocci in the *Staphylococcus* and *Micrococcus* genera.

DISTINGUISHING *STREPTOCOCCUS-ENTEROCOCCUS* FROM *STAPHYLOCOCCUS* AND *MICROCOCCUS*

The *Streptococcus-Enterococcus* group and the *Staphylococcus* and *Micrococcus* genera all include ubiquitous, human-borne Gram-positive cocci. It is important to be able to distinguish one group from the others in a clinical setting, because identifying the cause of a disease is an important step in its treatment. The methods described below are used to identify and differentiate genera and species of Gram-positive cocci.

Gram staining

Once the cells have been stained, these groups can be separated on the basis of cell growth patterns. *Streptococcus* and *Enterococcus* cells grow in chains or pairs, while *Staphylococcus* and *Micrococcus* cells grow in clusters and tetrads.

Catalase assay

The catalase assay detects the presence of a specific enzyme, catalase, which is an important oxygen-detoxifying enzyme (see Lab 14). The enzyme is essential to cells that carry out aerobic respiration, and it is found in strict aerobes and facultative anaerobes, but usually not in aerotolerant species or strict anaerobes. Catalase eliminates reactive hydrogen peroxide (H_2O_2), which is a reactive oxygen species and a by-product of oxygen use, by converting it to water and oxygen.

$$2\,H_2O_2 \xrightarrow{\text{catalase}} 2\,H_2O + O_2$$

Oxygen, obviously, is a gas. And when an aqueous solution of hydrogen peroxide is exposed to catalase, the oxygen generated will form gas bubbles in the hydrogen peroxide solution. So, the production of catalase by bacteria can be revealed by pouring a hydrogen-peroxide solution over the cells, and that's what's done in the catalase test.

Positive reaction: Bubbles in the peroxide solution When catalase-positive organisms are exposed to hydrogen peroxide, the oxygen gas produced will result in a "foaming" action, and the peroxide solution will appear to bubble vigorously (**Figure 21.2**). In the context of this lab, this result means the cells are either *Staphylococcus* or *Micrococcus*, and that they cannot be either *Streptococcus* or *Enterococcus*.

Negative reaction: No bubbles in the peroxide solution When a hydrogen peroxide solution is poured over catalase-negative cells, the peroxide will not be converted to water and oxygen gas. No oxygen gas means no foaming action, so the peroxide solution will not bubble vigorously. In the context of this lab, this result means the cells are either *Streptococcus* or *Enterococcus*, and that they cannot be either *Staphylococcus* or *Micrococcus*.

As an aside, it should be noted that humans are strict aerobes and that human cells are, predictably, catalase-positive. You can do a catalase test on yourself next time you happen to cut yourself (we recommend that you do not do this deliberately). Pour a dilute hydrogen peroxide solution (3%, available at most drugstores) into the wound and you will see the same vigorous bubbling that you will observe when hydrogen peroxide is added to *Staphylococcus* and *Micrococcus* cells. The bubbling occurs because your blood and most of your living cells produce a catalase enzyme that will generate oxygen gas and water from peroxide. In addition to providing hours of entertainment, this may actually have some benefit, as hydrogen-peroxide has a mild antiseptic effect, and more significantly, the foaming action is effective at removing microbes and debris from the injured area, a process called "debriding the wound."

IDENTIFYING SPECIES WITHIN THE *STREPTOCOCCUS-ENTEROCOCCUS* GROUP

Detecting hemolysis patterns with blood agar plates

When a *Streptococcus* species is suspected to be the cause of an infection, one of the first steps in identifying the bacterium is to streak the cells onto a blood agar plate to observe the effect the bacterial cells have on mammalian red blood cells. **Blood agar** is a nutrient-rich medium composed of peptones, yeast extract, and liver or heart extracts with 5% sheep's blood added to the medium after autoclaving. The medium is opaque or cloudy due to the presence of whole, intact red blood cells, and it is not very selective, as its nutrients can support the growth of a wide range of organisms. Blood agar is very valuable as a differential medium because bacterial species differ from one another in terms of how they break apart, or fail to break apart, red blood cells. The act of breaking red blood cells is called **hemolysis**, and the pattern of hemolysis produced by bacterial cells is useful in the differentiation and identification of *Streptococcus* and *Enterococcus* species.

Alpha hemolysis

Species that carry out **alpha (α) hemolysis** damage and cause a partial breakdown of the red blood cells in a blood agar plate (**Figure 21.3**, bottom streak). The opacity ("cloudiness") of the blood agar is due to the presence of red blood cells, and since the lysis of red blood cells is incomplete, the medium around the colonies remains opaque. While the medium stays cloudy, there is a distinct color change in the medium immediately surrounding the colonies due to the conversion of red-colored hemoglobin to a

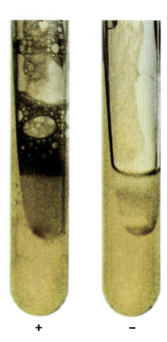

FIGURE 21.2 Results of the catalase assay.
The positive reaction on the left is produced by *Staphylococcus epidermidis*. The negative reaction on the right is produced by *Enterococcus faecalis*.

greenish-brown-tinted pigment called **biliveridin**. This same pigment is partly responsible for the greenish-brown color of bruises, which result from the partial breakdown of red blood cells in damaged tissues. The increased concentration of biliveridin produced by alpha-hemolytic species **changes the color of the medium adjacent to the colonies from bright red to opaque dark greenish brown**; for this reason, alpha hemolysis is also known as "green hemolysis." The exact shade and color of the alpha-hemolytic zone around the colonies depends on culture conditions, the species streaked, and the degree of crowding on the plate. So the tint of the hemolytic zone may be more brown than green, especially if some of the biliveridin is converted to a yellow-tinted pigment called **bilirubin**.

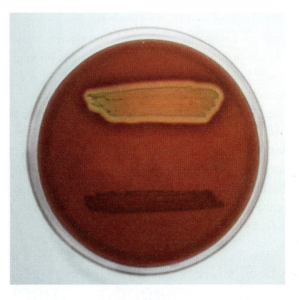

FIGURE 21.3 Beta-hemolysis.
Caused by *Streptococcus equisimilis*, beta-hemolysis is visible on the upper part of this blood agar plate, and **alpha-hemolysis** caused by *Streptococcus sanguis* can be seen on the lower part.

"Viridans" streps Most strep species found in the oral cavity are alpha-hemolytic. These numerous and diverse *Streptococcus* species are also referred to as the "viridans streps"; *viridis* is the Latin word for "green," as in the greenish color of alpha hemolysis. The viridans species are constantly present in our mouths in great numbers, so a typical throat swab of a healthy individual will produce large numbers of alpha-hemolytic colonies. Despite their abundance, they normally live on us without causing much harm, although there are a few exceptions. For example, *S. mutans* can convert sucrose into a gummy plaque that anchors the bacteria to the teeth, and causes tooth decay when the cells secrete lactic acid onto the surface of the teeth.

Streptococcus pneumoniae There is one major pathogenic strep species that is alpha-hemolytic: *Streptococcus pneumoniae*. Although it is alpha-hemolytic, *S. pneumoniae* is usually placed in a separate category or group apart from the viridans strep species. Cells of *S. pneumoniae* are usually arranged in pairs called diplococci, an exception to the rule that strep species grow in chains. Colonies of this species can be recognized or identified by their shiny, mucoid appearance, a product of the protein-polysaccharide capsule around the cells. *S. pneumoniae* can also be distinguished from other alpha-hemolytic streps by the fact that it is sensitive to the chemical optochin; alpha-hemolytic viridans streps are almost always resistant to optochin. It is important to properly identify *S. pneumoniae* because this species is the most common cause of pneumonia and a common cause of bacterial meningitis; these infections are referred to as pneumococcal infections.

Beta hemolysis

Species that carry out **beta (β) hemolysis** secrete toxic polypeptides called streptolysins, small proteins that lyse or split open and destroy red blood cells completely and totally. This reaction produces a **clear, transparent, slightly yellow-tinted zone of clearing** around the colonies (**Figure 21.3**, top streak). Most of the common, serious-disease-causing species in the genus *Streptococcus* are beta-hemolytic (an exception is *S. pneumoniae*), and most of the common beta-hemolytic streps are pathogens. So, generally speaking, "b" or "B" = BAD. One beta-hemolytic pathogen, *Streptococcus pyogenes*, is definitely worth further mention.

Streptococcus pyogenes *Streptococcus pyogenes* is a causative agent of a specific type of pharyngitis called **strep throat**. The prefix "pyo-," as in *pyogenes*, is derived from the Greek *puo-* or *puon*, meaning "pus," and "pyogenes," literally translated, means "pus-generating." The formation and appearance of pus in the throat and on the tonsils is a useful way to distinguish an infection caused by *S. pyogenes* from a more common one caused by a virus. It is very important to identify and treat strep throat cases because these infections occasionally lead to more serious complications, such as **scarlet fever** and **rheumatic fever**. Rheumatic fever can cause permanent damage to the heart and may be fatal. In addition, *S. pyogenes* can cause life-threatening blood infections, wound infections, and pneumonia. Jim Henson, the creator of the Muppets, died in 1990 at the age of 54 of pneumonia caused by *S. pyogenes*.

Gamma hemolysis

Gamma-hemolytic species do not lyse red blood cells, so **gamma (γ) hemolysis** equals no hemolysis. No clear zones appear around the colonies, and the **medium remains bright red and opaque (Figure 21.4)**. Some of the normal oral strep species

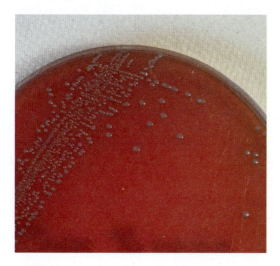

FIGURE 21.4 Gamma-hemolysis.
Caused by *Staphylococcus epidermidis*, Gamma-hemolysis result on a blood agar plate.

are gamma-hemolytic, and this trait is normally associated with nonpathogenic species. In addition, many of the gut-dwelling "fecal streps," now classified as *Enterococcus* species, are usually gamma-hemolytic.

Enterococcus faecalis *E. faecalis*, which is found in the large intestine or feces, is one of the *Enterococcus* species once classified as *Streptococcus*. Strains of this species are typically gamma-hemolytic, but some are capable of alpha hemolysis, so the hemolysis pattern for the species is somewhat variable. *E. faecalis* cells are usually found at high densities in the large intestine, and if found in drinking water, may be taken as an indicator of fecal contamination of the water. This species is considered to be a part of the normal flora of the large intestine, and under most conditions, *E. faecalis* does not cause human disease. However, if the cells relocate to the urinary tract, then this species may cause urinary tract infections such as cystitis (bladder infection). Trauma that ruptures the large intestine may enable *E. faecalis* cells to gain entry into the bloodstream, travel to and colonize the heart valves, and cause a life-threatening **endocarditis** (inflammation of the inner lining of the heart).

Lancefield Groups

Rebecca Lancefield

Hemolysis patterns are very useful for a preliminary sorting of *Streptococcus* and *Enterococcus* species into alpha-, beta-, and gamma-hemolytic species. But there are hundreds of species in these two genera, so there was a need for a system that could further subdivide the genera into multiple, distinct groups. In the 1920s, Dr. Rebecca Lancefield, a microbiologist at the Rockefeller Institute for Medical Research, developed an identification system based on the observation that different types of *Streptococcus* carry different types of carbohydrates—specifically, different types of sugars and amino sugars—on the surface of their cells. Each type of surface "marker" has a unique three-dimensional shape, and these unique shapes can act as unique antigens. Lancefield used letter names to differentiate the various groups of *Streptococcus* strains associated with specific types of antigens. Group A strains carry group A antigens, group B strains carry group B antigens, and so on; the list of Lancefield groups now goes up to group U.

To distinguish one type of *Streptococcus* from another, Lancefield and her colleagues at the Rockefeller Institute injected a given type of antigenic carbohydrate from a given type of *Streptococcus* into mammals. The animals produced antibodies that would bind specifically to the particular carbohydrate marker found on the surface of that particular strain of *Streptococcus*. Because each carbohydrate surface marker is specific, the antibodies would not bind to antigens found on other types of strep. For example, anti–group A antibodies would bind only to group A antigens and would react only with or against cells that were group A cells. This specificity allowed unidentified *Streptococcus* cells to be easily identified when screened with a panel of different anti-group antibodies.

After the Lancefield system based on group-specific differences in antigens was developed, it was found that some of the groups were well correlated with a single *Streptococcus* species name. For instance, the term "group B streps" describes a group of related strains that are also known as *Streptococcus agalactiae*. Another important example is the group A streps, a group of related strains in which the cells carry a group A antigen composed of the sugar rhamnose and the amino sugar N-acetylglucosamine. The group A strains are basically the same set of strains that we also call *Streptococcus pyogenes*, so group A strep and *Streptococcus pyogenes* are essentially synonymous terms. In addition to possessing group A antigens, these strains can be identified by their sensitivity to the antibiotic bacitracin, which is assessed by applying an "A disk" ("A" as in group A strep) to blood agar plates. Almost all other types of *Streptococcus* are resistant to bacitracin.

In other cases, microbiologists discovered that a given Lancefield group designation encompassed many different species of *Streptococcus*; for example, the group D streps are a collection of dozens of different species. All of the species originally placed in group D carry the group D antigen on the surface of their cells, and all were once classified as *Streptococcus* species. However, DNA studies showed that many of the large intestine–adapted or gut-dwelling members of this group were genetically different enough to warrant placement in a separate genus called *Enterococcus*. *E. faecalis* and *E. faeceum* are two widespread group D species now classified as "enterococci" or *Enterococcus* species. While some group D species are now in a different genus, there are still many group D "non-enterococci" that remain in the genus *Streptococcus*.

While the Lancefield system was successful at dividing most of the beta-hemolytic species into several subgroups, it turned out that many of the alpha- and gamma-hemolytic *Streptococcus* species lacked the carbohydrate antigens used in this system. So, there is no Lancefield group designation for species such as *Streptococcus pneumoniae* or for species in the viridans group.

Lancefield herself went on to a 60-year career in microbiology, during which time, she would author or co-author more than 50 scientific papers. Her research included further study of Group A streptococci and rheumatic fever as she continued to work at the Rockefeller Institute, and later, as a professor at Columbia University. She was elected to the National Academy of Science in 1970, and died in 1981 at the age of 86.

SUMMARY

Bacteria Specics	Hemolysis Pattern	Lancefield Group	Main Location in Body	Diseases and Significance
"Viridans" streptococcus species	Alpha (greenish brown)	No Lancefield antigens	Mouth, upper respiratory	Normal oral flora, rarely causes disease except tooth decay by *S. mutans*
*Streptococcus pneumoniae**	Alpha (greenish brown)	No Lancefield antigens	Upper respiratory tract	Common cause of pneumonia, meningitis ("pneumococcal diseases")
Streptococcus pyogenes	Beta (clear)	Group A	Upper respiratory tract	Strep throat, scarlet fever, rheumatic fever, pneumonia, septicemia (blood infection)
Streptococcus agalactiae	Beta (clear)	Group B	Large intestine, vagina	Neonatal infections, including septicemia (blood infection), meningitis, and pneumonia
Enterococcus faecalis	Gamma (red) or Alpha	Group D	Large intestine	Normal intestinal flora, usually harmless, may cause urinary tract infections, endocarditis

*S. pneumoniae can be distinguished from other alpha-hemolytic *Streptococcus* species by the optochin test. *S. pneumoniae* is sensitive to optochin while alpha-hemolytic *Streptococcus* viridans species are almost always resistant to this chemical.

PROTOCOL

> **IMPORTANT:** *S. pyogenes* and certain other oral *Streptococcus* species are biosafety level 2 species, so gloves and safety glasses must be worn throughout both days of this lab.

Day 1

Materials

- 2 pairs of safety glasses
- 2 pairs of disposable gloves
- 2 yeast extract tryptone (YET) broths
- 2 YET slants
- 2 blood agar plates
- 2 sterile swabs
- Beaker with disinfectant

Cultures:

- *Enterococcus faecalis*
- *Staphylococcus epidermidis*

CATALASE ASSAY OF KNOWN SPECIES

1. Use a sterile loop to inoculate one YET broth and one YET slant with *E. faecalis*.

2. Inoculate one YET broth and one YET slant with *S. epidermidis*.

THROAT SWAB ON BLOOD AGAR

1. Use a sterile cotton swab to swab the back of your throat and/or along your gum lines.

2. Use the swab to inoculate a quarter of a blood agar plate in a manner similar to the initial streaking of a plate when you wish to streak for isolated colonies **(Figure 21.4A)**.

3. Discard the swab in a beaker of disinfectant or directly into the receptacle designated for trash to be autoclaved.

Do not put the used swab down on the benchtop. Do not put the used swab in the regular trash.

A.

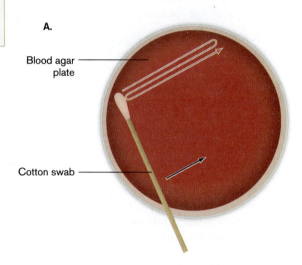

Blood agar plate

Cotton swab

B.

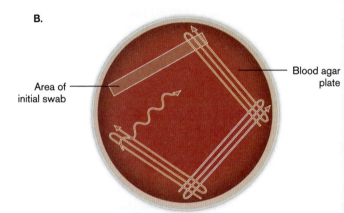

Area of initial swab

Blood agar plate

FIGURE 21.4 The swab-streak method.

4. Using your sterile inoculating loop, finish streaking the blood agar plate as you would do if your goal was to produce isolated colonies **(Figure 21.4B)**.

5. Incubate all broths, slants, and plates at 37°C for 24 hours.

Day 2

Materials

- YET broths and slants from day 1
- Blood agar plates from day 1
- Bottle of 3% hydrogen peroxide (H_2O_2)
- Demonstration plates showing alpha, beta, and gamma hemolysis

CATALASE ASSAY OF KNOWN SPECIES

Add about five drops of H_2O_2 to each of the two YET broths and the two YET slants, and record the reaction (bubbles/no bubbles) for *E. faecalis* and *S. epidermidis*.

DEMONSTRATION PLATES: DESCRIPTION OF HEMOLYSIS PATTERNS

Examine the demo plates to observe the different types of hemolysis patterns. Record your observations.

THROAT SWAB ON BLOOD AGAR

1. Using two small pieces of tape, secure the lid of the streaked blood agar plate to the base of the petri dish.

2. **Without opening the plate at any time,** examine the blood agar plate you streaked with your throat/mouth swab and describe the plate in the Results section. Note various colony types and the hemolysis patterns associated with the different colonies. Again, do not open this plate at any time.

> **IMPORTANT:** If you see more than a few beta-hemolytic colonies, **be sure the plate is taped shut** and take it to your lab instructor. It's not unusual to see a few beta-hemolytic colonies; these are mostly *S. aureus* or *Bacillus* colonies. But large areas of beta hemolysis could mean *S. pyogenes* and strep throat.

3. Discard the throat swab plates in the receptacle designated for trash to be autoclaved.

RESULTS

Name: _____ Section: _____

Course: _____ Date: _____

CATALASE ASSAY

Species and Medium	Reaction (Bubbles/No Bubbles)	Catalase (Positive or Negative?)
E. faecalis in YET Broths		
E. faecalis on YET Slants		
S. epidermidis in YET Broths		
S. epidermidis on YET Slants		

DEMONSTRATION PLATES: DESCRIPTION OF HEMOLYSIS PATTERN

Alpha hemolysis plate:

Beta hemolysis plate:

Gamma hemolysis plate:

THROAT SWAB ON BLOOD AGAR:

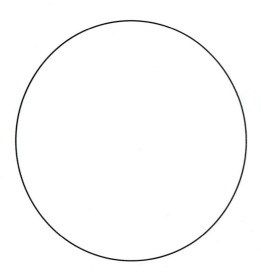

Written description: (Describe the different colony types seen on the plate, and for each colony type, be sure to include information about the relative size, color, abundance and hemolysis pattern—alpha, beta, or gamma.)

Name: _____   Section: _____

Course: _____   Date: _____

1. Did it matter whether the cells for the catalase test were grown in broth or on slants? That is, did the medium affect the catalase reaction for a given species? Why or why not?

2. What was the most common hemolysis pattern for the colonies on your plate? Was this what you expected? Why or why not?

3. After seeing what grew on the blood agar plate, are you concerned about the types of *Streptococcus* species that are living in your mouth? Why or why not?

4. If you found a beta-hemolytic colony on your plate, and you were concerned that it might be *S. pyogenes*, would you feel better or worse if the colony bubbled when you added hydrogen peroxide? Explain your choice.

Staphylococcus and *Micrococcus* bacterial genera are salt-tolerant bacteria found on the human body, notably on our skin. Some of the odor we associate with our sweat is thanks to *Micrococcus*, who metabolize chemicals in sweat, producing smelly products.

Learning Objectives

- Describe the traits that characterize the *Staphylococcus* and *Micrococcus* genera, and understand how these genera can be distinguished from the *Streptococcus-Enterococcus* group through cell morphology, the catalase assay, and salt tolerance.

- Become familiar with and describe diseases caused by *Staphylococcus* and *Micrococcus* species.

- Understand how fermentation, coagulase assays, and novobiocin resistance are used to distinguish *Staphylococcus* cells from *Micrococcus* cells and to identify species within the *Staphylococcus* genus.

- Use skin swab samples streaked onto mannitol salt agar plates and various assays to examine, isolate, and identify skin flora.

LAB 22

Staphylococcus and *Micrococcus*

Staphylococcus and *Micrococcus* are classified in two different Gram-positive phyla due to differences in the guanine-plus-cytosine content of their genomes. In this lab, however, we will consider these genera together because they share many morphological traits, and because salt-tolerant species from both genera can be found on the human body, especially the skin.

Common traits of these two genera include the following:

1. Cells are Gram-positive cocci.
2. Cells remain attached after cell division, so they are often observed in tetrads (*Micrococcus*) or clusters (*Staphylococcus*) **(Figure 22.1)**.
3. Cells do not produce endospores.
4. Cells are adapted to living on human skin, which is very salty. So, species of *Staphylococcus* and *Micrococcus* can tolerate high salt concentrations; specifically, they can grow on media containing 7.5% NaCl.

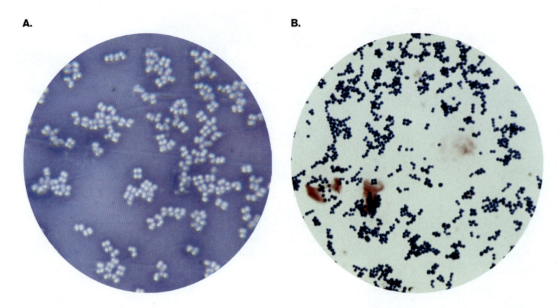

A.

B.

FIGURE 22.1 *Micrococcus* and *Staphylococcus*.
A. *Micrococcus mucilaginosis,* negatively stained and viewed under bright-field microscopy. Note the many tetrads of cells.
B. *Staphylococcus epidermidis,* stained and viewed under bright-field microscopy. Almost all of the cells are clumped together in bunches.

5. Cells use oxygen in aerobic respiration, a process that breaks down sugar to CO_2 and H_2O and releases the energy needed to synthesize ATP. *Micrococcus* cells are strictly aerobic and *Staphylococcus* cells are facultative anaerobes.

6. Cells show a positive response to the catalase assay, as is typical for cells that can use oxygen in aerobic respiration. As noted in Lab 21, the catalase assay is especially useful for separating *Staphylococcus* and *Micrococcus* from Gram-positive cocci in the *Streptococcus* and *Enterococcus* genera.

IMPORTANCE OF *STAPHYLOCOCCUS* AND *MICROCOCCUS* SPECIES

Staphylococcus species

Staphylococcus aureus Wild-type *S. aureus* colonies are golden yellow in color (in Latin, *aurum* and *aureus* mean "gold" and "golden"), especially on blood agar, where the colonies are also often beta-hemolytic (lab strains tend to be paler in color). This species is easily the biggest, baddest pathogen in the *Staphylococcus* genus. Infections caused by *S. aureus* include serious skin infections, such as boils and carbuncles; wound infections; **septicemia** (blood infections); and **pneumonia**. Some strains produce a toxin, called the **TSST-1 toxin** or toxic shock syndrome toxin, that triggers a fever and causes a potentially fatal drop in blood pressure. And when *S. aureus* cells grow in certain types of food, they can cause food poisoning, because the cells secrete an enterotoxin that causes vomiting and diarrhea.

S. aureus infections can spread rapidly in hospitals because the cells can be carried easily from one patient to another on the hands and clothing of people moving between patients. If a patient's immune system is functioning below normal effectiveness, that patient may be more vulnerable to infections than the average person. In addition, hospitals are home to many antibiotic-resistant strains of this pathogen, which are difficult to treat. In particular, methicillin-resistant *S. aureus*, or **MRSA** (pronounced "mer-sa"), strains have become a life-threatening problem in many clinical settings.

Outside of a hospital setting, most *S. aureus* infections are probably **autoinfections**, or infections from a person's own *S. aureus* cells. That is because about 10%–20% of healthy individuals carry this pathogen in their nasal passages. These cells usually don't cause trouble until a break in the skin or a weakening of the immune system gives them an opportunity to cause infections. Cells of this species can survive outside of the body for a long time, so people can also become infected when they come in contact with surfaces, such as sweat-soaked gym equipment, that are contaminated with *S. aureus* cells. Yuck. So, disinfect the equipment before and after use.

Staphylococcus epidermidis *S. epidermidis* is probably the most common type of *Staphylococcus* found on human skin, and almost all of us carry huge numbers of this type of bacteria on our bodies. This species usually does not cause disease unless it is introduced into the body by a catheter or other medical device. When this happens, *S. epidermidis* can be an opportunistic pathogen, and it has been known to cause infections of heart valves.

Staphylococcus saprophyticus *S. saprophyticus* may also be found on the skin, and it is a part of the normal microflora of the female genital tract. This species is not likely to be the source of a life-threatening disease, but it can be a cause of urinary tract infections.

Micrococcus species

Micrococcus species are often found on the skin and in the soil. Many species are pigmented; *M. luteus*, for example, produces bright yellow colonies, a fact reflected in the species name, *luteus*, which is one of several Latin words for yellow (*flavus* is another). These species are usually not considered to be pathogenic, although they may cause opportunistic blood infections in immunosuppressed patients.

DISTINGUISHING *STAPHYLOCOCCUS* AND *MICROCOCCUS* FROM THE *STREPTOCOCCUS-ENTEROCOCCUS* GROUP

Staphylococcus, *Micrococcus*, and the *Streptococcus-Enterococcus* group all include ubiquitous, human-borne Gram-positive cocci. It is important to be able to distinguish one group from the others in a clinical setting, because this is valuable in disease treatment. The methods described in this lab are used to identify species in the *Staphlylococcus* and *Micrococcus* genera.

Gram staining

Once the cells have been stained, these groups can be separated on the basis of cell growth patterns. *Staphylococcus* and *Micrococcus* cells grow in clusters and tetrads, respectively, and *Streptococcus* and *Enterococcus* cells grow in chains or pairs.

Catalase assay

The catalase assay is described in Lab 21. *Staphylococcus* and *Micrococcus* species are catalase-positive and *Streptococcus* and *Enterococcus* species are catalase-negative (see Figure 21.2).

Salt tolerance

Staphylococcus and *Micrococcus* species are adapted to the skin, so cells in these genera are very salt tolerant. If you wish to select for species in these genera or inhibit the growth of species in other genera, including *Streptococcus* and *Enterococcus*, you can use an agar-based medium with a high salt concentration. On such media, *Staphylococcus* and *Micrococcus* species will grow and form colonies, while *Streptococcus* and *Enterococcus* species, as well as many other types of bacteria, will grow poorly or not at all.

One example of such a medium is **mannitol salt agar**, a highly selective medium that contains the following components (per liter):

Proteose peptone	10.0 g
Beef extract	1.0 g
D-Mannitol	10.0 g
Sodium chloride	75.0 g
Agar	15.0 g
Phenol red	25.0 mg

The key selective component is, of course, sodium chloride. Mannitol salt agar is a selective medium that favors the growth of *Staphylococcus* and *Micrococcus* because it contains 7.5% NaCl (75 g NaCl per 1,000 ml). It is also a differential medium because it contains the sugar **mannitol** and the pH indicator phenol red. These components allow us to test isolates for their ability to ferment mannitol, a trait that helps to identify or differentiate species within the *Staphylococcus* genus.

DISTINGUISHING *MICROCOCCUS* FROM *STAPHYLOCOCCUS*

While they are in different phyla and are not considered closely related, species in the genera *Micrococcus* and *Staphylococcus* have many traits in common. So, how can we tell species from these two different genera apart?

Differences in ability to ferment sugars

Perhaps the simplest way to tell these two genera apart is to focus on differences in how they handle the absence of oxygen. *Micrococcus* species are strict aerobes: they must have oxygen to carry out aerobic respiration, which is essentially the only way in which these cells can produce enough ATP to grow and divide. In the absence of oxygen, *Micrococcus* cells cannot switch to fermentation or produce organic acids from sugars. By contrast, *Staphylococcus* species are facultative anaerobes: when oxygen is present, they will employ aerobic respiration, but in the absence of oxygen, *Staphylococcus* cells switch to fermentation and produce organic acids from sugars. So, all we have to do is check a given isolate for the ability to ferment sugars, and we can distinguish *Micrococcus* (nonfermenters) from *Staphylococcus* (fermenters).

Detecting fermentation: Phenol red carbohydrate assays

Phenol red carbohydrate assays are discussed in detail in Lab 10. Remember:

- If the broth has a bright yellow color, this is considered a positive reaction for the fermentation of a carbohydrate to organic acids.

- Any orange to red color is considered negative for the fermentation of a carbohydrate to acids.

IDENTIFYING SPECIES WITHIN THE GENUS *STAPHYLOCOCCUS*

There are many *Staphylococcus* species, but only a few are common. The common "staph" species can be differentiated from one another on the basis of several characteristics:

1. Fermentation of mannitol
2. Resistance to the antibiotic novobiocin
3. Production of coagulase

Mannitol fermentation

Recall that mannitol salt agar is both selective and differential. As described above, this medium is selective for growth of *Staphylococcus* and *Micrococcus* species because it contains 7.5% NaCl, which inhibits the growth of many other types of bacteria. The presence of the sugar mannitol and phenol red in this medium also allows us to determine an isolate's ability to ferment mannitol. The principle of the assay is the same as with phenol red carbohydrate broths (see Lab 10), but in this case, the medium is a solid, agar-based medium in a plate instead of a liquid medium in a tube.

Positive reaction If cells growing on mannitol salt agar ferment mannitol, then they will produce organic acids as end products. The acids will diffuse into the medium, lowering the pH of the agar in the immediate vicinity of the colonies to a point well below pH 6.8. At these pH levels, there are "excess" protons (H^+), and a high percentage of the phenol red molecules are uncharged and yellow in color. So, the medium around the colonies will be yellow; that is, there will be a yellow zone around and extending out from any colonies containing cells that can ferment the sugar mannitol **(Figure 22.2)**. The size of the yellow zone will depend on the amount of organic acids produced and the rate at which the acids diffuse.

If the medium around the colonies has a bright yellow color, this is considered a positive reaction for the fermentation of mannitol to organic acids.

In the context of identifying *Staphylococcus* and *Micrococcus*, a positive reaction means that the cells could be *S. aureus* or *S. saprophyticus*; (some *S. saprophyticus* strains do not ferment mannitol, so ask your lab instructor if you have any doubts). A positive reaction eliminates the possibility that the cells could be *S. epidermidis* or any *Micrococcus* species.

Negative reaction If the cells growing on the plate do not ferment mannitol, then no organic acids are produced, and the medium around the colonies remains at pH 6.8 or higher. At these pH levels, a high percentage of the phenol red molecules have lost a proton and are negatively charged, and the medium around the colonies is orange-red to pink-red in color.

If the medium immediately adjacent to the colonies is orange-red to pink-red in color—that is, if there are no yellow zones at all—this is considered a negative reaction for the fermentation of mannitol to organic acids.

FIGURE 22.2 Mannitol salt agar results.
The yellow bacterial colonies on the upper part of the
plate are mannitol fermenters. The colonies on the lower
part of the plate do not ferment mannitol.

A negative reaction does not necessarily mean that the cells didn't use the sugars at
all. Remember, if the sugars were used in an aerobic respiratory pathway, then there
won't be any organic acids accumulating in the agar. And if the respiring cells also used
the amino acids derived from the peptones in the agar as an additional energy source,
the amino group ($-NH_2$) will be converted to as an ammonia (NH_3) molecule, the alkaline ammonia will actually raise the agar pH, and the color may be a deeper red than
the orange-red of the uninoculated medium. But don't forget, this is a fermentation
assay, so this is still considered a negative result for this test.

In the context of identifying *Staphylococcus* and *Micrococcus*, a negative reaction
means that the cells could be either *S. epidermidis* or any of a number of *Micrococcus*
species, because neither *S. epidermidis* nor any of the *Micrococcus* species ferment mannitol or produce organic acids from the use of mannitol. In fact, as strict aerobes, all
Micrococcus species will be negative for any fermentation assay. A negative reaction also
eliminates *S. aureus* and some strains of *S. saprophyticus*.

Novobiocin sensitivity and resistance

Novobiocin is an antibiotic that is effective against some, but not all, *Staphylococcus*
species. As a result, *Staphylococcus* species may be separated and differentiated by their
response to the drug. In particular, *S. saprophyticus* is resistant to novobiocin, while *S.
epidermidis* and *S. aureus* cells are sensitive to it. In addition, *Micrococcus* species are
sensitive to novobiocin. So, this assay is particularly useful for separating or differentiating *S. saprophyticus* from other *Staphylococcus* and *Micrococcus* species.

A given isolate's reaction to novobiocin can be assessed with a method similar to the
Kirby-Bauer assay used in Lab 17. Cells of an unknown bacterium are streaked onto a
plate, and a paper disk containing 5 micrograms of novobiocin is added to the streak.

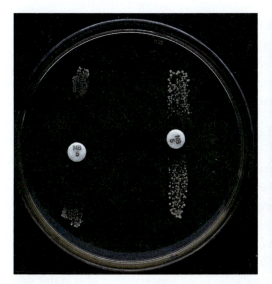
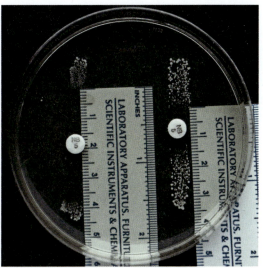

FIGURE 22.3 Novobiocin assay.
In both plates, *S. epidermidis* (left-hand streak) is sensitive to the antibiotic (zone of inhibition = 37 to 38 mm) and *S. saprophyticus* (right-hand streak) is resistant to the antibiotic (zone of inhibition = 14 to 15 mm).

The drug diffuses into the medium, and **if the cells are sensitive to the antibiotic, it will form a zone of inhibition of at least 18 to 20 mm in diameter (Figure 22.3)**. Resistant cells will be able to grow within a zone or circle that is 18 to 20 mm in diameter across the center of the disk. In this lab, such cells will be *S. saprophyticus* cells.

Coagulase assay

Staphylococcus aureus, the major pathogen of the *Staphylococcus* genus, has many traits—known as **virulence factors**—that enhance its ability to cause disease. Some of these are secreted toxins, such as the TSST-1 toxin and the enterotoxins that cause vomiting and diarrhea. Other virulence factors include two types of **coagulases**: a bound or cell-surface coagulase that is also known as "clumping factor," and a free or secreted coagulase.

1. The **bound coagulase** or **clumping factor** acts as a receptor for **fibrinogen**, a soluble blood plasma protein normally involved in blood clotting. When this coagulase latches onto fibrinogen, it can cause clumping of blood plasma proteins and promote the attachment of *S. aureus* cells to blood clots, traumatized tissue, and implanted medical devices such as catheters.

2. The free or **secreted coagulase** catalyzes a reaction that converts fibrinogen into insoluble, "stringy" **fibrin** protein. Normal blood clotting mechanisms involve a number of clotting factors found in the bloodstream that interact sequentially to produce fibrin from fibrinogen. Fibrin forms a meshwork that traps platelets and other blood cells to form a clot to stop bleeding. However, when *S. aureus* cells are present, the fibrin that they create from fibrinogen is for the benefit of the bacteria. The fibrin coats the bacteria, and this may protect the cells from certain parts of the immune system, including microbe-destroying phagocytic white blood cells.

The bottom line is this: Coagulases are a part of what makes *S. aureus* a potent pathogen. Not only is *S. aureus* the only major human pathogen in the genus *Staphylococcus*, but probably not coincidentally, it is also the only species that carries coagulases, so staph isolates that are coagulase-positive are almost always *S. aureus*. (We say "almost always," because biology is a science of exceptions.)

Several coagulase detection methods have been developed, but among the most rapid are assays that use colored microscopic latex beads coated with fibrinogen to detect bound coagulase. The coated beads can be of any color; as it happens, the Staphyloslide kit used in this lab contains blue beads. If bound coagulase is present in the liquid containing the fibrinogen-coated beads, it will interact with the fibrinogen to link the beads together.

Positive reaction When a loopful of coagulase-positive cells is mixed with a reagent containing blue fibrinogen-coated latex beads on a cardboard card, the bound coagulase on the cell surface acts as a fibrinogen receptor and interacts with the fibrinogen on the beads. The bacteria's coagulase links blue latex beads to bacterial cells to more beads to more cells to more beads to more cells … and the net effect is to clump, or aggregate, the blue particles. As the size of the bead-cell clusters increases, the reagent changes in appearance from a homogeneous, opaque blue suspension of beads to a "grainy" suspension with visible blue clumps in a nearly colorless liquid after 30–60 seconds of mixing **(Figure 22.4)**.

So, a positive reaction for the coagulase assay is the appearance of blue grains suspended in a colorless liquid within 30–60 seconds of mixing.

If an unidentified or unknown *Staphylococcus* isolate produces a positive coagulase test, then it is no longer unidentified. It's *S. aureus*.

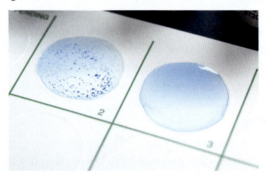

FIGURE 22.4 Coagulase assay results.
Coagulase assay results: The positive reaction (left) shows visible "grains" of clumped blue beads while the negative reaction (right) shows a homogeneous suspension of blue beads.

Negative reaction When a loopful of coagulase-negative cells is mixed with a drop of coagulase reagent on a cardboard card, there is no bound coagulase present on the bacterial cell surface to interact with the fibrinogen on the beads. Without coagulase present, blue latex beads and bacterial cells will not be linked together to produce clumps of cells and beads. The appearance of the reagent will remain unchanged; it will remain a homogeneous, opaque blue suspension of beads without any visible blue clumps after 30–60 seconds of mixing (Figure 22.4).

So, the absence of visible blue aggregates or the retention of a homogeneous, opaque blue appearance after 30–60 seconds of mixing is a negative reaction.

Caution: False positive results Note the importance of time in this assay. When coagulase is present on the surface of the bacterial cells, aggregation occurs within 30–60 seconds, so the assay should be read within a minute or two of the mixing of cells and reagent. If you wait too long to make your observations, then the liquid in the drop of reagent will evaporate. With no liquid to keep them suspended, the beads will have no choice but to stick together, producing visible clumps. So, even if the cells are actually coagulase-negative, you may see blue grains on the card, and you will come to the false conclusion that the cells are coagulase-positive.

SUMMARY

The summary table shows the typical results of the tests described in this lab for various bacterial species. Based on the tests that you will run in this lab, **Figure 22.5** shows you the steps for identifying various clinically important Gram-positive bacteria. In this lab, you'll attempt to identify bacteria in an unknown culture provided by your instructor and in an environmental sample. You will also run the tests using several known organisms so that you can see what the expected results look like. The first two lab periods are devoted to isolating a salt-tolerant pure culture that is also unable to ferment mannitol. The remaining lab periods are spent setting up all of the tests and analyzing the results. Have fun!

Bacterial species	Catalase	Salt Tolerance (on mannitol salt agar)	Mannitol Fermentation (on mannitol salt agar)	Glucose Fermentation (on phenol red broth)	Novobiocin	Coagulase
Micrococcus luteus	+	+	−	−	Sensitive	−
Staphylococcus epidermidis	+	+	−	+	Sensitive	−
Staphylococcus saprophyticus	+	+	+	+	Resistant	−
Staphylococcus aureus	+	+	+	+	Sensitive	+
Streptococcus species	−	−	N/A	+	N/A	N/A
Enterococcus species	−	−	N/A	+	N/A	N/A

PROTOCOL

Day 1

Materials

- Mannitol salt agar plates (1 per student)
- Sterile cotton swabs (1 per student)
- Sterile saline
- Beaker with disinfectant

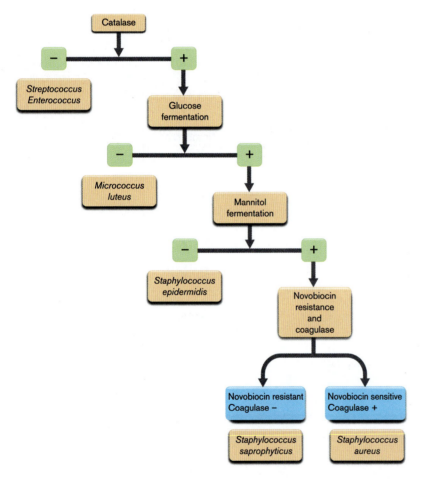

FIGURE 22.5 Steps for distinguishing clinically important Gram-positive bacteria.

SKIN SWABS ON MANNITOL SALT AGAR PLATES

1. Moisten a sterile cotton swab with a little sterile saline, and use it to swab a patch of **unbroken, uninfected** skin that is usually damp, dark, and unwashed . . . like behind your ears. We emphasize, **do not swab cut or broken skin**.

2. Use the swab to inoculate a quarter of a mannitol salt agar plate in a manner similar to the initial streaking of a plate when you wish to streak for isolated colonies.

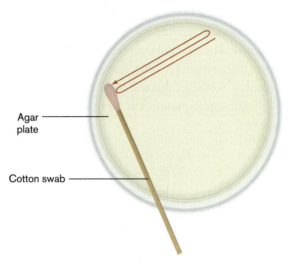

3. Discard the swab in a beaker of disinfectant or directly into the receptacle designated for trash to be autoclaved.

   Do not put the used swab down on the bench top. Do not put the used swab in the regular trash.

4. Using a sterile inoculating loop, finish streaking the mannitol salt agar plate with your goal being to produce isolated colonies.

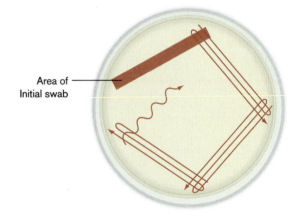

5. Incubate the plates at 37°C for 48 hours.

Day 2

> **IMPORTANT:** *S. aureus* is a biosafety level 2 species, so gloves and safety glasses must be worn throughout the Day 2 activities.

Materials
- Safety glasses
- Disposable gloves
- Tubes of nutrient broth (1 per student)
- Inoculated mannitol salt agar plates from day 1

SKIN SWABS ON MANNITOL SALT AGAR PLATES

1. Examine the mannitol salt agar plate streaked with your skin swab, and describe this plate in the Results section. Note the various colony types and the color of the medium associated with the different colonies.

2. Pick an isolated mannitol fermentation–negative colony (the color of the medium around the colony should be pink or red), and use a sterile loop to transfer this colony to a tube of nutrient broth.

> **IMPORTANT:** Do not pick colonies surrounded by medium that is yellow in color; these colonies are positive for mannitol fermentation and could be *S. aureus* colonies, and we do not want to culture wild-type *S. aureus*.

3. Incubate the broths at 37°C for 48 hours.

Day 3

> **IMPORTANT:** *S. aureus* is a biosafety level 2 species, so gloves and safety glasses must be worn throughout the Day 3 activities.

Materials
- 2 nutrient agar plates
- 4 nutrient agar slants
- 3 mannitol salt agar plates
- 4 phenol red glucose broths
- 4 novobiocin antibiotic disks
- Safety glasses
- Disposable gloves

Cultures:
- *Escherichia coli*
- *Micrococcus luteus*
- *Staphylococcus aureus*
- *Staphylococcus epidermidis*
- *Staphylococcus saprophyticus*
- *Streptococcus mitis*
- Broths inoculated with skin isolates on day 2
- Unknown Gram-positive bacterial isolates

CATALASE AND COAGULASE ASSAYS

1. Label two nutrient agar slants with your name. Then label one of the two tubes *S. aureus* and the other *S. epidermidis*. Inoculate the slants with the correct pure cultures.

2. Select one Gram-positive culture from the set of numbered Gram-positive unknowns. Record the identification number of your unknown culture. Label one nutrient agar slant with your name and "Gram-positive unknown no. ___" (fill in the identification number of your unknown). Inoculate the "Gram-positive unknown" nutrient agar slant with cells from your Gram-positive unknown culture.

3. Label one nutrient agar slant with your name and "Skin isolate." Inoculate the "Skin isolate" nutrient agar slant with cells from the nutrient broth inoculated with the isolated colony on the mannitol salt agar skin swab plate.

At this point, each student should have a total of four inoculated nutrient agar slants:

a. *S. aureus*
b. *S. epidermidis*
c. Skin isolate
d. Gram-positive unknown

Incubate the slants at 37°C for 48 hours.

GLUCOSE FERMENTATION ASSAY

1. Label two phenol red glucose broths with your name. Then label one of the two tubes *M. luteus* and the other *S. epidermidis*. Inoculate the broths with the correct pure cultures.

2. Label one phenol red glucose broth with your name and "Gram-positive unknown no. ___" (fill in the identification number of your unknown). Inoculate the "Gram-positive unknown" phenol red glucose broth with cells from your Gram-positive unknown culture.

3. Label one phenol red glucose broth with your name and "Skin isolate." Inoculate the "Skin isolate" phenol red glucose broth with cells from the nutrient broth inoculated with the isolated colony on the mannitol salt agar skin swab plate.

At this point, each student should have a total of four inoculated phenol red glucose broths:

a. *M. luteus*
b. *S. epidermidis*
c. Skin isolate
d. Gram-positive unknown

Incubate the broths at 37°C for 48 hours.

MANNITOL SALT AGAR PLATES

1. Label two mannitol salt agar plates with your name and the date, and divide each plate in half by drawing a line on the bottom with a wax pencil or Sharpie.

2. Label one side of the first plate "*E. coli*" and the other side "*S. mitis*." Label one side of the second plate "*S. aureus*" and the other side "*S. epidermidis*." Inoculate the plates with single streaks of the appropriate bacteria.

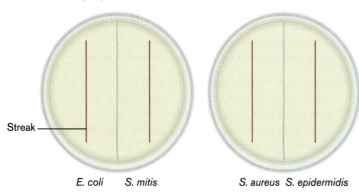

Streak

E. coli   *S. mitis*   *S. aureus*   *S. epidermidis*

3. Incubate the plates at 37°C for 24 hours.

4. Streak a mannitol salt agar plate with your unknown culture using streaking techniques to produce isolated colonies (see day 1, steps 2 and 4).

5. Incubate the plate at 37°C for 48 hours.

NOVOBIOCIN ASSAY

1. Label a nutrient agar plate with your name and the date, and divide the plate in half by drawing a line on the bottom with a wax pencil or Sharpie.

2. Label one side of the plate *S. epidermidis* and the other side *S. saprophyticus*. Inoculate the plate with single streaks of the appropriate bacteria.

3. Place a single novobiocin antibiotic disk somewhere along each of the streaks.

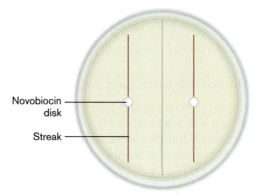

Novobiocin disk

Streak

S. epidermidis   S. saprophyticus

4. Label a nutrient agar plate with your name and the date, and divide the plate in half by drawing a line on the bottom with a wax pencil or Sharpie.

5. Label one side of the plate "Skin isolate," and label the other side of the plate "Gram-positive unknown no. ___" (fill in the identification number of your unknown).

6. Inoculate the "Skin isolate" side of the plate with a single streak of cells from the nutrient broth inoculated with the isolated colony on the mannitol salt agar skin swab plate.

7. Inoculate the "Gram-positive unknown" side of the plate with a single streak of cells from your selected Gram-positive unknown culture.

8. Place a single novobiocin antibiotic disk somewhere along each of the streaks.

9. At this point, each pair of students should have two inoculated nutrient agar plates with novobiocin disks:

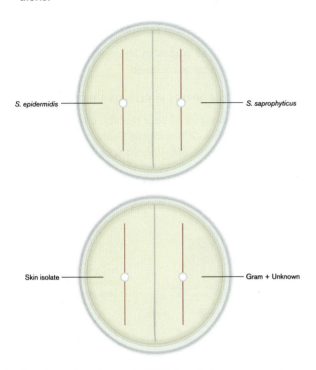

10. Incubate the plates at 37°C for 48 hours.

Day 4

MATERIALS

- Safety glasses
- Disposable gloves
- Metric rulers
- Staphyloslide coagulase reagents and cards
- Bottles of 3% hydrogen peroxide (H_2O_2)
- Plates and slants inoculated on day 3

MANNITOL SALT AGAR PLATES

1. Observe the colonies and the color of the medium for all of the mannitol salt agar plates streaked with known species. Record your observations in the table in the Results section. Discard the plates in the receptacle designated for trash to be autoclaved.

2. Examine the mannitol salt agar plates streaked with Gram-positive unknown cells for growth and determine whether the cells fermented mannitol. Record your observations in the table in the Results section. Discard the plates in the receptacle designated for trash to be autoclaved.

COAGULASE ASSAY

1. Pick up a coagulase assay card and mark the circles as follows:

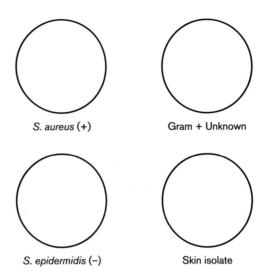

2. Add one small drop of coagulase reagent to the center of each circle. The reagent drops should not exceed 0.5 cm, or about ¼ inch in diameter.

3. Using a sterile loop, transfer a large amount of each type of cell from the nutrient agar slants to the drop in the appropriate circle. When we say "a large amount," we mean that the amount of colonial material on the loop should be enough that you can clearly see it as you transfer it. Use the loop to mix cells and reagent.

4. Once all cell types have been transferred to each circle, immediately begin rocking the card and continue for about 60 seconds.

5. Look for a change to a "grainy" pattern in the *S. aureus* circle, and compare its appearance with that of the *S. epidermidis* circle. Using these two circles as "positive" and "negative" controls, determine coagulase reactions (+ or −) for the other cell types. Cards should be read and results should be recorded immediately following the end of the 60 second rocking period.

   Do not record any observations made more than 3 minutes after the rocking has ended, because the evaporation of the liquid can clump the beads and make coagulase negative cells appear to be coagulase positive.

6. Dispose of the cards in the receptacle designated for trash to be autoclaved. Do <u>not</u> put them in the regular trash.

CATALASE ASSAY

After removing cells for the coagulase assay, add about five drops of H_2O_2 to the nutrient agar slants inoculated with your skin isolate and Gram-positive unknown cells, and record the reaction (bubbles or no bubbles) for each culture.

You don't have to add H_2O_2 to the *S. aureus* and *S. epidermidis* cultures. But you should understand why you do not need to add the reagent to these bacteria.

GLUCOSE FERMENTATION ASSAY

Record the colors of the phenol red glucose broths inoculated with *M. luteus* and *S. epidermidis* and the colors of the phenol red glucose broths inoculated with your skin isolate and Gram-positive unknown cells to determine whether a given culture ferments glucose. Record your observations in the Results section.

NOVOBIOCIN ASSAY

1. Measure the diameters of the zones of inhibition around the novobiocin disks for the nutrient agar plate that was inoculated with *S. epidermidis* and *S. saprophyticus*. Zones of inhibition should be measured from the ends of the heavy growth (see figure to the right); ignore any tiny isolated colonies near the disk.

2. Measure the diameters of the zones of inhibition around the novobiocin disks for the nutrient agar plate that was inoculated with your skin isolate and Gram-positive unknown cells. Record your observations in the Results section.

3. Dispose of all plates in the receptacle designated for trash to be autoclaved.

IDENTIFICATION OF GRAM-POSITIVE BACTERIA

After recording all of your results, use your data to identify both your numbered Gram-positive unknown and your skin isolate. Record your conclusions in the Results section.

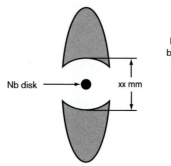

Nb disk

xx mm

Measure zone as distance between the middles of the arcs at the ends of heavy growth

Name: _____   Section: _____

Course: _____   Date: _____

SKIN SWABS ON MANNITOL SALT AGAR PLATES (DAY 2)

Written description of the morphology of the different types of colonies:

(Include colony color, shape, relative size and color of medium around the colonies)

MANNITOL SALT AGAR PLATE RESULTS

Species or Unknown type	Growth or No Growth	Color of Medium Around Colonies	Mannitol Fermentation (positive or negative?)
Colony picked for skin isolate (day 2 results)			
E. coli (day 4 results)			
S. mitis (day 4 results)			
S. aureus (day 4 results)			
S. epidermidis (day 4 results)			
Gram-positive unknown (day 4 results)			

COAGULASE ASSAYS

Species or Unknown Type	Appearance of Reagent (grainy or not grainy?)	Coagulase (positive or negative?)
S. aureus		
S. epidermidis		
Colony picked for skin isolate		
Gram-positive unknown		

CATALASE ASSAYS

Unknown Type	Reaction (bubbles or no bubbles)	Catalase (positive or negative?)
Colony picked for skin isolate (day 4 results)		
Gram-positive unknown (day 4 results)		

GLUCOSE FERMENTATION ASSAYS

Species or Unknown Type	Color of Medium	Positive or Negative for Fermentation of Glucose?
M. luteus		
S. epidermidis		
Colony picked for skin isolate		
Gram-positive unknown		

NOVOBIOCIN ASSAYS

Species or Unknown Type	Diameter of Zone of Inhibition (mm)	Greater or Less than 18 to 20 mm?	Novobiocin Resistant or Sensitive?
S. saprophyticus			
S. epidermidis			
Colony picked for skin isolate			
Gram-positive unknown			

IDENTIFICATION OF SKIN ISOLATE AND GRAM-POSITIVE UNKNOWN

Fill in the table with positive or negative, resistant or sensitive

Unknown Type	Catalase	Salt Tolerant (grows on mannitol salt agar)	Mannitol Fermentation	Glucose Fermentation	Novobiocin	Coagulase
Skin Isolate						
Gram-positive Unknown						

1. **Skin isolate species name:**

2. **Gram-positive unknown no._____ species name:**

1. What is different about the oxygen requirements of *Micrococcus* and of *Staphylococcus*, and how can we use these differences to distinguish *Micrococcus* from *Staphylococcus*?

2. Is mannitol salt agar a selective medium, a differential medium, or both? Explain why mannitol salt agar is or is not selective, and explain why mannitol salt agar is or is not a differential medium.

3. After inoculating a mannitol salt agar plate with a skin swab from behind your ear, would you rather find mannitol fermentation–negative colonies or mannitol fermentation–positive colonies on the plate? Explain your answer.

4. Why is the novobiocin assay so useful for the identification of *S. saprophyticus*?

5. If you wanted to isolate *S. aureus* from a throat swab, what could you add to blood agar to improve your chances, and what traits would you look for among the colonies growing on the plate? Explain your answer.

6. If you were specifically told to try to isolate *S. aureus* by swabbing some part of your body followed by inoculation of a mannitol salt agar plate, where would you swab, and what would you look for on the mannitol salt agar plate inoculated with that swab?

7. Why is the coagulase assay so useful for the identification of *S. aureus*?

8. Is it a coincidence that *S. aureus* cells can both produce coagulase and cause disease? Explain your answer.

Pseudomonas is the cause of the leaf spot disease seen in this hibiscus plant. *Pseudomonas* species are mainly non-pathogenic, free-living bacteria found in all types of soil and in water. They often inhabit the surfaces of plants without causing disease.

LAB 23

Pseudomonas, a Gram-Negative Genus

Learning Objectives

- Describe the traits that characterize species in the *Pseudomonas* genus.

- Consider the role and significance of *Pseudomonas aeruginosa* in human disease.

- Understand the principles of the cytochrome *c* oxidase and oxidative/fermentative (O/F) glucose assays, and explain how the metabolism of *Pseudomonas* species allows us to use these assays to distinguish these species from other Gram-negative microbes.

- Perform the cytochrome *c* oxidase and O/F glucose assays to differentiate *P. aeruginosa* from other Gram-negative species, including Gram-negative enterics.

Pseudomonas species are primarily free-living, nonpathogenic organisms often found in water, soils, and on the surfaces of plants. Species in the *Pseudomonas* genus have the following traits:

1. Cells are Gram-negative rods (bacilli; **Figure 23.1**).
2. Cells do not produce endospores.
3. Cells possess flagella and are motile.
4. Many of the species produce blue, yellow, green, or red pigments, many of which are water-soluble and diffuse into bacteriological media.
5. Cells use respiratory metabolic pathways to extract energy from glucose, that is, the cells rely on oxidizing electron transport systems; these systems use a type of cytochrome *c* and cytochrome *c* oxidase.
6. Cells do not ferment glucose.
7. Some species are strict aerobes, relying completely on aerobic respiration and oxygen as the terminal electron acceptor in their electron transport systems. Other species can live anaerobically if nitrate (NO_3) is available. The nitrates

replace oxygen as the terminal electron acceptor, and nitrogen (N_2) and nitrous oxide (N_2O) gases are produced as end products. This type of metabolism, called **anaerobic respiration**, is distinct from fermentation. It denitrifies environments under anaerobic conditions because the nitrogen that is present is converted into forms that are lost from the soil or water.

PSEUDOMONAS AERUGINOSA

Clinical significance of P. aeruginosa

Pseudomonas species rarely cause human infections. There is one notable exception: *Pseudomonas aeruginosa*. This species of *Pseudomonas* is usually free-living and can be found in soil and water, but it can also cause infections in plants, humans, and other animals. Most of the human diseases it causes are **opportunistic infections**; that is, they occur after an individual has been weakened, injured, or made vulnerable by a preexisting condition or decline in the immune system. For example, *P. aeruginosa* is a major cause of infections in tissue that has been badly burned. In addition, in patients with **cystic fibrosis**, *P. aeruginosa* often causes pneumonia and other lung infections because cystic fibrosis causes a buildup of gummy mucus in the lungs, which leaves individuals vulnerable to infection.

Although it is considered "only" an opportunistic pathogen, once an infection has become established, *P. aeruginosa* can be quite dangerous. The cells produce various toxins that enhance the bacterium's virulence (that is, its ability to cause disease). These toxins can destroy tissue and may assist in the potential spread of *P. aeruginosa* into the blood, where it can cause a life-threatening septicemia. Even worse, this species is already resistant to many antibiotics, which makes it difficult to treat infections once the cells are entrenched in the body.

P. aeruginosa does not have to be transmitted person-to-person because cells of this species can survive in a wide range of locations under a variety of conditions. Cells can grow at temperatures from 20°C to 42°C, and if necessary, they can survive with just one type of molecule as their carbon source and with NH_4^+ (ammonium) as their sole nitrogen source. Thus, cells can endure in environments with little in the way of organic nutrients. In addition, this species is resistant to many chemical control agents and has been known to remain alive in disinfectant solutions in clinical settings. It can also survive in contact lens solutions and in hot tubs, subsequently causing eye and skin infections, respectively. So, while, hot tubs have their appeal, they also have their hazards, and not everyone is a fan.

Some people sit in the Jacuzzi,
Every night and every day,
'Cause they think that boiling water,
Will take all their blues away.

But not me, [pause]
I like my lobsters in the sea.
I get my power from the shower,
Cause I'm a hot tub refugee.

—Steve Goodman, "Hot Tub Refugee," on *Santa Ana Winds* (1984)

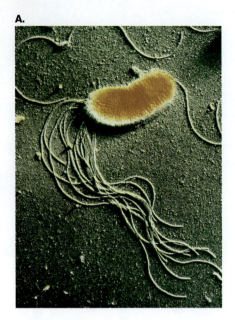

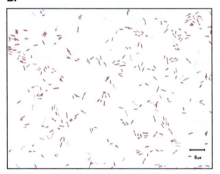

FIGURE 23.1 *Pseudomonas.*
A. Electron micrograph of *Pseudomonas fluorescens* showing numerous flagella.
B. Gram-stained *Pseudomonas aeruginosa* under light microscopy.

Distinguishing *P. aeruginosa* from other Gram-negative bacilli

Given the clinical significance of *P. aeruginosa*, it is important to be able to identify it in clinical samples, and in particular, it is essential to distinguish it from other Gram-negative rods.

There are certain non-fermenting aerobes, including *Alcaligenes*, *Moraxella*, and *Campylobacter* species, that can be mistaken for *P. aeruginosa*. Like *Pseudomonas* species, *Alcaligenes* species are common in soil and water, but they rarely cause disease (except for an occasional opportunistic bladder infection). We will use *Alcaligenes faecalis* in this lab as a relatively safe stand-in for other aerobic Gram-negative rod genera, such as *Moraxella* (cause of eye, ear, and respiratory infections) and *Campylobacter* (cause of intestinal infections).

It is also important to differentiate *P. aeruginosa* from fermenting facultative anaerobes, such as the Enterobacteriaceae or enteric family, a group encompassing several genera of gut-adapted bacteria. These species can use oxygen, if it is present, for aerobic respiration, but they can also convert sugars to organic acids and gases by fermentation if oxygen is unavailable. There are many, many medically important genera in this family, such as *Escherichia* (as in *E. coli*) and *Salmonella*, and we will visit some of these genera in other labs.

Two widely used assays can distinguish *Pseudomonas aeruginosa* from other Gram-negative groups: the cytochrome *c* oxidase assay and the oxidative/fermentative (O/F) glucose assay. The cytochrome *c* oxidase assay separates *P. aeruginosa* and many other aerobic Gram-negative genera from species in the medically important enteric family. The O/F glucose assay reveals a metabolic trait of *Pseudomonas* species that is unique enough to differentiate *P. aeruginosa* from most other medically important Gram-negative groups.

CYTOCHROME *C* OXIDASE ASSAY

Pseudomonas aeruginosa produces the enzyme cytochrome *c* oxidase, while species in the enteric family (Enterobacteriaceae) do not. Enteric species are facultative anaerobes and are quite capable of aerobic respiration if oxygen is available, as we'll see in Lab 24. However, they carry out aerobic metabolism using cytochromes and enzymes other than cytochrome *c* oxidase. So the cytochrome *c* oxidase assay can be used to differentiate *P. aeruginosa* from enteric species, because *P. aeruginosa* will be oxidase-positive, while the enteric species will be oxidase-negative.

There are, however, other Gram-negative bacteria that are also oxidase-positive, because they too carry out aerobic metabolism using cytochrome *c* and cytochrome *c* oxidase. These bacteria include species in the genus *Neisseria*, which contains species that cause gonorrhea and meningitis. Since *Neisseria* and *Pseudomonas* species are both oxidase-positive, the oxidase assay is not useful for distinguishing *Pseudomonas* from *Neisseria*.

The oxidase assay detects the presence of cytochrome *c* oxidase, an enzyme that transfers electrons from cytochrome *c* to molecular oxygen (O_2). This electron transfer is the last step in a series of electron transfers carried out by the electron transport systems that are essential to oxygen-dependent respiration.

A quick review of redox reactions

The oxidase assay involves a couple of oxidation-reduction (redox) reactions, so it may be useful to refresh our memory of how redox reactions work.

A **redox reaction** involves the transfer of electrons from one molecule to another. The molecule that gains the electrons becomes reduced, while the molecule that loses the electrons becomes oxidized.

This principle can be summed up in the mnemonic OIL RIG:

Oxidation, It Loses electrons
Reduction, It Gains electrons

A redox reaction is actually composed of two tightly coupled reactions in which molecules switch between two states: oxidized and reduced. In **Figure 23.2**, when molecule B goes from oxidized to reduced, it does so because molecule A goes from reduced to oxidized. Since molecule B is responsible for molecule A becoming oxidized (B takes the electrons from A), we say that molecule B is an oxidizing agent. Similarly, molecule A is called a reducing agent because it causes molecule B to become reduced.

In a cell, such as a *Pseudomonas* bacterium, the electron transport chain carries out a series of redox reactions that oxidizes the reduced NADH and $FADH_2$ produced in glycolysis and the Krebs cycle. The process sequentially reduces and oxidizes a series of electron carriers until, in the last step, the electrons are passed from reduced cytochrome *c* to O_2. This causes cytochrome c to become oxidized and O_2 to become reduced in the form of H_2O **(Figure 23.3)**. The oxidized cytochrome *c* will be reduced again when more electrons are passed down the chain from the oxidation of more NADH and $FADH_2$.

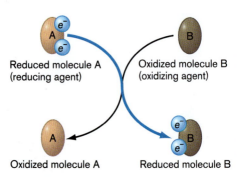

FIGURE 23.2 An oxidation-reduction reaction.
Reducing agent A donates electrons to molecule B, which becomes reduced. Since molecule A is losing electrons, it is being oxidized by the oxidizing agent, molecule B.

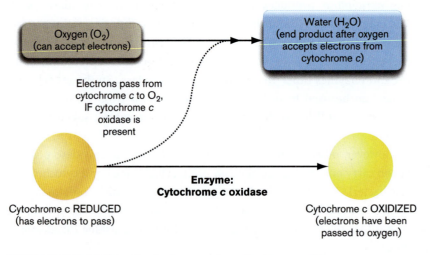

FIGURE 23.3 Oxidation of cytochrome *c* by cytochrome oxidase.
Oxidation of cytochrome *c* by cytochrome oxidase: Electrons are transferred from reduced cytochrome *c* to oxygen by the oxidase enzyme, reducing the oxygen to water and oxidizing cytochrome *c*.

The assay itself

The oxidase assay substitutes a chromogenic reducing reagent, N,N,N',N'-tetramethyl-*p*-phenylenediamine dihydrochloride as a source of electrons for the reduction of oxidized cytochrome *c*. "Chromogenic reducing agent" means that the reagent changes color when it becomes oxidized. In the case of tetramethyl-*p*-phenylenediamine, it is colorless when it is reduced and turns a dark blue-purple when oxidized by the donation of its electrons to an oxidized cytochrome *c* molecule **(Figure 23.4)**.

Although this method is called an "oxidase assay," the reducing reagent tetramethyl-*p*-phenylenediamine does not interact directly with cytochrome *c* oxidase. Instead, it detects the presence of oxidized cytochrome *c*, a product of cytochrome *c* oxidase activity. Since oxidized cytochrome *c* will not be present unless the cytochrome *c* oxidase enzyme is present to produce it, a reagent that reveals the presence of oxidized cytochrome *c* indirectly demonstrates that the oxidase enzyme is also present.

Positive reaction: If cells have an electron transport chain that includes cytochrome *c* and active cytochrome *c* oxidase, then as we've noted, reduced cytochrome *c* is oxidized by cytochrome *c* oxidase, producing oxidized cytochrome *c*. The electrons lost from cytochrome *c* are picked up by an oxygen molecule, and with the addition of two protons (H^+) per oxygen atom, the reduction of the oxygen produces water.

In the assay, the oxidized cytochrome *c* molecule can now accept electrons from the reduced tetramethyl-*p*-phenylenediamine reagent. This reaction reduces cytochrome *c* and oxidizes the reagent. When the reagent is oxidized by the loss of electrons to cytochrome *c*, it changes from colorless to a dark blue-purple color.

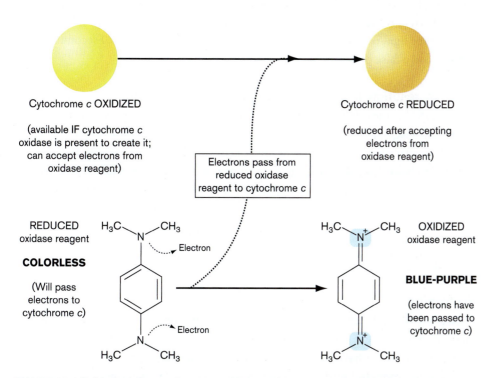

FIGURE 23.4 Oxidation of cytochrome *c* oxidase reagent, tetramethyl-*p*-phenylenediamine. Electrons are transferred from the colorless reduced reagent to oxidized cytochrome *c*, creating a blue-purple oxidized reagent and a reduced cytochrome *c*.

A.

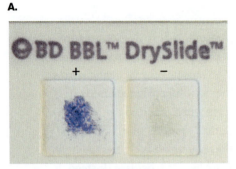

After 20 seconds

B.

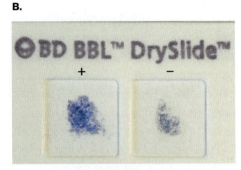

After 40 seconds

FIGURE 23.5 Positive and negative oxidase reactions on DrySlides.

So, when oxidase-positive cells are exposed to the oxidase reagent, the very rapid appearance of a dark blue-purple color demonstrates that oxidized cytochrome *c* is present and, by extension, reveals the presence of active cytochrome *c* oxidase. That is, the rapid appearance of a dark blue-purple color when cells are mixed with the oxidase reagent is a positive reaction for the oxidase assay **(Figure 23.5A)**.

The oxidase assay can be performed using fresh oxidase reagent solutions applied to filter paper or a swab, but in this lab, we will use the DrySlide oxidase test kit (Becton Dickinson and Company). Cells to be tested are applied to filter paper infused with tetramethyl-*p*-phenylenediamine, and color changes are observed for a maximum of 15–20 seconds. A **dark blue-purple color at 20 seconds** is considered a **positive oxidase reaction**. Note that reaction times are critical, and that **any color change after 20 seconds must be ignored**. This is because the reagent is unstable, and once dampened by applied cells, it will begin to slowly oxidize on its own, even in the absence of oxidized cytochrome *c*. By 30–40 seconds, oxidase-negative cells may appear to be producing some light blue-purple color, so any color that appears after 20 seconds should be viewed as a false positive **(Figure 23.5B)**.

Negative reaction If cells lack cytochrome *c* oxidase, then they will be unable to oxidize cytochrome *c*. In the absence of oxidized cytochrome *c* to accept electrons from the oxidase reagent, the reduced, colorless form of the reagent cannot be oxidized to the dark blue-purple form. So when cells are mixed with the oxidase reagent, **the absence of a rapid change to a dark blue-purple color is a negative reaction for the oxidase assay**. Again, when using the DrySlide test kit, **any color change after 20 seconds must be ignored**.

OXIDATIVE/FERMENTATIVE ASSAY

Principles

The **oxidative/fermentative (O/F) glucose assay** focuses on a metabolic trait that is useful for distinguishing *P. aeruginosa* from many other Gram-negative human pathogens. *Pseudomonas* species have the ability to oxidize glucose in such a way that it produces organic acid products, but this is a process that occurs only under aerobic conditions. That is, *Pseudomonas* species oxidatively produce organic acids from glucose.

By contrast, many other medically important Gram-negative aerobic bacterial species do not metabolize glucose, either oxidatively or by fermentation; in other cases, they oxidize glucose, but produce little organic acid product. And when facultative anaerobic species, such as those in the enteric family, ferment glucose, or use glucose fermentatively, they do produce organic acids, but they do so in pathways that do not use oxygen. Again, it is critical to note that oxygen is required for the production of organic acids from glucose by *Pseudomonas* species, so this metabolic phenomenon is not fermentation and should not be confused with fermentation.

O/F glucose medium

The medium used for the O/F assay has the following composition (per liter):

Peptone	2.0 g
Sodium chloride	5.0 g
Dipotassium phosphate	0.3 g
Bromthymol blue	0.1 g
Glucose	10.0 g
Agar	2.0 g

The O/F glucose medium is designed to detect the conversion of glucose to organic acids, either by oxidative or fermentative means. To that end, it contains 1% glucose, as a substrate for oxidation and fermentation, and a pH indicator, bromthymol blue. This pH indicator is close to phenol red in terms of the pH values at which it changes color, but the color transitions for bromthymol blue occur at pH values that are about 0.2–0.3 pH units below those for phenol red. Bromthymol blue is yellow in color when the molecule is uncharged and blue in color when the molecule loses a proton (H^+) and becomes negatively charged. The color of a given tube of O/F medium depends on the ratio of yellow-colored molecules to blue-colored molecules.

At pH 6.0, uncharged bromthymol blue molecules outnumber negatively charged molecules by ten to one, so there are ten times more yellow-colored molecules than blue-colored molecules, and the broth is clearly yellow in color. So if significant quantities of organic acids are produced from glucose, the medium will be yellow in color.

Between pH 6.5 and pH 7.5, the bromthymol blue indicator will go through a transition from yellowish green to green to bluish green. At pH 7.0, or neutral pH, there are equal numbers of yellow uncharged and blue charged molecules, and the O/F medium appears <u>green</u> in color, because yellow and blue make green.

At pH 8.0, there's a relative shortage of free protons. So it's the negatively charged blue form of bromthymol blue that is ten times more common than the yellow form, and the medium is blue.

Since our goal is to observe the production of acids from glucose, we don't want any other biochemical products "masking" the presence of the organic acids. Amino acid metabolism can generate ammonia (NH_3), and ammonia can make the medium alkaline, raising the pH. If too much ammonia were produced, then the medium might have a net alkaline pH, and a green or blue color, even if organic acids had been churned out by oxidation or fermentation. To reduce the probability of false negatives, the medium contains a low concentration of peptones (an amino acid source) compared with nutrient agar. Specifically, the O/F medium contains 2.0 grams per liter of peptones, in contrast to the 5.0 grams of peptones and 3.0 grams of meat extract in nutrient broths and agars. In addition, a small amount of dipotassium phosphate is added to act as a buffer against pH changes caused by production of low and diagnostically insignificant levels of either acids or ammonia.

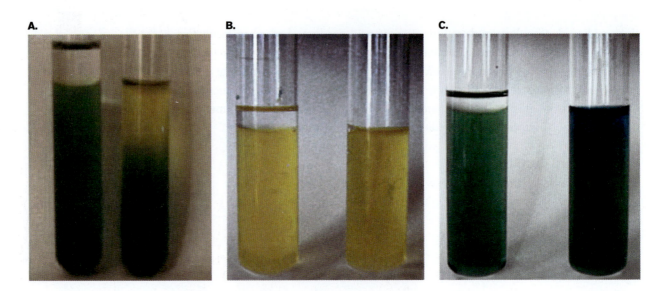

FIGURE 23.6 Oxidative/fermentative assay results.
A. *P. aeruginosa*, sealed (left) and open (right). **B.** *E. coli*, sealed (left) and open (right).
C. *A. faecalis*, sealed (left) and open (right).

Finally, we must differentiate between organic acids produced from glucose by oxidation and acids produced by fermentation. So, the assay requires that two tubes of O/F glucose medium be inoculated with each isolate tested. After inoculation, the medium in one tube is left open to the air to allow for oxidation of glucose, and the medium in the other tube is sealed with mineral oil. Sealing prevents oxygen from reaching the cells in the medium, and the only option available to the bacteria is to use the glucose by fermentative pathways.

Interpreting results of the O/F assay

Possibility 1: Open tube is yellow at the top and green at the bottom. Closed tube is green from top to bottom. The presence of organic acids, and the resulting yellow color at the top of the open tube, demonstrates the ability of the bacteria to produce acids when metabolizing glucose. But the pH at the bottom of the open tube remains unchanged, and the medium remains green, because the cells require oxygen to produce acids oxidatively **(Figure 23.6A)**. Thus, the acids have not been produced by fermentation. Furthermore, in the tube sealed with oil, the entire medium remains green, indicating that the cells are not capable of anaerobic or fermentative use of glucose.

The ability to produce organic acids when using glucose oxidatively is a trait that is almost unique to *Pseudomonas* species, and in the context of this lab and in most clinical cases, **this result would identify the bacteria as being *Pseudomonas aeruginosa*.**

Possibility 2: Open tube is yellow from top to bottom. Closed tube is yellow from top to bottom. The presence of organic acids, and the resulting yellow color throughout the closed tube, demonstrates the ability of the bacteria to produce acids from glucose under anaerobic conditions. In other words, a lowered pH in a closed tube demonstrates that glucose has been metabolized by fermentation. Glucose is also fermented in the low-oxygen environment of the lower part of the open tube, and the subsequent diffusion of the resulting acids from the bottom to the top of the tube lowers the pH throughout the medium. Thus, the entire open tube is yellow, despite the fact that the glucose may have been used by aerobic respiration with little acid production at the top of the tube **(Figure 23.6B)**.

The ability to produce organic acids when using glucose fermentatively is a trait that is found in many, many genera of Gram-positive and Gram-negative bacteria. But, in the context of this lab, this result would identify the isolate as a Gram-negative rod-shaped bacterium from the family Enterobacteriaceae.

Possibility 3: Open tube is green to blue at the top and green at the bottom. Closed tube is green from top to bottom.

The absence of organic acids, and the resulting lack of change in the initial green color throughout the closed tube, shows that, under anaerobic conditions, the cells cannot ferment glucose, cannot produce acids from glucose, and cannot lower the pH of the medium. The green or blue color at the top of the open tube, where oxygen is available, shows that the cells either cannot metabolize glucose or cannot produce significant quantities of organic acids oxidatively **(Figure 23.6C)**. If the medium is actually bluer at the top, that is most likely due to the release of ammonia from amino acids: if that happens, the pH will rise, especially in the absence of acid production, and the indicator will turn blue.

The inability to metabolize glucose at all, or the inability to produce much organic acid from glucose, either oxidatively or fermentatively, means that the isolate is neither a *Pseudomonas* species nor an enteric family species. If the isolate is a Gram-negative aerobic rod, this still leaves many possibilities, including *Alcaligenes*, *Moraxella*, and *Campylobacter*.

PROTOCOL

Day 1

Materials

- Safety glasses
- Disposable gloves
- 1 DrySlide with four windows
- 8 oxidative/fermentative glucose tubes
- Wax pencils or Sharpies for labeling DrySlides
- Platinum wire inoculating loops or sterile swabs
- Bunsen burner and striker (if using loops)
- Disinfectant in beakers (if using swabs)
- Inoculating needles
- 1-ml transfer pipette with bulb end
- Sterile mineral oil

Cultures (on slants):

- *Alcaligenes faecalis*
- *Escherichia coli*
- *Proteus vulgaris*
- *Pseudomonas aeruginosa*

> **IMPORTANT:** *Pseudomonas aeruginosa* and *Proteus vulgaris* are biosafety level 2 species, so gloves and safety glasses must be worn throughout this lab.

OXIDASE ASSAY

1. Cut open a pouch containing DrySlides, remove one slide, and place the slide on the bench with the filter paper windows facing up.

2. Use a wax pencil or Sharpie to label the plastic frames adjacent to the filter paper windows as follows:

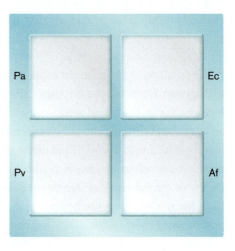

Af = *Alcaligenes faecalis*
Ec = *Escherichia coli*
Pv = *Proteus vulgaris*
Pa = *Pseudomonas aeruginosa*

3. Using either a sterile swab or a flame-sterilized platinum wire inoculating loop, gently touch the colonial growth on the *Pseudomonas aeruginosa* slant.

   The type of metal used in the loop matters because iron wire may cause a false positive reaction with the oxidase assay.

4. Transfer the *Pseudomonas* cells onto the filter paper in the slide window marked "Pa" by gently wiping the paper with the swab or wire. If the cell transfer was done with a swab, then discard the swab in a beaker of disinfectant or an autoclave bag. If the transfer was done with a platinum wire loop, then flame-sterilize the loop after the transfer.

5. As soon as the cells touch the filter paper, begin timing the reaction. You should record the color that you see in the slide window at 15–20 seconds. Do not record any color change observed after 20 seconds. As noted above, timing is very important with this assay.

6. Repeat steps 3 through 5 using cells from *E. coli*, *P. vulgaris*, and *A. faecalis* slants. In each case, transfer cells to the appropriate filter paper window and record the color after 15–20 seconds.

7. After all species have been tested and the results recorded, discard the used slide in an autoclave bag.

OXIDATIVE/FERMENTATIVE ASSAY

1. Label eight O/F glucose tubes with your name, the date, and the following descriptions:

   a. *E. coli*: Open
   b. *E. coli*: Sealed

   c. *P. vulgaris*: Open
   d. *P. vulgaris*: Sealed

   e. *A. faecalis*: Open
   f. *A. faecalis*: Sealed

   g. *P. aeruginosa*: Open
   h. *P. aeruginosa*: Sealed

2. For each of the following transfers, use a flame-sterilized inoculating needle to transfer cells from the slant culture to the appropriate O/F glucose tubes. Be sure that the needle is well coated with cells, and don't forget to sterilize the entire length of the needle between culture transfers.

3. Inoculate the O/F glucose tubes with the appropriate species by stabbing the cell-coated inoculating needle straight down into the agar **to the bottom of the tube**. Do not wiggle the needle or stir the agar before pulling the needle straight up out of the agar.

4. Using a 1-ml transfer pipette with a bulb end, add about 1 cm of mineral oil to each of the four tubes labeled "Closed." Oil should be added without touching the sides of the tube; if you touch the side of a tube with the pipette or contaminate the tip of the pipette in any other way, discard the pipette and use a new, sterile pipette.

5. Incubate all tubes at 37°C for 48 hours.

Day 2

Observe the colors in all of the O/F glucose tubes and record them in the Results section.

RESULTS

Name: _____   Section: _____

Course: _____   Date: _____

OXIDASE ASSAY

Bacterial Species	Color on Slide at 20 Seconds	Cytochrome *c* Oxidase: Positive or Negative?
E. coli		
P. vulgaris		
A. faecalis		
P. aeruginosa		

OXIDATIVE/FERMENTATIVE ASSAY

Bacterial Species	Open/ Sealed	Color of Top Half of Tube	Is the Top Acidic, Neutral, or Alkaline?	Color of Bottom Half of Tube	Is the Bottom Acidic, Neutral, or Alkaline?
E. coli	Open				
E. coli	Sealed				
P. vulgaris	Open				
P. vulgaris	Sealed				
A. faecalis	Open				
A. faecalis	Sealed				
P. aeruginosa	Open				
P. aeruginosa	Sealed				

Name: _____   Section: _____

Course: _____   Date: _____

1. Did any of the species produce a dark blue-purple color on the DrySlide filter paper within 20 seconds, and if so, which species did this?

2. Did the oxidase assay distinguish *P. aeruginosa* from species in the enteric family? If it did so, what is the underlying biochemical or metabolic explanation for the assay's ability to separate these taxonomic groups?

3. Why is it important clinically to have assays that distinguish between *P. aeruginosa* and enteric family species?

4. Did the oxidase assay distinguish *P. aeruginosa* from *A. faecalis*? If it did not, what is the underlying biochemical or metabolic explanation for the assay's inability to separate or distinguish between these species?

5. Did any of the species produce a blue color at the top of the open O/F glucose tube, and if so, which species did this? Why might the top of an inoculated O/F glucose tube be blue in color? What biochemical reactions would lead to the blue color?

6. Why wasn't the bottom of the open *P. aeruginosa* tube the same color as the top of the tube? What biochemical reactions create this color pattern?

7. What is the underlying biochemical reason why the O/F assay fails to distinguish *E. coli* from *Proteus vulgaris*? Why does it fail to distinguish species within the enteric group?

8. Let's say that you knew in advance that a given species was oxidase-positive. Predict what you would see in a closed O/F glucose tube inoculated with this species, and explain the reasoning behind your prediction.

9. What set of symptoms might prompt a doctor to ask a clinical lab to run an oxidase assay, an O/F glucose assay, or both on a pure bacterial culture isolate from a patient?

10. Some types of *Staphylococcus aureus* infections may resemble *Pseudomonas aeruginosa* infections. In addition to the O/F glucose assay, what lab procedures could be used to distinguish *S. aureus* from *P. aeruginosa*?

11. Both *Bacillus cereus* and *Pseudomonas aeruginosa* are rod-shaped bacteria that live in the soil. In addition to the O/F glucose assay, what lab procedures could be used to distinguish *B. cereus* from *P. aeruginosa*?

The Enterobacteriaceae is a large and enormously important family of bacteria. Many genera in this Gram-negative family cause serious human diseases. Since the mode of transmission for these pathogens is often fecal-oral, crowded unsanitary conditions (like in this refugee camp) are breeding grounds for epidemics of dysentery, typhoid fever, and other serious diarrheal and systemic illnesses.

LAB 24

The Family Enterobacteriaceae

Learning Objectives

- Describe the characteristics of the bacteria in the Enterobacteriaceae family (enteric family).

- Understand the significance of lactose fermentation in the differentiation of species within the enteric family.

- Explain how differential and selective media, such as MacConkey agar and EMB agar, can be used to begin the process of isolating and identifying species in the enteric family.

- Examine the colony morphologies of several enteric species growing on MacConkey and EMB agar plates.

One of the most medically important groups of Gram-negative bacteria is the family **Enterobacteriaceae**, or the **enteric family**. Enteric genera include *Escherichia*, *Enterobacter*, *Citrobacter*, *Klebsiella*, *Proteus*, *Salmonella*, *Shigella*, *Serratia*, and *Providencia* **(Figure 24.1)**. These species are most often found in the large intestines as part of the normal flora of healthy people. However, some of the species in this group are pathogens, and the disease-causing enterics are responsible for millions of cases of illness every year.

Enteric species have the following characteristics:

1. Cells are Gram-negative rods.

2. Cells do not produce spores.

3. All species are facultative anaerobes: cells can use oxygen in aerobic respiration pathways, but if oxygen is unavailable, they will switch to a type of fermentation that typically produces organic acids and gases.

4. With only a few exceptions, enterics ferment glucose to produce acids and gases.

A.

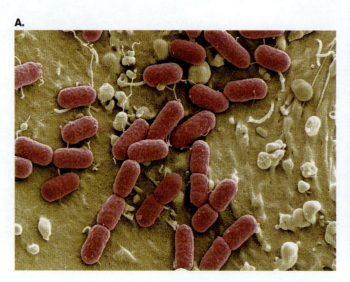

B.

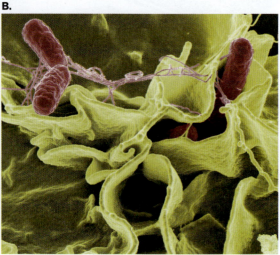

FIGURE 24.1 Enterobacteriaceae.
A. *E. coli* 0157:H7 detected in ground beef (scanning electron micrograph).
B. *Salmonella* spp. (scanning electron micrograph).

DISTINGUISHING AMONG THE ENTERICS

While all enteric species ferment glucose, only some ferment lactose, because the breakdown of the disaccharide lactose to the monosaccharide glucose requires additional enzymes. Utilization of lactose requires production of the **beta-galactosidase enzyme**, which splits lactose into the monosaccharides glucose and galactose. Cells must also produce additional enzymes to convert galactose into glycolysis intermediates. In the case of *E. coli*, galactose is converted, in a multi-step process, to the glycolysis intermediate glucose-6-phosphate. In addition, some enteric species can be differentiated from one another by their ability to ferment sucrose, a process that requires **beta-fructosidase**, an enzyme that hydrolyzes sucrose, producing glucose and fructose as end products.

So, testing enteric isolates for their ability to ferment lactose can be a very useful procedure when we are trying to differentiate one enteric species from another. Species that are capable of fermenting lactose are sometimes referred to as **coliform** species. The coliform subgroups of enterics include almost all species and strains in the genera *Escherichia* (including *E. coli*), *Enterobacter*, and *Klebsiella*. In addition, most common species and strains in the genus *Citrobacter* ferment lactose. **Non-coliform enterics**, or species in the enteric family that are unable to ferment lactose, include almost all the species and strains in the genera *Proteus*, *Salmonella*, *Shigella*, *Serratia*, and *Providencia*. Because the distinction between coliform and non-coliform is well correlated with specific sets of genera, information about the ability of an isolate to ferment lactose usually allows us to determine that an unidentified organism belongs in one set of genera or another **(Figure 24.2)**.

Identification of enteric species starts with the use of selective and differential media designed to favor species in the enteric family (selection) and to begin the process of distinguishing one enteric species from another (differentiation). Two media that are widely used to isolate species in this family are MacConkey and eosin methylene blue (EMB) agars.

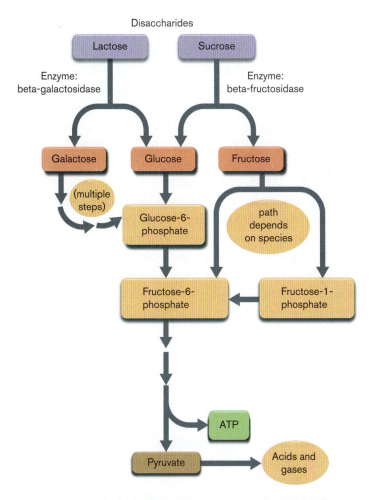

FIGURE 24.2 Pathways for the use of carbohydrates by some enteric bacteria.
All enterics can ferment the monosaccharide glucose, but only some species produce the specific enzymes needed for fermentation of the disaccharides lactose and sucrose.

MACCONKEY AGAR

MacConkey agar contains peptones and sodium chloride to meet basic nutritional needs as well as components that allow it to act as both a selective and a differential medium. MacConkey agar has the following composition (per liter):

Peptones	20.0 g
Lactose	10.0 g
Bile salts	1.5 g
Sodium chloride	5.0 g
Neutral red	30.0 mg
Crystal violet	1.0 mg
Agar	13.5 g

Selection by MacConkey agar

MacConkey agar is specifically designed to select for Gram-negative enterics and other Gram-negative species and to inhibit or prevent the growth of Gram-positive species. Gram-positive cells are specifically inhibited by the presence of crystal violet and bile salts, while most Gram-negative species can tolerate these compounds. Crystal violet was once widely used as a topical antiseptic to treat skin, wound, and oral cavity infections caused by members of the Gram-positive *Staphylococcus* and *Streptococcus* genera. It also has antifungal properties. Bile salts are produced by the liver and are present in the digestive tracts of animals. It makes sense that gut-adapted bacteria, such as the enterics, can grow in the presence of these liver products. Note, however, that many nonenteric Gram-negative species, such as those in the *Pseudomonas* and *Alcaligenes* genera, can also tolerate the addition of bile salt to their media.

Differentiation by MacConkey agar

In addition to being selective for Gram-negative species, MacConkey agar also allows separation of enteric isolates into lactose fermenters and nonfermenters—that is, coliforms and non-coliforms, respectively—because it contains the components needed to detect lactose fermentation. The medium contains 1% lactose as a substrate for fermentation and neutral red as a pH indicator. **Neutral red** passes through a transition from red to orange to yellow as the pH rises from 6.4 to 7.4 to 8.4. So, if the pH of the medium or the fluid in or around the cells is below about 6.5, then neutral red will be bright red to brick red in color.

Positive reaction for lactose fermentation
The neutral red indicator in the medium is water-soluble, and it can diffuse from the medium into the fluid surrounding the cells and into the cells themselves. If cells growing on MacConkey agar can ferment lactose, then the organic acids they produce will diffuse into the fluids surrounding the cells and into the nearby medium. As a result, as the acids accumulate, the pH in all these locations will drop below pH 6.5, and the majority of the neutral red molecules will be in the red form.

So, if the medium surrounding the colonies and the colonies themselves are bright red in color, then this is considered a positive reaction for the fermentation of lactose to organic acids (Figure 24.3, left); if the cells are known to be enteric species, then we can conclude that they belong to a coliform species.

In addition, if a lactose fermenter produces a large quantity of organic acids, then the pH of the medium around the colonies may decrease to the point at which the bile salts precipitate out of the agar-based medium. If this happens, the medium will become more opaque or hazy or cloudy in the regions surrounding the colonies. This reaction is typical of species that produce a high concentration of acids, such as *E. coli* and *Citrobacter freundii*. Other coliform species do not produce as much acid from lactose, so while their colonies will be red, there will be little of the hazy bile salt precipitate; this reaction is typical of *Enterobacter aerogenes*.

Negative reaction for lactose fermentation
If the cells growing on MacConkey agar cannot ferment lactose, then they do not produce organic acids. Without any acids to diffuse into the fluids surrounding the cells within the colonies and into the medium surrounding the colonies, the pH in all these locations will remain above 6.5, and the majority of the neutral red molecules will be in the pale yellow form.

So, if the medium surrounding the colonies remains unchanged or becomes pale yellow in color, or if the colonies themselves remain essentially uncolored or pale

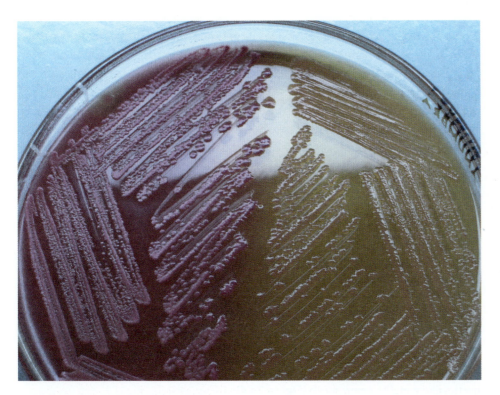

FIGURE 24.3 MacConkey agar results.
Positive (left) and negative (right) results for lactose fermentation.

pink in color (as opposed to brick-red in color), then this is considered a negative reaction for the fermentation of lactose to organic acids (Figure 24.3, right); if the cells are known to be enteric species, then we can conclude that these cells do not belong to a coliform species.

It should be noted that lactose-negative colonies may appear slightly, pale pink because the medium itself has a pale pink color and the colonies may be tinted by the light passing through the bottom of the plate.

EOSIN METHYLENE BLUE (EMB) AGAR

Eosin methylene blue (EMB) agar contains a digest of gelatin protein to meet basic nutritional needs as well as other components that allow it to act as both a selective and a differential medium. EMB agar has the following composition (per liter):

Pancreatic digest of gelatin	10.0 g
Lactose	10.0 g
Dipotassium phosphate	2.0 g
Eosin Y	0.4 g
Methylene blue	65.0 mg
Agar	15.0 g

Selection by EMB agar

Like MacConkey agar, EMB agar is specifically designed to select for Gram-negative enterics and other Gram-negative species and to inhibit or prevent the growth of

Gram-positive species. Gram-positive cells are specifically inhibited by the presence of two dyes, eosin and methylene blue, whereas most Gram-negative species can tolerate these compounds. EMB agar is slightly more inhibitory than MacConkey agar, and some of the bacterial species that we would like to isolate and grow from a clinical sample may not do well on this medium. A clinically important species could be missed if we started the process of isolating species from a mixed clinical sample with the EMB medium instead of the MacConkey medium. For this reason, many microbiologists prefer MacConkey agar over EMB agar, at least for the initial isolation steps. But as we'll see, EMB agar has its advantages.

Differentiation by EMB agar

In addition to being selective for Gram-negative species, EMB agar, like MacConkey agar, allows separation of enteric isolates into lactose fermenters (coliform species) and nonfermenters (non-coliform species) because it contains the components needed to detect lactose fermentation. The medium contains 1% lactose as a substrate for fermentation, and the eosin and methylene blue dyes act as pH indicators. When the pH drops into the acidic range, the dyes form dark purple-black or blue-black precipitating complexes. If the pH drops low enough—that is, if large quantities of acids are produced—then further changes in the eosin and methylene blue produces a Japanese beetle–like metallic green sheen **(Figure 24.4)**.

Positive reaction for lactose fermentation The eosin and methylene blue dyes are both water-soluble and can diffuse from the medium into the fluid surrounding the cells and into the cells themselves. If cells growing on EMB agar ferment lactose, then the organic acids produced by the cells will diffuse into the fluids surrounding the cells and into the nearby medium. As the acids accumulate in various locations, the pH in

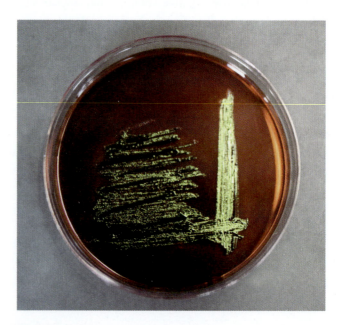

FIGURE 24.4 EMB agar results.
The green metallic sheen shows that this species of bacteria produces a high concentration of acids during lactose fermentation. Other types of lactose fermenters produce dark purple-black colonies.

all of these locations will drop below pH 6.5, and the two dyes will combine to form a dark, purple-black to blue-black precipitate.

So, if the medium surrounding the colonies and the colonies themselves become dark purple-black to blue-black in color, then this is considered a positive reaction for the fermentation of lactose to organic acids. If the cells are known to be enteric species, then we can conclude that these cells belong to a coliform species.

All species that ferment lactose, including all coliform species, produce dark-colored colonies on EMB agar. However, different coliform species vary in the amount of acid they produce. Some coliforms produce large amounts of acid, and others produce smaller amounts of acid mixed with neutral fermentation products. These differences can be detected by the EMB agar, and we can use them to further differentiate bacteria within the coliform group. The colonies of species that produce a high concentration of acids, such as *E. coli* and some *Citrobacter* species, have a metallic green sheen, especially in locations on a plate where the cells are crowded together and there are no isolated colonies (Figure 24.4). So, EMB agar may be particularly useful if we are looking specifically for *E. coli* in clinical or environmental samples because colonies with a green sheen on EMB agar may be presumptively identified as *E. coli*. Coliform species that do not generate high concentrations of organic acids, such as those in the *Enterobacter* and *Klebsiella* genera, still produce dark purple colonies, but the colonies lack a green sheen, and the rims of those colonies are often lighter in color.

So, if the colonies have a metallic green sheen, this indicates that lactose fermentation produced a high concentration of acids, and that the cells are from a specific subtype of coliform species that includes *E. coli*.

Negative reaction for lactose fermentation If the cells growing on EMB agar cannot ferment lactose, then they do not produce organic acids. Without any acids to diffuse into the fluids surrounding the cells within the colonies and into the medium surrounding the colonies, the pH in all these locations will remain above pH 6.5, and the majority of the dye molecules will not produce dark-colored precipitates.

So, if the medium surrounding the colonies remains unchanged in color or if the colonies themselves remain essentially uncolored or translucent light yellow-purple in color (as opposed to dark purple or metallic green), then this is considered a negative reaction for the fermentation of lactose to organic acids. If the cells are known to be enteric species, then we can conclude that these cells probably do not belong to a coliform species.

It should be noted that lactose-negative colonies may appear pale pink because the medium itself has a pale red color and the colonies may be tinted by the light passing through the bottom of the plate.

PROTOCOL

Day 1

Materials

- Disposable gloves

- Safety glasses

- 2 Eosin methylene blue (EMB) agar plates

- 2 MacConkey agar plates

Cultures:

- *Enterobacter aerogenes*

- *Escherichia coli*

- *Salmonella* Typhimurium

- *Staphylococcus epidermidis*

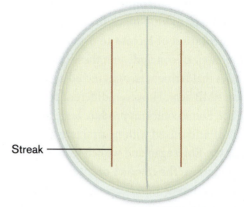

E. coli   S. epidermidis

> **IMPORTANT:** *Salmonella* Typhimurium is a biosafety level 2 species, so gloves and safety glasses must be worn throughout this lab.

1. Label two EMB agar plates and two MacConkey agar plates with your names and the date, and divide each plate in half by drawing a line on the bottom with a wax pencil or Sharpie.

2. Label one side of one of each type of plate "*E. coli*" and the other side "*S. epidermidis*." Label one side of each of the remaining two plates "*E. aerogenes*" and the other side "*S.* Typhimurium."

3. Inoculate the plates with single streaks of the appropriate bacteria.

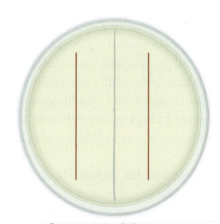

S. aerogenes S. Typhimurium

4. Incubate the plates at 37°C for 24 hours.

Day 2

Observe the growth and color of the colonies, and the color and cloudiness of the medium, for all species streaked onto the EMB and MacConkey agar plates. Record your observations in the Results section. Discard the plates in the receptacle designated for trash to be autoclaved.

RESULTS

Name: _____   Section: _____

Course: _____   Date: _____

MACCONKEY AGAR PLATES

Bacterial Species	Growth/ No Growth	Color of Colonies	Color of Agar	Cloudiness in the Agar? (yes/no)	Lactose Fermented? (yes/no)
Escherichia coli					
Staphylococcus epidermidis					
Enterobacter aerogenes					
Salmonella Typhimurium					

EMB AGAR PLATES

Bacterial Species	Growth/ No Growth	Color of Colonies	Color of Agar	Green Metallic Sheen? (yes/no)	Lactose Fermented? (yes/no)
Escherichia coli					
Staphylococcus epidermidis					
Enterobacter aerogenes					
Salmonella Typhimurium					

Name: _____   Section: _____

Course: _____   Date: _____

1. Are MacConkey and EMB agars selective media, differential media, or both? Explain why these media are or are not selective, and why they are or are not differential.

2. Did the *S. epidermidis* cells grow well on the EMB or MacConkey agar? Explain why the cells did or did not grow well.

3. How did *E. coli* and *S.* Typhimurium differ in terms of how they appeared on EMB and on MacConkey agar? What is the underlying biochemical basis for these observed differences?

4. How did *E. coli* and *E. aerogenes* differ in term of how they appeared on EMB and on MacConkey agar? What is the underlying biochemical basis for these observed differences?

5. Typhoid fever is a life-threatening intestinal infection caused by a species in the genus *Salmonella*. If a patient was thought to have typhoid fever, and if a fecal sample from the patient was streaked onto MacConkey and EMB agar, which of the following colonies should be subcultured and identified? (There may be more than one answer.) Explain your answer.

a. Metallic green colonies growing on EMB agar
b. White colonies growing on MacConkey agar
c. Nearly colorless colonies growing on EMB agar
d. Purple colonies growing on EMB agar
e. Red colonies growing on MacConkey agar

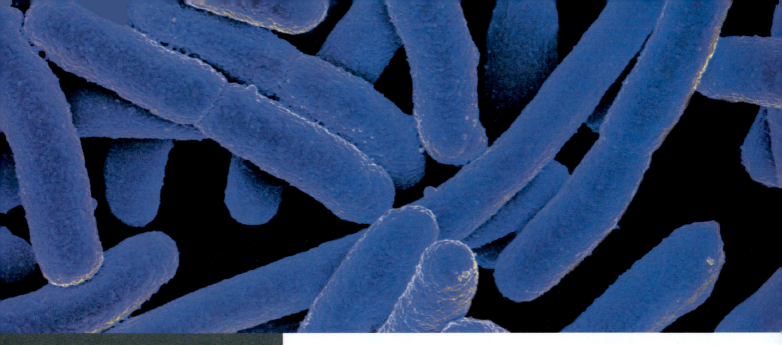

Enteric bacteria are a large family within the class of bacteria known as gammaproteobacteria. The enterics are, evolutionarily, a tightly-knit group, so many diagnostic tests have been designed to distinguish them from one another, and to allow for identification of a single species in a patient sample that is the cause of an illness. The photo above is a scanning electron micrograph of the best known enteric bacteria, *Escherichia coli*, or *E. coli* for short.

LAB 25

Assays for Identification of Enteric Species

Learning Objectives

- Understand the components, reactions, and applications of assays used to identify enteric species, including fermentation assays, the methyl red assay, the citrate assay, the urease assay, and the SIM assays.

- Use the various assays to distinguish species within the Enterobacteriaceae family.

The family Enterobacteriaceae is a very large family with hundreds of Gram-negative species. The MacConkey and eosin methylene blue (EMB) agars are useful for the initial isolation of enteric species and the preliminary division of unidentified isolates into coliform and non-coliform enteric groups (Lab 24). Two assays that were described in Lab 10 are also useful for distinguishing among enteric species:

- Phenol red carbohydrate assays are useful for differentiating bacteria based on their ability to ferment sugars.
- The lysine decarboxylase assay is useful for differentiating bacteria based on the presence of specific enzymes that act on amino acids.

However, many additional assays are needed to complete the identification of a specific enteric species.

METHYL RED ASSAY

The **methyl red assay** is designed to differentiate bacterial species that produce small amounts of organic acids by the fermentation of glucose from species that generate high concentrations of acids by glucose fermentation. Note that we are not separating species on the basis of whether they ferment glucose, as we do when we use the phenol red carbohydrate assay; all enteric species ferment glucose. Instead, we are separating them on the basis of the **quantity** of certain end products of fermentation. This approach is useful in distinguishing coliform species in genera such as *Enterobacter* and *Klebsiella* from other coliform species such as *Escherichia coli*. For example, while both *Enterobacter aerogenes* and *E. coli* ferment glucose, *E. aerogenes* produces less total organic acid than *E. coli* because *E. aerogenes* converts acidic fermentation products such as α-acetolactic acid into neutral molecules such as acetoin (acetylmethylcarbinol) and butanediol, which raise the environmental pH (**Figure 25.1**).

Methyl red Voges Proskauer (MRVP) broth

The **methyl red Voges Proskauer (MRVP) broth** is used for the methyl red (MR) assay as well as for a test called the Voges-Proskauer (VP) assay. In almost all cases, methyl red–negative species will be positive for the VP assay, and vice versa, because the VP assay tests for the presence of acetoin, which is produced by fermentation from pyruvate via an organic acid intermediate. So, more acetoin (organism will be VP-positive) usually means correspondingly lower concentrations of organic acids (organism will be methyl red–negative).

The composition of MRVP broth is (per liter):

Buffered peptones	7.0 g
Dipotassium phosphate	5.0 g
Glucose	5.0 g

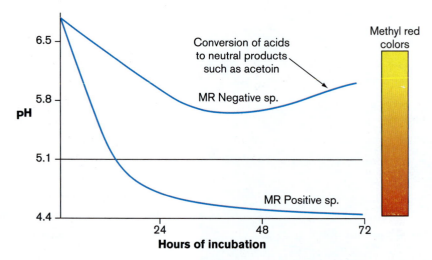

FIGURE 25.1 Changes in pH during the methyl red assay for two glucose fermenting enteric species.
Methyl red positive species produce large quantities of organic acids (lower final pH) while methyl red negative species produce a mix of organic acids and neutral products (higher final pH).

The peptones provide carbon and nitrogen, while glucose acts as a substrate for fermentation and as a source of carbon and energy. Dipotassium phosphate is a buffer that prevents large changes in medium pH unless the cells produce a high concentration of acids. Methyl red is not a component of the medium itself; instead, it will be added to the tubes after the medium is inoculated and incubated. Once the glucose has been fermented, the addition of methyl red will reveal whether cells have produced large amounts of acid by fermentation.

Methyl red indicator

Methyl red is a pH indicator that is yellow in color when the molecule is uncharged and red in color when the molecule gains a proton (H$^+$) and becomes positively charged (**Figure 25.2**). Methyl red gains a H$^+$ under highly acidic conditions. The color of a medium after methyl red is added depends on the ratio of yellow-colored molecules to red-colored molecules (see Figure. 25.1):

At pH 4.1, positively charged methyl red molecules outnumber negatively charged molecules by ten to one, so there are ten times more red-colored molecules than yellow-colored molecules, and the indicator is clearly red in color.

Between pH 4.6 and pH 5.6, the methyl red indicator will go through a transition from reddish orange to orange to yellowish orange. At pH 5.1, there are equal numbers of both types of molecules, and the indicator appears orange in color.

At pH 6.1, there's a relative shortage of free protons (H$^+$). So, it's the uncharged or unprotonated yellow form that is ten times more common than the red form, and the indicator is yellow.

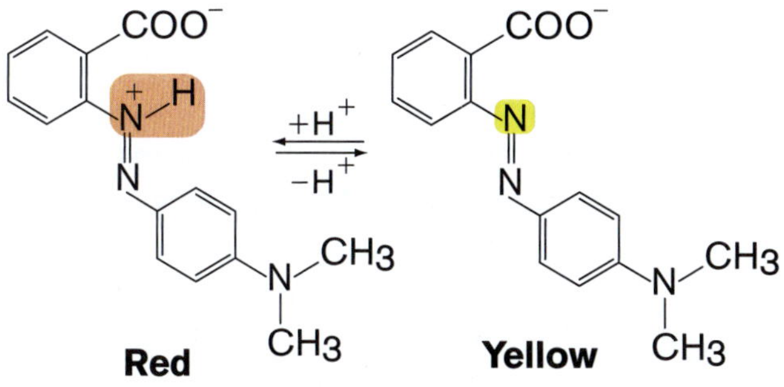

FIGURE 25.2 Methyl red reaction.
As the pH changes, the addition and loss of a proton causes methyl red to change color between red and yellow.

Positive and negative reactions for the methyl red assay

Figure 25.3 displays both positive and negative reactions for the methyl red assay as described below.

Positive for high concentration of acids produced by fermentation
If the cells in the MRVP broth ferment glucose and produce large quantities of organic acids, then the medium pH will be below pH 5.0 and may be as low as pH 4.0. If the pH is 5.0 or lower, then a majority of the methyl red molecules will pick up protons (H$^+$) from the acidic medium, and the indicator molecules will become positively charged and orange to red in color. Remember, the indicator turns orange to red only at pH 5.0, when the proton concentration of the medium is 100 times greater than at neutral pH.

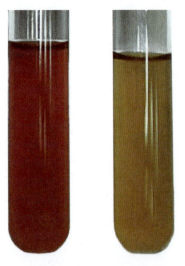

FIGURE 25.3 Results of methyl red assay. A positive (red) reaction is shown on the left and a negative (yellow) reaction is shown on the right.

So, if the medium to which methyl red has been added is orange to red in color, this is considered a positive reaction for fermentation of glucose in which a high concentration of organic acids is produced (Figure 25.3, left tube).

Notice that these colors are the opposite of those produced by phenol red, which turns yellow when the medium is more acidic and red when it is more alkaline. With methyl red, high levels of acidity are revealed by a red color; as the pH rises, methyl red turns yellow. It would be simpler if the color pattern for methyl red was not the reverse of the pattern for phenol red, but as the musical philosopher Sir Michael Philip Jagger has noted, you can't always get what you want.

In the context of separating the two coliform species *E. coli* and *E. aerogenes*, a positive reaction means that the cells must be *E. coli*, because these cells produce large quantities of organic acids when they ferment glucose. *E. aerogenes* also ferments glucose, but the cells produce a mix of organic acids and neutral products such as acetoin. Differentiating *E. coli* from *E. aerogenes* in environmental samples is important because *E. coli* is almost always of intestinal origin, and its presence in a sample almost always indicates fecal contamination. By contrast, *E. aerogenes* can be isolated from soil and water even if there is no fecal contamination.

Negative for high concentration of acids by fermentation If the cells in the MRVP broth ferment glucose and produce some organic acids, but also significant quantities of neutral products such as acetoin, then the medium pH will be above pH 5.1 and may well be above pH 6.1. If the medium pH is above about pH 5.6–6.1, then a large majority of the methyl red molecules will fail to pick up protons (H^+) from the medium, and the indicator molecules will be uncharged and yellow in color. The medium may be below pH 7.0 and thus acidic, but unless the pH is below about pH 5.6, then the indicator will still be more yellow than red.

So, if the medium to which methyl red has been added is yellow in color, this is considered a negative reaction for fermentation of glucose in which a high concentration of organic acids is produced (Figure 25.3, right tube).

In the context of separating the two coliform species *E. coli* and *E. aerogenes*, a negative reaction means that the cells must be *E. aerogenes*, because these cells produce a mix of organic acids and neutral products when they ferment glucose.

CITRATE ASSAY

The **citrate assay** detects the ability of bacteria to use the organic compound citrate (citric acid), an intermediate in the Krebs or citric acid cycle, as its sole source of carbon and energy. Such bacteria can metabolize citrate and synthesize all other essential carbohydrates, fats, and proteins, as well as meet their energy needs, without any additional source of carbon atoms. To pull off this trick, cells must have genes for the citrate permease enzyme, which is important for the uptake of citrate into the cells.

Simmons citrate agar

The medium used to determine if a cell can use citrate as its sole carbon source was invented by J. S. Simmons and S. A. Koser in the 1920s as a means of differentiating *Escherichia coli* (citrate-negative) and *Enterobacter aerogenes* (citrate-positive). The ability to separate these species was valuable because approximately 90%–95% of coliforms isolated from human feces are citrate-negative (*E. coli*, etc.), while 90% of coliforms found in soil samples are citrate-positive (*Enterobacter aerogenes*, etc.).

The medium, which now carries Simmons's name, has the following composition (per liter):

Ammonium dihydrogen phosphate	1.0 g
Dipotassium phosphate	1.0 g
Sodium chloride	5.0 g
Sodium citrate	2.0 g
Magnesium sulfate	0.2 g
Agar	15.0 g
Bromthymol blue	0.08 g

In this medium, citrate is the only source of carbon and energy, and the ammonium (NH_4^+) in ammonium dihydrogen phosphate is the sole source of nitrogen. Sodium chloride improves the osmotic balance, and dipotassium phosphate acts as a buffer against pH changes due to small quantities of acidic or alkaline products. Magnesium sulfate provides magnesium, which is a cofactor in certain metabolic reactions, and sulfur, for amino acids such as methionine and cysteine. **Bromthymol blue** is the pH indicator; it is the same indicator used in the O/F glucose medium and is covered in depth in the *Pseudomonas* lab (Lab 23).

Detecting the use of citrate

When cells that can use citrate as their sole carbon source and ammonium as their sole nitrogen source are transferred to citrate agar, they grow and metabolize the carbon- and nitrogen-containing components in the medium. Citrate is broken down into oxaloacetate and acetate by the action of an enzyme called citrase or citrate lyase. Then the oxaloacetate, a four-carbon molecule, is converted to pyruvate, a three-carbon molecule, and carbon dioxide (CO_2) in a reaction catalyzed by oxaloacetate decarboxylase. The smaller molecules are then used to synthesize all the other carbon-containing molecules that the cell needs to survive. So, the presence of specific enzymes is the key to survival when citrate is the only carbon source involved in metabolizing citrate.

The ammonium in ammonium dihydrogen phosphate is converted to ammonia (NH_3) and ammonium hydroxide (NH_4OH) as part of the process of using the nitrogen to synthesize amino acids. During amino acid synthesis, the nitrogen in the ammonia or ammonium will become the nitrogen in the amino ($-NH_2$) groups. As these alkaline ammonia or ammonium compounds begin to accumulate, the pH of the medium will begin to rise. Initially, the pH of the medium is near neutral, and at pH values around 7.0, bromthymol blue will be green in color. But when the medium becomes more alkaline as a result of the accumulation of ammonia, the indicator will turn blue.

Positive and negative reactions for the citrate assay

Positive for ability to use citrate as the sole carbon source

If the cells on Simmons citrate agar can use citrate as their sole carbon source and ammonium salts as their only nitrogen source, they will begin to grow and divide, eventually producing visible colonies. Metabolism of citrate and the ammonium salts will

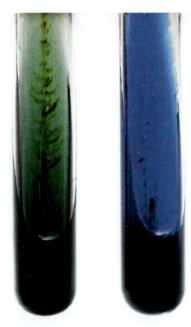

FIGURE 25.4 Results of the citrate assay.
A negative (green) reaction is shown on the left and a positive (blue) reaction is shown on the right.

produce ammonia as described above, and the rise in medium pH will turn the indicator blue.

So, if there is visible, macroscopic evidence of cell growth and if the medium is deep or royal blue in color, this is considered a positive reaction for the ability to use citrate as the sole carbon source (Figure 25.4, right tube).

In the context of separating the two coliform species *E. coli* and *E. aerogenes*, a positive reaction means that the cells must be *E. aerogenes*, because these cells can grow in an environment in which citrate is the only carbon source. *E. coli* cannot survive on citrate alone, and *E. coli* cells will not grow on the citrate medium.

Negative for high concentration of acids by fermentation

If the cells on Simmons citrate agar cannot survive on citrate as their sole carbon source, they will not grow, and they will not produce alkaline end products. The pH of the medium will remain near neutral, and the indicator will remain green in color.

So, if there is no evidence of cell growth, and if the medium is green in color, this is considered a negative reaction for the ability to use citrate as the sole carbon source (Figure 25.4, left tube). In the context of distinguishing between the two coliform species *E. coli* and *E. aerogenes*, a negative reaction means that the cells must be *E. coli* and are likely of fecal origin.

Note that a heavy initial inoculum may leave behind a visible layer of "slime," which can lead to a false positive if this material is confused with colonies generated by actual growth on the medium. If there is any doubt about what you are seeing, let the final color of the medium be your guide.

UREASE ASSAY

The **urease assay** is designed to detect the ability of bacteria to produce an enzyme called **urease** that breaks down **urea** into **ammonia (NH_3)** and carbon dioxide (CO_2). The ability to produce urease is relatively uncommon among the enterics, so when an unidentified isolate tests positive for the urease assay, it considerably narrows down the list of possible suspects. *Proteus* species, in particular, are noted for their ability to quickly generate large amounts of urease and then rapidly hydrolyze urea to ammonia and carbon dioxide, especially in broth-based media. Positive reactions among *Proteus* species may be seen as soon as 12–18 hours after inoculation. Perhaps not coincidentally, *Proteus* species can cause urinary tract and kidney infections; that is, the cells of *Proteus* species do well in locations where urea is present in high concentrations. When these species infect the kidneys, their production of ammonia from urea may result in considerable damage to the organs.

There are a few other enteric genera, such as *Klebsiella* and *Citrobacter*, in which some species can produce urease. But in many cases, the production of the enzyme is less vigorous than in *Proteus* species, and it may take longer for these urease-producing species to produce a positive reaction in urea broth. A quicker reaction may be seen when these "slow" or "late-positive" species are grown in an agar-based urea medium such as the type found in one of the compartments of the EnteroPluri tube assays (Lab 27). In this lab, however, we are using a broth-based urea medium and short incubation times, so for our purposes, the only urea-positive species will be *Proteus* species.

Urea broth

Urea broth contains the following components (per liter):

Yeast extract	0.1 g
Monopotassium phosphate	9.1 g
Dipotassium phosphate	9.5 g
Urea	20.0 g
Phenol red	0.01 g

The broth contains 2% urea as a substrate for urease. Yeast extract is added to provide carbon, nitrogen, and vitamins. However, the concentration of yeast extract is deliberately kept low, which reduces the ability of late-positive or slow urease enzyme–producing species to produce urease. In addition, the medium is buffered by the mono- and dipotassium phosphate, so species that produce only small quantities of ammonia will not be able to significantly change the pH of the broth. The pH will change dramatically only if a given isolate is able to produce large quantities of urease and, as a result, is able to generate a high concentration of ammonia. Changes in pH are detected by phenol red, an indicator discussed in earlier labs in the context of phenol red carbohydrate broths and fermentations that produce organic acids (see Lab 10). In this case, we are looking for a color shift from orange in the uninoculated, near-neutral pH broth to a bright pink when ammonia is produced and the pH rises.

Positive and negative reactions for the urease assay

Positive for ability to produce large amounts of urease rapidly If the cells in the urea broth can produce large amounts of the urease enzyme rapidly, even under low-nutrient conditions, then the urea will be quickly converted to ammonia and carbon dioxide. The accumulation of large amounts of ammonia will raise the pH of the medium, despite the presence of phosphate buffers. Under these conditions, most of the phenol red pH indicator molecules will be in the red form, and the medium will change from an initial orange color (neutral pH) to a Day-Glo hot pink color (alkaline pH).

So, if the urea broth turns bright pink within 24 hours after inoculation, this is considered a positive reaction for the ability to produce large amounts of urease rapidly (Figure 25.5, right tube).

In the context of identifying enteric species, a positive reaction means that the cells are most likely *Proteus* species. While many *Klebsiella* and *Citrobacter* species express low levels of urease, the combination of low nutrient levels and high buffering capacity in urea broth is designed to prevent these species from producing positive reactions. In other words, this medium differentiates *Proteus* and a few other closely related species from those species that produce little or no urease at all.

Negative for ability to produce large amounts of urease rapidly If the cells in the urea broth do produce urease at all, or if they only slowly produce small amounts of the enzyme, then little or no urea will be converted to ammonia and carbon dioxide. In the absence of high concentrations of ammonia, the high buffering capacity of the urea broth will prevent significant changes in pH, and the medium pH will remain near neutral. At these pH levels, the number of phenol red molecules in the yellow form is

FIGURE 25.5 Results of urease assay. A negative (tan-orange) reaction is shown on the left and a positive (bright pink) reaction is shown on the right.

similar to the number of molecules in the red form, and the indicator will be orange in color.

So, if the urea broth remains unchanged or orange in color 24 hours after inoculation, this is considered a negative reaction for the ability to produce large amounts of urease rapidly (Figure 25.5, left tube).

A negative reaction could mean that the cells did not produce any urease, or that urease levels were too low to cause a pH change sufficient to change the indicator color to bright pink. Some types of urea media contain higher nutrient levels and lower buffering capacities, and these media can differentiate between species incapable of producing any urease at all from those in genera such as *Klebsiella* that produce "late-positive" reactions due to low levels of the enzyme.

SULFIDE INDOLE MOTILITY (SIM) ASSAYS

The **SIM assays** are a set of tests for three independent traits. All of the assays can be done at once by inoculating cells into a single tube containing a single medium, called **SIM agar**. The acronym "SIM" is an abbreviation for sulfide production, indole production, and motility, which are the three characteristics that can be examined using SIM agar. Since one tube gives us information about three different traits, SIM agar is quite useful for the separation and identification of many enteric species. The composition of SIM agar is (per liter):

Pancreatic digest of casein (peptones)	20.0 g
Peptic digest of animal tissue (peptones)	6.1 g
Ferrous ammonium sulfate	0.2 g
Sodium thiosulfate	0.2 g
Agar	3.5 g

The significance of each component will be discussed in the context of the individual assays that constitute the SIM assay.

Sulfide production assay

The **sulfide production assay** is designed to detect the generation of hydrogen sulfide (H_2S) from sulfates (SO_4^{2-}), thiosulfates ($S_2O_3^{2-}$), or from sulfur-containing amino acids such as cysteine. The sulfates, thiosulfates, and cysteine are provided in the medium by ferrous ammonium sulfate, sodium thiosulfate, and the amino acids in the peptones, respectively. Sulfide production from inorganic salts involves the reduction of sulfur from the +6 (SO_4^{2-}, sulfate) or +2 ($S_2O_3^{2-}$, thiosulfate) oxidation state to the –2 (S^{2-}, sulfide) oxidation state. When cysteine is the source of the sulfur, the reaction entails the removal of the sulfhydryl (–SH) group from the amino acid, producing H_2S and the amino acid alanine **(Figure 25.6)**.

Hydrogen sulfide is a toxic gas with a strong rotten egg smell; in theory, we could detect it by sniffing the SIM tubes … but we're not going to do that. Instead, the medium contains ferrous iron (Fe^{2+}), provided by the ferrous ammonium sulfate salt. Ferrous ions will readily combine with any highly reactive sulfide to form a black precipitate composed of ferrous sulfide ($Fe^{2+} + S^{2-} \rightarrow FeS$). So, the appearance of a black precipitate reveals the production of hydrogen sulfide.

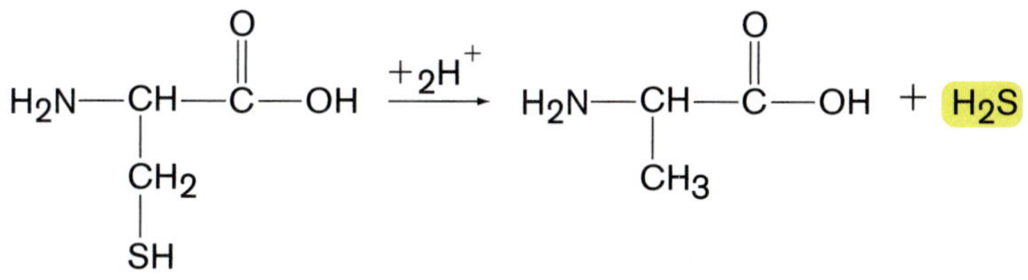

$$H_2N-CH-\overset{\overset{\displaystyle O}{\|}}{C}-OH \xrightarrow{+2H^+} H_2N-CH-\overset{\overset{\displaystyle O}{\|}}{C}-OH + \boxed{H_2S}$$

with CH₂–SH below the left CH, and CH₃ below the right CH

Cysteine   Alanine

FIGURE 25.6 Production of hydrogen sulfide from cysteine.

Positive and negative reactions for the sulfide assay

Positive for the ability to produce hydrogen sulfide If the cells in the SIM agar tube can produce hydrogen sulfide from the inorganic sulfate salts, cysteine molecules, or both, then the sulfide will react rapidly with the ferrous iron ions to produce a black ferrous sulfide precipitate. Note that the black color is not actually the hydrogen sulfide itself, but a product of the reaction between hydrogen sulfide molecules and iron.

So, if the SIM agar darkens or turns black where the cells were deposited by an inoculating needle, this is considered a positive reaction for the ability to produce hydrogen sulfide (Figure 25.7, right tube).

While there are exceptions, hydrogen sulfide–producing species within the enterics are most often from the *Salmonella*, *Proteus*, and *Citrobacter* genera.

Negative for the ability to produce hydrogen sulfide If the cells in the SIM agar tube cannot produce hydrogen sulfide either from the inorganic sulfate salts or from cysteine molecules, then there won't be any sulfide to react with the ferrous iron ions. No reaction with the ferrous iron means no black precipitate.

So, if the SIM agar does not darken or turn black where the cells were deposited by an inoculating needle—that is, if it retains its initial pale tan to yellow color—this is considered a negative reaction for the ability to produce hydrogen sulfide (Figure 25.7, left tube).

While there are exceptions, species within the enteric group that do not produce hydrogen sulfide are most often in the *Escherichia*, *Enterobacter*, *Klebsiella*, *Providencia*, *Serratia*, and *Shigella* genera.

Indole production assay

The **indole production assay** detects the breakdown of the amino acid **tryptophan** to **indole**, pyruvate, and ammonia **(Figure 25.8)**.

The tryptophan is provided in the medium by the pancreatic digest of casein and the peptic digest of animal tissue, both of which are rich in amino acids, including tryptophan. The breakdown of tryptophan requires an enzyme called **tryptophanase** or, more formally, L-tryptophan indole-lyase. The gene for this enzyme is carried by some, but not all, enteric species. For example, most strains of *E. coli* can convert tryptophan to indole, but the species in two other coliform genera, *Enterobacter* and *Klebsiella*, cannot.

The indole production assay uses a reagent called **Kovac's reagent** to reveal the production and presence of indole in SIM agar. The active ingredient in Kovac's reagent, **dimethyl-aminobenzaldehyde**, is dissolved in an organic solvent, butanol. Indole dissolves in organic solvents, so when Kovac's reagent is added to the top of the SIM agar, the

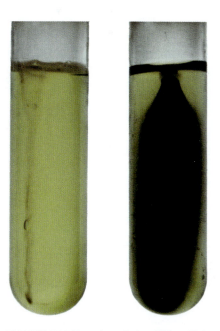

FIGURE 25.7 Results of the SIM sulfide production assay.
A negative (no black precipitate) reaction is shown on the left and a positive (black precipitate) reaction is shown on the right.

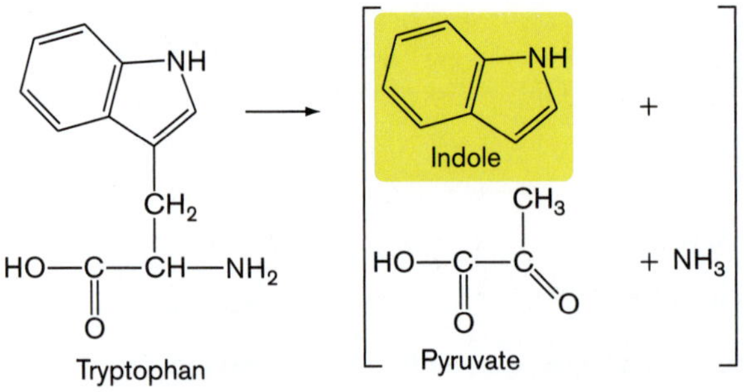

FIGURE 25.8 SIM indole production.

indole is extracted from the medium by the butanol. The dimethyl-aminobenzaldehyde then combines with the indole via the reagent's aldehyde group to produce a cherry-red end product, a di-dimethyl ammonium product **(Figure 25.9)**.

The test is designed to detect indole because that compound is specific for the degradation of tryptophan. The other two products of tryptophan degradation are pyruvate and ammonia, but these compounds are also produced by many other chemical reactions common to all enterics. Pyruvate is produced when glucose is broken down by glycolysis, and ammonia is a frequent product of amino acid degradation. Thus, by using a reagent that reacts specifically with indole, we can distinguish among certain enteric species.

Positive and negative reactions for the indole assay

Positive for the ability to produce indole from tryptophan

If the cells in the SIM agar tube can produce indole by breaking down tryptophan, then the indole will be extracted from the medium by butanol when the Kovac's reagent is added to the tube. The extracted indole will then quickly react with the dimethyl-aminobenzaldehyde in the Kovac's reagent to form a bright red end product in the butanol layer on top of the SIM agar.

So, if the layer of Kovac's reagent on top of the medium turns bright cherry red, this is considered a positive reaction for the ability to break down tryptophan to produce indole (Figure 25.10, right tube).

As a general rule, there are fewer indole-positive enteric species than indole-negative species. But among the indole-positive species is *E. coli*, so this is another example of an assay that can distinguish *E. coli* from *E. aerogenes*. It can also separate *E. coli* from *Klebsiella pneumoniae*, another important coliform species that is indole-negative. *K. pneumoniae* can cause a form of pneumonia that is frequently resistant to antibiotics.

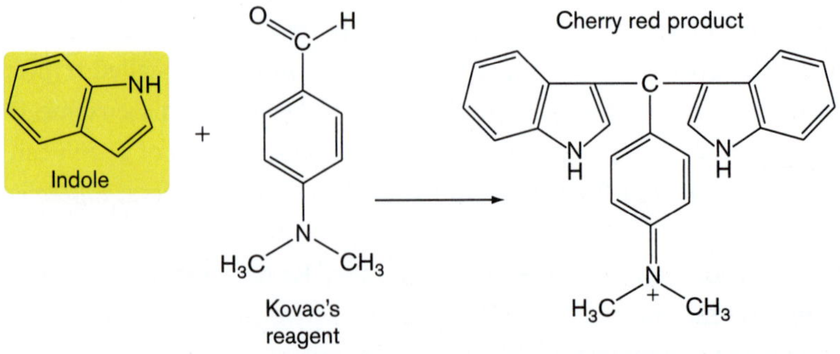

FIGURE 25.9 Detecting indole with Kovac's reagent (dimethyl-aminobenzaldehyde).

Negative for the ability to produce indole from tryptophan

If the cells in the SIM agar tube cannot produce indole by breaking down tryptophan, then there will be no indole to extract from the medium. The dimethyl-aminobenzaldehyde in the Kovac's reagent will not be converted to a bright red end product, and the layer of butanol on top of the SIM agar will remain unchanged or yellowish in color.

So, if the layer of Kovac's reagent on top of the medium remains yellow or unchanged in color, this is considered a negative reaction for the ability to break down tryptophan to produce indole (Figure 25.10, left tube).

E. aerogenes and *Klebsiella pneumoniae*, as well as the majority of enteric species, are indole-negative.

Motility assay

Many species of bacteria possess **flagella** that enable them to move from one location to another at speeds of up to 50 cell lengths per second. Most enteric species carry **peritrichous flagella**, meaning that these cells have multiple flagella distributed uniformly over the cell surface **(Figure 25.11)**. These structures are only 20–30 nm (0.02–0.03 μm) in diameter and are normally too thin to be seen with a light microscope. The **flagella stain** makes it possible to visualize these structures by "thickening" the flagella with stain in a process akin to applying mascara to eyelashes. But these delicate structures are often stripped off during the staining process, so this stain may fail to reveal the presence of flagella.

The SIM agar offers another way to test for the presence of flagella because bacteria with flagella are **motile**; that is, they can swim through the medium. In the **motility assay**, cells are deposited deep in a SIM agar tube by making a neat, straight stab with an inoculating needle. The medium in the tube contains 0.35% agar, as opposed to the 1.2%–1.5% agar concentrations used in plate media, so the SIM agar is a semisolid agar, and flagella-bearing motile cells can move into the agar from the central stab. As the motile cells disperse, grow, and divide, they will accumulate to the point at which the medium well beyond the stab becomes cloudy or turbid with cells. By contrast, if the cells cannot move, then all the cell division will be confined to the stab, and the rest of the medium will remain fairly transparent **(Figure 25.12)**.

Positive and negative reactions for the motility assay

When examining SIM agar for turbidity, always use an uninoculated control tube to compare the slight cloudiness that is due to the medium itself with the turbidity produced by dispersing motile cells. The medium is not crystal clear to begin with, and this slight background opacity should be mentally subtracted when you look at the inoculated tubes. What you are looking for is an increase in turbidity or a cloudiness that goes beyond the turbidity inherent in the uninoculated medium.

Positive for motility and the presence of flagella
If the cells in the SIM agar tube possess flagella and are motile, they will disperse from the stab, and they will make the medium around the stab more turbid or cloudy than that in the uninoculated control tube. And the stab will be less distinct than in a

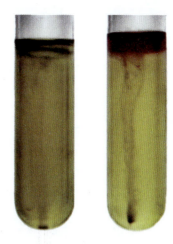

FIGURE 25.10 Results of the SIM indole production assay.
A negative (yellow or unchanged) reaction is shown on the left and a positive (cherry red) reaction is shown on the right.

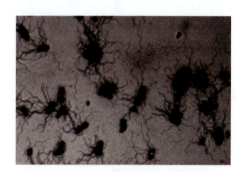

FIGURE 25.11 Peritrichous flagella.

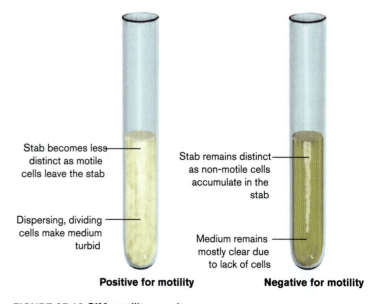

Stab becomes less distinct as motile cells leave the stab

Dispersing, dividing cells make medium turbid

Stab remains distinct as non-motile cells accumulate in the stab

Medium remains mostly clear due to lack of cells

Positive for motility   **Negative for motility**

FIGURE 25.12 SIM motility results.

tube inoculated with nonmotile species, because many of the multiplying cells will have moved out of the stab and into the agar.

So, if there is a bloom of cloudiness or turbidity extending out from the area of the stab, or if the stab is less clearly defined when compared with the stab for a known nonmotile species, this is considered a positive reaction for motility (Figure 25.12, left tube).

With the exception of *Klebsiella* and *Shigella* species, all the major enteric genera are composed of species with peritrichous flagella, so almost all enteric species are motile.

Note that when cells produce hydrogen sulfide, the black precipitate produced by its reaction with ferrous iron tends to mask the cells in the SIM agar. It can be difficult to distinguish between the light-colored cloudiness due to motility and the dark-colored cloudiness due to the diffusion of hydrogen sulfide from the cells followed by precipitation of ferrous sulfide. Fortunately, almost all sulfide producers are also motile, so if you see a positive reaction for the sulfide test, you can assume that the species is also motile.

Negative for motility and the presence of flagella If the cells in the SIM agar tube do not possess flagella, then they will not be motile. The cells will remain in the stab, and as they accumulate due to cell division, the stab will become distinct and bright white to tan in color. The medium around the stab will not contain any cells and will remain unchanged in transparency compared with an uninoculated control tube.

So, if there is no bloom of cloudiness or turbidity extending out from the area of the stab, if the medium around the stab remains relatively transparent, or if the stab itself is clearly defined by tan to white colonial growth, this is considered a negative reaction for motility (Figure 25.12, right tube).

If the isolate is a nonmotile enteric species, then there is a reasonably good chance that it is either a *Klebsiella* or a *Shigella* species.

SUMMARY

Assay	Medium	Reagent Added	Positive Reaction	Negative Reaction
Phenol red carbohydrate	Phenol red carbohydrate broth	None	Yellow	Red
Methyl red	MRVP broth	Methyl red	Red	Yellow
Lysine decarboxylase	Lysine decarboxylase broth	None	Purple	Yellow
Citrate	Citrate agar	None	Blue	Green
Urease	Urea broth	None	Bright pink	Orange, no change in color
Sulfide	SIM agar	None	Black precipitate	No black precipitate
Indole	SIM agar	Kovac's	Red	Yellow, no change in color
Motility	SIM agar	None	Medium turbid, stab indistinct	Medium not turbid, stab clearly defined

PROTOCOL

Day 1

MATERIALS

- Safety glasses
- Disposable gloves
- 4 phenol red glucose broths with Durham tubes
- 4 phenol red lactose broths with Durham tubes
- 4 phenol red sucrose broths with Durham tubes
- 3 methyl red broths
- 5 SIM agar deeps
- 3 citrate agar slants
- 2 urea broths
- Inoculating needle

Cultures:

- *Citrobacter freundii*
- *Enterobacter aerogenes*
- *Escherichia coli*
- *Proteus vulgaris*
- *Staphylococcus epidermidis*

In this lab, we will be using *Staphylococcus epidermidis* as an example of a nonmotile organism. *S. epidermidis* is <u>not</u> an enteric species; cells of this species are Gram-positive cocci. We're using *S. epidermidis* because it gives us a relatively safe example of a nonmotile bacterium, so it's a good species to use as a negative control.

Table 25.1 provides a handy chart of the inoculations performed on day 1 of this lab.

CARBOHYDRATE FERMENTATION ASSAYS

> **IMPORTANT:** *Proteus vulgaris* is a biosafety level 2 species, so gloves and safety glasses must be worn throughout this lab.

1. Collect four tubes each of the phenol red glucose, lactose, and sucrose broths, and label them with your name and the date. Note that all of these different media look alike, so be sure to **label the tubes immediately** or take any and all other steps needed to keep track of the sugar type in each tube of broth.

2. For each of the three types of phenol red broths, label four tubes as follows:

   a. *Escherichia coli*
   b. *Enterobacter aerogenes*
   c. *Proteus vulgaris*
   d. Uninoculated control

3. Inoculate tubes a–c of each broth type with pure-culture material of the appropriate species using a flame-sterilized loop. Do not add cells to the control tubes.

4. Incubate all tubes at 37°C for 48 hours.

METHYL RED ASSAY

1. Label three methyl red broths with your name and the date, and then label the tubes as follows:

   a. *Escherichia coli*
   b. *Enterobacter aerogenes*
   c. *Proteus vulgaris*

2. Inoculate all tubes with the correct species and incubate at 37°C for 48 hours.

SIM ASSAYS

1. Label five SIM agar deeps with your name and the date, and then label the tubes as follows:

   a. *Escherichia coli*
   b. *Citrobacter freundii*
   c. *Proteus vulgaris*
   d. *Staphylococcus epidermidis*
   e. Uninoculated control

2. Use a flame-sterilized inoculating needle to transfer cells from the pure cultures of the appropriate species to SIM agar deeps a–d. Inoculate each tube by stabbing the cell-coated inoculating needle **straight** down into the SIM agar. Stab to a depth of about 1 inch, and draw the needle straight back up and out of the agar without wiggling the needle from side to side. Do not add cells to the control tube.

3. Incubate all SIM tubes at 37°C for 48 hours.

CITRATE ASSAY

1. Label three citrate agar slants with your name and the date, and then label the tubes as follows:

   a. *Escherichia coli*
   b. *Enterobacter aerogenes*
   c. *Proteus vulgaris*

2. Use a flame-sterilized inoculating loop to inoculate all tubes with the correct species. Use a heavy inoculum

Table 25.1 | SUMMARY OF ENTERIC INOCULATIONS

Assay	Medium	No. of Tubes	Species Used
Glucose fermentation	Phenol red glucose broth	4	*Escherichia coli, Enterobacter aerogenes, Proteus vulgaris,* Uninoculated control
Lactose fermentation	Phenol red lactose broth	4	*Escherichia coli, Enterobacter aerogenes, Proteus vulgaris,* Uninoculated control
Sucrose fermentation	Phenol red sucrose broth	4	*Escherichia coli, Enterobacter aerogenes, Proteus vulgaris,* Uninoculated control
Methyl red	MRVP broth	3	*Escherichia coli, Enterobacter aerogenes, Proteus vulgaris*
Citrate	Citrate agar	3	*Escherichia coli, Enterobacter aerogenes, Proteus vulgaris*
Urease	Urea broth	2	*Escherichia coli, Proteus vulgaris*
Sulfide, indole, and motility	SIM agar	5	*Escherichia coli, Citrobacter freundii, Proteus vulgaris, Staphylococcus epidermidis,* Uninoculated control

(lots of slime on the loop), and streak the entire length of the citrate agar slant.

3. Incubate all tubes at 37°C for 48 hours.

UREASE ASSAY

1. Label two urea broths with your name and the date, then label the tubes as follows:

   a. *Escherichia coli*
   b. *Proteus vulgaris*

2. Use a flame-sterilized inoculating loop to inoculate the tubes with the correct species.

3. Incubate the tubes at 37°C for 48 hours.

Day 2
MATERIALS

- Inoculated tubes from day 1
- Disposable gloves
- Kovac's (indole) reagent
- Methyl red reagent
- Uninoculated SIM tube (control)
- Uninoculated phenol red broths (controls)

CARBOHYDRATE FERMENTATION ASSAYS

Examine the phenol red broths. Record (in the Results section) the color of the broth and the presence or absence of bubbles in the Durham tube for each phenol red broth tube.

METHYL RED ASSAY

Add four or five drops of the methyl red reagent to each inoculated methyl red broth. Gently shake the

tube to mix the broth and reagent. Record the colors of the broths in the Results section.

HYDROGEN SULFIDE ASSAY (SIM AGAR)

Observe the color of the agar around the stabs in all the inoculated SIM deeps. Note any blackening of the medium, and record your observations in the Results section.

MOTILITY ASSAY (SIM AGAR)

Observe the agar around the stabs in all the inoculated SIM deeps to determine if the cells are clearly confined to a sharply defined stab or if the cells have moved out from the stab, producing a less distinct stab and cloudiness (turbidity) around the stab. Use an uninoculated SIM agar tube to distinguish motility-caused turbidity from the slight cloudiness or opacity of the SIM agar itself. Record your observations in the Results section.

INDOLE ASSAY (SIM AGAR)

Put on gloves, and after recording results for the hydrogen sulfide and motility assays, add four or five drops of Kovac's reagent to each of the inoculated SIM deeps, gently shake the tube to swirl the reagent around the top of the agar, and record any changes in the color of the Kovac's reagent in the Results section.

CITRATE ASSAY

Record the colors (green/blue) of the citrate agar slants in the Results section.

UREASE ASSAY

Record the colors (orange/pink) of the urea broths in the Results section.

Name: _____   Section: _____

Course: _____   Date: _____

CARBOHYDRATE FERMENTATION (PHENOL RED GLUCOSE BROTH)

Bacterial Species	Broth Cloudy?	Broth Color	Broth pH*	Were Acids Produced by Fermentation?	Bubble (Yes/No)	Were Gases Produced by Fermentation?
E. coli						
E. aerogenes						
P. vulgaris						
Control						

CARBOHYDRATE FERMENTATION (PHENOL RED LACTOSE BROTH)

Bacterial Species	Broth Cloudy?	Broth Color	Broth pH*	Were Acids Produced by Fermentation?	Bubble (Yes/No)	Were Gases Produced by Fermentation?
E. coli						
E. aerogenes						
P. vulgaris						
Control						

CARBOHYDRATE FERMENTATION (PHENOL RED SUCROSE BROTH)

Bacterial Species	Broth Cloudy?	Broth Color	Broth pH*	Were Acids Produced by Fermentation?	Bubble (Yes/No)	Were Gases Produced by Fermentation?
E. coli						
E. aerogenes						
P. vulgaris						
Control						

*Broth pH: Choose from (a) pH < 6.8, (b) 6.8–7.8, or (c) pH > 7.8.

METHYL RED ASSAY

Bacterial Species	Broth Color	Broth pH*	Did Fermentation Produce High Acid Concentration?	Positive or Negative?
E. coli				
E. aerogenes				
P. vulgaris				

*Broth pH: Choose from (a) pH < 4.6, (b) 4.6–5.6, or (c) pH > 5.6.

HYDROGEN SULFIDE ASSAY (SIM AGAR)

Bacterial Species	Color of Medium Around the Stab	Positive or Negative?
E. coli		
C. freundii		
P. vulgaris		
S. epidermidis		

MOTILITY ASSAY (SIM AGAR)

Bacterial Species	Is Medium Cloudy Around Stab?	Positive or Negative?
E. coli		
C. freundii		
P. vulgaris		
S. epidermidis		

INDOLE ASSAY (SIM AGAR)

Bacterial Species	Color of Added Kovac's Reagent	Positive or Negative?
E. coli		
C. freundii		
P. vulgaris		
S. epidermidis		

CITRATE ASSAY

Bacterial Species	Agar Color	pH (acidic, neutral, or alkaline)	Positive or Negative?
E. coli			
E. aerogenes			
P. vulgaris			

UREASE ASSAY

Bacterial Species	Broth Color	pH (acidic, neutral, or alkaline)	Positive or Negative for this Assay?
E. coli			
P. vulgaris			

Name: _____   Section: _____

Course: _____   Date: _____

1. If the only assays you had available were the phenol red glucose, lactose, and sucrose assays, would you be able to distinguish *E. coli* from *E. aerogenes*? Explain your answer.

2. If the only assays you had available were the phenol red glucose, lactose, and sucrose assays, would you be able to distinguish *E. coli* from *P. vulgaris*? Explain your answer.

3. The sugar in the methyl red broth is glucose, and all enteric species (family Enterobacteriaceae) can ferment glucose. So why are some enteric species methyl red–positive and some species methyl red–negative? That is, how does the methyl red assay separate or distinguish one group of enterics from another?

4. If the only assay you had available was the methyl red assay, would you be able to distinguish *E. coli* from *E. aerogenes*? Explain your answer.

5. What is the value of an uninoculated SIM control tube when you are trying to determine if a species is motile or not?

6. Why is it difficult to determine motility (in SIM agar) when a species is positive for hydrogen sulfide production?

7. If you just wanted to determine motility for a H_2S-positive species, how would you change the formula of the SIM agar to prevent problems with H_2S production? Explain your reasoning.

8. In the Kovac's reagent used in the indole assay, what are the roles of the butanol and dimethyl-aminobenzaldehyde, respectively?

9. Say you started with a mixed culture of *Escherichia coli* and *Enterobacter aero-genes*. If you were going to choose one of the media described in this lab to favor the growth of *Enterobacter aerogenes* cells over that of *E. coli* cells (i.e., if you were using the medium as a selective medium for *Enterobacter*), which medium would you chose? Explain your answer.

10. Why is the urease assay so useful for the identification of *Proteus* species, and why is the answer relevant to *Proteus* species' ability to cause kidney infections?

LAB 26

Identification of Unknown Enteric Species

Learning Objectives

- Use your knowledge of various assays described in previous labs to identify the genus and species of a bacterium in an unknown enteric culture.

In this lab, you will perform a series of assays described in Lab 25 and use the results to identify an unknown Gram-negative bacterium from the Family Enterobacteriaceae. This is similar to identifying Gram-positive unknowns in Lab 22, but in this case, you will generate and maintain your own pure cultures, perform many more assays and identify an unknown from a longer list of possible answers. To succeed, you will need to apply a number of the methods used throughout this lab course. It is especially important that you use all of your talents to prevent contamination of the pure culture of your unknown bacterium, both when you create it and as you inoculate the various media. This exercise is a challenge, but many students find identifying unknown bacteria to be one of the most anticipated and exciting experiences in an introductory microbiology course. So, apply your skills and you will find the answer to one of life's persistent questions. Of course, to answer the Great Question of Life, the Universe and Everything, you will need to consult the Hitchhiker's Guide to the Galaxy.

Note that Lab 27 provides a protocol for running miniaturized versions of these assays and can be run in conjunction with Lab 26 to identify your unknown bacterium. **Figure 26.1** provides a flowchart of the steps in this lab.

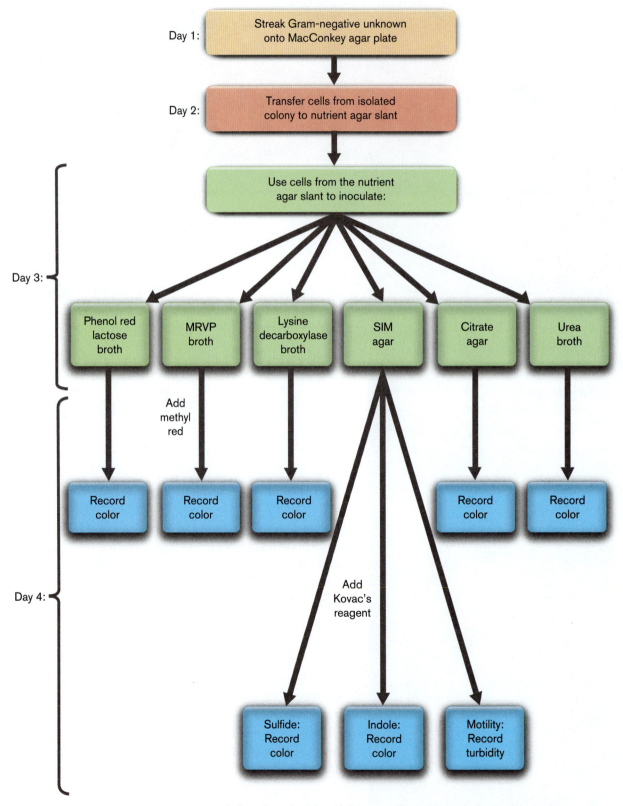

FIGURE 26.1 A flowchart for identifying an unknown enteric bacteria species.

PROTOCOL

Day 1
MATERIALS

- MacConkey agar plate (1 per student)
- Bunsen burner
- Striker
- Inoculating loop
- Wax pencils or Sharpies for labeling plates (or masking tape and pens)
- Disposable gloves
- Safety glasses

Cultures:
- Unknown Gram-negative enteric species

> **IMPORTANT:** Some of the unknown enterics are biosafety level 2 species, so gloves and safety glasses must be worn throughout all four days of this lab.

1. Select one Gram-negative culture from the set of numbered Gram-negative unknowns. Record the ID number of your unknown culture on your MacConkey agar plate.

2. Streak the MacConkey agar plate with the unknown culture following the streak plating protocol used to produce isolated colonies (see Lab 3).

3. Incubate the plate at 37°C for 24 hours.

Day 2
MATERIALS

- Nutrient agar slant (1 per student)
- Streaked MacConkey agar plate from day 1
- Disposable gloves
- Safety glasses

1. Show your MacConkey agar plate to your lab instructor. It is important to verify that you have isolated colonies.

2. Record the appearance of the colonies in the Results section.

3. Pick one isolated colony growing on the MacConkey agar plate. Use a sterile loop to transfer cells from this colony to a nutrient agar slant to produce a pure culture of your Gram-negative unknown. Label the slant with your name and the ID number of your unknown.

Be sure to run your loop up and down the face of the slant several times so that cells grow over the entire slant. This is going to be your stock culture for the next day's work, and you're going to want to have plenty of cells available to inoculate six different types of media.

4. Incubate the slant at 37°C for 24 hours.

Day 3
MATERIALS

Per student:
- 1 MRVP broth
- 1 SIM agar deep
- 1 citrate agar slant
- 1 urea broth
- 1 phenol red lactose broth
- 1 lysine decarboxylase broth
- Sterile mineral oil
- Slant of Gram-negative unknown from day 2
- Disposable gloves
- Safety glasses

1. Label tubes of the six types of media listed above with your name and the ID number of your Gram-negative unknown.

2. Inoculate the six types of media with cells from your stock culture (on the nutrient agar slant you inoculated on day 2), following the procedures described in Lab 10 for lysine decarboxylase broth and in Lab 25 for all the other media.

   Don't forget to add mineral oil on top of the lysine decarboxylase broth after inoculating the tube.

3. Incubate all media at 37°C for 48 hours.

Day 4
MATERIALS

- Inoculated media from day 3
- Methyl red reagent
- Kovac's reagent
- Disposable gloves
- Safety glasses

1. Use the procedures described in Labs 10 and 25 to interpret your results from the media inoculated on day 3 with your Gram-negative enteric unknown. Record your observations and results in the table provided in the Results section.

   To help determine positive and negative results for each assay, use photos and your observations of the reactions in media inoculated with known species in Lab 25. For lysine decarboxylase results, refer to Lab 10, Figure 10.9.

2. Use all the results to identify your unknown. Write your answer in the table in the Results section. Table 26.1 can assist you in identifying your unknown.

TABLE 26.1 TYPICAL RESULTS OF TESTS FOR IDENTIFYING ENTERIC BACTERIA.

Bacterial Species	Lactose Fermenter	Indole	Methyl Red	Citrate*	H_2S	Motility	Urea	Lysine Decarboxylase[†]
Salmonella Typhimurium	–	–	+	+	+	Weak + (H_2S interferes)	–	+
Shigella flexneri	–	–	+	–	–	–	–	–
Escherichia coli	+	+	+	–	–	+	–	+
Citrobacter freundii	+	–	+	+	+	Weak + (H_2S interferes)	–	–
Klebsiella pneumoniae	+	–	+	+	–	–	+[‡] (late)	+
Enterobacter aerogenes	+	–	–	+	–	+	–	+
Serratia marcescens	–	–	–	+	–	+	–	+
Proteus vulgaris	–	+[§]	+	+	+	+ (H_2S interferes)	+	–
Providencia alcalifaciens	–	+	+	+	–	+	–	–

*Any blue-green is positive; compare to *E. coli* to see true negative.

[†]Positive tubes should be deep purple; dark yellow is negative.

[‡]"Late" means that it takes 3–7 days to develop a positive reaction.

[§]Most *Proteus* strains are positive for indole, but some strains are negative, so check with your instructor about what is to be expected with your particular lab strain.

Name: _____   Section: _____

Course: _____   Date: _____

Gram-negative unknown culture ID number:

Did your streak produce isolated colonies on the MacConkey agar plate?

Colony morphology on MacConkey agar plate, including colony and medium color:

Assay	Observations (Describe what you see)	Interpretation (positive/negative)
Lactose fermentation		
Indole production		
Methyl red		
Citrate use		
H_2S production		
Motility		
Urease		
Lysine decarboxylase		
The genus and species name of my Gram-negative unknown no. _____ is:		

Name: _____

Section: _____

Course: _____

Date: _____

1. I came to my conclusion about my unknown because:

2. Do a little research and write a short paragraph describing where your particular Gram-negative species might be found outside the human body, and describing how it interacts with us when it's in the human body. As part of this description, consider which biochemical assays might be relevant to the activity of the organism inside the body. Determine if your organism is normally found in humans or if it is found in the body only in cases of disease. If your organism can cause illness, for each of the diseases caused by your unknown, describe (a) the modes of transmission, (b) the nature and symptoms of the diseases, and (c) how the diseases would be treated.

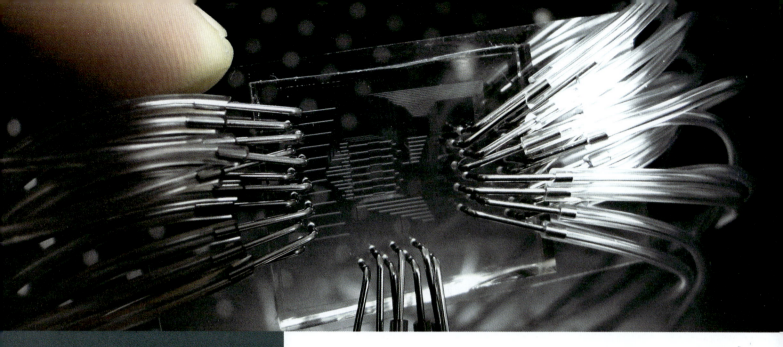

The tests performed in the previous two labs have been miniaturized and assembled into plastic tubes about the size of a pen. Using these products in a clinical diagnostic lab saves time and money, and sometimes patients' lives. You will use one of these products in this lab. But tests are shrinking ever smaller. Above is a lab-on-a-chip—a dynamic system about the size of a quarter, in which fluids flow through tubes etched into the chip in a manner analogous to the printing of a circuit board onto a tiny computer chip.

LAB 27

Identification of Enteric Bacteria Using the EnteroPluri System

Learning Objectives

- Define the terms bacterial strain and bacterial species.

- Understand that variation among strains must be considered when identifying bacterial species and when differentiating one species from another.

- Describe how identification tools such as the EnteroPluri system incorporate multiple assays into a single device to quickly identify unknown bacteria.

- Identify enteric bacteria using results from inoculated EnteroPluri tubes and the *EnteroPluri Code Book*.

IDENTIFYING BACTERIAL SPECIES AND STRAINS

In clinical and research lab settings, people often need to rapidly determine the species names for a very large number of unidentified pure bacterial cultures. This task often requires the use of many different types of media, and it can be tedious and time-consuming to produce and inoculate many sets of tubes of different media, as we have done in several labs. It is made even more difficult by the reality that there is often variation in test results among the different strains of a given bacterial species.

A **bacterial strain** or **variant** is a population of cells derived from a single cell. So, potentially, all of the cells of a given strain would be genetically identical to each other, that is, they would all be exact genetic clones of the original cell. In practice, due to independent mutations in different descendant cells, there will be some minor genetic differences among cells within a strain. What's significant in the context of this lab is that all of the cells of a given strain or variant will be very nearly identical to each other in their genes and very similar in their responses in biochemical assays. **Bacterial species**, on the other hand, is a term that applies to collections of different strains—in some cases, collections of hundreds of different strains. Though not genetically identical, all of the strains share many genetically stable traits, and they differ significantly from other collections of strains or other species. The relationship of "strain" to "species" is roughly similar to the relationship between breeds of dogs and the species *Canis familiaris*. All breeds of dogs are called "dogs" (*C. familiaris*), and no breeds of dogs are said to be of the "jackal" species (*Canis mesomelas, Canis aureus*), but there is still a considerable difference within the *C. familiaris* species between a Chihuahua and a Saint Bernard.

As a general rule, strains within a species are similar enough to one another that they usually produce the same results for most of the assays used to identify bacteria. But since the strains within a species are not genetically identical to one another, two strains within a species may produce different results for a given assay. For example, about 95% of *E. coli* strains are strong lactose fermenters, but about 5% are weak fermenters or do not ferment lactose **(Figure 27.1)**. So if we happen to have an odd *E. coli* strain, and if we rely too heavily on one assay (the lactose fermentation assay), we might fail to properly identify our culture as *E. coli*.

So, in summary, what do we need to rapidly identify a large number of unknown cultures?

1. We need to perform many different biochemical assays in a short time, and we don't want to spend a lot of time making media.

2. We need to know something about the variation in biochemical or metabolic activity among many different strains within a species. That is, we need a "database of variation" among the strains of a given species **(Figure 27.1)**.

Table of Biochemical Reactions

Genus	Species	Glucose	Gas	Lysine	Ornithine	H2S	Indole	Adonitol	Lactose	Arabinose	Sorbitol	VP	Dulcitol	PA	Urea	Citrate
Citrobacter	C. freundii	100	95	0	20	80	5	0	50	99	98	0	55	0	70	95
	C. koseri	100	98	0	99	0	100	98	35	98	99	0	50	0	75	99
	C. amalonaticus	100	97	0	95	0	100	0	50	99	97	0	0	0	80	85
	C. farmeri	100	93	0	100	0	100	0	19	99	100	0	4	0	45	1
Enterobacter	E. aerogenes	100	100	98	98	0	0	98	95	100	99	98	5	0	2	95
	E. asburiae	100	95	0	95	0	0	0	75	100	100	2	0	0	65	100
	E. cloacae	100	100	0	96	0	0	25	93	99	95	99	15	0	65	98
	E. gergoviae	100	98	90	100	0	0	0	55	100	0	100	0	0	93	99
	E. sakazakii	100	98	0	91	0	11	0	99	100	0	100	5	50	1	99
Escherichia	E. coli	100	95	90	65	1	98	5	95	99	94	0	60	1	1	1
	E. fergusonil	100	95	95	100	0	98	98	0	98	0	0	60	0	0	0
	E. hermanii	100	97	6	100	0	99	0	45	100	0	0	19	0	0	0
	E. vulneris	100	97	85	0	0	0	0	15	100	1	0	0	0	0	0
	E. blattae	100	100	100	100	0	0	0	0	100	0	0	0	0	0	50
Klebsiella	K. pneumoniae	100	97	98	0	0	0	90	98	99	99	98	30	0	95	98
	K. oxytoca	100	97	99	1	0	99	99	100	98	99	95	55	1	90	95

FIGURE 27.1 Percentage of strains that produce a positive assay reaction.
For example, 60% of *E. coli* strains produce a positive reaction for the dulcitol carbohydrate fermentation assay (number is highlighted in red). VP is Voges-Proskauer; PA is Phenylalanine. *Source*: Adapted from EnteroPluri Code Book, Liofilchem S.l.r., Roseto, Italy.

3. We need an identification system that uses such a database to take variation within species into account. This way, when a given strain produces a reaction that is unusual or atypical for a given species for one assay out of many, this does not lead to a failure to properly identify the bacterium.

Fortunately, there is a great deal of money to be made in supplying microbiologists and medical technologists with what they need. To supply what is needed to identify enteric bacteria, at least two systems have been developed: the EnteroPluri system and the API 20E system. In this lab, we will use the EnteroPluri system, but the API 20E system works in a similar manner.

THE ENTEROPLURI SYSTEM

The **EnteroPluri system** is a rapid, multi-test system used in the identification of the Gram-negative rod species in the family Enterobacteriaceae (enterics). It consists of a tube with twelve chambers containing different media, all of which can be inoculated by touching a needle at one end of the tube to a single isolated colony (**Figure 27.2**). The needle is then drawn through the twelve chambers in the tube, inoculating the twelve different media in a few seconds. A total of fifteen different assays can be performed using these twelve different kinds of media. These assays include seven carbohydrate fermentation assays (glucose, gas from glucose fermentation, lactose, adonitol, arabinose, sorbitol, and dulcitol) and two different amino acid decarboxylase assays (lysine and ornithine), as well as the indole, hydrogen sulfide, urea, citrate, phenylalanine deamination, and Voges-Proskauer assays.

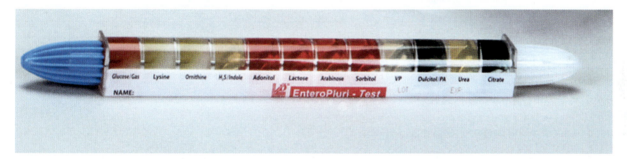

FIGURE 27.2 An EnteroPluri system tube.

After incubation, each biochemical assay is scored either positive or negative, based on the color of the medium in each chamber (**Figure 27.3**). If a given assay is positive, then the number below that assay is circled on a table provided by the manufacturer; if the assay is negative, the number below the assay remains unmarked. After all assays have been scored, the circled numbers are added together in five separate clusters. The result is a five-digit number referred to as a **biocode**.

For example, if the unknown isolate has a positive reaction for:

1. acid from glucose fermentation
2. gas from glucose fermentation
3. lysine decarboxylase
4. ornithine decarboxylase
5. indole production
6. lactose fermentation
7. arabinose fermentation

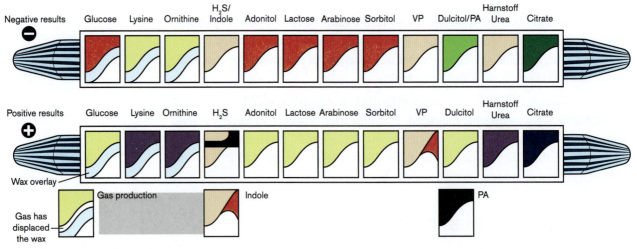

Sector	Biochemical Reactions	Sector Color	
		Positive Reaction	**Negative Reaction**
Glucose/Gas	Glucose fermentation	yellow	red
	Gas production	lifted wax	overlaying wax
Lysine	Lysine decarboxylation	violet	yellow
Ornithine	Ornithine decarboxylation	violet	yellow
H₂S/Indole	Hydrogen sulfide production	black-brown	beige
	Indole test	cherry-red	colorless
Adonitol	Adonitol fermentation	yellow	red
Lactose	Lactose fermentation	yellow	red
Arabinose	Arabinose fermentation	yellow	red
Sorbitol	Sorbitol fermentation	yellow	red
VP	Acetoin production	red	colorless
Dulcitol/PA	Dulcitol fermentation	yellow	green
	Phenylanine deamination	dark brown	green
Urea	Urea hydrolysis	purple	beige
Citrate	Citrate utilization	deep blue	green

FIGURE 27.3 EnteroPluri tube test results.
As in Lab 25, for each assay designed to identify enteric species, there are colors that correspond to positive and negative reactions, respectively.

8. sorbitol fermentation

9. dulcitol fermentation

Then the five-digit biocode generated will be **7 5 3 5 0 (Figure 27.4)**.

Test	Group 1			Group 2			Group 3			Group 4			Group 5		
	Glucose	Gas	Lysine	Ornithine	H₂S	Indole	Adonitol	Lactose	Arabinose	Sorbitol	VP	Dulcitol	PA	Urea	Citrate
Value if positive	④	②	①	④	2	①	4	②	①	④	2	①	4	2	1
Results	+	+	+	+	−	+	−	+	+	+	−	+	−	−	−
Code	4 + 2 + 1 = 7			4 + 0 + 1 = 5			0 + 2 + 1 = 3			4 + 0 + 1 = 5			0 + 0 + 0 = 0		

FIGURE 27.4 *E. coli* strain results entered into an EnteroPluri system data table.
Values for a given assay are circled if the reaction is positive; all positive (circled) numbers within a group are added to produce the five digits of a biocode.

The biocode is a product of all the results from all the assays, and it can be matched to a specific enteric species using the *EnteroPluri Code Book*, which contains hundreds of bio-codes corresponding to dozens of enteric species. If you look up biocode 75350, you will find that the number tells you that the unknown isolate is *Escherichia coli*. In this case, there are no "atypical results"; that is, this strain does exactly what we expect *E. coli* to do.

But the code book also takes strain variation into account, and there are many, many other biocodes that correspond to other related strains of *E. coli*, and entering these other biocodes will also lead to a species identification of *E. coli*. These biocodes lead to *E. coli* strains in which the assay results are almost, but not exactly, identical to those for the "typical" *E. coli* with its biocode of 75350. In the case of these strains, one or two assay results are atypical, and so the set of circled numbers will be slightly different, and ultimately, the biocode will be slightly different.

For example, there is an atypical strain of *E. coli* that is capable of producing urease. This strain will produce a "urease positive" result, and the number "2" will be circled in the last set of assays. All other assay results are the same as those seen with typical *E. coli*, so this strain is still behaving more like *E. coli* than any other enteric species. When all the numbers are added up, they generate the biocode **7 5 3 5 2 (Figure 27.5)**. This is obviously different from the biocode of the typical *E. coli*, but there's no need to fear, the *EnteroPluri Code Book* is here. The code book knows all, including the fact that some strains of *E. coli* produce urease. The code book takes this variation into account, and the number 75352 will still give you the correct identification of *E. coli*.

Test	Group 1			Group 2			Group 3			Group 4			Group 5		
	Glucose	Gas	Lynine	Ornithine	H₂S	Indole	Adonitol	Lactose	Arabinose	Sorbitol	VP	Dulcitol	PA	Urea	Citrate
Value if positive	④	②	①	④	2	①	4	②	①	④	2	①	4	②	1
Results	+	+	+	+	−	+	−	+	+	+	−	+	−	+	−
Code	4 + 2 + 1 = 7			4 + 0 + 1 = 5			0 + 2 + 1 = 3			4 + 0 + 1 = 5			0 + 2 + 0 = 2		

FIGURE 27.5 Results for a second *E. coli* strain entered into an EnteroPluri system data table. Assay reactions yield a different biocode from Figure 27.4, but the code book still identifies the unknown as *E. coli*.

We present two versions of the EnteroPluri lab here. In the first protocol, you will look at a number of EnteroPluri tubes previously inoculated by your instructor with different Gram-negative organisms. You will then determine the biocode for each tube and use the EnteroPluri codebook to identify these enteric bacteria. In the second protocol, you will use the EnteroPluri tube in conjunction with the same Gram-negative Unknown and the tests you ran in Lab 26.

You may be wondering, if the EnteroPluri system is so great, why didn't we use it in Lab 26 for identifying enteric unknowns? There are two reasons. First, we want you to focus on and learn each individual assay, including the key medium components, how each assay works, positive and negative reactions, what the assay is particularly useful for, and so forth. When a single tube contains twelve different media and fifteen different biochemical assays, it is more difficult to focus on the details of each individual assay. Second, each EnteroPluri tube costs over $15, and to put it simply, $15 per student for a single experiment can become very expensive. So, you had to do things the hard way in Lab 26.

PROTOCOL 1: INSTRUCTOR-INOCULATED ENTEROPLURI TUBES

> **IMPORTANT:** Some of the unknown enterics are bio-safety level 2 species, so gloves and safety glasses must be worn throughout this lab.

Day 1
MATERIALS

- Safety glasses
- Disposable gloves
- *EnteroPluri Code Book*
- Pre-inoculated EnteroPluri tubes

1. Examine each of the EnteroPluri tubes inoculated with a different Gram-negative unknown species. There will be four to six different Gram-negative species among all of the tubes, and for each unknown, there are two or three tubes inoculated with that species.

2. Results for some of the assays will be provided on sheets next to the tubes. Indole and VP results are provided because adding the Kovac's and VP reagents to the tubes creates a mess when tubes are to be shared among many students. Record indole, VP, and any other provided assay results in the EnteroPluri tables in the Results section.

3. For all other assays, examine the tubes and score each assay as either positive or negative. In the Results section, circle the number below the assay if positive and do not circle the number if negative.

4. Add up the numbers for positive assays as indicated on the EnteroPluri tables provided in the Results section, and use the *EnteroPluri Code Book* to determine the identity of each unknown.

PROTOCOL 2: USING THE ENTEROPLURI SYSTEM

The EnteroPluri system can be used to identify the unknown Gram-negative species assigned in Lab 26. Students can use the Lab 26 protocol and this protocol concurrently to identify their unknown culture if the same culture and MacConkey plate are used.

> **IMPORTANT:** Some of the unknown enterics are bio-safety level 2 species, so gloves and safety glasses must be worn throughout all three days of this lab.

Day 1
MATERIALS

- Safety glasses
- Disposable gloves
- MacConkey agar plate (1 per student)

Cultures:
- Unknown Gram-negative enteric species

1. Select one Gram-negative culture from the set of numbered Gram-negative unknowns. Record the ID number of your unknown culture on your MacConkey agar plate.

2. Streak the MacConkey agar plate with the unknown culture following the streak plating protocol used to produce isolated colonies (see Lab 3).

3. Incubate the plate at 37°C for 24 hours.

Day 2
MATERIALS

- Safety glasses
- Disposable gloves
- EnteroPluri tube
- Streaked MacConkey plate from day 1

1. Show your MacConkey agar plate to your lab instructor. It is important to verify that you have isolated colonies.

2. Record the appearance of the colonies in the Results section.

3. Label an EnteroPluri tube with your name and the ID number of your unknown.

4. Holding the tube in the middle, remove both caps of the EnteroPluri tube.

Figure 27.2 illustrates the procedures in steps 5 through 11.

5. Continuing to hold the tube in the middle, touch the tip of the inoculating needle (under the white cap) to one and only one isolated colony growing on the MacConkey agar plate to transfer cells from the colony to the needle (**Figure 27.6A**).

   Your goal is to pick up as much of the colony as possible, but avoid digging into the agar, and do not touch any other colonies with the needle.

6. Hold the looped-handle end of the needle (under the blue cap) and inoculate all the chambers in the EnteroPluri tube by turning the needle as you withdraw it from the tube (**Figure 27.6B**).

7. Once the tip of the needle has deposited cells in the last chamber, push the needle back through all of the chambers, turning the needle as you go, to add more cells to each chamber.

8. Again, slowly withdraw the needle with a turning motion until you see the breakage notch in the needle, located about a quarter to a third of the distance from the looped-handle end. The notch is a small groove carved into the shaft of the needle.

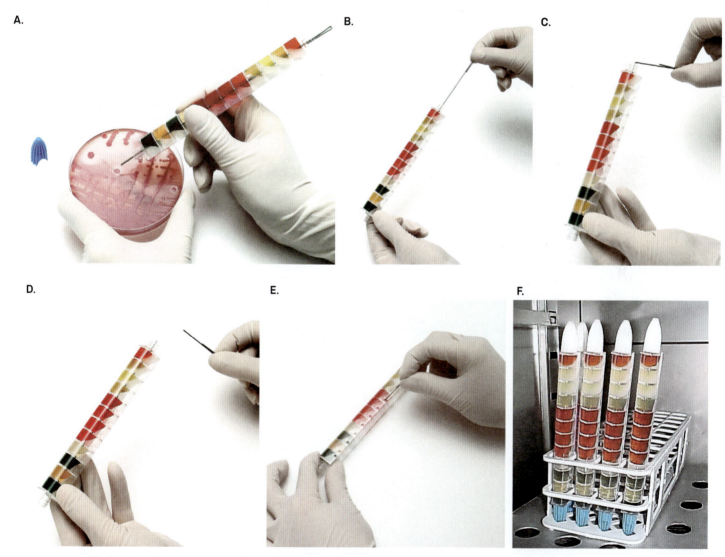

FIGURE 27.6 Inoculating the EnteroPluri tubes: (A) A single isolated colony is touched by the inoculating needle, **(B)** the needle is slowly drawn through all of the chambers, **(C and D)** the needle is snapped at the breakage notch, leaving most of the length of the needle in the tube, **(E)** the broken end of needle is used to punch holes in the plastic film through small slots located on side of the tube, and **(F)** the tubes are incubated by standing the tubes on the blue (citrate medium) ends of the tubes.

9. With the bulk of the needle still inserted in the tube, position the breakage notch such that it is just sticking out of the end of the tube, and bend the needle back and forth until it breaks off at the notch (**Figures 27.6C and D**).

The portion of the needle remaining inside the tube helps to maintain the anaerobic conditions needed for the glucose/gas, lysine, and ornithine assays.

10. Use the broken end of the needle to punch holes in the plastic film through the small slots located on the side of the tube in the adonitol, lactose, arabinose, sorbitol, VP, dulcitol/phenylalanine, urea, and citrate chambers (**Figure 27.6E**). **Immediately dispose of this piece of needle in the receptacle designated for trash to be autoclaved without touching the end of the needle and without putting the needle down on your bench; this needle is contaminated with live enteric bacteria.**

11. Screw both caps back on the ends of the inoculated tube and incubate the tube at 37°C for 48 hours. Incubate horizontally in an upside-down position (with words upside down) or vertically in a test tube rack with the Glucose/Gas chamber at the top of the tube (**Figure 27.6F**).

Day 3
MATERIALS

- Safety glasses
- Disposable gloves
- Inoculated EnteroPluri tube

1. Examine your inoculated EnteroPluri tube.

2. For some of the EnteroPluri assays, the results will be provided for your unknown Gram-negative culture. Indole and VP results will be provided because adding the Kovac's and VP reagents to the tubes tends to create a leaky mess. If a particular assay is listed as positive, then circle the number under that assay on the EnteroPluri table provided in the Results section.

As an alternative, students can add reagents to the indole and VP chambers themselves. To add the reagents, lay the EnteroPluri tube on a bench with its flat surface pointing upward, punch a tiny slit in the cellophane with a razor blade or similar instrument, and add a few drops of the appropriate reagents with a micropipetter. After reading the results, when disposing of the tube, be sure to carry it so that the reagents do not run out of the chambers.

3. For all the other assays, examine the tubes, and score each assay as either positive or negative. If positive, circle the number below the assay on the EnteroPluri table provided in the Results section; do not circle the number if the assay is negative.

4. Add up the numbers for positive assays as indicated on the EnteroPluri table, and use the *EnteroPluri Code Book* to determine the identity of your unknown.

5. Discard the EnteroPluri tubes in the receptacle designated for trash to be autoclaved.

RESULTS

Name: _____   Section: _____

Course: _____   Date: _____

PROTOCOL 1

ENTERIC UNKNOWN 1

Test	Group 1			Group 2			Group 3			Group 4			Group 5		
	Glucose	Gas	Lysine	Ornithine	H_2S	Indole	Adonitol	Lactose	Arabinose	Sorbitol	VP	Dulcitol	Phenyl alanine	Urea	Citrate
Value if positive	4	2	1	4	2	1	4	2	1	4	2	1	4	2	1
Results															
Code															

Enteric unknown 1 is: _____

ENTERIC UNKNOWN 2

Test	Group 1			Group 2			Group 3			Group 4			Group 5		
	Glucose	Gas	Lysine	Ornithine	H_2S	Indole	Adonitol	Lactose	Arabinose	Sorbitol	VP	Dulcitol	Phenyl alanine	Urea	Citrate
Value if positive	4	2	1	4	2	1	4	2	1	4	2	1	4	2	1
Results															
Code															

Enteric unknown 2 is: _____

ENTERIC UNKNOWN 3

Test	Group 1			Group 2			Group 3			Group 4			Group 5		
	Glucose	Gas	Lysine	Ornithine	H₂S	Indole	Adonitol	Lactose	Arabinose	Sorbitol	VP	Dulcitol	Phenyl alanine	Urea	Citrate
Value if positive	4	2	1	4	2	1	4	2	1	4	2	1	4	2	1
Results															
Code															

Enteric unknown 3 is: _____

ENTERIC UNKNOWN 4

Test	Group 1			Group 2			Group 3			Group 4			Group 5		
	Glucose	Gas	Lysine	Ornithine	H₂S	Indole	Adonitol	Lactose	Arabinose	Sorbitol	VP	Dulcitol	Phenyl alanine	Urea	Citrate
Value if positive	4	2	1	4	2	1	4	2	1	4	2	1	4	2	1
Results															
Code															

Enteric unknown 4 is: _____

ENTERIC UNKNOWN 5

Test	Group 1			Group 2			Group 3			Group 4			Group 5		
	Glucose	Gas	Lysine	Ornithine	H₂S	Indole	Adonitol	Lactose	Arabinose	Sorbitol	VP	Dulcitol	Phenyl alanine	Urea	Citrate
Value if positive	4	2	1	4	2	1	4	2	1	4	2	1	4	2	1
Results															
Code															

Enteric unknown 5 is: _____

PROTOCOL 2

Gram-negative unknown culture ID number: _____

Did streaking produce isolated colonies?

Colony morphology on MacConkey agar plate, including colony and medium color:

Test	Group 1			Group 2			Group 3			Group 4			Group 5		
	Glucose	Gas	Lysine	Ornithine	H$_2$S	Indole	Adonitol	Lactose	Arabinose	Sorbitol	VP	Dulcitol	Phenyl alanine	Urea	Citrate
Value if positive	4	2	1	4	2	1	4	2	1	4	2	1	4	2	1
Results															
Code															

Enteric Unknown is: _____

Name: _____   Section: _____

Course: _____   Date: _____

1. If an unknown isolate produced a score of 0 (zero) for the first digit of the biocode, is there a reason why you might not bother to score the rest of the chambers or assays in the EnteroPluri tube?

2. How or why can two different isolates or pure cultures produce two different biocodes and yet be identified by the EnteroPluri system as being the same species?

3. If your class followed Protocol 2, then compare the Lab 26 assay data for your Gram-negative unknown with the data from Lab 27. Did both labs produce the same answer with respect to species name? For assays that were used in both labs (lactose fermentation, lysine decarboxylase, sulfide production, urease, etc.), did both methods produce the same results? If they did not, why might the results have been different?

4. What are advantages and disadvantages of using the EnteroPluri system to identify Gram-negative enteric bacteria compared with the "traditional" methods used in Lab 26?

5. Which sections of the protocol for Lab 26 must be carried out first, regardless of whether the Gram-negative enteric bacteria are subsequently identified using the EnteroPluri system or the traditional methods of Lab 26? Why must these procedures be done for both the Lab 26 and Lab 27 approaches to identification? That is, what do all of these methods require for proper identification of unknown bacteria?

6. What symptoms or types of illnesses might prompt a doctor to order an EnteroPluri assay of a sample taken from a patient?

7. Clinical labs pride themselves on the accurate identification of pathogenic bacteria. What problems might result from the misidentification of a bacterium in a hospital setting?

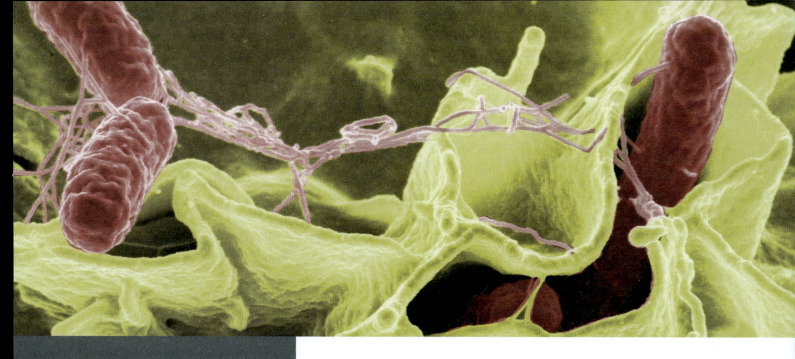

Food- and water-borne pathogens, like the rod-shaped *Salmonella* shown in this digitally-colored scanning electron micrograph, cause disease by secreting toxins or, as in this case, being taken up by host cells. When someone becomes sick from a bacterial pathogen, identifying the culprit quickly can literally be a matter of life and death. *Salmonella*-specific antibodies form the basis of some rapid-identification immunoassay tests.

LAB 28

Immunoassay Identification of *Salmonella*

Learning Objectives

- Understand how the specificity of antibodies and antibody-antigen interactions allows us to use antibodies to identify bacteria.

- Review *Salmonella* taxonomy to understand current practices in naming *Salmonella* strains.

- Explain the principles of an antibody-based rapid identification test and use this test to identify *Salmonella* in a pool of unknown enteric species.

In Labs 22 and 26, we saw how bacteria can be identified on the basis of the biochemical reactions they can perform. However, these assays can take days to complete, and identification may require the results from ten to twenty different assays. **Immunoassays** offer a faster way to identify bacteria. Immunoassays rely on specific antibodies that bind to specific antigen molecules associated with particular bacterial strains or species. If you have the right antibody in hand, you can get results in minutes.

ANTIBODIES AND ANTIGENS

Antibodies are antimicrobial immune system proteins produced and secreted into the plasma (the liquid part of the blood) and other body fluids by B lymphocytes. These proteins are roughly Y-shaped. The amino acid sequences in the stem of the

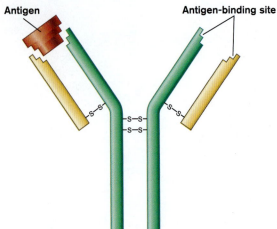

Antigen

Antigen-binding site

—S—S—

FIGURE 28.1 Antibody structure.
Different antibodies have different
shapes at the antigen-binding sites.
The antibody can bind only an antigen
with a complementary shape; in this
case, it's the red shape. —S—S— is a
disulfide bond.

Y are similar in different antibodies, so there's not much difference in the stem shape among different antibodies. In contrast, the amino acid sequences at the tips of the arms of the Y are variable, **so the shape or conformation of the tip varies considerably among different antibodies (Figure 28.1)**. The tips of the Y arms bind to specific molecules, called **antigens**, so these tip regions are referred to as **antigen-binding sites.** Each type of antibody has a unique, specific shape at its antigen-binding sites. The shape of a particular, unique antigen-binding site will very closely fit, or complement, the shape of a unique, specific antigen. If the fit between antigen and antibody is close enough, then after the initial binding, the conformation of these molecules will change slightly to make the match even closer or tighter. The key point is that **a given antibody will bind to, or lock onto, only a very narrow range of antigen shapes**.

Antigens are molecules that stimulate antibody production and that fit the produced antibody's antigen-binding sites. In the case of bacterial cells, these antigens could be proteins, glycoproteins, or lipopolysaccharides at the bacterial cell surface, which vary among Gram-negative strains and species in terms of protein structure or the types and sequences of the sugars that are present at the ends of carbohydrate side chains. What matters to us in the context of immunoassays is that **different bacterial strains and species produce and carry different sets of antigens**. In many cases, a given antigen is unique to, or specific for, a given strain or species. So if we identify various strain-specific and species-specific antigens, and if we produce antibodies against these antigens (and we can do this), then we might be able to use those antibodies to identify many different organisms.

For example, we could produce a set of antibodies that binds to and identifies only bacterial species A, and another set that binds to and identifies only bacterial species B. We could even produce a set of antibodies that binds only to strains 1, 2, and 3 of species B, and so we might be able to differentiate and separate these strains of species B from hundreds of other species B strains. To put it in the context of this lab, if we have an antibody that binds to an antigen that is unique to *Salmonella* species, and if this antibody and this antigen join in molecular matrimony, then we know that we are in the presence of *Salmonella*.

However, there is a problem here. All of this lovely binding and linking occurs on a submicroscopic level, and a single antibody linked to a single antigen is invisible to anyone peering into a light microscope. So, how do we know when an antibody has bound to antigen? How can we reveal the linkage of a strain-specific or species-specific antibody to a strain-specific or species-specific antigen? How can we tell when no connection is made, an observation that suggests that an unknown microbe is not what we thought it might be? There are actually several ways to do this, and thus several different types of immunoassays. In this lab, we will examine the Singlepath Salmonella Rapid Test to see how the invisible can be made visible.

Vision is the art of seeing what is invisible to others.

—Attributed to Jonathan Swift

But first, a word or two about *Salmonella*.

SALMONELLA TAXONOMY

Salmonella is a genus name, and historically, almost every new strain of bacteria that seemed to fit into this genus was given a new and different species name, such as *Salmonella enterica*, *Salmonella cholerasuis*, *Salmonella typhimurium*, or *Salmonella typhi*. At one point in the twentieth century, there were literally hundreds and hundreds of different species in the genus *Salmonella*. Then newly developed techniques in molecular genetics revealed that almost all "species" of *Salmonella* that caused illness in humans were so genetically similar that they should be grouped under a single species name—specifically, the species *Salmonella enterica*. Further study led taxonomists to split the species *Salmonella enterica* into about six or seven subspecies, placing almost all human-associated salmonellae in just one of those subspecies, the subspecies *enterica*.

So, in summary, almost all *Salmonella* strains that infect humans belong to taxon *Salmonella enterica* subspecies *enterica*. But what about all of those old species names (*cholerasuis*, *typhimurium*, *typhi*, etc.)? Most of those names live on as strain or **serovar** names within the taxon *Salmonella enterica* subspecies *enterica* (**Figure 28.2**). For example, the *Salmonella* strain that was once called *Salmonella typhimurium* is now most accurately and fully written as *Salmonella enterica* subspecies *enterica* serovar Typhimurium. This is almost as bad as Prince William's actual full name (you know, Kate Middleton's husband), which is William Arthur Philip Louis Mountbatten-Windsor. No wonder the British press refers to him as just "Wills."

Well, *Salmonella enterica* subspecies *enterica* serovar Typhimurium is an awfully long name. So in this lab, we'll generally just use *Salmonella* Typhimurium for short. In addition, references in this manual to *Salmonella* should be taken as shorthand for the many, many strains or serovars of *Salmonella enterica* subspecies *enterica*.

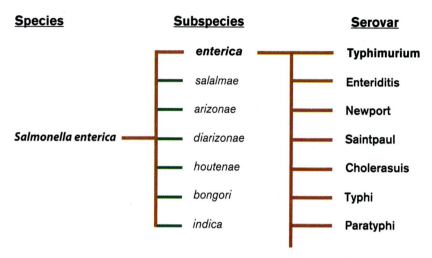

FIGURE 28.2 *Salmonella enterica* tree.

SINGLEPATH *SALMONELLA* RAPID TEST ASSAY AND ASSAY DEVICE

As discussed in Lab 31, in which we will look for enterics in contaminated meat, *Salmonella* strains cause intestinal infections resulting in a type of gastroenteritis called **salmonellosis**, which features diarrhea, cramps, vomiting, and a low fever. *Salmonella* cells can be identified by growing the bacteria on a variety of selective and differential media followed by a series of biochemical assays, such as the lactose fermentation and sulfide assays. However, these methods can be very time-consuming, and with food-borne pathogens, time matters. In contrast, immunoassays can identify these cells in minutes; the Singlepath *Salmonella* assay is an example of such an approach. The **Singlepath *Salmonella* Rapid Test** (produced by EMD Millipore) relies on antibodies that bind to bacterial antigens specific for, or unique to, the numerous *Salmonella* strains capable of causing intestinal diseases in humans.

The test device consists of chromatography medium, enclosed in a plastic case, to which three types of antibodies have been attached. The first set of antibodies is located in a reaction zone near the base of the sample well **(Figure 28.3A)**. These antibodies can bind to a specific *Salmonella* antigen; in addition, they carry tiny colloidal gold particles (which are linked to the antibodies during the manufacturing of the device). When broth containing *Salmonella* cells is added to the well, the broth and antigen-displaying cells begin to spread by capillary action through the chromatography medium at the base of the well. As the *Salmonella* cells pass through the zone filled with antibodies, the *Salmonella* antigen binds to the antibodies to form complexes of gold-labeled antibody bound to antigens on the *Salmonella* cells.

The complexes migrate along the chromatography medium from the reaction zone into the binding zone in the test area (marked on the device by the letter T). This binding zone contains a band of anchored antibodies that are capable of attaching to a different *Salmonella*-specific antigen **(Figure 28.3B)**.

The complexes flow past the anchored antibodies, and this second set of antibodies binds to the complexes and retain them in the test area **(Figure 28.3C)**. As more and more gold-labeled *Salmonella* complex accumulates in the test area, a faint red band appears, aligned with the "T" mark on the test device.

In contrast, cells of non-*Salmonella* species lack the antigen that fits the binding site of the anchored antibodies in the test area. These cells carry antigens, but they are of different shapes. So, non-*Salmonella* cells flow past the test area without binding to the attached antibodies, and no red line appears aligned with the "T" mark **(Figure 28.3D)**.

When *Salmonella* cells are present in the sample, the red color of the band containing bound complexes is due to the reflection of light by the tiny gold particles. We're used to thinking of gold as being, well, gold in color. However, this assay uses colloidal gold nanoparticles, which are about 20–50 nanometers in size (a nanometer is 1/1,000,000 of a millimeter or 1/1,000 of a micrometer). When gold particles are this tiny, they tend to absorb blue, green, and yellow wavelengths of light while reflecting or allowing the passage of red wavelengths. So, a collection of these particles in the test area of the device will appear to be red in color instead of golden or yellow.

You can see the same phenomenon in some medieval stained-glass windows. The artisans who created these windows mixed tiny amounts of gold or gold chloride (a gold salt) with molten glass. Under these conditions, the gold formed spherical colloidal nanoparticles, and these particles reflected and transmitted sunlight in a

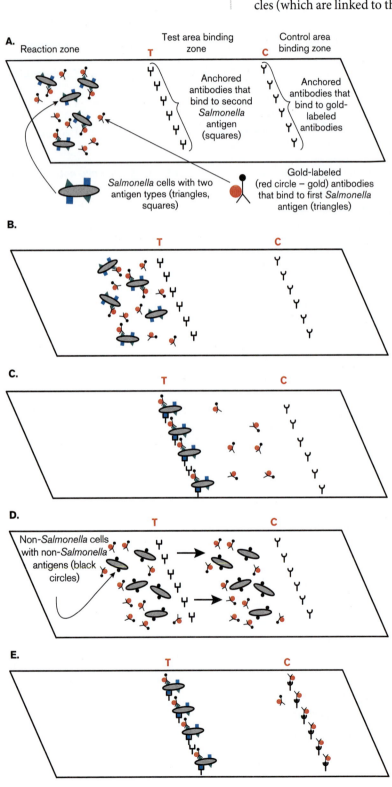

FIGURE 28.3 How the Singlepath *Salmonella* Rapid Test works.

way that produced a ruby red color. Red-colored stained glass could also be made using certain copper compounds. But we digress, so back to the assay.

Beyond the test area, the rest of the sample continues to flow toward a second binding zone in the control area (marked on the device by the letter C). This binding zone contains a different set of anchored antibodies; this is the third set of different antibodies in the device. These antibodies can bind to regions found on the gold-labeled antibodies; that is, these are antibodies that bind to other antibodies. The sample flowing by this second binding zone will include gold-labeled antibody that failed to latch onto *Salmonella* antigen in the reaction zone, as well as some of the *Salmonella* antigen-antibody complexes that missed their chance to be bound and immobilized in the test area. Or if the cells being tested are not from a *Salmonella* species, then the flowing sample will be composed entirely of gold-labeled antibodies that are not bound to any bacterial cells.

As excess gold-labeled antibodies flow past the anchored anti-antibody antibodies, this antibody set binds the gold-labeled antibodies and holds them in the control area **(Figure 28.3E)**. As bound antibodies accumulate in the control area, a faint red band appears, aligned with the "C" mark on the test device. The appearance of a red line in the control area tells us the device is working properly because it indicates that the flow of broth covered the length of the device and that anchored antibodies are capable of stopping enough gold-labeled antibody molecules to create a visible red line. Even if the bacteria being tested are not *Salmonella* species, a red line should always appear in the control area, because the antibodies anchored to the device will capture the passing gold-labeled antibodies even if the gold-labeled antibodies are not bound to *Salmonella* or any other antigen.

In summary, **if *Salmonella* cells are present, then there will be a faint red line in the test area and another red line in the control area (Figure 28.4A)**.

If *Salmonella* cells are absent, then there will be a red line in the control area but no line in the test area **(Figure 28.4B)**.

If no lines at all appear, then either mistakes were made or the device is defective.

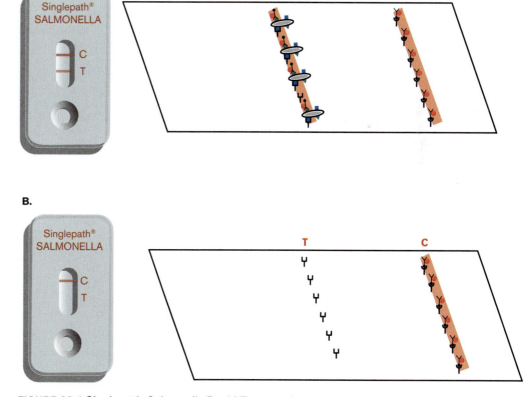

FIGURE 28.4 Singlepath *Salmonella* Rapid Test results.
A. Sample contains *Salmonella*. **B.** There is no *Salmonella* in this sample.

PROTOCOL

MATERIALS

- 5 empty sterile petri dishes
- 5 sterile 1.0-ml pipettes
- 2-ml (blue-barrel) pipette pump
- 5 Singlepath *Salmonella* Rapid Test devices
- Safety glasses
- Disposable gloves

Cultures in BHI broths:

- *Citrobacter freundii*
- *Enterobacter aerogenes*
- *Escherichia coli*
- *Proteus vulgaris*
- *Salmonella* Typhimurium

> **IMPORTANT:** *Salmonella* Typhimurium and *Proteus vulgaris* are biosafety level 2 species, so gloves and safety glasses must be worn throughout this lab.

The broth cultures you receive will be labeled A, B, C, D and E, so that the identity of the bacteria in each broth will not be known to you. The point of the lab, after all, is to identify *Salmonella* species using an immunoassay. It wouldn't be as much fun if you knew at the start which tube contains *Salmonella* Typhimurium.

1. Label five Singlepath *Salmonella* Rapid Test devices with your names and the date. The first device should then also be labeled "A," the second labeled "B," and so forth, until the fifth device is labeled "E."

2. Place each device in a separate empty petri dish.

   The devices are placed in petri dishes to prevent the spread of bacteria from inoculated devices.

3. Use a sterile 1.0 ml pipette to transfer 0.2 ml of the broth culture in Tube A to the sample well in the test device labeled A. Add the broth slowly but steadily, one drop at a time, so that the broth does not overflow the well, but neither does the well run dry before all the broth is added.

4. Incubate the devices for 20–30 minutes at room temperature.

5. After 20–30 minutes, check each test device for red-colored lines that mark deposits of gold-labeled antibodies. Look for lines aligned with the "C" (control) and "T" (test) marks on each device. Record your observations in the Results section.

   In the case of the *Salmonella* culture, the line at the test mark may be fainter than the line at the control mark, but it will still be darker than any test mark line for the non-*Salmonella* cultures.

6. Dispose of all inoculated materials in the receptacle designated for trash to be autoclaved.

RESULTS

Name: _____   Section: _____

Course: _____   Date: _____

DESCRIPTION (RED LINE/NO LINE)			
Culture type	Test Area	Control Area	Positive or Negative for *Salmonella*?
Culture A			
Culture B			
Culture C			
Culture D			
Culture E			

Name: _____   Section: _____

Course: _____   Date: _____

1. Which tube contained *Salmonella*?

2. How do you know that this tube contained *Salmonella*?

3. In the case of the tube containing *Salmonella*, what was happening on the molecular level to produce an observation that you interpreted as positive for *Salmonella*?

4. How do you know that the other tubes did not contain *Salmonella*?

5. Was this assay able to differentiate *E. coli* from *Enterobacter*? Explain why this assay can or cannot differentiate *E. coli* from *Enterobacter*.

6. Let's say that a broth culture of *Salmonella enterica* failed to produce a red line in the test area (produced a false negative), but did produce a red line in the control area. Give two possible explanations for this observation.

7. Now let's say that a broth culture of an unknown or unidentified bacterial species failed to produce red lines in either the test area or the control area. So, no red lines at all. Can you say conclusively that the culture does not contain *Salmonella*? Why or why not?

8. Finally, let's say that a broth culture containing something other than a *Salmonella* species did produce a red line in the test area (produced a false positive). What might explain this observation?

9. In addition to the non-*Salmonella* species, what other controls might be used to ensure that any red lines in the test area are due to *Salmonella* species?

10. **a.** According to current taxonomic systems, how many different species of *Salmonella* are responsible for almost all human cases of salmonellosis?

    b. What is the taxonomic difference between the names *Salmonella typhimurium* and *Salmonella* Typhimurium?

11. Which foods are most commonly associated with *Salmonellosis*?

12. Often the antibodies produced from an infection protect you from ever having that infection again. For example, I would only expect to have rubella once. But you can have food poisoning multiple times resulting from *Salmonella* infections. Explain why. You may need to do a little research on your own here.

Milk is rich in proteins and fats, making it a terrific growth medium for microorganisms. The milk you buy is not sterile, but the microbial load has been reduced by pasteurization. Refrigeration thereafter slows the increase in microbial numbers, but all milk eventually spoils.

LAB 29

Milk Microbiology

Learning Objectives

- Understand that nutrient-rich milk can become contaminated with microbes and that pasteurization is needed to control the growth of these microbes.

- Describe the value of semi-log paper in graphing the pasteurization-produced exponential changes in bacterial populations and in the calculation of decimal reduction times.

- Examine the role of beneficial bacterial species in the production of fermented dairy products such as yogurt.

- Employ pour plate methods to follow the effects of pasteurization, and use *Lactobacillus* and *Streptococcus* species to make yogurt.

OF MICROBES AND MILK

Animal milk is a rich medium. It contains biologically meaningful amounts of sugars and other carbohydrates, amino acids and proteins, fats, and vitamins and minerals because young mammals live solely on milk for the first part of their lives. And as a consequence of its nutritional richness, milk readily supports the growth of many types of microorganisms, including bacteria that will spoil the milk and bacteria that can cause human disease. The bacteria in question are usually not present inside the udders of healthy cows and other mammals that provide milk for human use. Instead, they are found on the outsides of the udders, in manure, and in other locations in the dairy, and they contaminate the milk after the milk leaves the animal. Proper sanitation in milking facilities can significantly reduce the level of contamination by bacteria and other microbes, but even under ideal conditions, it is almost inevitable that some bacterial cells will get into the milk before it is consumed.

Fortunately, most of the contaminants in milk are relatively harmless. These "mostly harmless" bacteria include species in the soil-dwelling *Bacillus* genus and **lactic acid bacteria,** a collection of species in genera such as *Lactococcus*, *Lactobacillus*, *Leuconostoc*, and *Streptococcus*. These species will eventually spoil the milk, but they usually don't present much of a threat to human health. But milk can also be contaminated by, and can transmit, dangerous pathogens, including *Salmonella* species, which can cause intestinal infections; *Listeria monocytogenes*, which causes high-mortality-rate infections of the blood and central nervous system; and *Mycobacterium bovis*, a species of *Mycobacterium* that can cause tuberculosis in cattle and humans.

Pasteurization

So, there is clearly a need to control the microbes in milk, both to prevent or delay spoilage and to reduce the transmission of pathogens and disease. One of the main methods of microbial control is to heat the milk in a process called **pasteurization.** Pasteurization is so important that it's a key part of the definition of the term **raw milk.** Before milk is pasteurized, it is referred to as raw milk; after pasteurization, it is called **pasteurized milk.**

While there are several combinations of heating times and temperatures that are considered acceptable for pasteurization, most of the milk that is sold for direct human consumption in the United States is treated by either the LTLT or the HTST method. **LTLT** stands for "low temperature, long time." In this case, the raw milk is heated to about 63°C (145°F) for 30 minutes. **HTST** is short for "high temperature, short time," and this method heats the raw milk to about 72°C (161°F) for 15–20 seconds. Notice that an increase in temperature of about 10°C reduces the time required for pasteurization about a hundredfold.

LTLT and HTST pasteurization is designed to reduce microbial populations in raw milk by 10,000- to 100,000-fold, but these methods are not intended to sterilize the milk. Some live bacteria will remain in the pasteurized product, so milk must be refrigerated to prevent rapid spoilage. Nevertheless, pasteurization usually eliminates all or almost all pathogen cells, rendering the milk safe for human consumption. It also significantly increases shelf life, or the time before the milk is spoiled by *Bacillus* and lactic acid bacteria. Once the density of these bacteria has been greatly reduced by pasteurization, it takes time for their populations to recover to a point where there are enough cells making enough lactic acid and other fermentation products to spoil the milk.

Why not simply sterilize milk? In part, it's because any heating of the milk that is sufficient to kill all bacteria also alters the taste and nutritional content of the milk. The LTLT and HTST times and temperatures are a compromise between the need to slow spoilage and prevent disease versus the desire to maintain the original flavor and as much nutritional value as possible. In short, these methods do change the milk, but they also save lives.

Now, if one wants to sterilize milk, it certainly can be done using other types of pasteurization. UHT (ultra-high temperature) pasteurization uses temperatures of around 135°C (275°F) to kill all microbes in the milk, including bacterial endospores. However, these very high temperatures will alter the taste and the nutritional content of the milk. Despite this drawback, UHT pasteurization has its uses, because as long as it remains sealed, sterilized milk doesn't have to be refrigerated and has a shelf life that is measured in months, not weeks.

Microbial death rates and semi-log paper

When microbial populations are being reduced by methods such as exposure to high temperatures, then these populations usually show an exponential response, or exponential

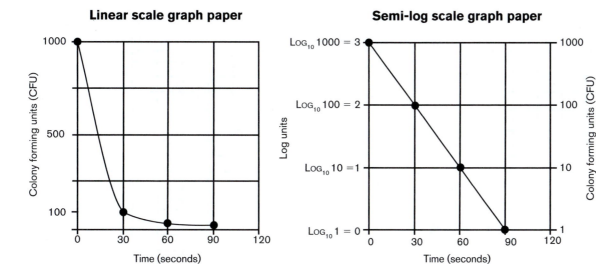

FIGURE 29.1 Plotting microbial death rates.
These plots show data for a population starting with an initial density of 1,000 cells/ml in which 90% of cells are killed every 30 seconds. Numbers of cells (CFU) are plotted on linear, non-log scale graph paper (left) and on semi-log scale paper (on which the y-axis uses a log scale) (right).

decline, in cell density. In an **exponential response,** the same fraction or percentage of the remaining population is eliminated in any given time interval. If this decline is plotted on regular, linear, non-log graph paper, the curve describing the relationship between time and population size takes the form of a concave curve. But if the cell densities are plotted on the y-axis of graph paper with a log scale on the y-axis (called **semi-log paper**), and time is plotted on a linear x-axis, the result is a nice, mathematically simple, easy-to-work-with, straight line **(Figure 29.1).** The line is exactly the same as the line that you would get if you plotted the $\log_{10}$ of the cell counts on the y-axis of regular graph paper, but the semi-log paper saves you the trouble of entering the number in a calculator. And who doesn't like that?

So, how do we use semi-log paper? Semi-log paper uses a log or exponential scale on the y-axis, on which each major division, sometimes marked as "1___" or "10," represents one log unit, or a tenfold change, from the next major division. The major divisions on the y-axis should be marked with whole log units, or powers of ten, that are appropriate for your data, such that the top line has a higher value than any of your CFU/ml counts, but is not more than ten times your highest count. That is, the exponent for the top line should be one log unit higher than the exponent of your highest count. For example, if your highest CFU/ml count is 5×10^3, or 5,000, or 3.7 log units, then the top line should be set at 1×10^4, or 10,000, or 4.0 log units **(Figure 29.2).** Then the next major division down will be tenfold, or one log unit, lower than the top division, so in this example, the next major division would be marked as 1×10^3, or 1,000, or 3.0 log units. Continue this pattern until you get to the bottom line.

Having marked the major divisions, you need to understand what the minor divisions mean as well. Minor divisions are marked 2, 3, 4, 5 … 9, and it's important to realize that these numbers do not refer to values between 1 and 10. For example, the lines marked 2, 3, 4, 5 … 9 that are above the 1×10^3, or 1,000, line do not equal "1,002," "1,003," "1,004," and so on. Instead, the line designated by the number X marks the point where the value on the y-axis equals $X \times 10^A$, where A is the exponent of the next lowest major division. So, if the next lowest major

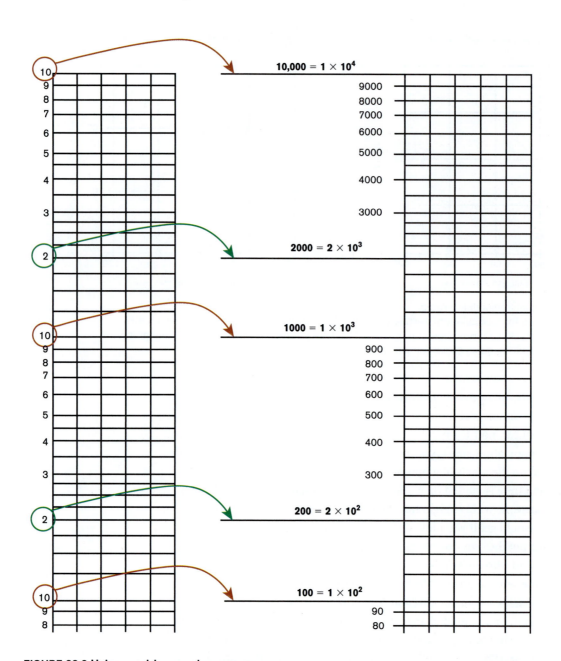

FIGURE 29.2 Using semi-log graph paper.
Note that the major divisions, marked as 10, represent one log unit interval or a ten-fold change from the next major division. A line designated with a number other than 10 marks the value equal to that number times 10^E, where E is the exponent of the closest major division below that line. So, the lowest 2 along the left side represents 2×10^2, which equals 200.

division is 1×10^3, or 1,000, then the minor division line above this division marked "2" is equal to 2×10^3, or 2,000. The line does not mark the value "1,002." The minor division line (above the 1×10^3 line) that is marked "3" is 3×10^3, or 3,000, and so on.

Decimal reduction time

When the use of microbial control methods (such as pasteurization) produces an exponential death rate, then the curve describing the relationship of time to the number of live cells remaining will be a straight line when plotted on semi-log paper.

And that straight-line curve can be a very useful thing if one wishes to calculate a value that is widely used in the food industry, namely, the decimal reduction time. **Decimal reduction time** is the time required to:

1. reduce a microbial population to a CFU/ml count that is one decimal point to the left of the starting count, or
2. reduce a microbial population to a CFU/ml count that is one power of ten less, or one exponent unit less, or one log unit less than the starting count, or
3. reduce a microbial population to a CFU/ml count that is 90% less than the starting count; this is the same as reducing the population to 10% of its original size.

For example, if the starting population is 1,000 cells/ml, then the decimal reduction time is the time required to:

1. reduce the population from 1,000.0 to 100.00, or
2. reduce the population from (1×10^3) or $(\log_{10} = 3)$ to (1×10^2) or $(\log_{10} = 2)$, or
3. reduce the population by 90% from 1,000 to 100, because 100 is 10% of 1,000.

Calculation of decimal reduction time

1. Plot the change in cells or CFU/ml upon exposure to pasteurization (y-axis) versus time (x-axis) on semi-log paper **(Figure 29.3)**.
2. Draw the straight-line curve that is the "best fit" to the data: Due to experimental error, the plotted points are unlikely to fall exactly along any single straight line, so draw the line such that the different points are as close to the drawn line as possible.
3. Select two major divisions on the y-axis (1×10^3, 1×10^2, etc.) that are both crossed by the curve and that are one log unit (tenfold) apart.
4. Draw horizontal lines at the two major divisions selected in step 3; note that the reduction from the top line to the bottom line in cell or CFU/ml density is 90%, or one decimal point to the left.
5. For each horizontal line, find the point where the curve crosses the line. Draw vertical lines from these two points down to the horizontal axis.
6. To determine decimal reduction time, subtract the time on the x-axis crossed by the vertical line from the upper major division line from the time on the x-axis crossed by the vertical line from the lower major division line. The result is equivalent to the time required to reduce the population from $1 \times 10^{X+1}$ to 1×10^X, or the time required to reduce the population by one log unit, or one power of ten.

Decimal reduction time example: The data shown in the table to the right have been plotted on semi-log paper in Figure 29.3. The relationship between heating time and cells/ml or CFU/ml has been drawn as a straight-line curve. The difference in time between the point at which the population equaled 1,000 and the point at which the population equaled 100 was 60 seconds; this is the decimal reduction time.

Time to reduce populations to 1 cell or CFU/ml

Once the decimal reduction time has been calculated, this information can be used to calculate other values, such as the time required to reduce the microbial population from some initial density to a target or end-point density. This is useful information because food producers may have to meet a certain standard or limit in terms of the maximum number

Heating time (sec)	CFU/ml
0	5,000
30	1,700
60	400
90	160

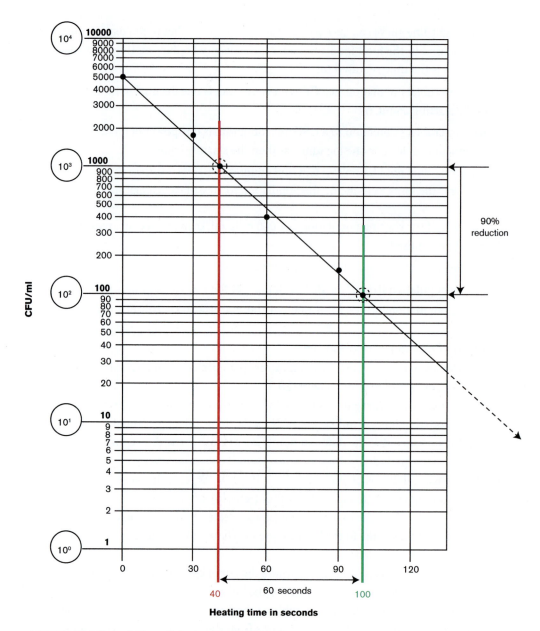

FIGURE 29.3 Calculation of decimal reduction time.

of viable microbial cells permitted in food as a condition for the sale of the food. What if regulations required that a particular food item have no more than 1 cell (CFU)/ml before it could be sold? How long would we have to heat or pasteurize this item before we reached our target population of 1 cell/ml?

If our target density is 1 cell/ml, then we need to reduce the population to 10^0, or a $\log_{10}$ value of 0 ($\log_{10}$ of 1.0 = 0). So, starting with an initial population described in log units, we can think of this problem in terms of the number of times the population must be reduced by one log unit until we get to zero log units. That number of times is equal to the $\log_{10}$ of the initial population density. For example, if the initial population was 1,000 cells/ml, or 1×10^3, or $\log_{10}$ = 3.0, then we would need 3.0 rounds of reduction by one log unit to reach our target. One round of a one-log-unit reduction takes us to 100 cells/ml, or 1×10^2, or $\log_{10}$ = 2; a second round reduces the population to 10 cells/ml, or 1×10^1, or $\log_{10}$ = 1; and a third round takes us to 1 cell/ml, or 1×10^0, or $\log_{10}$ = 0.

How long does each round of one-log-unit reduction take? Remember that decimal reduction time is the time required to reduce a microbial population by tenfold, or one log unit, so the time required for a one-log-unit reduction is equal to the decimal reduction time. And, therefore, the total time needed to reduce a population from an initial density to a density of 1 cell/ml is:

$$\log_{10} \text{ of the initial cell density} \times \text{decimal reduction time}$$

To use the example given above:

$$\text{Time to 1 cell/ml} = (\log_{10} \text{ of } 5,000) \times 60 \text{ seconds (per log unit reduction)}$$

$$= 3.7 \times 60 \text{ seconds} = 222 \text{ seconds}$$

POUR PLATE QUANTITATIVE PLATE COUNTS

In this lab, we will determine the density of bacteria in various milk samples using pour plate counting (described in Lab 4). In this method, a given sample is serially diluted, and measured amounts (usually 1 ml) of the diluted cell suspensions are mixed with melted agar and poured into sterile plates. Cell division by viable cells produces visible colonies, so the number of colonies reflects the number of live cells in the diluted samples mixed with the agar. Countable plates are those with 30 to 300 colonies per plate, and multiplying colony counts by the dilution factor for a counted sample yields the cell density of the starting sample. As in spread plating, counts are expressed in terms of colony-forming units per milliliter.

Lactic acid bacteria often produce relatively small colonies, and so to make these colonies easier to spot, we will use a 3 × phenol red glucose agar. When lactic acid bacteria ferment the glucose, the lactic acid they produce will lower the pH around the colonies. The phenol red indicator will change from orange-red to yellow, and this will create a small, pale yellow halo around the colony against an orange-red background **(Figure 29.4)**. Phenol red is present in this medium at three times the concentration found in phenol red broths to make these colors more intense. Small, agar-embedded colonies are best seen by **holding the plate up to a light** so that the light passes through the medium in the plate. When this technique is used, colonies will appear as dark dots against a yellow halo background.

A.   **B.**

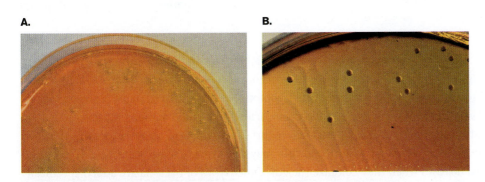

FIGURE 29.4 Lactic acid bacteria colonies on 3× phenol red glucose agar.
A. Top view of plate. **B.** Colonies back lit by plate held to the light.

Note, however, that at high colony densities, the medium may be yellow throughout the plate, and you may not be able to see separate yellow halos. When colonies are numerous and close together, the zones of lactic acid (and lowered pH) that are generated by the individual colonies will overlap, so the yellow halos will merge together.

"GOOD BUGS" IN DAIRY PRODUCTS

Milk products may contain or transmit many kinds of spoilage and pathogenic microbes, necessitating the use of pasteurization to reduce the microbial load and prevent disease. But not all microbes are bad; in fact, some "good bug" species may improve our intestinal health, and others are very useful in the manufacture of fermented dairy products. So, in many cases, we deliberately add various microbes—mostly certain species of bacteria—to milk and milk products.

For example, there has been a recent upturn in interest in using milk and dairy products as a means of "inoculating" humans with certain types of gut bacteria that may improve our health. These **probiotic** species, or **normal flora,** colonize the large intestine and protect us against disease by manufacturing vitamins that can be absorbed into the bloodstream and by literally taking up space, a process that creates a habitat that may be too crowded and too competitive for pathogenic species. That is, the pathogens cannot establish an infection and multiply to the point of causing disease because they cannot outcompete the probiotic species for space and nutrients. Think of your colon as a "tangled bank," in the Darwinian sense.

It is interesting to contemplate a tangled bank, clothed with many plants of many kinds, with birds singing on the bushes, with various insects flitting about, and with worms crawling through the damp earth, and to reflect that these elaborately constructed forms, so different from each other, and dependent upon each other in so complex a manner, have all been produced by laws acting around us. These laws, taken in the largest sense, being Growth with reproduction; Inheritance which is almost implied by reproduction; Variability from the indirect and direct action of the conditions of life, and from use and disuse; a Ratio of Increase so high as to lead to a Struggle for Life, and as a consequence to Natural Selection, entailing Divergence of Character and the Extinction of less improved forms.

—Charles Darwin, *Origin of Species*

While we often acquire these beneficial species naturally and without specifically trying to eat them, we can aid the acquisition or inoculation process by deliberately adding probiotic species to dairy products. Probiotic species added to dairy products include *Lactobacillus acidophilus,* used to create acidophilus milk, and several species that are added to either produce or enhance yogurt. The lactic acid bacteria (described below) are the key to making yogurt, and they may also provide some protection from intestinal diseases when they colonize the gut. Other types of bacteria are added specifically for their possible health benefits, but are not important in the creation of the yogurt itself. For example, Activia brand yogurts contain a proprietary strain of *Bifidobacterium animalis,* a strain that the company has dubbed (and trademarked as) Bifidus Regularis.

YOGURT PRODUCTION

For many centuries, long before we knew about microorganisms, people manipulated the microbe-mediated spoilage of milk to produce tasty products with longer shelf lives, such as yogurts and cheeses. One group of bacteria, the Gram-positive lactic acid bacteria, is particularly important in producing yogurt and cheese. The lactic acid bacteria are aerotolerant and rely solely on fermentation pathways to release the energy in sugars for the production of ATP.

All lactic acid bacteria produce large amounts of lactic acid during fermentation, and this organic acid adds flavor and aroma to the fermented dairy product. Further, by lowering the pH of milk, lactic acid inhibits spoilage microbes. Some types of lactic acid bacteria produce additional fermentation end products that also contribute unique tastes and odors to cheeses and yogurts. These additional molecules include small amounts of ethanol, CO_2, acetic acid, acetaldehyde, and diacetyl. The diacetyl molecule is of particular interest because it gives foods and beverages a buttery flavor; its production is favored when citrate is available during fermentation.

In the case of yogurt, two species from the lactic acid bacteria group are particularly important: *Streptococcus thermophilus* and *Lactobacillus bulgaricus* (full taxonomic name: *Lactobacillus delbrueckii* subspecies *bulgaricus*) **(Figure 29.5A** and **B)**. These species interact in a symbiotic manner to convert milk into yogurt. Both can grow alone or independently of each other in milk, but each species will grow better in the presence of the other.

Isn't it warm?
Isn't it rosy?
Side by side!
Ports in a storm,
Comfy and cozy,
Side by side!

—Stephen Sondheim, *Company*

Of the two, *Streptococcus thermophilus* is usually the faster-growing species in milk. At the start of yogurt production, its activity quickly raises the lactic acid concentration to about 0.6%–0.8% and lowers the milk pH to around 4.5 to 5. Speed matters: The sooner the milk pH is lowered, the sooner the growth of undesirable spoilage bacteria

A.

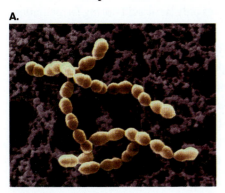

B.

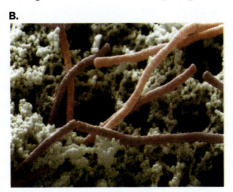

FIGURE 29.5 Bacteria in milk and yogurt.
A. *Streptococcus thermophilus* isolated from yogurt. **B.** *Lactobacillus bulgaricus* is another bacterium that is used to produce yogurt.

is inhibited. *S. thermophilus* cells can grow alone, but in the presence of *Lactobacillus bulgaricus*, the *S. thermophilus* cells may grow better and produce more lactic acid in less time. This is because *L. bulgaricus* cells are more proteolytic than *S. thermophilus* cells; the activity of the *L. bulgaricus* proteases yields amino acids such as valine that promote the growth of *S. thermophilus*. So in the absence of amino acids provided by *Lactobacillus*, it takes *S. thermophilus* cells more time to produce inhibitory levels of lactic acid.

In turn, *Lactobacillus bulgaricus* benefits from the presence of *Streptococcus thermophilus*. *Streptococcus* cells rapidly create the acidic conditions preferred by *L. bulgaricus*, and in addition, *S. thermophilus* produces carbon dioxide and formate (formic acid) from pyruvate. The *L. bulgaricus* cells grow better when CO_2 concentrations are higher, and these cells use formate as an important nutrient. In the presence of the *S. thermophilus*–supplied formate, the *L. bulgaricus* cells produce lactic acid more rapidly than in the absence of *Streptococcus*. This becomes more important when the milk pH drops below 4.5 to 5, because at that point *S. thermophilus* cells are inhibited by the accumulating organic acids. *L. bulgaricus* then takes over as the primary acid producer, and ultimately, the additional lactic acid produced by *Lactobacillus* yields a final lactic acid concentration of 1.0%–1.3%, and the final pH drops to about 4.

As noted, lactic acid acts as a preservative and accounts for much of the flavor and aroma of yogurt. However, it has one more role to play in yogurt production (and also in cheese production): It is very important in curd formation.

Curds are formed when casein molecules clump or stick together. The term **caseins** refers to a family of milk proteins that have phosphates attached to some of the amino acids. At normal milk pH (6.3–6.7), caseins are present in the form of tiny aggregates, or micelles, that are suspended in a cloudy colloid system. At near-neutral pH values, the phosphates give the casein molecules a net negative charge, and the colloidal system is maintained by the mutual repulsion of the negatively charged casein micelles. When milk pH is lowered by any acid (such as the lactic acid in yogurt), the excess protons (H^+) in the acidic environment interact with the negative charges on the phosphates and produce net neutral micelles. Now, under acidic conditions, the neutral caseins begin to clump together, forming semisolid clots or curds, a process that will give the yogurt its semisolid texture. This same phenomenon occurs when lactic acid bacteria in the old milk in the back of your refrigerator creates chunky spoiled milk.

While we've focused on lactic acid in this discussion of yogurt, it's worth noting that the *L. bulgaricus* and *S. thermophilus* cells also produce small amounts of acetaldehyde and diacetyl. These organic molecules give yogurt a unique taste and aroma different from those of other fermented dairy products such as cheddar cheese. Cheddars are created by a type of lactic acid bacterium that produces only lactic acid during fermentation.

It is only recently that we've been able to inoculate milk with pure cultures of *Lactobacillus bulgaricus* and *Streptococcus thermophilus* for the production of yogurt. So how were yogurt makers in the past able to favor these species in the production of yogurt when they were starting with raw milk contaminated with a mix of microbe types? Fortunately for yogurt lovers, *L. bulgaricus* and *S. thermophilus* are thermophilic: They grow well at temperatures of 45°C–50°C (113°F–122°F), and they can handle temperatures up to 55°C (130°F). So, as long as raw milk is incubated at these higher temperatures as the yogurt is made, the thermophilic *L. bulgaricus* and *S. thermophilus* can outcompete any heat-inhibited contaminating mesophiles.

PROTOCOL

IMPORTANT: Some of the bacteria in raw milk may be pathogenic, so gloves and safety glasses must be worn throughout both days of this lab.

Day 1

Work in pairs for all exercises.

MATERIALS

Per pair of students:

- 8 melted 3× phenol red glucose agar, 15 ml agar per tube
- 8 sterile petri dishes
- 2 empty sterile test tubes
- 6 sterile 9-ml water blanks
- 2 sterile 1-ml transfer pipettes
- 10 sterile 1-ml graduated pipettes
- 2-ml (blue-barrel) pipette pump
- Paper towels
- 2 150–250-ml beakers
- 1 teaspoon
- 100-ml graduated cylinder
- Aluminum foil
- pH paper
- Safety glasses
- Disposable gloves

Milk samples:

- Raw milk
- Pasteurized milk
- Pasteurized milk left out at room temperature for a day
- Pasteurized milk cooked immediately prior to lab in a 75°C water bath for 20 minutes, to be used in making yogurt

Cultures:

- *Streptococcus thermophilus*
- *Lactobacillus bulgaricus*
- Commercial yogurt with active cultures

BACTERIAL POPULATIONS IN MILK

1. Each pair of students will be assigned one milk type (of three possibilities) to test. Once given your assignment, be sure to write down immediately which milk type you are to test so that you don't forget.

2. Label four empty sterile petri dishes with your names and the date. Label the first plate 10^0, the second plate 10^{-1}, the third plate 10^{-2}, and the last plate 10^{-3}.

3. Label three 9-ml sterile water blanks with your names and the date. Label the first tube 10^{-1}, the second tube 10^{-2}, and the third tube 10^{-3}.

4. Transfer about 3 ml of your assigned milk sample from the stock beaker to an empty sterile test tube using a 1-ml transfer pipette.

   The line at the top of the shaft of the pipette is marked 1.0 ml, so if you fill the pipette to the base of the bulb, you have about 1 ml of liquid in the pipette.

   In steps 5 through 7 you will perform a serial dilution of the milk stock **(Figure 29.6)**.

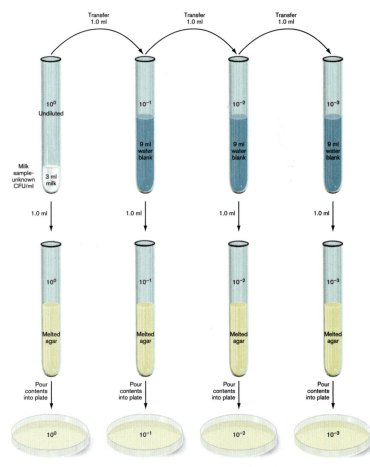

FIGURE 29.6 Dilution scheme for determination of bacterial populations in milk and pasteurization of raw milk.

5. Use a sterile 1.0-ml graduated pipette to transfer 1.0 ml of milk from the test tube to the 10^{-1} water blank tube to create a 10^{-1} milk dilution. Carefully draw the water up into the pipette to a line about halfway between the 1.0 ml mark and the top of the pipette, or to a total volume of about 1.2 ml, and then empty the contents of the pipette back into the tube. Repeat several times to finish the process of mixing the cells into the water. Do not discard the pipette.

6. Use the pipette from step 5 to transfer 1.0 ml of cells from the 10^{-1} milk dilution tube to the 10^{-2} water blank tube to create a 10^{-2} milk dilution. Repeat the draw-and-empty step to mix the cells, and do not discard the pipette.

7. Using the same pipette as in step 6, transfer 1.0 ml of cells from the 10^{-2} milk dilution tube to the 10^{-3} water blank tube to create a 10^{-3} milk dilution. Repeat the draw-and-empty step to mix the cells. Discard the pipette in the receptacle designated for trash to be autoclaved.

   Do **not** put the contaminated pipette down on the benchtop.

8. Prepare four labels as follows and set them aside: (a) 10^0, (b) 10^{-1}, (c) 10^{-2}, and (d) 10^{-3}.

   Assemble your labeled petri dishes, serially diluted milk samples, and labels, and be ready to work fast once you have removed your tubes of agar from the water bath.

9. Remove four tubes of melted $3\times$ phenol red glucose agar from the water bath and wipe off any excess water with a paper towel.

10. Apply a masking tape label to each of the four melted $3\times$ phenol red glucose agar tubes.

    In steps 11 through 18, you will transfer the serial dilutions of the milk stock to melted agar and pour them into petri dishes (see Fig. 29.6). One student will mix the milk samples with the melted agar. The other student will pour the inoculated agar into the petri dishes.

11. Use a new sterile 1.0-ml graduated pipette to transfer 1 ml of milk from the original undiluted milk sample in your test tube (from step 4) to the 10^0 melted agar tube (Tube 1). Gently stir the agar with the pipette to mix the cells into the agar, but don't stir for more than about 10–15 seconds, as the agar is cooling quickly at this point. Discard the pipette in the receptacle designated for trash to be autoclaved.

    Do **not** put the contaminated pipette down on the benchtop.

12. After stirring the agar for 10–15 seconds, hand the tube to your lab partner. This partner should carefully pour the contents of the 10^0 melted agar tube (Tube 1) into the petri dish labeled 10^0 while the "stirrer" quickly goes on to step 13.

13. Use a new sterile 1.0-ml graduated pipette to transfer 1 ml of cells from the 10^{-1} milk dilution tube to the 10^{-1} melted agar tube (Tube 2). Gently stir the agar with the pipette to mix the cells into the agar, but again, don't stir for more than about 10–15 seconds, as the agar is cooling quickly at this point. Discard the pipette in the receptacle designated for trash to be autoclaved.

14. Hand the tube to your lab partner, who will carefully pour the contents of the 10^{-1} melted agar tube (Tube 2) into the petri dish labeled 10^{-1} while the "stirrer" goes on to step 15.

15. Use a new sterile 1.0-ml graduated pipette to transfer 1 ml of cells from the 10^{-2} milk dilution tube to the 10^{-2} melted agar tube (Tube 3). Gently stir the agar with the pipette to mix the cells into the agar, but again, don't stir for more than about 10–15 seconds, as the agar is cooling quickly at this point. Discard the pipette in the receptacle designated for trash to be autoclaved.

16. Hand the tube to your lab partner, who will carefully pour the contents of the 10^{-2} melted agar tube (Tube 3) into the petri dish labeled 10^{-2} while the "stirrer" goes on to step 17.

17. Use a new sterile 1.0-ml graduated pipette to transfer 1 ml of cells from the 10^{-3} milk dilution tube to the 10^{-3} melted agar tube (Tube 4). Gently stir the agar with the pipette to mix the cells into the agar, but again, don't stir for more than about 10–15 seconds, as the agar is cooling quickly at this point. Discard the pipette in the receptacle designated for trash to be autoclaved.

18. Hand the tube to your lab partner, who will carefully pour the contents of the 10^{-3} melted agar tube (Tube 4) into the petri dish labeled 10^{-3}. Discard the pipette in the receptacle designated for trash to be autoclaved.

19. After all of the plates are poured, give them about 10–15 minutes to solidify. When the agar is clearly firm, invert the plates and incubate them at 30°C for 48 hours.

PASTEURIZATION OF RAW MILK

1. Each pair of students will be assigned one time point (30, 60, 120, or 240 seconds). Once given your assignment, be sure to write down immediately

which time point you are to test so that you don't forget.

2. Label an empty and capped sterile test tube with your names, the date, and your assigned time point. Transfer about 3 ml of raw milk from the stock beaker to the empty sterile test tube using a 1-ml transfer pipette. (In this part of the lab, everyone uses the raw milk.)

The line at the top of the shaft of the pipette is marked 1.0 ml, so if you fill the pipette to the base of the bulb, you have about 1 ml of liquid in the pipette.

3. Place your tube of raw milk in the rack provided in the 75°C water bath and incubate the tube for the assigned length of time (30, 60, 120, or 240 seconds). Warning: A 75°C bath is HOT!

4. After heating the milk for the assigned length of time, move the test tube to the room-temperature water bath to bring the temperature of the milk back down to room temperature as fast as possible.

5. After cooling the milk for 1 minute, take the tube back to your bench, and follow steps 2 and 3, and then steps 5–19, of the Bacterial populations in milk protocol in order to determine the number of viable cells remaining in the raw milk after heating.

In this case, the milk sample to be serially diluted is the heated milk and not the assigned milk type.

YOGURT PRODUCTION

1. Each pair of students will be assigned one of the following pure-cultures:
   a. *L. bulgaricus*
   b. *S. thermophilus*
   c. *L. bulgaricus* and *S. thermophilus*

   Once given your assignment, be sure to write down immediately which pure culture you are to test so that you don't forget.

2. Label two 150–250-ml beakers with your names and the date. Label one of the two beakers "Yogurt Cultures" and label the other beaker with the name of the pure culture(s) that you were assigned.

3. Use a 100-ml graduated cylinder to measure and add 100 ml of cooked milk to each of the beakers.

   Milk will be cooked by your instructor for 20 minutes at 75°C prior to lab.

4. Use pH paper to determine the initial pH of the cooked milk (either beaker will do). Record the pH value in the Results section under "Initial pH of cooked milk."

5. Add a teaspoon of the commercial yogurt with active cultures to the milk in the beaker labeled Yogurt Cultures.

6. Use a sterile 1.0-ml pipette to add 1.0 ml of broth culture of your assigned pure-culture species to the beaker labeled with the pure culture's name. If you were assigned "*L. bulgaricus* and *S. thermophilus*," then add 1.0 ml of broth culture of each species to the milk.

7. Cover the beakers with aluminum foil and incubate the milk at 45°C for 24 hours, then move the beakers to 4°C until day 2.

Day 2
MATERIALS

- Fresh commercial yogurt
- Pour plates and beakers from day 1
- Safety glasses
- Disposable gloves

BACTERIAL POPULATIONS IN MILK

1. Count all of the colonies on all of your plates and record the results in the table in the Results section. Look closely, because *Lactobacillus* and *Lactococcus* colonies can be quite small.

   The small, agar-embedded colonies are best seen by holding the plate up to a light so that the light passes through the medium in the plate. This way, colonies will appear as dark dots against a yellow halo background.

   If in the process of counting the colonies on a plate, you reach a count of about 400 colonies or more, then you can stop counting that plate at that point, and you can write "TNTC" ("too numerous to count") on the appropriate line in the Results section. There is no need to count all the colonies on plates with more than 400 colonies per plate because you won't be using those plates to do your calculations. But go ahead and record results if there are between 300 and 400 colonies, in case the next dilution produces plates with fewer than 30 colonies.

2. After counting the colonies on your plates, calculate the CFU/ml for your particular milk sample (see Lab 4). If there are no plates with counts between 30 and 300, you can use plates with 300–400 colonies to calculate CFU/ml. If no plates have more than 30 colonies, use the lowest dilution that produced colonies to calculate CFU/ml. For example, if both the 10^0 and the 10^{-1} plates have a few colonies, use the 10^0 plate to do the calculations.

3. Record the CFU/ml for your particular milk type in the Results section. Post your results with those of the entire class and record the pooled class results.

4. Use the class data to calculate mean CFU/ml for each milk type.

PASTEURIZATION OF RAW MILK

1. Count all of the colonies on all of your plates and record the results in the table in the Results section. Again, hold the plates up to the light for best results.

   If in the process of counting the colonies on a plate, you reach a colony count of about 400 colonies or more, then you can stop counting that plate at that point, and you can write "TNTC" ("too numerous to count") on the appropriate line in the Results section. There is no need to count all the colonies on plates with more than 400 colonies per plate because you won't be using those plates to do your calculations. But go ahead and record results if there are between 300 and 400 colonies, in case the next dilution produces plates with fewer than 30 colonies.

2. After counting the colonies on all plates, calculate the CFU/ml for your heating time. If there are no plates with counts between 30 and 300, you can use plates with 300–400 colonies to calculate CFU/ml. If no plates have more than 30 colonies, use the lowest dilution that produced colonies to calculate CFU/ml. For example, if both the 10^0 and the 10^{-1} plates have a few colonies, use the 10^0 plate to do the calculations.

3. Record the CFU/ml for your particular heating time in the Results section. Post your results with those of the entire class and record the pooled class results.

4. Use the class data to calculate mean CFU/ml for each heating time.

5. Copy the results for raw milk from the Bacterial populations in milk part of the lab to determine CFU/ml for raw milk at time = 0 seconds.

6. Plot the mean values for each heating time on the semi-log graph paper in the Questions section, and sketch in the line that is the best fit to the points.

YOGURT PRODUCTION

1. **Do not eat the yogurt!**

   Look around the lab. Does your lab look like a USDA-approved food production facility? The correct answer is NO! No eating in lab!

2. Use pH paper to measure the pH of fresh commercial yogurt. Record the value in the Results section.

3. Record your impressions of the odor, texture, and appearance of the fresh commercial yogurt.

4. Use pH paper to measure the pH of the milk you inoculated with commercial yogurt on day 1, and record the value in the Results section.

5. Record your impressions of the odor, texture, and appearance of the milk inoculated with commercial yogurt. Does this milk now look and smell like yogurt?

6. Repeat steps 2 and 3 for the milk inoculated with your assigned pure culture(s).

7. Borrow examples of milk inoculated with the other two pure-culture options, and again, repeat steps 2 and 3.

8. When finished, place all beakers and petri dishes in the receptacles designated for materials to be autoclaved.

Name: _____   Section: _____

Course: _____   Date: _____

BACTERIAL POPULATIONS IN MILK

Individual results

Assigned milk type:

Record the number of colonies for each of your plates and calculate the CFU/ml for your assigned milk sample using the equation at the bottom of the table.

CALCULATION OF CFU/ML			
	INDIVIDUAL DILUTION	TOTAL DILUTION	DILUTION FACTOR	COLONIES
Tube 1	Undiluted	10^0	$10^0 = 1$	
Tube 2	10^{-1}	10^{-1}	$10^1 = 10$	
Tube 3	10^{-1}	10^{-2}	$10^2 = 100$	
Tube 4	10^{-1}	10^{-3}	$10^3 = 1,000$	

CFU/ml = (Colonies from plate with 30 to 300 colonies) × (Dilution factor)

Pooled results

Post your CFU/ml calculation on the board in the appropriate column, record all of the data for all other pairs of students, and calculate the mean CFU/ml for all milk types.

POOLED RESULTS			
	RAW MILK	PASTEURIZED	KEPT WARM FOR A DAY
Pair 1			
Pair 2			
Pair 3			
Pair 4			
Pair 5			
Mean			

PASTEURIZATION OF RAW MILK

Individual results

Assigned heating time:

Record the number of colonies for each of your plates and calculate the CFU/ml for your assigned milk sample using the equation at the bottom of the table.

CALCULATION OF CFU/ML				
	INDIVIDUAL DILUTION	DILUTION AFTER ...	DILUTION FACTOR	COLONIES
Tube 1	Undiluted	10^0	$10^0 = 1$	
Tube 2	10^{-1}	10^{-1}	$10^1 = 10$	
Tube 3	10^{-1}	10^{-2}	$10^2 = 100$	
Tube 4	10^{-1}	10^{-3}	$10^3 = 1,000$	

CFU/ml = (Colonies from plate with 30 to 300 colonies) × (Dilution factor)

Pooled results

Post your CFU/ml calculation on the board in the appropriate column, record all of the data for all other pairs of students, and calculate the mean CFU/ml for all milk types.

POOLED RESULTS					
	T = 0 SEC*	*T* = 30 SEC	*T* = 60 SEC	*T* = 120 SEC	*T* = 240 SEC
Pair 1					
Pair 2					
Pair 3					
Pair 4					
Mean					

*Use data from raw milk column from the Bacterial populations in milk section on page 427 for *T* = 0 seconds.

PRODUCTION OF YOGURT

Initial pH of cooked milk:

pH, appearance, odor, and texture of commercial yogurt:

OBSERVATIONS OF YOGURT MADE IN LAB		
BACTERIA	pH AT FINISH	APPEARANCE, ODOR, TEXTURE, DESCRIPTION OF CURDS
Yogurt cultures		
Lactobacillus bulgaricus		
Streptococcus thermophilus		
L. bulgaricus and *S. thermophilus*		

Name: _____   Section: _____

Course: _____   Date: _____

1. Use the mean CFU/ml values from the pooled class data, and compare the CFU/ml values for raw milk and pasteurized milk when answering (a) and (b).

   a. Use the equation below to express CFU/ml for pasteurized milk as a percentage of the CFU/ml for raw milk:

   $$\frac{\text{CFU/ml pasteurized milk}}{\text{CFU/ml raw milk}} \times 100$$

   b. What do these numbers tell you about the value of pasteurization?

2. Use the mean CFU/ml values from the pooled class data, and compare the CFU/ml values for pasteurized milk and for milk that was pasteurized, but then kept at room temperature for a day when answering (a) and (b).

   a. What do these numbers tell you about the value of refrigeration?

   b. Is pasteurization the same thing as sterilization? Cite evidence from this lab as part of your answer.

Graph the results from the **Pasteurization of raw milk** section and use the graph to answer Questions 3 and 4.

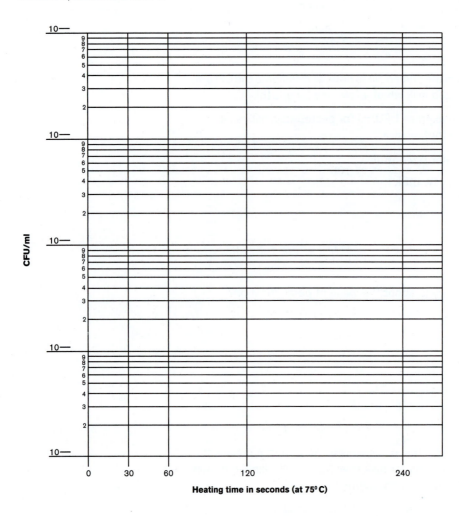

Heating time in seconds (at 75° C)

CFU/ml

3. What is the decimal reduction time for raw milk at 75°C?

4. Given the initial bacterial density of our raw milk sample, at 75°C, how long will it take to reduce the population of raw milk to 1 cell/ml (1 CFU/ml)?

5. How might improved sanitation at the point of collection in a dairy change the time required to reduce the bacterial population in raw milk to one cell (at 75°C)? Explain your answer.

6. In recent years, there has been increased consumer interest in raw or unpasteurized milk.

   a. List a few reasons why someone might chose to consume raw milk as opposed to pasteurized milk (you may need to do some research here).

   b. What are the risks of consuming raw milk, and why do you think that people may choose to take these risks?

7. Look over all the procedures you used for making yogurt and describe two things that were done (including things done to the milk prior to the start of the lab) to improve the odds that the vast majority of the cells growing in the milk inoculated with yogurt cultures would be either *Lactobacillus* cells or *Streptococcus* cells, or both. Be specific about how a particular procedure favored these genera.

8. Did the milk inoculated with a teaspoon of yogurt look and smell like yogurt (after incubation)? What changes occurred in the milk inoculated with yogurt cultures?

9. Why does yogurt have a longer shelf life than raw milk?

10. After incubation, how did the milk inoculated with only *L. bulgaricus* compare with the milk inoculated with yogurt cultures? Explain what happened in the *L. bulgaricus*-inoculated milk with respect to pH and curd formation.

11. After incubation, how did the milk inoculated with only *S. thermophilus* compare with the milk inoculated with yogurt cultures? Explain what happened in the *S. thermophilus*-inoculated milk with respect to pH and curd formation.

12. After incubation, how did the milk inoculated with both *L. bulgaricus* and *S. thermophilus* compare with the milk inoculated with yogurt cultures? Explain what happened in this milk with respect to pH and curd formation.

13. Offer evidence that *L. bulgaricus* and *S. thermophilus* interacted in a symbiotic manner in the milk inoculated with both species by comparing this milk with the milk samples inoculated with a single species of lactic acid bacteria. What did each species do for the other, and how did this change what you saw in the milk inoculated with both species versus the single-species milk samples?

14. Yeast infections sometimes result from taking antibiotics, due to the death of "good" bacteria along with the death of the "bad" bacteria. A person is sometimes advised to eat yogurt while taking prescribed antibiotics. Why would this be beneficial? Explain the process behind its benefit.

LAB 30

Water Analysis

Learning Objectives

- Consider the problems created by fecal contamination of water.

- Understand the use of indicator species, water analysis protocols, and most probable numbers (MPN) tables to detect fecal contamination of water.

- Explain the methods used to differentiate "total coliforms" from "fecal coliforms."

- Test a variety of water samples for fecal contamination.

THE PROBLEM OF FECAL CONTAMINATION OF WATER

There is a very long list of diseases caused by microbes that can be transmitted from one human host to another via fecal contamination of drinking water (**Figure 30.1**). Waterborne or water-transmitted microbes include the bacteria *Salmonella* Typhi and *Vibrio cholerae*, species responsible for typhoid fever and cholera, respectively. These two diseases alone killed literally millions of Americans in the nineteenth century. Other diseases often transmitted by consumption of fecally contaminated water include viral diseases such as polio and hepatitis A and protozoan diseases such as giardiasis, caused by *Giardia intestinalis*.

Typically, these microbes establish intestinal infections, cause vomiting and diarrhea, and in some cases, spread to other parts of the body such as the blood, kidneys and nervous system, where they cause serious and sometimes fatal diseases. The pathogens then leave the body with the host's intestinal waste (feces) and enter another host when that individual drinks fecally contaminated water. So, if we can detect fecal contamination of water, we can prevent or reduce the incidence of a large number of diseases. Besides, who wants to drink water "enhanced" with *merde*?

A.

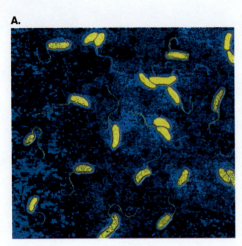

B.

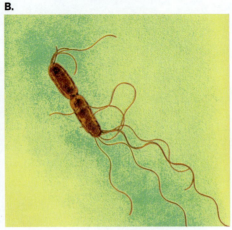

C.

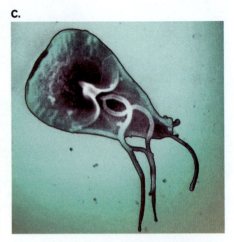

FIGURE 30.1 Waterborne microbial pathogens: all cause diarrhea and other intestinal symptoms.
A. *Vibrio cholerae.* **B.** *Salmonella* Typhimurium. **C.** *Giardia lamblia.*

WHAT IS NOT A PART OF THE ROUTINE MICROBIOLOGICAL ANALYSIS OF WATER?

How do we detect fecal contamination of water? Since it's the pathogens that we're concerned about, you might expect that this would be done by routinely testing for a range of disease-causing intestinal microbes such as *Salmonella* or poliovirus. But this is not what is done. There are several reasons why this is not the approach taken:

1. Pathogenic cells and virus particles are often present at low concentrations in water.
2. In most cases, pathogenic bacteria probably won't multiply in the water, so their numbers will remain low.
3. Many types of waterborne pathogens, especially viruses, are difficult to isolate and culture.
4. There are many, many types of waterborne pathogens, and testing for all of them would be expensive and time-consuming, although new methods have significantly reduced these costs.

INDICATOR SPECIES

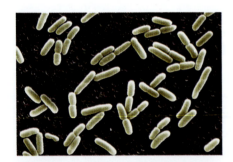

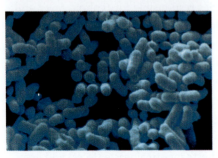

FIGURE 30.2 Coliform bacteria.
Coliforms are all Gram-negative, rod-shaped bacteria like *E. coli* (top) and *E. aerogenes* (bottom).

So, what do we look for in our drinking water to detect microbial contamination? Routine water testing looks for coliform bacteria **(Figure 30.2)**. Recall that coliform bacteria are Gram-negative species in the family Enterobacteriaceae (enteric family) that are capable of rapidly fermenting lactose to produce acid and gas (see Labs 24–27). As we look for coliforms in water, we will focus on the fermentation of lactose as a key distinguishing trait for this group of bacteria.

Most coliform species are usually not pathogenic; that is, most don't routinely cause disease in healthy humans. In fact, we carry these bacteria in our guts throughout our lives, and they usually maintain a relatively benign relationship with their human hosts (though there are exceptions to this rule). So, our search for coliforms isn't particularly motivated by the direct risks they pose to us. Instead, we look for coliforms because these bacterial species are well adapted to the intestines of humans and other mammals, and as a result, their presence in water indicates that the water may have been contaminated with fecal matter. If fecal contamination is present, then intestinal pathogens could be in the water, too, although the assays described in this lab are not designed to directly confirm their presence. What we're trying to do is use indicator organisms to show fecal contamination of water, and that's all.

Coliforms are good indicators of fecal contamination for several reasons:

1. They are adapted to the gut environment.
2. They are relatively easy to detect and quantify.
3. They are uncommon in uncontaminated water, especially the "fecal coliforms."
4. They are almost always present where intestinal pathogens are present.
5. Their population densities are usually related to the degree of fecal contamination.

DISTINGUISHING FECAL COLIFORMS FROM TOTAL COLIFORMS

As noted, coliforms are good indicators of fecal contamination. However, as a group, coliforms are not solely or exclusively found in mammalian colons. Some coliforms, such as *Enterobacter* species, can be found in soil or water when there is no fecal contamination at all. Ideally, we would like to be able to distinguish the subgroup of coliforms that are specifically of fecal origin from other coliforms that may come from non-fecal sources, because it's the fecal coliforms that are real indicators of fecal contamination. For example, the fecal coliform species *E. coli* is not usually found in water, so if *E. coli* cells are found in water in significant numbers, then it's very likely that intestinal waste of some type has found its way into the water.

So how can we select for fecal coliforms such as *E. coli* over non-fecal coliforms such as *Enterobacter*? Mammalian intestines are maintained at near-constant temperatures ranging from 36°C to nearly 40°C (97°F–104°F), so fecal coliforms are adapted to warmer environments than are non-fecal coliforms. In fact, fecal coliforms can grow and multiply even at temperatures as high as 44.5°C, a temperature at which most coliforms of non-fecal origin will not grow at all. Thus, fecal coliforms can be defined as the subgroup of all coliforms that rapidly ferment lactose to acids and gases at 44.5°C. And we can select for the fecal coliform subgroup within the total coliform group by incubating cells in lactose-containing broths at 44.5°C.

Analysis of water for total coliforms

A series of three assays is used to measure total coliforms in water. The assays are performed in the following order: the presumptive test, the confirmed test, and the completed test.

Presumptive test The **presumptive test** uses a medium specifically designed to favor the growth of coliforms and to detect the ability of microbes to rapidly produce gas as one end product of the fermentation of lactose. The medium used for this assay is a lactose broth containing beef extract and peptones to meet basic nutritional needs and 0.5% lactose as a substrate for fermentation. Each tube of medium contains a Durham tube to collect gases produced by fermentation.

Unlike the phenol red lactose broths used elsewhere in the course, this lactose broth does not contain a pH indicator. While the fermentation of lactose by coliforms produces organic acids and lowers the pH of the medium, the production of acids by lactose metabolism is a trait seen in many other bacterial groups, including the Gram-positive lactic acid bacteria (see Lab 21). So, it would not be unusual if a feces-free water sample contained non-coliform bacteria able to produce a positive reaction for the production of acids from lactose. In other words, the production of acids from lactose is not definitive or specific for the presence of coliforms in a water sample. By contrast, the production of gas from lactose is a much less common trait. It is more specific to the coliforms, so when gas production is detected, it is a significant and meaningful observation.

In the presumptive test, water samples are added to three sets of three lactose broth tubes in varying volumes: 10 ml to the first set of three tubes, 1.0 ml to the second set of three tubes, and 0.1 ml to the last set of three tubes. In effect, this creates three different dilutions of the initial water sample. In the case of the tubes receiving 10 ml of the water sample, the medium begins at double, or 2×, strength, because adding 10 ml of water to 10 ml of 1× medium will cut the concentration of the medium components in half. Thus, after the addition of the water sample, the final concentrations of the medium components will be back to "normal," or "single," or "1×" strength for the incubation step.

Once the tubes are inoculated, they are incubated at 37°C for 24 hours to determine total coliforms. If there is just one live cell in the water added to a given tube, then that one cell will multiply, and the population will build rapidly. If that one cell was a coliform, then lactose will be fermented, producing acids and gases as end products. Obviously, the same thing will happen (gas will be produced) if there is more than one coliform cell in the water added to the tube, but the key point is that it takes only a single coliform cell at the start to produce gas in the medium.

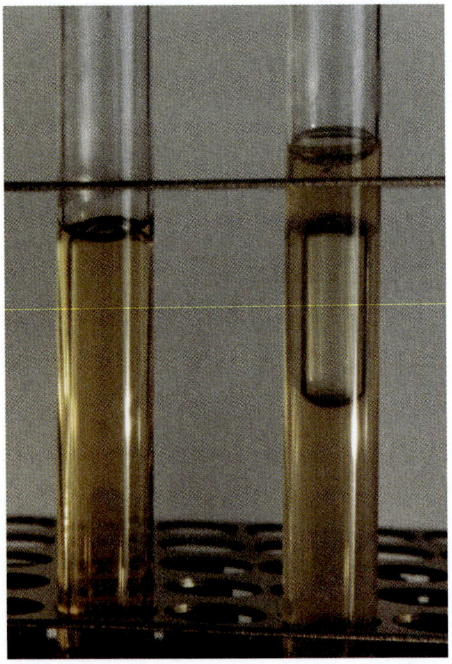

After 24 hours of incubation, each tube is checked for a positive or negative result. **The tube is scored as positive for rapid fermentation of lactose to gas if there is a large bubble at least 5 mm in diameter in the Durham tube. If there are only tiny bubbles or no bubble, that is a negative result (Figure 30.3)**. Remember, fermentation of lactose to acid alone isn't that unusual in the bacterial world, but fermentation that produces both acid and gas is more particular to the coliform group. In each set of three tubes (0.1-ml group, 1.0-ml group, 10-ml group), there may be 0, 1, 2, or 3 positive tubes. The number of positive tubes in each group is used in a most probable numbers table **(Table 30.1)** to determine the number of coliform cells per 1 ml or per 100 ml of the original water sample.

The **most probable number (MPN)** is an estimate of coliform population density that is based on probability theory. The MPN or density value generated by the table is the coliform density that makes the observed outcome of a given number of positive reactions (0, 1, 2, or 3) in each of the three sets of tubes (0.1 ml, 1.0 ml, 10 ml) the "most probable" outcome. To put it another way, it is based on the probability of getting a certain number of positive tubes in each of the three sets of tubes when the initial population of coliforms is X. Since we're dealing with probabilities, one should always remember that the MPN value is an estimate of density (a "most probable number"), and the actual density is unlikely to be exactly, precisely the value given in the table.

FIGURE 30.3 Results of presumptive test.

Table 30.1 | MOST PROBABLE NUMBERS TABLE

0.1 ml	1.0 ml	10 ml	MPN per 1 ml	MPN per 100 ml	0.1 ml	1.0 ml	10 ml	MPN per 1 ml	MPN per 100 ml
0	0	0	<0.03	<3	2	0	0	0.09	9
0	0	1	0.03	3	2	0	1	0.14	14
0	0	2	0.06	6	2	0	2	0.20	20
0	0	3	0.09	9	2	0	3	0.26	26
0	1	0	0.03	3	2	1	0	0.15	15
0	1	1	0.06	6	2	1	1	0.20	20
0	1	2	0.09	9	2	1	2	0.27	27
0	1	3	0.12	12	2	1	3	0.34	34
0	2	0	0.06	6	2	2	0	0.21	21
0	2	1	0.09	9	2	2	1	0.28	28
0	2	2	0.12	12	2	2	2	0.35	35
0	2	3	0.16	16	2	2	3	0.42	42
0	3	0	0.09	9	2	3	0	0.29	29
0	3	1	0.13	13	2	3	1	0.36	36
0	3	2	0.16	16	2	3	2	0.44	44
0	3	3	0.19	19	2	3	3	0.53	53
1	0	0	0.04	4	3	0	0	0.23	23
1	0	1	0.07	7	3	0	1	0.39	39
1	0	2	0.11	11	3	0	2	0.64	64
1	0	3	0.15	15	3	0	3	0.95	95
1	1	0	0.07	7	3	1	0	0.43	43
1	1	1	0.11	11	3	1	1	0.75	75
1	1	2	0.15	15	3	1	2	1.20	120
1	1	3	0.19	19	3	1	3	1.60	160
1	2	0	0.11	11	3	2	0	0.93	93
1	2	1	0.15	15	3	2	1	1.50	150
1	2	2	0.20	20	3	2	2	2.10	210
1	2	3	0.24	24	3	2	3	2.90	290
1	3	0	0.16	16	3	3	0	2.40	240
1	3	1	0.20	20	3	3	1	4.60	460
1	3	2	0.24	24	3	3	2	11.00	1100
1	3	3	0.29	29	3	3	3	>11.00	>1100

Number of Positives In Each Group of Tubes With a Given Volume

A.

B.

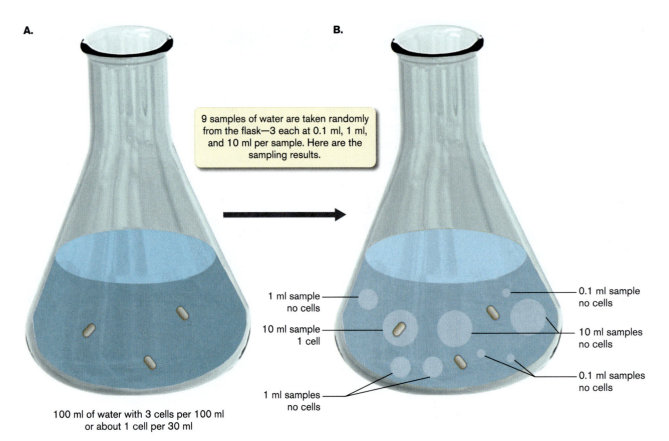

9 samples of water are taken randomly from the flask—3 each at 0.1 ml, 1 ml, and 10 ml per sample. Here are the sampling results.

1 ml sample no cells

10 ml sample 1 cell

1 ml samples no cells

0.1 ml sample no cells

10 ml samples no cells

0.1 ml samples no cells

100 ml of water with 3 cells per 100 ml or about 1 cell per 30 ml

FIGURE 30.4 An illustration of most probable number sampling.
Note that one of the three 10-ml samples netted a cell while none of the 1-ml or 0.1-ml samples included any cells.

For example, let's say the initial population in a water sample is 3 coliform cells per 100 ml of water, or about 1 cell per 30 ml **(Figure 30.4A)**.

In this case, there is a very low probability that a given 1.0-ml or 0.1-ml sample will contain any cells, so all of the 1.0-ml-inoculated and 0.1-ml-inoculated tubes will almost always be negative **(Figure 30.4B)**. That is, the number of positive tubes in these two sets of tubes will be zero for both sets. However, if three independent 10-ml samples are drawn from a 3 cells/100 ml population, then there is a very good chance that one of those 10-ml samples will contain a single cell, because the cell density of the sample is about 1 cell per 30 ml of sample (3 cells/100 ml), and a total volume of 30 ml was taken in the three 10-ml samples **(Figure 30.4B)**. So, if the coliform density is 3 cells/100 ml, you would expect one of the three 10-ml tubes to produce a positive result. And a glance at the MPN table shows that a result of 0 positive out of 3 for 0.1 ml—0 positive out of 3 for 1.0 ml— 1 positive out of 3 for 10 ml gives a most probable number of 3 cells/100 ml. That is, the coliform density that would make the observed outcome (0 0 1) the "most probable" outcome would be 3 cells/100 ml. This reasoning is used for all possible combinations of positive and negative tubes to give estimates of population densities for a given combination. In the end, you don't have to worry about the statistical theory behind the table; all you have to do is enter your results and determine the most probable number for your sample.

It should be noted that if all the tubes are positive, the result will be "greater than 1,100 coliform cells per 100 ml." This could mean that the concentration is 1,200 cells per 100 ml, or 1,200,000 cells per 100 ml, or any other number in between or above. The problem is that all of these population densities (>1,100 cells/100 ml) will produce the same result; namely, all nine tubes will be positive. To get an accurate measurement in such a sample, one must first dilute the sample until the assay has at least one negative tube. Then the MPN value can be multiplied by the dilution factor to get the coliform density in the original sample.

So to recap, the MPN method uses a series of inoculated Durham tubes to capture carbon dioxide gas from the rapid fermentation of lactose. The dilutions are set up and gas bubbles are measured at 24 hours to determine which tubes are positive. A statistical analysis is then used to determine the most probable number of bacterial cells present based on the number and distribution of positive tubes.

Once an MPN value has been determined, decisions can be made about the proper use of the water. For example, the standard for drinking water is very strict: Drinking water is not considered to be **potable**, or safe for human consumption, unless the coliform count is less than 1 cell/100 ml. By contrast, "recreational water," such as the water in a lake or river, is generally considered acceptable for swimming and boating as long as the coliform count is below 125 cells/100 ml. And when an area is to be used for commercial shellfishing (oysters, clams, etc.), the standards are usually set at 50–100 cells/100 ml.

Confirmed test The **confirmed test** is designed to confirm that the gas producers in the lactose broth used in the presumptive test are really coliforms and not some other type of lactose-fermenting microbes. An EMB agar plate is inoculated with material from each positive tube produced in the presumptive test, and the resulting colonies are examined to confirm that they are coliforms. The composition of EMB agar is discussed in Lab 24. The medium inhibits the growth of Gram-positive bacteria, and **Gram-negative lactose fermenters will produce darkly pigmented colonies**. Gram-negative species that do not ferment lactose will produce colonies with little or no color; these species include non-coliform enterics such as *Salmonella* and *Shigella*.

Further, *E. coli* and other strong acid producers will generate colonies with a distinct, metallic green sheen, while *Enterobacter aerogenes* and other lactose fermenters that produce a mix of acids and neutral end products will generate dark purple colonies (**Figure 30.5**). This difference is actually quite significant in the context of this assay, because while *E. coli* is almost always of intestinal origin and an excellent indicator of fecal contamination, *Enterobacter* species may be found in uncontaminated water and

A.

B.

FIGURE 30.5 Results of the confirmed test.
A. Colonies of the coliform species *Escherichia coli* are metallic green due to the large amounts of acid from lactose fermentation. **B.** Colonies of the coliform species *Enterobacter aerogenes* are dark purple due to smaller amounts of acid from lactose fermentation.

A.   **B.**

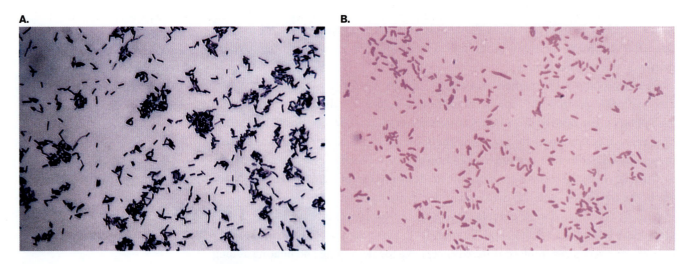

FIGURE 30.6 Results of the completed test.
A. Gram-positive cells (negative for the completed test). **B.** Gram-negative cells (positive for the completed test).

soil. Thus, the presence of metallic green colonies is a stronger indicator of fecal contamination than dark purple colonies.

Completed test The **completed test** clinches the case that the lactose-broth-positive microbes are coliforms. A darkly pigmented colony is picked from the EMB plates of the confirmed test and used to inoculate (1) another lactose broth with a Durham tube and (2) a nutrient agar slant to produce cells to be Gram-stained. **If this isolate is (1) positive for gas from lactose fermentation and (2) a Gram-negative rod, it's almost certainly a coliform (Figure 30.6B).**

Analysis of water for fecal coliforms

Water can be analyzed for fecal coliforms using a procedure similar to the one used for total coliforms, except that the presumptive test is performed with a different medium and at a different incubation temperature. A broth called A-1 medium is used in the presumptive test. This medium contains lactose to differentiate cells that ferment lactose and produce gas from those that do not, as well as peptones as an amino acid source, salicin as an additional energy and carbon source, and sodium chloride for osmotic balance. The medium itself is not highly selective, but inoculated tubes are incubated at 44.5°C to favor the growth of fecal coliforms. Remember, **at this temperature, fecal coliforms will produce gas from lactose, while most coliforms of nonfecal origin will not grow at all.**

The steps of this lab are summarized in **Figure 30.7**.

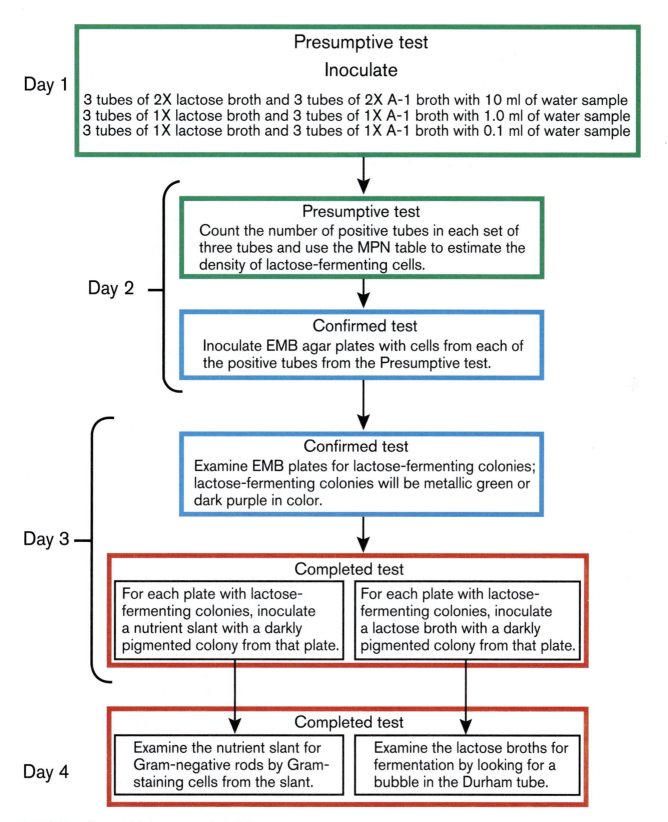

FIGURE 30.7. Steps of the water analysis lab.

PROTOCOL

> **IMPORTANT:** Some of the bacteria in environmental water samples may be pathogenic, so gloves and safety glasses must be worn throughout all four days of this lab.

Day 1
Materials

- 6 1× lactose broths with Durham tubes
- 3 2× lactose broths with Durham tubes
- 6 1× A-1 medium (broths) with Durham tubes
- 3 2× A-1 medium (broths) with Durham tubes
- 2 sterile 1-ml pipettes
- 2 sterile 10-ml pipettes
- 2-ml (blue-barrel) pipette pump
- 10-ml (green-barrel) pipette pump
- 2 sterile 125-ml flasks
- Safety glasses
- Disposable gloves

Cultures:
- Drinking fountain water
- Duck pond water (or stream water)
- Toilet water

PRESUMPTIVE TEST (TOTAL COLIFORM COUNT)

1. Each pair of students will be assigned water from one source (of three possibilities) to test. Once given your assignment, be sure to write down immediately which water type you are to test so that you don't forget.

2. Fill a 125-ml flask with 50 ml of water from your assigned water source.

3. Label three 2× lactose broths and six 1× lactose broths with your names and the date. Mark the three 2× lactose tubes "10 ml," mark three 1× lactose tubes "1.0 ml," and mark three 1× lactose tubes "0.1 ml."

4. Use a sterile 10-ml pipette to transfer 10 ml of your water sample to each of the 2× lactose broth tubes labeled 10 ml, and use a 1-ml pipette to transfer the correct volume of water sample to each of the 1.0 ml and 0.1 ml 1× lactose broth tubes.

As long as your pipette tip doesn't touch any unsterile surfaces besides the water itself, you can use one pipette for all 10-ml tubes and one pipette for all 1.0-ml and 0.1-ml tubes. But if the end of the pipette touches anything unsterile at any point in time, discard the pipette and start over with a new one.

5. Discard all pipettes in the receptacle designated for trash to be autoclaved.

6. Incubate all nine tubes at 37°C for 24 hours.

PRESUMPTIVE TEST (FECAL COLIFORM COUNT)

1. Using the same water source that you were assigned for the total coliform count, repeat steps (2) through (5) except, in this case, inoculate the 1× and 2× A-1 medium broths instead of the 1× and 2× lactose broths.

2. Incubate all A-1 medium tubes at 37°C for 3 hours. Then transfer all tubes to a water bath or incubator set at 44.5°C, and incubate for an additional 21 hours.

   A water bath will do a better job of holding the temperature of the media at 44.5°C, but a regular incubator will do in a pinch.

Day 2
MATERIALS

- 2 eosin methylene blue (EMB) agar plates
- Tubes inoculated with water on day 1
- Safety glasses
- Disposable gloves

PRESUMPTIVE TEST

1. Examine each of the inoculated lactose and A-1 broths for the presence of a large gas bubble in the Durham tube. Ignore any tiny bubbles at the top of the tube that are smaller than about 5 mm in diameter. True coliform cells make bubbles greater than 5 mm in diameter in Durham tubes.

2. Record the number of tubes with large bubbles (positive reaction) in each volume category (10 ml, 1.0 ml, 0.1 ml) for both lactose and A-1 broths in the Results section.

3. Use results from the lactose broths and the most probable numbers table (Table 30.1) to determine the most probable number of total coliform cells per 1 ml and per 100 ml in your water sample.

4. Use the results from the A-1 broths and the most probable numbers table to determine the most probable number of fecal coliform cells per 1 ml and per 100 ml in your water sample.

5. For both total and fecal coliforms, one member of each of the day 1 student pairs should post the MPN values on the board for 100 ml only for the purpose of pooling class results. Then record the results of all students in the Results section, and use the class data to calculate mean MPN for all water samples of each type.

CONFIRMED TEST

6. Label two EMB plates with your names and the date. Label one of the plates Total Coliforms and the other plate Fecal Coliforms.

7. Pick any lactose broth tube with a large bubble, and using cells from this tube, streak the Total Coliforms EMB agar plate following the streak plating protocol used to produce isolated colonies (see Lab 3). If you don't have any positive tubes (which we hope will be the case with the drinking fountain water), just use a positive tube from any other student in the class.

8. Repeat step 7, using any A-1 broth tube with a large bubble to inoculate the Fecal Coliforms EMB plate.

9. Incubate all plates at 37°C for 24 hours.

Day 3
MATERIALS

- 2 1× lactose broths with Durham tubes
- 2 nutrient agar slants
- EMB plates inoculated on day 2
- Safety glasses
- Disposable gloves

CONFIRMED TEST

1. Examine both EMB plates inoculated on day 2 for any coliform colonies. Record your observations in the Results section.

COMPLETED TEST

2. Label your two 1× lactose broth tubes with your names and the date. Label one of the tubes Total Coliforms and the other tube Fecal Coliforms.

3. Label your two nutrient agar slants with your names and the date. Label one of the slants Total Coliforms and the other slant Fecal Coliforms.

4. Find an isolated coliform colony (a shiny green or dark purple colony) on the EMB plate inoculated from the lactose broth tube (on day 2). Use a sterile loop to transfer cells from that colony to the 1× lactose broth labeled Total Coliforms.

5. Using the same isolated colony as in step 4, use a sterile loop to transfer cells from that colony to the nutrient agar slant labeled Total Coliforms.

6. Find an isolated coliform colony on the Fecal Coliform EMB plate inoculated from the A-1 broth tube. Use a sterile loop to transfer cells from that colony to the 1× lactose broth labeled Fecal Coliforms.

7. Using the same isolated colony as in step 6, use a sterile loop to transfer cells from that colony to the nutrient agar slant labeled Fecal Coliforms.

8. Incubate all four tubes at 37°C for 24 hours.

Day 4
MATERIALS

- Lactose broths inoculated on day 3
- Nutrient agar slants inoculated on day 3
- Clean glass microscope slides
- Gram stain reagents
- Safety glasses
- Disposable gloves

COMPLETED TEST

1. Examine both 1× lactose broths, note the presence or absence of a large bubble in the tubes, and record your observations in the Results section.

2. Use a flame-sterilized inoculating loop to transfer cells from each nutrient agar slant to a glass slide, and Gram-stain these cells (see Lab 6 for instructions and materials). Record your observations in the Results section.

Name: _____   Section: _____

Course: _____   Date: _____

TOTAL COLIFORMS: PRESUMPTIVE TEST WITH LACTOSE BROTH

Results for your water sample

Assigned water type:

CALCULATION OF MPN				
NUMBER OF POSITIVE TUBES IN EACH GROUP OF THREE TUBES OF A GIVEN VOLUME			MPN per 1 ml	MPN per 100 ml
0.1 ML	1.0 ML	10 ML		

Post your MPN calculation on the board in the appropriate column, record all of the data for all other pairs of students, and calculate the mean MPN for all water types.

POOLED RESULTS			
	Drinking Fountain Water MPN per 100 ml	Duck Pond Water MPN per 100 ml	Toilet Water MPN per 100 ml
Pair 1			
Pair 2			
Pair 3			
Pair 4			
Pair 5			
Pair 6			
Mean			

FECAL COLIFORMS: PRESUMPTIVE TEST WITH A-1 MEDIUM

Results for your water sample

Assigned water type:

CALCULATION OF MPN				
NUMBER OF POSITIVE TUBES IN EACH GROUP OF THREE TUBES OF A GIVEN VOLUME			MPN per 1 ml	MPN per 100 ml
0.1 ML	1.0 ML	10 ML		

Post your MPN calculation on the board in the appropriate column, record all of the data for all other pairs of students, and calculate the mean MPN for all water types.

POOLED RESULTS			
	Drinking Fountain Water MPN per 100 ml	Duck Pond Water MPN per 100 ml	Toilet Water MPN per 100 ml
Pair 1			
Pair 2			
Pair 3			
Pair 4			
Pair 5			
Pair 6			
Mean			

TOTAL COLIFORMS: CONFIRMED TEST

EMB agar plate inoculated with cells from positive lactose broth tube

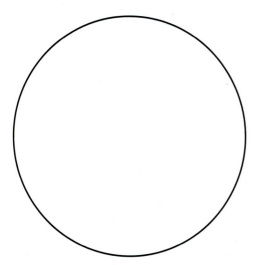

Written description of colonies: (Include colony size, shape, color, and whether or not the colonies ferment lactose.)

FECAL COLIFORMS: CONFIRMED TEST

EMB agar plate inoculated with cells from positive A-1 medium tube

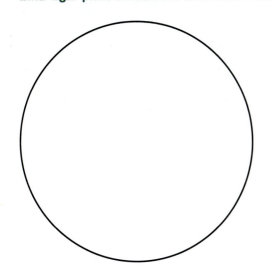

Written description of colonies: (Include colony size, shape, color, and whether or not the colonies ferment lactose.)

TOTAL COLIFORMS: COMPLETED TEST

(From EMB agar plate inoculated with cells from positive lactose broth tube)

Was there a large bubble in the Durham tube inside your 1× lactose broth tube?

Gram stain of nutrient agar subculture of colony from EMB agar plate

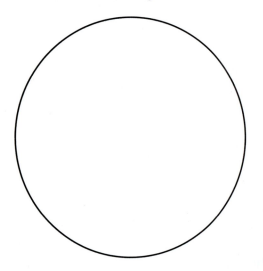

Written description of cell morphology: (Be sure to note the Gram reaction—positive or negative.)

FECAL COLIFORMS: COMPLETED TEST

(From EMB agar plate inoculated with cells from positive A-1 medium tube)

Was there a large bubble in the Durham tube inside your 1× A–1 medium tube?

Gram stain of nutrient agar subculture of colony from EMB agar plate

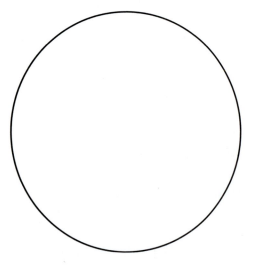

Written description of cell morphology: (Be sure to note the Gram reaction—positive or negative.)

Name: _____   Section: _____

Course: _____   Date: _____

1. a. What were the average total and fecal coliform counts (MPN) for toilet water?

    b. Was coliform population density consistent with what you'd expect to find
        for this water sample? Explain your answer. If this result is not consistent
        with what you'd expect to find, why do you suppose the results are not what
        was expected?

    c. Was there any difference between the total coliform count and the fecal
        coliform count? What does a comparison of the two counts tell you about
        the coliform population in the toilet water?

2. a. What were the average total and fecal coliform counts (MPN) for duck pond
        water?

    b. Was coliform population density consistent with what you'd expect to find
        for this water sample? Explain your answer. If this result is not consistent
        with what you'd expect to find, why do you suppose the results are not what
        was expected?

    c. Was there any difference between the total coliform count and the fecal
        coliform count? What does a comparison of the two counts tell you about
        the coliform population in the duck pond water?

3. Use lab data to support your conclusions in answering the following questions:

   a. Would you consider duck pond water safe for drinking? Why or why not?

   b. Would you consider the duck pond safe for recreational uses such as swimming? Why or why not?

   c. Would you consider the duck pond safe for commercial freshwater shellfishing? Why or why not?

4. a. What were the average total and fecal coliform counts (MPN) for drinking fountain water?

   b. Was coliform population density consistent with what you'd expect to find for this water sample? Explain your answer. If this result is not consistent with what you'd expect to find, why do you suppose the results are not what was expected?

   c. Was there any difference between the total coliform count and the fecal coliform count? What does a comparison of the two counts tell you about the coliform population in the drinking fountain water?

5. Did the lactose broth tube that you chose for the total coliform confirmed test produce a positive confirmed test? Explain your answer.

6. Did the A-1 medium tube that you chose for the fecal coliform confirmed test produce a positive confirmed test? Explain your answer.

7. Compare the colonies on the EMB plate streaked with cells from the lactose broth with those on the EMB plate streaked with cells from the A-1 medium. Are there differences in the types of colonies found on these two plates? If so, explain why there are differences.

8. Did the colony on the EMB plate streaked with cells from the lactose broth produce a positive completed test? Explain your answer.

9. Did the colony on the EMB plate streaked with cells from the A-1 medium produce a positive completed test? Explain your answer.

10. As noted in the introduction, in the 1800s, millions of people in the United States died from waterborne *Salmonella typhi* and *Vibrio cholerae* infections. Today, such infections are very rare in the United States. Describe three specific discoveries, technologies, or practices that have played, and still play, a major role in this dramatic change.

11. While typhoid fever and cholera are very rare in the United States, other types of intestinal illnesses linked to fecal contamination of water still occur in this country. Describe a couple of scenarios that might result in waterborne intestinal illness, despite the discoveries, technologies, and practices that you described in the previous question. In other words, how could fecal coliforms find their way into drinking water, and why is fecal contamination of water still a problem?

For thousands of years humans have raised animals for food, hides, and fur. Many of the same commensal bacteria and pathogenic ones that interact with humans are also found in animals like these sheep and this cow on a farm in South Africa. Because the slaughtering process can contaminate animal meat with enteric bacteria, it is important to test the meat for pathogens before it is consumed or sold for consumption.

LAB # 31

Enterics in Meat

Learning Objectives

- Understand how enteric bacteria such as *E. coli* and *Salmonella* species may be spread by contaminated meat.

- Describe diseases that may be caused by certain strains of meat-borne *E. coli* and *Salmonella* species.

- Understand methods for isolating and identifying enteric bacteria in meat, including the use of EMB and HE agars.

- Use selective and differential media to test meat samples for enteric bacteria.

B acterial species in the family Enterobacteriaceae are not just adapted to living in the human digestive tract, where the temperature is 37°C; they are also well adapted to the intestines of all warm-blooded vertebrates, where temperatures are between 35°C and 38°C. So, if we consume foods such as meat, eggs, and milk from warm-blooded cows, pigs, sheep, chickens, or turkeys, then we can potentially be infected by enteric species from these sources. In this lab, we're going to focus on enteric contamination of meat as a cause of disease.

Meat is most often "contaminated" with enteric bacteria during the slaughtering and butchering process. This is when the sterile muscle tissue of animals (i.e., meat) is exposed to any traces of manure present in the meat processing facility or to bacteria spilling out of ruptured, enteric-rich animal guts. Yummy. With most cuts of meat, the highest concentration of bacteria will be found on the surface of the meat. But in the case of ground products such as hamburger, any bacteria on the exposed surface of the muscle tissue prior to grinding will be mixed throughout the meat during processing, so it is particularly important to cook ground beef thoroughly, until all parts of the meat reach a temperature of 160°F–165°F (71°C–74°C).

PATHOGENIC ENTERICS IN MEAT

Most of the enteric species living in livestock and poultry are relatively harmless, but there are several intestinal pathogens that can be transmitted by foods of animal origin.

Escherichia coli

Most strains of *Escherichia coli* are relatively benign, even beneficial. Not long after birth, we begin to acquire *E. coli* from our environment, and for the rest of our lives, our digestive tracts will be colonized by *E. coli* to one degree or another. But some specific strains of *E. coli* can be deadly, and great effort goes into preventing the transmission of these strains via the foods we eat. For example, **E. coli O157:H7** cells have extra genes that can cause a life-threatening disease. This strain can produce a toxin that is capable of destroying red blood cells (hemolytic) and causing fatal kidney damage (uremic), resulting in a condition known as **hemolytic-uremic syndrome**.

E. coli O157:H7 can be spread from a variety of food sources, including unpasteurized dairy products, fruits, vegetables, and meat. Although this strain does not normally colonize fruits and vegetables, produce can be contaminated with *E. coli*–rich fecal material by food handlers or by animals living in close contact with these foods. In past outbreaks, O157:H7 has been traced to unpasteurized apple cider as well as lettuce, spinach, and clover spouts.

But one of the most common modes of transmitting pathogenic *E. coli* strains has been undercooked hamburger. In a 1993 epidemic linked to Jack-in-the-Box restaurants in four states in the western United States, hundreds of people were severely sickened, and four people died. The aftermath of this event provided momentum to efforts to tighten regulations regarding the temperatures at which hamburger must be cooked in restaurants. As a result, eateries must now cook hamburger to a higher temperature than in the past. More recent *E. coli* O157:H7 outbreaks have been traced to ground beef sold in grocery stores. To prevent infection, the CDC recommends cooking ground beef to a temperature of at least 160°F (71°C). The temperature should be checked with a thermometer, as color is not a very reliable indicator of safeness.

Salmonella species

The genus *Salmonella* contains many species able to cause intestinal infections resulting in a type of gastroenteritis called salmonellosis, featuring diarrhea, cramps, vomiting, and a low-grade fever. If the infected individual is otherwise healthy, most salmonellosis cases are relatively mild and self-limiting; that is, the patient recovers on his or her own in a few days. The *Salmonella* bacteria may be spread from many sources, but transmission is often associated with poultry and poultry products, including both the meat and the eggs of chickens and turkeys. However, it should be noted that, like *E. coli*, *Salmonella* can also be spread by other foods, such as fruits and vegetables, following fecal contamination by food handlers. Recent *Salmonella* outbreaks have been linked to cantaloupes, alfalfa sprouts, cucumbers, and peanut butter.

The severity of salmonellosis is related to the pathogen's ability to spread beyond the intestinal tract. In most cases, the bacteria remain in the gut and cause relatively mild symptoms. However, if *Salmonella* cells spread to and multiply in the blood, then the infection can quickly become life-threatening. This is more likely if the infected individual is elderly or immunosuppressed. One highly pathogenic and often deadly *Salmonella* strain, *Salmonella* Typhi, is infamous for its capacity to penetrate the lining of the intestines and spread to other organs. Fortunately, *Salmonella* Typhi is extremely rare in the United States today.

Detecting *E. coli* and *Salmonella* in meat

Since all the species of interest in this lab are in the Gram-negative enteric family, we will use selective and differential media designed to isolate and identify species in this group. Typical media used to identify enterics contain chemicals that inhibit the growth of Gram-positive organisms. In addition, these media detect traits that distinguish species within the enteric group, including lactose fermentation and hydrogen sulfide production. The two media to be used in this lab are eosin methylene blue (EMB) agar, described in detail in Lab 24, and Hektoen enteric (HE) agar, described below.

SELECTION AND DIFFERENTIATION BY HEKTOEN ENTERIC AGAR

Hektoen enteric (HE) agar is used primarily for the isolation and identification of two enteric family genera, *Salmonella* and *Shigella*; *Shigella* is another genus in which there are species capable of causing intestinal infections and illnesses. Isolating these species can be a challenging task because cells of these two groups may be a tiny portion of the total microbial population present in the gut. HE agar contains peptones and yeast extract to meet basic nutritional needs as well as other components that allow it to act as both a selective and a differential medium. The composition of HE agar is as follows (per liter):

Peptones	12.0 g
Yeast extract	3.0 g
Bile salts	9.0 g
Lactose	12.0 g
Sucrose	12.0 g
Salicin	2.0 g
Sodium chloride	5.0 g
Sodium thiosulfate	5.0 g
Ferric ammonium citrate	1.5 g
Bromthymol blue	65 mg
Acid fuchsin	0.1 g
Agar	14.0 g

HE agar was developed in the 1960s by Sylvia King and William Metzger of the Hektoen Institute in order to increase the frequency of isolation of *Shigella* and *Salmonella* organisms over what was possible with other media available in clinical laboratories at that time. Like EMB agar, HE agar is specifically designed to select for Gram-negative enterics and other Gram-negative species and to inhibit or prevent the growth of Gram-positive species. Gram-positive cells are specifically inhibited by the addition of bile salts and by the presence of two dyes, bromthymol blue and acid fuchsin. The dyes are also somewhat inhibitory for enteric species, but the high concentration of sugars and peptones allows the enterics to grow in the dye's presence.

The bile salts, in particular, select for and encourage the growth of enterics such as *Salmonella* and *Shigella* even if they are a tiny minority of the original gut bacterial population. The differential components in HE agar are designed to detect *Salmonella* and *Shigella* species and to differentiate them from each other.

Sugar fermentation

In addition to being selective for Gram-negative bacteria, HE agar allows separation of enteric isolates into lactose or sucrose fermenters and non-fermenters. As we have seen, the ability to ferment lactose usually distinguishes coliform enterics from non-coliform enterics. This trait is worth detecting because most of the pathogenic enterics are non-coliforms that do not ferment lactose or sucrose. To differentiate fermenters from non-fermenters, HE agar contains 1.2% lactose and 1.2% sucrose as fermentation substrates. The concentration of the sugars in this medium is higher than in most fermentation-detecting media, which aids in the visualization of fermenters and minimizes the problem of delayed lactose fermentation. The pH indicators are bromthymol blue (described in Lab 23) and **acid fuchsin**. When the pH is neutral or above, the combination of dyes produces a green to blue-green color **(Figure 31.1A)**, and when the pH drops into the acidic range, the indicators produce a pink-orange color—you could call it salmon.

Positive reaction for lactose or sucrose fermentation If cells growing on HE agar ferment lactose or sucrose, or both, then the organic acids produced by the cells will lower the pH of the medium into the acidic range. This drop in pH will change the color of the combined pH indicators bromthymol blue and acid fuchsin to a pink-orange/salmon color. While in theory, the drop in pH could be due to the fermentation of sucrose alone, it is rare that enteric species can ferment sucrose, but not lactose. Thus, a change to a pink-orange color almost always indicates the ability to ferment lactose, independent of any ability to ferment sucrose. In addition, if the pH drops low enough—that is, if large quantities of acids are produced—then the bile salts will precipitate, producing a haze in the medium.

So, if the colonies and the medium surrounding the colonies become pink-orange/salmon in color, with or without the addition of a hazy bile salt precipitate in the medium, this is considered a positive reaction for the fermentation of lactose (and perhaps sucrose, too) to organic acids.

If the cells we are testing are enteric species, then we can conclude that these cells belong to a coliform species or genus such as *E. coli*, *Enterobacter*, or *Citrobacter* **(Figure 31.1B)**. In other words, if the colonies are pink-orange, they are probably not *Salmonella* or *Shigella*. And generally speaking, species producing pink-orange colonies are less likely to be pathogenic than those producing a negative reaction for lactose fermentation.

Negative reaction for lactose fermentation If the cells growing on HE agar cannot ferment lactose or sucrose, then they will not produce organic acids from these sugars, and there won't be any acids to lower the pH of the medium or the colonial material. As a result, the pH throughout the medium will remain near or above a neutral pH, and the combination of dye molecules (pH indicators) will produce a green to blue-green color. Furthermore, without acid production, the bile salts will not precipitate, and there won't be a haze in the medium around the colonies.

A.

B.

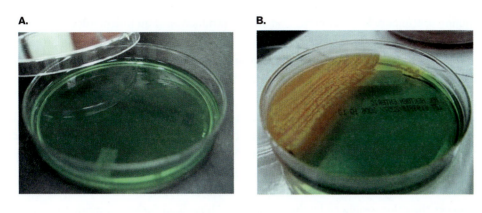

C.

FIGURE 31.1 Hektoen enteric agar.
A. Uninoculated HE agar. **B.** *E. coli* colonies produce a positive reaction for lactose and/or sucrose fermentation on HE agar. **C.** *Salmonella enterica* colonies produce a negative reaction for lactose and/or sucrose fermentation on HE agar.

So, if the colonies and the medium surrounding the colonies become green to blue-green in color, and there is no hazy bile salt precipitate in the medium, this is considered a negative reaction for the fermentation of lactose or sucrose to organic acids.

If the species being tested are enteric species, then we can conclude that these cells belong to a non-coliform species or genus such as *Salmonella*, *Shigella*, or *Proteus* (Figure 31.1C). In other words, if the colonies are green to blue-green in color, they could be the product of one of several pathogenic non-coliform enteric species. As noted above, HE agar was designed to detect this very thing; in particular, the presence of pathogenic *Salmonella* or *Shigella* cells.

Hydrogen sulfide (H₂S) production

HE agar not only reveals the ability to ferment lactose, but can also show when hydrogen sulfide is produced, which is useful for separating *Shigella* species from *Salmonella* and *Proteus* species. Sodium thiosulfate ($Na_2S_2O_3$) is added to the medium as a substrate for sulfide production. If the cells are sulfide producers, they will reduce the thiosulfate ($S_2O_3^{2-}$) to sulfides (S^{2-}). The sulfides will then react with the ferric (Fe^{3+}) ions derived from another medium component, ferric ammonium citrate, producing a black iron sulfide precipitate. Keep in mind that sulfide production is an independent trait; that is, it is not related to lactose fermentation. A lactose fermenter (coliform) can be either positive or negative for sulfide production, and a non-fermenter (non-coliform) can be either positive or negative for sulfide production.

Positive reaction for sulfide production   If cells growing on HE agar can produce sulfides by reduction of thiosulfates, then the sulfides will react with ferric ions, and black iron sulfide precipitate will begin to accumulate within the colony. Eventually, this reaction will produce a visible black center in the colony, and if sulfide production rates are high, the entire colony may look black. In fact, the black precipitate may obscure the color produced by the bromthymol blue–acid fuchsin combination, making it difficult to determine if the colonies fermented lactose.

So, if the colonies have a black center, this is considered a positive reaction for sulfide production.

HE agar is designed to select for, isolate, and identify *Salmonella* and *Shigella* species. *Salmonella* produces sulfides from thiosulfate, so a positive sulfide result combined with no lactose fermentation—that is, a green colony with a black center—raises the possibility that the colonies are from a *Salmonella* species and probably eliminates the possibility that an enteric isolate is a *Shigella* species.

Negative reaction for sulfide production   If the cells growing on HE agar do not produce sulfides by reduction of sulfates, then there won't be any sulfide to react with ferric ions. The center of the colonies will not turn black, and the colonies will be either pink-orange or blue-green, depending on the cell's ability to ferment lactose.

So, if the colonies do not have a black center, this is considered a negative reaction for sulfide production.

Shigella does not produce sulfides from thiosulfate, so a negative sulfide result combined with no lactose fermentation—that is, a green colony without a black center—raises the possibility that the colonies are from a *Shigella* species and reduces the possibility that an enteric isolate is a *Salmonella* species.

PROTOCOL

> **IMPORTANT:** Some of the bacteria in raw meat may be pathogenic and *Salmonella* Typhimurium and *Proteus vulgaris* are biosafety level 2 species, so gloves and safety glasses must be worn throughout both days of this lab.

Day 1

MATERIALS

- 6 Eosin methylene blue (EMB) agar plates
- 6 Hektoen enteric (HE) agar plates
- 4 Test tubes of sterile water with 0.1% Tween
- Beakers with disinfectant
- Sterile swabs
- Safety glasses
- Disposable gloves

Cultures:

- *Escherichia coli*
- *Citrobacter freundii*
- *Salmonella* Typhimurium
- *Proteus vulgaris*
- Cooked hamburger
- Cooked ground turkey
- Raw hamburger
- Raw ground turkey

KNOWN SPECIES ON EMB AND HE PLATES

1. Label two EMB plates and two HE plates with your name and the date, and divide each plate of both media types in half by drawing a line on the bottom with a wax pencil or Sharpie.

2. Label one side of one of each type of plate "*E. coli*" and the other side "*C. freundii*." Label one side of each of the remaining two plates "*S.* Typhimurium" and the other side "*P. vulgaris*."

3. Inoculate the plates with single streaks of the appropriate bacteria.

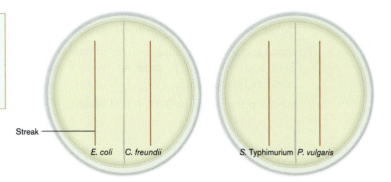

Streak — | *E. coli* | *C. freundii* | | *S.* Typhimurium | *P. vulgaris*

4. Incubate plates at 37°C for 24 hours.

RAW MEAT SAMPLES ON EMB AND HE PLATES

1. Label four EMB plates and four HE plates with your name and the date. Then label one EMB and one HE plate "Raw turkey," one EMB and one HE plate "Cooked turkey," one EMB and one HE plate "Raw hamburger," and one EMB and one HE plate "Cooked hamburger."

2. Label one tube of sterile water with Tween "Raw turkey," a second tube "Cooked turkey," a third tube "Raw hamburger," and a fourth tube "Cooked hamburger." (Tween is a mild detergent that helps to remove the bacterial cells from the meat.)

3. Add about 1 cc (about a ¼-inch cube) of each type of meat to the appropriate tube.

4. Use a separate sterile swab to mash and stir each meat sample in its test tube.

5. After mashing, wipe each swab over about a quarter of the appropriate EMB and HE plates, in a manner similar to the initial streaking of a plate when you wish to streak for isolated colonies.

6. Discard the swabs in a beaker of disinfectant or in the receptacle designated for trash to be autoclaved.

   Do not put the used swabs down on the bench top. Do not put the used swabs in the regular trash.

7. Using a sterile inoculating loop, finish streaking the EMB and HE plates in order to produce isolated colonies (see Lab 3).

8. Incubate the plates at 37°C for 24 hours.

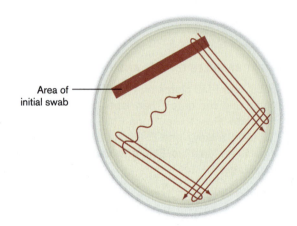

Area of initial swab

Day 2

MATERIALS

- EMB and HE plates from day 1
- Safety glasses
- Disposable gloves

KNOWN SPECIES ON EMB AND HE PLATES

Observe the colonies and the color of the medium for all species streaked onto EMB and HE plates. Record your observations in the Results section.

MEAT SAMPLES ON EMB AND HE PLATES

1. Use two pieces of masking or Scotch tape placed opposite each other to tape each of the plates shut.

2. Examine all EMB and HE plates streaked with the meat samples. **Do not open these plates, as we don't know what species are present, and some of these species could be pathogenic.** Describe these plates in the Results section, noting the various colony types and the color of the medium associated with the colonies.

3. Dispose of the plates in the receptacle designated for trash to be autoclaved.

RESULTS

Name: _____   Section: _____

Course: _____   Date: _____

KNOWN SPECIES ON EMB AND HE PLATES

EMB PLATE RESULTS

Bacterial species	Color of Colonies	Do Cells Ferment Lactose?	Are These Cells Coliforms?
E. coli			
C. freundii			
S. Typhimurium			
P. vulgaris			

HE PLATE RESULTS

Bacterial species	Color and Appearance of Colonies	Color of Medium Around Colonies	Do Cells Ferment Lactose/Sucrose?	Do the Cells Produce H_2S?
E. coli				
C. freundii				
S. Typhimurium				
P. vulgaris				

MEAT SAMPLES

Raw turkey swab on EMB plate:

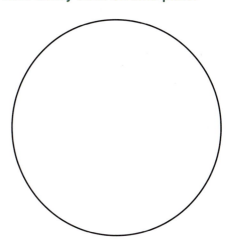

Written description of the morphology of the different types of colonies: (Include colony color, shape, relative size, and color of medium around the colonies.)

Raw turkey swab on HE plate:

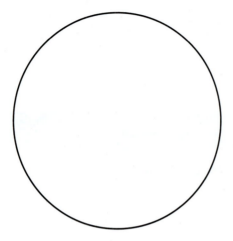

Written description of the morphology of the different types of colonies: (Include colony color, shape, relative size, and color of medium around the colonies.)

Cooked turkey swab on EMB plate:

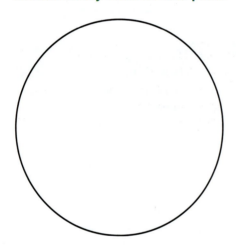

Written description of the morphology of the different types of colonies: (Include colony color, shape, relative size, and color of medium around the colonies.)

Cooked turkey swab on HE plate:

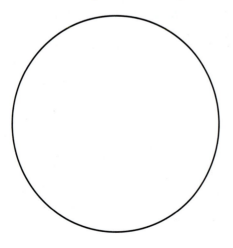

Written description of the morphology of the different types of colonies: (Include colony color, shape, relative size, and color of medium around the colonies.)

Raw hamburger swab on EMB plate:

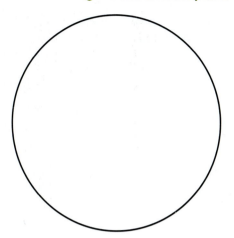

Written description of the morphology of the different types of colonies: (Include colony color, shape, relative size, and color of medium around the colonies.)

Raw hamburger swab on HE plate:

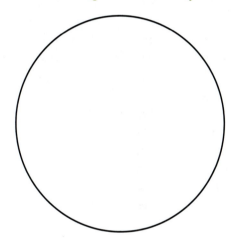

Written description of the morphology of the different types of colonies: (Include colony color, shape, relative size, and color of medium around the colonies.)

Cooked hamburger swab on EMB plate:

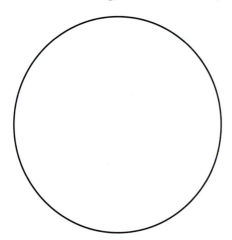

Written description of the morphology of the different types of colonies: (Include colony color, shape, relative size, and color of medium around the colonies.)

Cooked hamburger swab on HE plate:

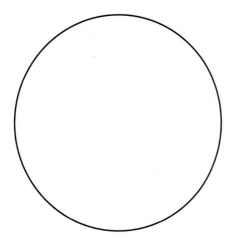

Written description of the morphology of the different types of colonies: (Include colony color, shape, relative size, and color of medium around the colonies.)

Name: _____

Section: _____

Course: _____

Date: _____

1. Can colonies on HE agar be both pink-orange/salmon in color and have black centers? Explain your answer.

2. Both *E. coli* and *Salmonella* species can be the cause of food-borne intestinal illness. If you streaked EMB and HE agar plates with a sample taken from meat suspected of causing a food-borne illness, how could you distinguish *E. coli* from *Salmonella*? Explain how this can be done with both the EMB and HE agar plates.

3. Did the EMB plates inoculated from meat sources have any colonies that might be *E. coli* colonies? Were these colonies on the hamburger or turkey plates, or on both? Why do you think that the colonies in question might be *E. coli* colonies?

4. Did the EMB plates inoculated from meat sources have any colonies that might be *Salmonella* or *Shigella* colonies? Were these colonies on the hamburger or turkey plates, or on both? Why do you think that the colonies in question might be *Salmonella* or *Shigella* colonies?

5. Did the HE plates inoculated from meat sources have any colonies that might be *E. coli* colonies? Were these colonies on the hamburger or turkey plates, or on both? Why do you think that the colonies in question might be *E. coli* colonies?

6. Did the HE plates inoculated from meat sources have any colonies that might be *Salmonella* or *Shigella* colonies? Were these colonies on the hamburger or turkey plates, or on both? Why do you think that the colonies in question might be *Salmonella* or *Shigella* colonies?

7. Were there any differences between the plates inoculated with raw meat samples and the plates inoculated with cooked meat samples in terms of the number and type of colonies on the various meats? Cite lab data to explain and support your answer.

8. What did you expect to find when you compared the raw meat and cooked meat plates, and why did you expect certain results? Did a comparison of raw meat and cooked meat plates produce the results that you expected? If not, what might explain the unexpected results?

9. Why are outbreaks of food-borne illnesses more likely to be due to consumption of hamburger than steak? What can you do to reduce the risk of being sickened by contaminated hamburger?

10. Which do you suppose would be riskier, consuming a raw hamburger contaminated with live *E. coli* cells or consuming one contaminated with live *Salmonella* cells? Assume equal numbers of bacterial cells and assume that you know nothing about the particular strains of *E. coli* or *Salmonella* growing in the meat. Explain your choice.

11. In an outbreak of intestinal illness, dozens of people experienced vomiting and diarrhea after eating at the same restaurant. Bacteria were cultured from several different food items at the salad bar, and the following results were observed:

	Nutrient Agar	EMB Agar	HE Agar
Green peppers	Numerous colonies	No colonies	No colonies
Lettuce	Numerous colonies	Pale pink colonies	Green colonies with black centers
Broccoli	Numerous colonies	No colonies	No colonies

Based on the above observations, which food item was contaminated with fecal bacteria, and what type of bacteria caused the intestinal illnesses? Explain your conclusions.

12. Fresh unpasteurized apple juice has been known to transmit enteric bacteria when consumed. List three ways that apples could become contaminated with these bacteria.

Large-scale production of wine dates back more than 6,000 years. This means that humans have been cultivating grapes for food and wine for at least that many years. Yeast and bacteria that normally live on the grapes and vines were the first fermenters. Today, although naturally occurring microbes still contribute, winemakers add strains of yeast that have been bred for turning grapes into wine.

LAB 32

Microbiology and Chemistry of Wine

Learning Objectives

- Understand the taxonomy and chemistry of the grapes used in wine production.

- Explain the chemistry and microbiology of the ethanol fermentation process used to produce wine.

- Describe methods for determining sugar and ethanol concentration before, during, and after primary ethanol fermentation.

- Examine the effects of variables such as sugar concentration, time, and temperature while creating a crude "wine."

This lab is designed to explore a few concepts in the microbiology and chemistry of wine making. It is not intended to create drinkable wine, and the products of this lab should not be consumed. Remember that undergraduate labs are not sanitary food production facilities, and food and drink should not be consumed in lab, regardless of the source. Further, this lab uses juice from grapes that are not typically used to make wine, and several key steps in the winemaking process have been omitted. If you are interested in producing your own wine, you should consult publications designed to teach you how to do this, and you should purchase winemaking supplies from reputable dealers.

GRAPES

Wine making typically begins with grape juice, although any juice with a high sugar content can be used in this process. Most grapes used in wine making are subspecies or varieties of a single grape species, *Vitis vinifera,* which translates as "vine (*vitis*) that is wine-bearing (*vinifera*)." This species is native to Europe and Asia, and it's the traditional winemaking grape of Europe. Several different varieties or sub-species of *Vitis vinifera* are used to make wine. Popular white grape varieties include Chardonnay, Riesling, and Sauvignon blanc, while widely used red or purple grape varieties include Merlot, Cabernet Sauvignon, and Pinot noir.

While European wines are traditionally made using *Vitis vinifera*, there are several North American *Vitis* species that have become important to wine making worldwide, including in Europe. In general, the main advantages of North American *Vitis* species are cold-hardiness and insect resistance. These traits have been introduced into *V. vinifera* grapes by grafting *vinifera* vines onto North American species rootstock and by the hybridization of *V. vinifera* with native North American wild grape species. One example of an important cold-hardy North American grape species is *Vitis labrusca* (fox grapes), a species that has been crossed with *V. vinifera* to produce hybrid grapes varieties such as Cayuga and DeChaunac. In addition, the fruit of *V. labrusca* is used for table grapes, for unfermented grape juice, and to create a few types of wines. Varieties of this species include Concord, used for purple grape juice, and Niagara, used for white grape juice. In this lab, we'll be fermenting white grape juice made from Niagara grapes.

At the start of fermentation, the percentage of sugar is high, the liquid has a higher density, and the hydrometer sits higher in the medium.

At fermentation's end, the percentage of sugar is low, the liquid has a lower density, and the hydrometer sits lower in the medium.

FIGURE 32.1 Use of a hydrometer to measure density.

CHEMISTRY OF GRAPES

The chemistry of grapes is critically important to wine production because ultimately, the chemistry, aroma, and taste of the wine are a direct product of the initial chemistry of the grapes. Grapes contain a complex mix of sugars, organic acids, flavonoids, and tannins, and grapes from the same vine produce different concentrations of each chemical component in different years, depending on the weather. In general, hot and dry conditions produce grapes with more sugars and less organic acids.

Grape chemistry can also be influenced by soil, geology, altitude, and direction of the slope with respect to the sun, so connoisseurs stress the importance of *terroir*, a French word that means "sense of place" or "the coming together of the climate, the soil, and the landscape." All of these factors can influence the chemistry and microbiology of the grapes and the process of wine production; thus, the *terroir* influences the taste and aroma of the final product.

Sugars

Mature grapes contain 15%–28% sugar by weight (15–28 g sugar/100 g of solution), and sugar is the main substrate for fermentation. In mature grapes, about half of the total sugar content is glucose and about half is fructose. Both sugars are derived from sucrose, a disaccharide composed of glucose and fructose. Sucrose is produced in leaves by photosynthesis, transported to the developing fruit, and almost immediately hydrolyzed into glucose and fructose. To produce a typical wine, the winemaker needs grapes with a concentration of at least 23%–25% sugar.

In the first step in making wine, grapes are harvested and pressed. When the juice has been pressed from the harvested grapes and is ready to be fermented, its initial sugar concentration is usually measured with a hydrometer. A **hydrometer** is a sealed glass tube, weighted on one end, that sinks to a certain depth in a given liquid depending on the specific gravity of that liquid **(Figure 32.1)**. **Specific gravity** is a measure of the density of the liquid and is defined as the ratio of the density of a test substance (such as grape juice or wine) to the density of a reference solution, usually water. Liquids with a high concentration of sugar, such as fresh grape juice, have a higher specific gravity than water, so the hydrometer floats higher in the denser liquid (Figure 32.1, left). As fermentation proceeds, the sugar is converted to alcohol, the specific gravity of the juice declines, and the hydrometer floats lower in the developing wine (Figure 32.1, right).

As noted, hydrometers measure specific gravity, and then, with the use of an adjacent scale, they can convert the specific gravity into a sugar concentration, measured in units called **degrees Brix (°Bx) (Figure 32.2)**. One degree Brix is equal to 1.0 g of sucrose in 100 g of solution (solution = solute + solvent). So, the Brix scale yields sucrose concentration as a percentage by weight (% w/w), or the percentage of total grams of solution that is composed of sucrose. For example, a sucrose solution with a specific gravity of 1.06 would measure 15 degrees Brix (15 °Bx; Figure 32.2). This solution could also be described as a 15% (w/w) sucrose solution, or as a solution containing 15 g of sucrose for every 100 g of solution.

Grape juice contains a mixture of different sugars, and relatively little of this sugar is sucrose. Since it's not a pure sucrose solution, a grape juice that measures, say, 15 °Bx is not really exactly, precisely a 15% (w/w) sucrose solution. However, for the purposes of this lab, the juice will be close enough to a 15% (w/w) sucrose solution that we can say that 15 °Bx means that the total sugar concentration of the juice is at

This grape juice has a specific gravity of 1.060.

If this specific gravity is a product of dissolved sugars, then this juice is 15 °Bx, or about 15% (w/w) sugar.

FIGURE 32.2 Converting specific gravity to degrees Brix.
Hydrometers have a specific gravity scale (left), which can be converted to a degrees Brix scale (right).

about 15% (w/w). That is, we can consider °Bx as being about the same as percentage of total sugar by weight.

Describing the sugar concentration in terms of percentage by weight, or the percentage of the total grams of solution that is composed of sugar, has its advantages. However, grape juice is usually measured by volume, not weight. So it's also useful to describe the percentage of sugar in the juice as grams of sugar per 100 ml of solution (juice); that is, as a percentage of sugar weight by volume. In the case of sugar in grape juice, you can convert the weight/weight percentage, or the measurement made in °Bx, into a weight/volume percentage using the following equation:

$$°Bx × \text{specific gravity} = \% \text{ weight/volume sugar concentration}$$

or

$$°Bx × \text{specific gravity} = \text{g sugar/100 ml juice}$$

For example, a juice with a specific gravity of 1.06 will measure 15 °Bx (see the hydrometer scale in Fig. 32.2). So, it contains 15 g of sugar for every 100 grams of juice (15% w/w). Now, 15 × 1.06 = 15.9, so the solution is 15.9% (w/v) sugar; that is, it contains 15.9 g sugar per 100 milliliters of grape juice.

Sugar concentration is important because it predicts the potential alcohol concentration at the end of fermentation. For every 100 g of sugar, fermentation yields 51 g of alcohol and 49 g of CO_2. So if all of the sugar is fermented, the potential final alcohol concentration will be approximately

$$\% \text{ Sugar (weight/volume)} × 0.51 = \% \text{ ethanol (weight/volume)}$$

For example, if a juice measures 15 °Bx, then it contains 15.9% sugar (w/v). Since 15.9 × 0.51 = 8.11, we can predict that the primary fermentation will yield a final ethanol

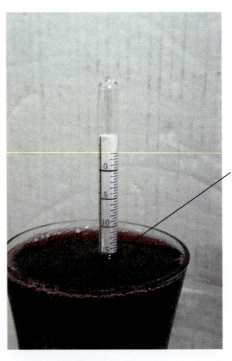

This grape juice is 15 °Bx (about 15% sugar).

If completely fermented, the final alcohol concentration will be about 8%.

FIGURE 32.3 Converting from degrees Brix to final alcohol concentration.
Brix scale (left) and predicted final alcohol concentration scale (right).

concentration of about 8%. The actual final ethanol concentration will probably be a little bit lower than this, because a little bit of the sugar that contributes to the density of the juice and degrees Brix is in the form of pentose sugars, and these sugars are usually not fermented to ethanol. To make life easier, most hydrometers designed for use in winemaking have a third scale, a scale that converts degrees Brix into predicted final ethanol concentration, so you can skip the math and just read the ethanol scale **(Figure 32.3)**.

Other components

In addition to sugars, grapes contain several different types of organic acids, including lactic, malic, acetic, and tartaric acids. Ideally, the juice will be between 0.5% and 0.9% organic acids. Too little organic acid, and the environment is not acidic enough to inhibit undesirable spoilage microbes. That is, the presence of the organic acids creates a selective environment favoring the right type of fermentation microbes. In contrast, if the juice is too acidic, then the final product may have a sharp and unpleasant taste.

Tannins are formed in plant cells by a combination of flavonoids, phenolics, and sugars. They are present in the skins, seeds, and stems of the grape plants, and they are extracted from the oak barrels often used to ferment and age wine. In wine, tannins play many roles, including adding yellow-to-brown colors and bitter and astringent flavors. They also bind to proteins and polysaccharides and can have antimicrobial properties as well.

PRIMARY FERMENTATION OF GRAPE JUICE

Saccharomyces cerevisiae and Saccharomyces bayanus

S. cerevisiae and *S. bayanus* are species of yeast that have been used for thousands of years to make wine. Traditional winemaking relied on a mix of wild yeast strains that naturally occurred on the surfaces of grapes, on the feet of grape stompers, and in previously made wines. However, the use of mixtures of wild yeast strains can give variable and unpredictable results, so most wine made today uses selected pure cultures of *S. cerevisiae* or *S. bayanus*, or both. In addition, as is true of all microbes, the activity of the yeast cells is affected by temperature, so the outcome of the winemaking process also depends on the temperature at which fermentation occurs.

Fermentation of sugar to ethanol

Yeast cells are facultative anaerobes. When they are first added to grape juice, they grow by oxidizing glucose and other sugars to carbon dioxide and water via aerobic respiration (*Saccharomyces* literally means "sugar-fungus"). Certain yeast types, including some *Saccharomyces* strains, will also produce some ethanol under aerobic conditions if the initial glucose concentration is high enough; this is known as the Crabtree effect. Respiration is more efficient than fermentation, and as long as oxygen is present, yeast cell populations can expand relatively rapidly. However, aerobic metabolism also consumes oxygen, so this type of metabolism soon leads to increasingly anaerobic conditions in the juice. As the oxygen is depleted, the yeast cells switch to fermentation pathways to survive, and they begin to convert glucose and fructose into the fermentation products ethanol and carbon dioxide **(Figure 32.4)**.

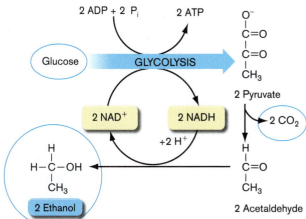

FIGURE 32.4 The chemistry of ethanol fermentation by yeast. Under anaerobic conditions, one glucose molecule will be converted to two CO_2 and two ethanol molecules.

The overall chemical reaction for ethanol fermentation is:

$$\text{1 glucose or 1 fructose} \rightarrow \text{2 ethanol (CH}_3\text{CH}_2\text{OH)} + \text{2 CO}_2$$

If there is a high starting concentration of sugar, fermentation will be vigorous, producing large amounts of ethanol and carbon dioxide in a relatively short time. In this lab, we'll monitor this process by putting a balloon over the mouth of a fermentation flask to collect the carbon dioxide gas as it bubbles out of the juice. As the yeast use the glucose and fructose as carbon and energy sources, the sugar concentration will decline, and the rate of fermentation will slow. Eventually, almost all the carbon dioxide will diffuse out of the liquid, and the "wine" will become still or flat like a carbonated soda that's been left uncapped. In this lab, you'll be able to see that fermentation is near or at an end when the balloon no longer inflates after it's been removed and replaced on the flask.

Residual sugar

The sugar that remains at the end of fermentation is called **residual sugar**. Residual sugar determines the dryness or sweetness of the final product. If the measured residual sugar concentration is less than about 0.2%–0.3% (2–3 g sugar per liter or 2,000–3,000 mg sugar per liter), then the wine has been fermented to "full dryness." Semi-dry wines contain 1%–2% sugar (10–20 g sugar per liter or 10,000–20,000 mg sugar per liter). Semi-sweet wines are about 2%–4% sugar (20–40 g sugar per liter or 20,000–40,000 mg sugar per liter), while the concentration of sugar in sweet wines is at or above 5% sugar (50 g sugar per liter or 50,000 mg sugar per liter).

Residual sugar concentrations can be measured by a hydrometer or by the use of enzyme-based assays.

Hydrometer method As the sugar in the juice is converted to ethanol, the specific gravity of the solution will decline, and this decline can be observed with a hydrometer. The specific gravity of pure water is 1.000 (at a standard temperature and pressure), so if the specific gravity of the juice drops to 1.000 or lower, we can conclude that the sugars have been removed and that there is little or no residual sugar in the final product. However, the ethanol produced by fermentation is less dense than water, and thus, when ethanol is dissolved in water, it lowers the specific gravity of the solution. As a result, the specific gravity at fermentation's end may actually be below 1.000 due to the ethanol, even if there is a still a little sugar in the wine, so a hydrometer provides only an approximate measure of the wine's residual sugar concentration.

Enzyme-based method Since ethanol affects specific gravity, we have other methods for measuring sugar in solutions (such as wine) that contain ethanol. One alternative approach measures sugar concentration using enzymes specific for fructose and glucose, the two dominant sugars in grape juice. These enzymes convert the sugars into end products that generate a color change in an end product–detecting reagent. Within a certain range of sugar concentrations, the intensity of the reagent color is directly related to the concentration of the sugars. Note that this method does not measure pentose sugars; however, these sugars are usually present in concentrations of less than1%–2%.

Measuring residual sugar is a basic part of winemaking, so companies have developed enzyme-based assays to do this. One example is a test strip assay produced by Accuvin, LLC **(Figure 32.5)**. In this residual sugar assay, a small, measured amount of wine is applied to a pad that contains a mix of enzymes and reagents. Over a period of 2 minutes, those enzymes and reagents interact with any glucose and fructose in the wine in a process that produces a purple color. The intensity of the color is related to the concentration of the sugars up to a final concentration of about 2,000 mg sugar per liter (0.2% w/v sugar). To convert mg/l to % w/v, multiply mg/l by 0.0001.

A series of reactions take place in the test strip:

$$\text{Reaction 1: fructose} \rightleftharpoons \text{glucose}$$

Fructose is converted to glucose by an isomerase enzyme. This reaction is reversible, but as glucose is converted to glucono-δ-lactone in reaction 2, more of the fructose is converted to glucose. With time, **almost all of the fructose is converted to glucose** as all of the glucose is consumed in reaction 2.

$$\text{Reaction 2: glucose} + O_2 + H_2O \rightarrow \text{glucono-}\delta\text{-lactone} + H_2O_2$$

Glucose is converted to glucono-δ-lactone and hydrogen peroxide (H_2O_2) by the enzyme glucose oxidase. Notice that **the number of H_2O_2 molecules is equal to the number of glucose molecules** (including glucose produced from isomerization of fructose to glucose in reaction 1).

$$\text{Reaction 3: } H_2O_2 + \text{colorless reagent} \rightarrow 4\,H_2O + \text{pink-purple colored reagent}$$

In an oxidation-reduction reaction catalyzed by a peroxidase enzyme, the **hydrogen peroxide molecules react with a colorless reagent to produce a pink-purple-colored molecule**. The intensity of the purple color is directly related to the concentration of sugar (glucose + fructose) via the H_2O_2 intermediate. The sugar concentration can be determined by comparison of the test strip color with a standard color scale.

FIGURE 32.5 The Accuvin, LLC residual sugar test strip assay.

Estimate of final percentage of ethanol

As we have seen, sugar concentrations can be used to estimate the final ethanol concentration of the wine. When using sugar concentrations for this purpose, we assume (1) that all the removed glucose and fructose molecules have been metabolized by fermentation to CO_2 and ethanol, and (2) that all the ethanol produced remains in the wine. In reality, the actual ethanol concentration will probably be a little lower than the value produced by the equations below because some of the glucose will have been converted to CO_2 and H_2O (instead of ethanol) by aerobic respiration, and because some of the ethanol will have evaporated from the surface of the wine during the days of fermentation. We will also assume that there has been relatively little change in the initial volume of the juice as it's converted into wine.

It should be noted that we will be estimating ethanol concentration in terms of the number of grams of ethanol produced per 100 ml of grape juice, or a weight/volume concentration, and not as an "alcohol by volume," or ABV, concentration. ABV, the ethanol concentration value listed on commercial alcoholic beverages, is a volume/ volume concentration (milliliters of ethanol in 100 ml of solution). Calculating a final ABV requires more equations and conversions, which we're not going to do in this lab, although the concentration values produced by the methods below should be within 10%–20% of the actual ABV values.

Ethanol concentration can be estimated using hydrometer measurements alone or by combining hydrometer measurements with enzyme-based assay results.

Hydrometer alone To estimate ethanol concentration, we can subtract the final percentage of sugar (w/v) as determined by the hydrometer from the initial percentage of sugar (w/v) as determined by hydrometer. The difference between these two values tells us how much sugar was removed by yeast metabolism. If we assume that all the removed sugar was fermented to CO_2 and ethanol, then we can multiply this quantity by 0.51 to convert the percentage of sugar into the percentage of ethanol (w/v):

$$\% \text{ ethanol} = (\text{initial } \% \text{ sugar} - \text{final } \% \text{ sugar}) \times 0.51$$

There will be some error in the estimate due to the fact that the measurement of the final percentage of sugar by hydrometer is going to be affected by the ethanol in the solution, which lowers its specific gravity. So, the true final percentage of sugar may actually be a little bit higher than is calculated via hydrometer.

Hydrometer combined with enzyme-based assay results Ethanol concentration can also be estimated by subtracting the final percentage of sugar (w/v) as determined by the enzyme-based assay from the initial percentage of sugar (w/v) as determined by hydrometer. As noted above, the difference between these two values tells us how much sugar was removed by yeast metabolism. If we again assume that all removed sugar was fermented to CO_2 and ethanol, then we can use the equation above to convert the percentage of sugar into the weight/volume percentage of ethanol.

$$\% \text{ ethanol} = (\text{initial } \% \text{ sugar} - \text{final } \% \text{ sugar as determined by assay}) \times 0.51$$

There will be some error in the estimate due to the fact that the assay does not measure pentose sugars. These sugars will add a little bit to the initial specific gravity and percentage of sugar, but they will not contribute or add to the final sugar concentration as measured by the assay. So, the true final percentage of sugar may actually be a little bit higher than was determined via the enzyme-based assay.

AND FINALLY, THERE'S AGING

The primary ethanol fermentation takes place over a few weeks to a month or so, depending on numerous variables. However, this is often just the first step in winemaking. After the primary fermentation, the wine may be stored and aged for years to complete the winemaking process. During this time, there are slow changes in the chemistry of the wine as a result of low levels of microbial activity as well as nonbiological chemical reactions. If the wine is stored in oak barrels, then it will extract tannins from the oak wood, which will contribute to the flavor of the final product. So, over the years, the aroma and taste of the wine will slowly change, and much is made of the optimal time to consume a given wine.

I love how wine continues to evolve, how every time I open a bottle it's going to taste different than if I had opened it on any other day.

Because a bottle of wine is actually alive—it's constantly evolving and gaining complexity. That is, until it peaks—like your '61—and begins its steady, inevitable decline.

—Maya (Virginia Madsen), *Sideways*

PROTOCOL

Day 1
MATERIALS

- 6 250–300-ml flasks
- 700 ml white grape juice
- 200 ml sterile distilled water
- 6 balloons (1-liter capacity)
- 1 winemaking hydrometer
- 1 100-ml graduated cylinder
- Glucose
- Digital scales and weigh boats
- 1 sterile 1-ml pipette
- 2-ml (blue-barrel) pipette pump
- Incubator set to 28°C–30°C

Culture:

- *Saccharomyces cerevisiae* (overnight culture in glucose nutrient broth)

Figure 32.6 provides a flowchart of the steps in this lab.

1. Label six 250–300-ml flasks with your names and the date. Then label two flasks each with the following labels:

   a. 1:1 grape juice:water
   b. Undiluted grape juice
   c. Grape juice plus added glucose

2. Transfer about 65 ml of undiluted grape juice to each of the two flasks labeled 1:1 grape juice:water, and then transfer 65 ml of sterile distilled water to each of the same two flasks. Gently swirl the flasks to mix the juice and water.

3. To measure the specific gravity of this juice, pour 100 ml of juice from one of the two flasks into a 100-ml graduated cylinder, then slowly lower a hydrometer into the juice (hydrometers are fragile!). When the hydrometer has stopped bobbing in the juice, read the lines on the scales that are level with the juice surface and record the specific gravity, degrees Brix, and predicted percentage of alcohol from the various scales.

   The specific gravity scale on a wine hydrometer usually ranges from 0.990 to 1.150. Divisions are often marked with two-digit numbers such as 10, 20, 30, and so forth. Such divisions between 1.000 and 1.100 should be read 1.0<u>10</u>, 1.0<u>20</u>, 1.0<u>30</u>, and so on. Most grape juices will have a specific gravity around 1.060.

4. After recording your measurements in the Results section, pour the juice back into the flask.

5. Transfer about 130 ml of undiluted grape juice into each of the two flasks labeled Undiluted grape juice.

6. To measure the specific gravity of this undiluted juice, pour 100 ml of juice from one of the two flasks into a 100-ml graduated cylinder and then use the hydrometer to read and record specific gravity, degrees Brix, and predicted percentage of alcohol from the various scales.

7. After recording your measurements in the Results section, pour the juice back into the flask.

8. Transfer about 130 ml of undiluted grape juice to each of the two flasks labeled Grape juice plus added glucose, and then add 15 g of glucose to each of the same two flasks. Gently swirl these flasks to dissolve the sugar.

9. To measure the specific gravity of this juice, pour 100 ml of juice from one of the two flasks into a 100-ml graduated cylinder and then use the hydrometer to read and record specific gravity, degrees Brix, and predicted percentage of alcohol from the various scales.

10. After recording your measurements in the Results section, pour the juice back into the flask.

11. Inoculate the juices in all six flasks by transferring 0.1 ml of *Saccharomyces cerevisiae* culture (in glucose nutrient broth) to each of the flasks. Gently swirl the juice in each flask to distribute the yeast cells throughout the juice.

12. For each of the six flasks, stretch the opening at the base of a balloon over the opening at the top of the flask so that the flask is essentially sealed off from the surrounding environment by the balloon. Balloons should be large and flexible enough to hold about 1 liter of gas.

13. Incubate one flask of each juice type (three flasks total) at room temperature for 2 days. Be sure that flasks are incubated in such a way as to allow for balloon inflation.

14. Incubate the other flask of each juice type (three flasks total) at 28°C–30°C for 2 days. Be sure that flasks are incubated in such a way as to allow for balloon inflation.

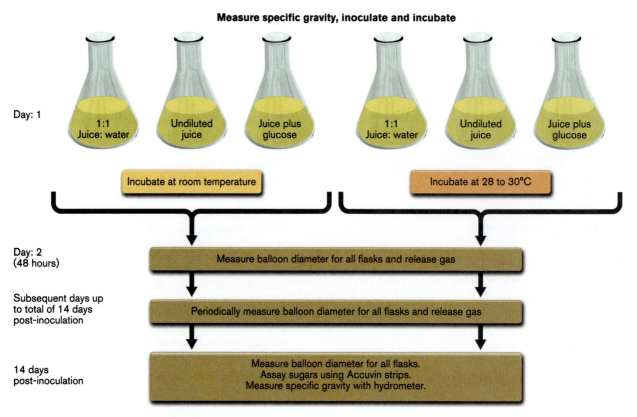

FIGURE 32.6 Flowchart of the steps in this lab.

Day 2
MATERIALS

- Inoculated flasks from day 1

- Flexible metric ruler or tape measure

  1. Use a flexible metric ruler or tape measure to measure the circumferences of the balloons attached to each of the inoculated flasks that have been incubated for 2 days. Measurements should be made at the widest point of the inflated balloon or at the point of the balloon's greatest diameter. If a balloon is not inflated, record the circumference as 0 cm. Record the data in the Results section.

     If a flexible ruler or tape measure isn't available, then wrap a string around the balloon, mark the circumference on the string, and use a metric ruler to measure the string and, hence, the circumference of the balloon.

  2. After measuring the circumference of the balloons, remove the balloons from the flasks, squeeze out all the gas, and reattach the balloons to the flasks.

  3. Return the flasks to the appropriate incubation temperatures.

Subsequent days
MATERIALS

- Inoculated flasks from day 1

- Flexible metric ruler or tape measure

  During subsequent incubation periods, the balloon circumferences can be measured every 2 to 3 days. Measurements can be made at day 4 or 5, at day 7, at day 9 or 10, and at day 14. Following each measurement, gases should be squeezed out of each balloon, and then the balloons should be reattached. After 2 weeks of incubation, fermentation should be finished, or nearly finished, in all of the flasks.

Final measurements (at 14 days)
MATERIALS

- Inoculated flasks from day 1

- Flexible metric ruler or tape measure

- Accuvin residual sugar assay strips

- Wine-making hydrometers

- 100-ml graduated cylinders

1. After 14 days of incubation, make a final measurement of the balloon circumferences.

2. Follow the directions provided with the Accuvin residual sugar assay strips to measure the residual sugar concentration in the wine in each of the six flasks.

   If the assay indicates a concentration of 2,000 mg/l or higher, you will have to dilute the wine 1:10 with water and repeat the assay. In this case, multiply the subsequent assay result by 10 to determine the residual sugar concentration.

3. Follow directions for day 1, step 3 to measure the specific gravity, degrees Brix, and predicted percentage of alcohol for the wine in each of the six flasks.

4. Autoclave all wine. Do not consume.

Name: _____   Section: _____

Course: _____   Date: _____

INITIAL SPECIFIC GRAVITY AND SUGAR CONCENTRATIONS OF GRAPE JUICES (DAY 1)

Juice Type	Specific Gravity (SG)	Degrees Brix (°Bx) or Percentage of Sugar by Weight/Weight	Percentage of Sugar by Weight/Volume (°Bx × SG)
1:1 mix of grape juice and water			
Undiluted grape juice			
Grape juice with added glucose			

BALLOON CIRCUMFERENCES

INCUBATED AT ROOM TEMPERATURE

	DAYS OF INCUBATION				
Juice Type	2 DAYS	__DAYS	7 DAYS	__DAYS	14 DAYS
1:1 mix of grape juice and water					
Undiluted grape juice					
Grape juice with added glucose					

Graph balloon circumference versus time for flasks incubated at room temperature. (Put lines for all three grape juice preparations on one graph—use colored pencils or different symbols to indicate juice types.)

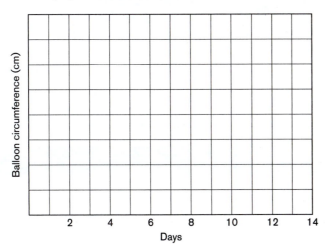

INCUBATED AT 28°C–30°C

Juice Type	DAYS OF INCUBATION				
	2 DAYS	__DAYS	7 DAYS	__DAYS	14 DAYS
1:1 mix of grape juice and water					
Undiluted grape juice					
Grape juice with added glucose					

Graph balloon circumference versus time for flasks incubated at 28°C–30°C. (Put lines for all three grape juice preparations on one graph—use colored pencils or different symbols to indicate juice types.)

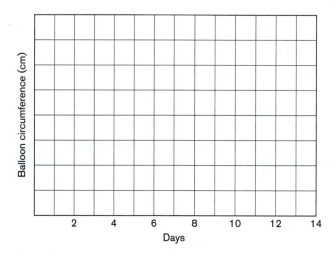

FINAL SPECIFIC GRAVITY AND SUGAR CONCENTRATION OF GRAPE JUICES (DAY 14)

INCUBATED AT ROOM TEMPERATURE

Juice Type	Specific Gravity* (SG)	Degrees Brix (°Bx) or Percentage of Sugar by Weight/Weight	Percentage of Sugar by Weight/Volume (°Bx × SG)
1:1 mix of grape juice and water			
Undiluted grape juice			
Grape juice with added glucose			

INCUBATED AT 28°C–30°C

Juice Type	Specific Gravity* (SG)	Degrees Brix (°Bx) or Percentage of Sugar by Weight/Weight	Percentage of Sugar by Weight/Volume (°Bx × SG)
1:1 mix of grape juice and water			
Undiluted grape juice			
Grape juice with added glucose			

*If specific gravity is below 1.000 (due to ethanol), enter zero for both degrees Brix (°Bx) and percentage of sugar by w/v.

FINAL RESIDUAL SUGAR BY ACCUVIN ASSAY AND DETERMINATION OF DRYNESS OR SWEETNESS

INCUBATED AT ROOM TEMPERATURE

Juice Type	Residual Sugar (mg/l)	Dry, Semi-dry, Semi-sweet, or Sweet?
1:1 mix of grape juice and water		
Undiluted grape juice		
Grape juice with added glucose		

INCUBATED AT 28°C–30°C

Juice Type	Residual Sugar (mg/l)	Dry, Semi-dry, Semi-sweet, or Sweet?
1:1 mix of grape juice and water		
Undiluted grape juice		
Grape juice with added glucose		

ESTIMATED FINAL ETHANOL CONCENTRATION BY HYDROMETER ALONE

INCUBATED AT ROOM TEMPERATURE

Juice Type	Initial % Sugar (°Bx × SG)	Final % Sugar (°Bx × SG)	Reduction in % Sugar (Initial/Final)	Estimated % Ethanol*
1:1 mix of grape juice and water				
Undiluted grape juice				
Grape juice with added glucose				

*Estimated % Ethanol = Reduction in % sugar × 0.51.

INCUBATED AT 28°C–30°C

Juice Type	Initial % Sugar (°Bx × SG)	Final % Sugar (°Bx × SG)	Reduction in % Sugar (Initial/Final)	Estimated % Ethanol*
1:1 mix of grape juice and water				
Undiluted grape juice				
Grape juice with added glucose				

*Estimated % Ethanol = Reduction in % sugar × 0.51.

ESTIMATED FINAL ETHANOL CONCENTRATION BY HYDROMETER AND ENZYME-BASED ASSAY

INCUBATED AT ROOM TEMPERATURE				
Juice Type	Initial % Sugar (°Bx × SG)	Final % Sugar (mg/l × 0.0001)	Reduction in % Sugar (Initial/Final)	Estimated % Ethanol*
1:1 mix of grape juice and water				
Undiluted grape juice				
Grape juice with added glucose				

*Estimated % Ethanol = Reduction in % sugar × 0.51.

INCUBATED AT 28°C–30°C				
Juice Type	Initial % Sugar (°Bx × SG)	Final % Sugar (mg/l × 0.0001)	Reduction in % Sugar (Initial/Final)	Estimated % Ethanol*
1:1 mix of grape juice and water				
Undiluted grape juice				
Grape juice with added glucose				

*Estimated % Ethanol = Reduction in % sugar × 0.51.

Name: _____   Section: _____

Course: _____   Date: _____

1. How did varying the incubation temperature affect the time required to complete the fermentation? Remember, once all of the sugar has been fermented, there will be little or no additional CO_2 produced.

2. How did varying the incubation temperature affect the residual sugar concentration?

3. How did varying the incubation temperature affect the final ethanol concentration?

4. How did varying the sugar concentration affect the time required to complete the fermentation? Remember, once all of the sugar has been fermented, there will be little or no additional CO_2 produced.

5. How did varying the sugar concentration affect the residual sugar concentration?

6. How did varying the sugar concentration affect the final ethanol concentration?

7. Which juice preparation had a sugar concentration similar to what is typically used in wine making, and what was the final ethanol concentration of the wine made from this juice preparation? How does this concentration compare with the alcohol concentration of typical commercial wines?

8. Were any of the six wines produced in this lab semi-dry, semi-sweet, or sweet? If so, why do you think that these particular wines contained more residual sugar than your dry wines?

9. Yeast cells are inhibited at about 15% ethanol. Based on that fact, if your goal was to produce a sweet wine with residual sugar concentration of 5% (w/v), then what should the sugar concentration be at the start of fermentation?

10. List several reasons why your calculated ethanol concentrations were probably higher than the actual ethanol concentrations.

11. Say you tried to do the fermentation with 130 ml of juice, but in a flask with a bottom diameter about twice that of the 250–300-ml flask you used in this lab (something like a 1,000-ml flask). How might that alter the course of events? You might want to do a little research into the role and effects of oxygen during wine production.

12. After the primary fermentation, during the aging process (which lasts months or years), wine barrels are filled all the way to the top and plugged. What might happen if barrels were only half-filled?

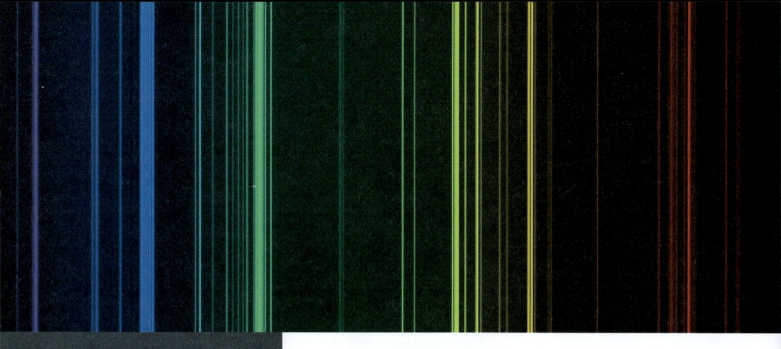

Nitrogen is the most abundant element in our atmosphere, where it exists primarily as dinitrogen gas, or N_2. This triple-bonded molecule is extremely stable and inert, and without it, there would be almost no life on Earth. Once N_2 is converted into forms of nitrogen that can be used by organisms, the nitrogen is incorporated into amino acids, nucleic acids, and a host of other compounds. The beautiful atomic spectrum above is elemental nitrogen's "fingerprint," a series of absorption and emission lines across the visible light spectrum.

LAB # 33

The Nitrogen Cycle

Learning Objectives

- Understand the microbe-mediated processes of the nitrogen cycle, including nitrogen fixation, ammonification, nitrification, and denitrification.

- Describe the microbes involved in the nitrogen cycle and their specific roles in the cycle.

- Culture, isolate, and detect bacteria involved in the nitrogen cycle by using selective media and assays that identify intermediates and end products of the cycle.

INTRODUCTION

The **nitrogen cycle** involves a series of steps or transitions from one form of nitrogen to another as this element moves through ecosystems from one type of organism to another **(Figure 33.1)**. Nitrogen is often a **limiting element** or **limiting nutrient** in an ecosystem; that is, a lack of nitrogen (in relation to the amounts of other elements that are available) limits the growth of microbial populations, as well as the sizes of both individual plants and plant populations. This, in turn, limits the sizes of plant-dependent animal populations. So, it is worth understanding how nitrogen enters an ecosystem, how it moves from one organism to another, and how it leaves an ecosystem. If nitrogen cannot enter an ecosystem, if it leaves too quickly, or if it becomes tied up for a long time in the organic remains of once-living organisms, the result may be a significant reduction in the sizes of populations or the biological diversity of that ecosystem.

The Nitrogen Cycle

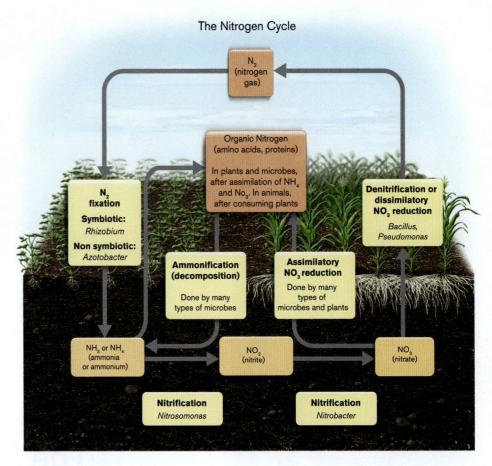

FIGURE 33.1 The nitrogen cycle.

NITROGEN FIXATION: NITROGEN ENTERS ECOSYSTEMS

Description

Nitrogen fixation is the reduction of atmospheric nitrogen (N_2) to ammonia (NH_3) or ammonium (NH_4^+). The ratio of NH_4^+ to NH_3 in an aqueous solution depends on pH. At pH 7, the ammonium form is by far the most common, accounting for 99% of the molecules in solution. Nitrogen fixation is very energy-expensive. It not only involves a reduction—and reductions usually require energy—but also requires breaking of the triple bond that holds the two nitrogen atoms in N_2 together. Regardless of the costs, nitrogen fixation is a crucial part of the nitrogen cycle because plants, animals, and most microbes cannot directly use the N_2 form of nitrogen.

After nitrogen is "fixed" as NH_3/NH_4^+, the nitrogen atoms are incorporated into amino acids as amino groups ($-NH_2$), and ultimately, amino acids are linked together to form proteins. Certain specific amino acids, such as glycine, glutamine, and aspartate, are also used to make nucleotides. At this point, the nitrogen can be called **organic nitrogen**; that is, the nitrogen atoms are a part of large organic molecules.

Symbiotic nitrogen fixation by *Rhizobium* and leguminous plants

Symbiotic nitrogen fixation is carried out by species of bacteria in the genus *Rhizobium*. These bacteria invade the roots of plants in the legume family (clover, alfalfa, peas, soybeans) and establish a **symbiotic,** or mutually beneficial, relationship with the plants. In response to the entry of the *Rhizobium* cells, the plants form root nodules around the bacteria **(Figure 33.2)**. The nodules provide the bacteria with a stable, protected environment, and the plants supply the bacteria with energy in the form of sugars created by photosynthesis. The *Rhizobium* cells release energy by metabolizing these sugars and use that energy to fix nitrogen within the nodules using an enzyme called nitrogenase. A good bit of the ammonium produced is used by the bacteria in synthesizing bacterial proteins and nucleic acids, and ultimately, for *Rhizobium* growth and cell division. Usually, however, excess ammonium is produced, which diffuses out of the cells, where it is absorbed and used by the legume sheltering the *Rhizobium*. For plants, ammonium is a very valuable nutrient; like bacteria, plants can use this form of nitrogen to build amino acids, proteins, and nucleic acids. So, in exchange for protection and sugars, the plants get their very own built-in fertilizer factories. Symbiosis!

There is another benefit to nitrogen fixation by *Rhizobium*. In addition to being used by the bacteria and their leguminous hosts, some of the excess ammonium is secreted into the surrounding soils, and additional ammonium is released after the plant dies and decays (see "Ammonification" below). So eventually, the presence of symbiotic nitrogen-fixing *Rhizobium* significantly improves the fertility of the soil for any plant growing in the same field. Centuries before this mechanism was understood, European farmers developed crop rotation systems in which legumes such as peas, beans, and lentils were planted in a given field once every 3 years to maintain the fertility of the soils. Medieval peasants didn't know why this system worked, but by expanding the food supply, *Rhizobium* quite literally kept many peasants alive. It's possible that the survival of one or more of your direct ancestors was a product of this crop rotation, in which case it can be said that you owe your existence today to *Rhizobium*.

Clover, alfalfa, soybeans, and other legumes that have been "infected" with *Rhizobium* have nodules on their roots that are visible to the unaided eye (see Figure 33.2A). If the nodules are crushed on a slide, the bacteria can be stained. While *Rhizobium* cells are usually clearly visible as Gram-negative rods before they enter legume roots, once the

A.

B.

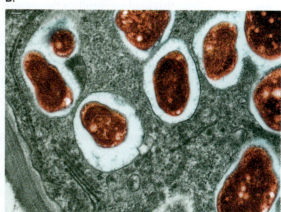

FIGURE 33.2 Symbiotic nitrogen fixers.
A. *Rhizobium* nodules on clover roots. **B.** *Rhizobium* bacteroids.

symbiotic relationship is established, the cells are surrounded by a plant-derived membrane and take on irregular, "Y" or "club-shaped" bacteroid forms. The bacteroids contain inclusions, or granules, composed largely of poly-beta hydroxylbutyrate, a small lipid.

Nonsymbiotic nitrogen fixation by free-living bacteria

Cyanobacteria Many kinds of bacteria that are capable of nitrogen fixation are free-living; that is, they do *not* live in a symbiotic relationship with other species. For example, there is a group of bacteria called **cyanobacteria** (once called blue-green algae) that live in water and moist soils. These bacteria are able to create ammonium from nitrogen gas and carry out photosynthesis to convert carbon dioxide (CO_2) gas into carbohydrates (carbon fixation). Thus, these cells can meet their nitrogen, carbon, and energy needs from sunlight and gases (N_2, CO_2) alone. Nice trick, that.

Azotobacter Other examples of free-living nitrogen-fixing bacteria include soil-dwelling species in the genus *Azotobacter*. These cells cannot photosynthesize, so they must acquire organic molecules from the environment as a source of carbon and energy. We can look for *Azotobacter* in soil using a selective medium called **mannitol salts broth**, which has the following composition (per liter):

Mannitol	15.0 g
Dipotassium phosphate	0.5 g
Magnesium sulfate	0.2 g
Calcium sulfate	0.1 g
Sodium chloride	0.2 g
Calcium carbonate	5.0 g

This medium contains a variety of salts to meet a cell's needs for phosphorus, sulfur, sodium, and calcium, but it does not contain nitrogen in any form, either organic or inorganic. Thus, this medium selects for nitrogen fixers, because when soil is added to the broth, only those cells that are able to create ammonium from the N_2 gas diffusing into the medium from the air will be able to survive and grow in the broth. In effect, this medium inhibits the growth of non-N_2 fixers by denying those species the nitrogen compounds they need to make amino acids, proteins, and nucleic acids. (In practice, the soil will add a tiny bit of nitrogen to the medium.)

In addition, the medium specifically selects for *Azotobacter* species over other nitrogen-fixing species by using the sugar mannitol as the sole source of carbon and energy. *Azotobacter* cells can survive on mannitol as their only source of carbon and energy. But many other nitrogen-fixing species need other organic molecules, such as other sugars or amino acids from peptones, in order to grow.

In this lab, soil is used as a source of *Azotobacter* species. Mannitol salts broth is inoculated with soil, then incubated for a week because *Azotobacter* cells grow slowly. Then the broth is examined with a microscope for the presence of oval *Azotobacter* cells; under certain conditions, these cells also form a spherical or ovoid-shaped resting stage called a cyst (**Figure 33.3**).

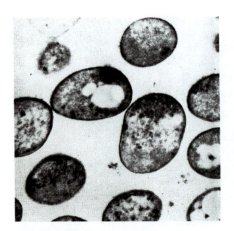

FIGURE 33.3 *Azotobacter* **cysts.**

AMMONIFICATION: NITROGEN COMPOUNDS ARE RELEASED FROM ORGANIC MATERIALS

Ammonification is a process that releases NH_3/NH_4^+ (ammonia/ammonium) from organic nitrogen sources such as amino acids, in which nitrogen is present in the form of amino ($-NH_2$) groups. Among other things, ammonification is a major part of the course of decomposition, in which a wide variety of bacterial and fungal species break down the remains of plants and animals. Once NH_3/NH_4^+ is released, it can be absorbed by microbes and plants and reassimilated into the amino acids, nucleic acids, and proteins of those organisms. Ammonium may also be used as an energy source by nitrifying bacteria, as described below.

The ability to release ammonia/ammonium from organic nitrogen compounds is a relatively common trait among microorganisms. We can encourage the growth of ammonifying bacteria by providing said microbes with broth containing 0.5% peptones (5 g peptone per liter). Remember that peptones are enzymatic digests of proteins such as casein and gelatin, so 0.5% peptone broths should provide pure-culture and soil bacteria with an overdose of amino group–rich ($-NH_2$-rich) amino acids.

In this lab, we will detect ammonia using an ammonia test strip that includes two pads at the end of the strip. One pad contains a base that raises the pH of the liquid to be tested to pH 10 or above, which converts all the ammonium (NH_4^+) to ammonia (NH_3). The second pad contains a mix of "proprietary indicators and dyes," which **change colors from yellow to green to green-blue to blue as the concentration of ammonia increases**. "Proprietary indicators and dyes" means that the Hach Company, the manufacturer of the strip, will not reveal the secret ingredients on the second pad. It's rumored that if you try to steal the mysterious formula for the test pad, that after a fair trial, you will be convicted and transported to the penal colony at Botany Bay for the term of your natural life.

NITRIFICATION: NITROGEN COMPOUNDS ARE USED FOR ENERGY, CHANGING THE CHARGE

Nitrification is a process that oxidizes ammonium (NH_4^+) to nitrite (NO_2^-) and then oxidizes nitrite to nitrate (NO_3^-). Not surprisingly, given all this oxidation, these reactions occur under aerobic conditions. The first oxidation is carried out by bacteria in the genus *Nitrosomonas*, and the second oxidation is carried out by *Nitrobacter* species **(Figure 33.4)**. These nitrifying bacteria are **autotrophic,** meaning that the cells are like plant cells: They can make sugars and all other needed organic molecules from CO_2 alone. Unlike plant cells, however, *Nitrosomonas* and *Nitrobacter* are **chemotrophic,** not photosynthetic; that is, they cannot use light energy, but instead must oxidize reduced molecules to acquire the energy needed for turning CO_2 into sugar (carbon fixation). And that brings us back to nitrification, because it's the oxidation of reduced nitrogen in the forms of ammonium (*Nitrosomonas*) and nitrite (*Nitrobacter*) that provides the energy for carbon fixation.

Nitrifying bacteria are common in soil, where they have the net effect of converting one form of nitrogen usable by plants and microbes into another form usable by

A.

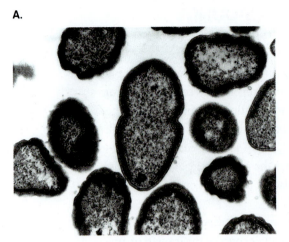

B.

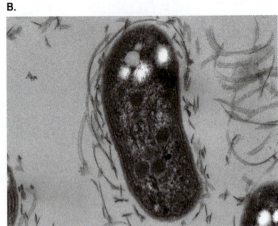

FIGURE 33.4 Nitrifying bacteria.
A. *Nitrosomonas.* B. *Nitrobacter.*

plants and microbes. Both ammonium and nitrate can be plant fertilizers because both are absorbed by plant roots, and when they are, their nitrogen atoms eventually show up in amino acids, proteins, DNA, and RNA. Thus, at first glance, the conversion of ammonium into nitrate by *Nitrosomonas* and *Nitrobacter* would appear to be a wash as far as soil fertility is concerned. However, the change from NH_4^+ to NO_3^- also changes the charge on the nitrogen-containing molecule from positive to negative. And since soil particles have a net negative charge, those particles bind ammonium, but repel nitrate. As a result, when water flows through soils, the nitrate is easily removed from the soil in a process called **leaching**. Removal of nitrogen compounds from soils reduces soil fertility, and if the concentration of nitrate in the water is too high, it can have toxic effects on aquatic organisms and on people drinking the water.

To select for nitrifying bacteria, we use **nitrite formation broth,** which has the following composition (per liter):

Ammonium sulfate	2.0 g
Dipotassium phosphate	1.0 g
Magnesium sulfate	0.5 g
Iron sulfate	0.4 g
Calcium carbonate	10.0 g

The various salts in the medium provide essential elements such as phosphorus, sulfur, magnesium, and iron. The calcium carbonate ($CaCO_3$) raises the CO_2 content of the broth, an important factor when selecting for autotrophic cells able to use CO_2 as a source of carbon for sugars and other organic molecules. And ammonium sulfate, $(NH_4)_2SO_4$, obviously adds the ammonium needed by *Nitrosomonas* as its energy and nitrogen source. If cells can use ammonium for energy and nitrogen, then they will produce nitrite as an end product, hence the name "nitrite formation broth." Then, once nitrites are produced, any *Nitrobacter* in the broth can also grow using the nitrites produced by *Nitrosomonas* as an energy source. Using ammonium and nitrites as an energy source is a tough way to get by, however, because the amount of energy released by the oxidation of one ammonium or nitrite molecule is relatively small. Thus, broths need to be incubated for several days before *Nitrosomonas* and *Nitrobacter* activity leads to the production of detectable levels of nitrite and nitrate.

Just as important is what is not in the medium: There are no organic molecules to be used as carbon and energy sources. There is no glucose, no amino acids, and no fats. Thus the broth selects against any heterotrophic cells; that is, it selects against any cells unable to use carbon dioxide as their sole carbon source. It also selects against any cells unable to use ammonium as their sole source of energy. And tubes are incubated in the dark to select against phototrophic cells able to use light for energy. So what can grow in nitrite formation broth? Well, initially, only cells able to do the specific trick of using carbon dioxide as a carbon source and ammonium as an energy source. In other words, all that will grow are nitrifying *Nitrosomonas* bacteria. Later, as nitrite levels rise, *Nitrobacter* cells can also grow, but that's about it.

The rise in nitrite can be detected using a nitrite/nitrate test strip with two pads at the end of the strip. In this case, both pads contain sulfanilic acid, which will react with nitrites (NO_2^-) to form a diazonium salt. The diazonium then reacts with one of those secret Hach Company indicators **to produce a pink to red color; red colors become deeper with increasing nitrite concentration (Figure 33.5).** The bottom pad contains an additional secret ingredient that will reduce any nitrates (NO_3^-) in the water to nitrites, and then the nitrites will subsequently react with the sulfanilic acid to produce a pink to red color. So, the bottom pad actually measures nitrites and nitrates combined.

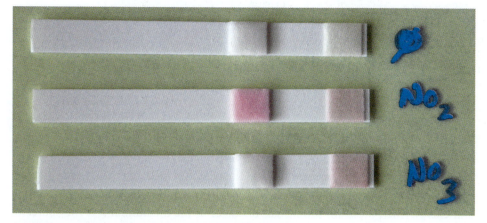

FIGURE 33.5 Nitrite detection using sulfanilic acid.
The diazonium salt that is formed is pink to red in color, depending on how much product is made; strips were dipped in nitrogen-free water (top), a nitrite solution (middle), or a nitrate solution (bottom). In this orientation, the pad on the left detects only nitrites, while the pad on the right detects both nitrites and nitrates.

DENITRIFICATION: NITROGEN LEAVES ECOSYSTEMS

Denitrification, also called **dissimilatory nitrate reduction,** is a process in which nitrates (NO_3^-) are reduced to either nitrogen (N_2) gas or nitrous oxide (N_2O) gas. Some facultative anaerobic soil bacteria, such as those in the genera *Bacillus* and *Pseudomonas,* can use nitrates in place of oxygen (O_2). If oxygen is available, these bacteria will use it as the terminal electron acceptor at the end of the electron transport chain as part of aerobic respiration. But in the absence of oxygen, in anaerobic environments such as wetlands and flooded agricultural fields, they use nitrates as terminal electron acceptors in a process called anaerobic respiration. As a result, depending on the pH of the environment, the nitrogen in the NO_3^- will be reduced to either nitrogen (N_2) gas or nitrous oxide (N_2O) gas.

Nitrate (NO_3^-) is a plant fertilizer; this form of nitrogen can be absorbed by plant roots, in which case the plants eventually use the nitrogen atoms in amino acids, proteins, DNA, and RNA. So regardless of which gas is produced, the denitrification process removes valuable nitrogen compounds from the ecosystem. Denitrification is one reason why water-saturated anaerobic wetland soils are often very nutrient poor. In addition, the process can create big problems for farmers who spend huge amounts of money to add nitrates to their fields. If fields are flooded for a significant length of time, much of that expensive nitrogen can be converted to nitrogen and nitrous oxide gas, and the investment in fertilizer goes up into the air.

To look for chemoheterotrophic denitrifiers such as *Bacillus* and *Pseudomonas* species, the following two media are prepared (amounts are per liter):

Nitrate Broth

Peptones	10.0 g
Magnesium sulfate	1.5 g
Dipotassium phosphate	1.5 g
Potassium nitrate	10.0 g
Sodium nitrite	1.0 g

Nitrate-Free Broth

Peptones	10.0 g
Magnesium sulfate	1.5 g
Dipotassium phosphate	1.5 g

Durham tubes are added to all tubes of both types of media.

Nitrate broth is essentially peptones, inorganic salts to supply phosphorus and magnesium, and 1.0 % potassium nitrate. The peptones meet the basic needs of many chemoheterotrophs; such cell types need organic molecules as sources of both carbon and energy. The potassium nitrate salt dissolves in the medium and is the source of NO_3^- for denitrification. At the bottom of the tube of inoculated broth, there is very little dissolved oxygen, so denitrifying bacteria will switch from using oxygen as a terminal electron acceptor to using nitrate in place of oxygen in anaerobic respiration. This process will produce nitrogen and nitrous oxide gases, and bubbles of these gases will collect in the Durham tubes. **So, the appearance of a bubble could indicate that denitrification has occurred, while the absence of a bubble tells us that denitrification did not occur.** That is, the absence of a bubble is negative for denitrification.

Why say "could indicate that denitrification has occurred?" Why not say that the presence of a bubble is positive for denitrification and be done with it? Remember that our nitrate broth medium favors the growth of chemoheterotrophs, and in this lab, we've seen many examples of chemoheterotrophs capable of fermenting sugars with the production of gas. And when sugars are fermented at the bottom of a tube, the gas bubbles can be trapped in a Durham tube. So, if you observe a bubble, it's possible that the bubble is the product of fermentation, not denitrification.

To reduce the risk of a "false positive" bubble from fermentation, not denitrification, nitrate broth does not contain any added sugars. This should significantly reduce the potential for gas production by fermentation. However, there may be traces of sugars or other fermentable substrates in the peptones, and we need the peptones in the medium to meet the basic nutritional needs of the chemoheterotrophs. As a result, we can't say for certain that fermentation is impossible. So, how can we tell how much of the gas is the product of denitrification and how much is due to fermentation?

Well, we can inoculate a control tube containing **nitrate-free broth** (nutrient broth with no nitrates) with the same bacteria or same soil sample used to inoculate the nitrate broth. The control tube will give us an idea of how much gas is produced by fermentation alone or by reactions other than denitrification. If the bubble in the nitrate broth tube is bigger than the bubble in the nitrate-free broth tube, then the "extra" gas in the nitrate broth is probably due to denitrification **(Figure 33.6). So, if there is a difference in bubble size, then we can say "positive for denitrification."**

This lab is carried out over a number of days because, as noted earlier, many of the bacteria involved in the nitrogen cycle grow slowly. A summary of the schedule for this lab is provided in **Table 33.1**.

FIGURE 33.6 Identification of denitrifying bacteria.
The tube on the left lacks nitrate in the medium, and thus no gas-producing denitrification has occurred. The tube on the right contains nitrate in the medium. The large gas bubble in the Durham tube is evidence that denitrifying bacteria such as *Pseudomonas* or *Bacillus* were able to grow.

Table 33.1	SUMMARY OF THE PROTOCOL SCHEDULE			
Nitrogen Cycle Step	Day 1	Day 2 2 days after day 1	Day 3 7 days after day 1	Day 4 9 days after day 1
Symbiotic N$_2$ fixation			Start/finish	
Non symbiotic N$_2$ fixation	Start			Finish
Nitrification	Start			Finish
Denitrification	Start	Subculture		Finish
Ammonification			Start	Finish

PROTOCOL

Day 1
MATERIALS

- 2 nitrite formation broths
- 2 nitrate broths with Durham tubes
- 2 nitrate-free broths with Durham tubes
- 1 mannitol salt broth

Cultures:
- *Bacillus cereus*
- Soil samples

NITRIFICATION

1. Label two nitrite formation broths with your name, the date, and the type of broth.

   Note that all of the broths you will be using in this lab look alike, so be sure to label the tubes immediately or take any and all other steps needed to keep track of the type of broth in each tube.

2. Inoculate one nitrite formation broth with a cubic centimeter (about a quarter-inch cube) of soil. Label a second tube "Uninoculated control."

3. Incubate at 30°C for 7–9 days.

DENITRIFICATION

1. Label two nitrate broths and two nitrate-free broths with your names, the date, and the type of broth. Then label one nitrate and one nitrate-free broth "*Bacillus cereus*" and one nitrate and one nitrate-free broth "Soil."

2. Inoculate all tubes with the appropriate material; for the soil broths, add about a cubic centimeter of soil.

3. Incubate the *Bacillus cereus*–inoculated tubes at 30°C for 9 days.

4. Incubate the tubes with soil suspensions at 30°C for 2 days (see day 2 protocol).

NONSYMBIOTIC NITROGEN FIXATION

1. Label one mannitol salt broth with your name, the date, and the type of broth. Then inoculate the broth with a cubic centimeter (about a quarter-inch cube) of soil.

2. Incubate at 30°C for 7–9 days.

Day 2
Do Day 2 procedures 2 days after day 1.
MATERIALS

- 1 nitrate broth
- 1 nitrate-free broth

- Nitrate and nitrate-free broths inoculated with soil on day 1

DENITRIFICATION

1. Transfer a loopful of broth from the soil-inoculated nitrate broth from day 1 to a second nitrate broth. Transfer a loopful of broth from the soil-inoculated nitrate-free broth from day 1 to a second nitrate-free broth.

   This subculture is performed because adding the soil to the nitrate-free broth can inadvertently add nitrates from the soil to the broth (depending on the quality of the soil). As a result, you can get a big bubble in the nitrate-free broth tube, and there's not much difference when you compare nitrate and nitrate-free broths. If you do a 2-day incubation, the denitrifiers have a chance to multiply, and then the subculturing should leave any soil-added nitrates behind, and the nitrate-free broth will be truly nitrate-free.

2. Incubate the two broths inoculated on day 2 at 30°C for another 7 days.

Day 3
Do day 3 procedures 7 days after day 1.
MATERIALS

- 4 0.5% peptone broths
- Demo slides of *Rhizobium* cells
- Clover or alfalfa plants with nodules (optional)

Cultures:
- *Bacillus cereus*
- *Staphylococcus epidermidis*
- Soil samples

AMMONIFICATION

1. Label four 0.5% peptone broths with your name and the date. Then label the tubes as follows:

   a. *Bacillus cereus*
   b. *Staphylococcus epidermidis*
   c. Soil sample
   d. Uninoculated control

2. Inoculate the labeled tubes with the appropriate species or with a cubic centimeter (about a quarter-inch cube) of soil, and leave the control tube uninoculated.

3. Incubate at 28°C–30°C for 48 hours.

SYMBIOTIC NITROGEN FIXATION

View demonstration slides of *Rhizobium* cells taken from alfalfa or clover nodules and record your observations in the Results section. If available, observe alfalfa or clover roots with *Rhizobium* nodules. You can see the root nodules without a microscope (although they are still pretty small).

Day 4

Do day 4 experiments 9 days after day 1.

MATERIALS

- 2 nitrite/nitrate test strips
- 4 ammonia test strips
- 4 sterile transfer pipettes with bulb end
- 4 sterile 9.0-ml water blanks
- 2 empty sterile petri dishes
- Glass slides
- Crystal violet solution
- Microscope
- Metric ruler
- Inoculated tubes from days 1, 2, and 3

NONSYMBIOTIC NITROGEN FIXATION

1. Transfer a loopful of cells from the soil-inoculated mannitol salt broth to a glass slide, and spread the cells over the slide.

2. Air-dry and heat-fix the slide, and then stain the cells with crystal violet, following the protocol for direct staining in Lab 5.

3. Look for the oval cells and the spherical to ovoid cysts characteristic of *Azotobacter* (see Figure 33.3). Record your observations in the Results section.

AMMONIFICATION

1. Label four 9.0 ml sterile water blanks as follows:

   a. *Bacillus cereus*
   b. *Staphylococcus epidermidis*
   c. Soil sample
   d. Uninoculated control

2. Use sterile transfer pipettes to transfer 1.0 ml of broth from each of the four 0.5% peptone broths (from day 2) to the labeled 9.0-ml sterile water blanks.

3. Label the non-pad ends of four ammonia test strips using a wax pencil or Sharpie or by writing on a small piece of tape as follows:

   a. *Bacillus cereus*
   b. *Staphylococcus epidermidis*

c. Soil sample
d. Uninoculated control

Timing is everything with the ammonia test strips, so lab partners should do steps 4–7 together at the same time because pad colors need to be compared, and color will change with time.

4. Each lab partner should choose two of the four labeled tubes created in steps 1 and 2 (the tubes with 1.0 ml of peptone broth diluted in 9.0 ml sterile water blanks). Remember, all tubes should be tested at the same time. Carefully tip a tube until the level of the water in the tube reaches the point where you can dip both pads on the test strip into the water. Swirl the strip in the water for about 15–30 seconds. Be sure the pads remain immersed.

5. After 15–30 seconds, pull the strip out of the water, tap it on the inside surface of the tube to remove excess water, and **place the strip in an empty petri dish**, with the label up and readable and the pad end in the dish.

   Remember that the pads used in inoculated tubes have live bacteria on them. Placing the strips in the dish will keep the bacteria from spreading throughout the lab.

6. Immediately after finishing with the first two tubes, as soon as the first two test strips are placed in the petri dish, test the last two tubes together, following steps 4 and 5 above.

7. At 30–60 seconds after the last two test strips are placed in the petri dish, record the colors of the pads at the end of all strips as either yellow to greenish yellow, green, light blue, or dark blue. Use the colors to estimate or rank the relative concentrations of ammonia by converting the color scale to a 0–3 scale. The yellow to greenish yellow color of the uninoculated sample should be considered the 0 point of the scale, and the sample with the deepest blue color should be considered a 3. Green to light blue colors would be 1s and 2s.

8. Leave the test strips in the petri dish and place it in the receptacle designated for materials to be autoclaved.

NITRIFICATION

1. Label the non-pad ends of two nitrite/nitrate test strips using a wax pencil or Sharpie or by writing on a small piece of tape as follows:

   a. Soil sample
   b. Uninoculated control

2. Test for nitrites and nitrates by dipping the pads on one test strip into the soil-inoculated nitrite formation broth and the pads on the other test strip into the uninoculated control nitrite formation broth for just 1 or 2 seconds.

3. Place the strips in an empty petri dish, with the label up and readable and the pad end in the dish.

    Remember that the pads used in inoculated tubes have live bacteria on them. Placing the strips in the dish will keep the bacteria from spreading throughout the lab.

4. About 30 seconds after the test strips are placed in the petri dish, record the color of the "nitrate plus nitrite pad" at the end of the strip and the color of the "nitrites only pad" farther up the strip **(Figure 33.7)** as either white, light pink, pink, light red, or dark red.

5. Leave the test strips in the petri dish and place it in the receptacle designated for materials to be autoclaved.

DENITRIFICATION

Examine the nitrate and nitrate-free broths inoculated with *Bacillus cereus* and the nitrate and nitrate-free broths with cells subcultured from the soil suspensions. Use a metric ruler to measure and record the length of any gas bubble in the Durham tubes in the Results section.

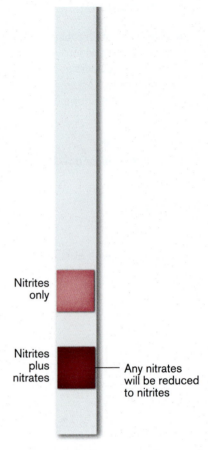

Nitrites only

Nitrites plus nitrates — Any nitrates will be reduced to nitrites

FIGURE 33.7 Nitrite/nitrate test strip.

Name: _____   Section: _____

Course: _____   Date: _____

SYMBIOTIC NITROGEN FIXATION

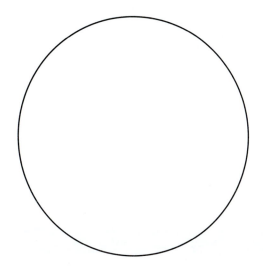

Rhizobium **cells from alfalfa or clover nodule**

Sketch of alfalfa or clover root nodules (if available):

Written description of cell morphology:

NONSYMBIOTIC NITROGEN FIXATION

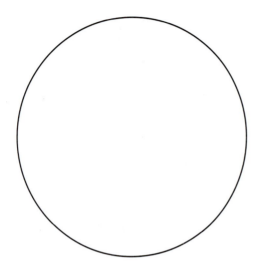

Written description of cell morphology:

Are *Azobacter* species present? How can you tell?

Cells from mannitol salt broth

AMMONIFICATION

Species or Sample Type	What Color is the Pad on the Ammonia Test Strip?	Numerical Scale (0–3)	Did Ammonification Occur?
B. cereus			
S. epidermidis			
Soil sample			
Control		0	—

NITRIFICATION

Species or Sample Type	Color of Nitrites Plus Nitrates Pad	Color of Nitrites Only Pad	Did Nitrification Occur?
Soil sample			
Control			—

DENITRIFICATION

Species or Sample Type	Broth	Length of bubble	Did dentrification occur?
Bacillus cereus	Nitrate broth		
Bacillus cereus	Nitrate-free broth		
Soil sample	Nitrate broth		
Soil sample	Nitrate-free broth		

Name: _____ Section: _____

Course: _____ Date: _____

1. Why are *Rhizobium* cells so important to humans?

2. How does the morphology of *Rhizobium* differ from that of typical rod-shaped bacteria such as *Bacillus*? Why is its morphology different? (Consider the immediate environment surrounding the different types of cells.)

3. How do *Rhizobium* and *Azotobacter* cells differ from each other in terms of where they obtain essential carbohydrates? How is this difference reflected in the formulation of the medium used to select for *Azotobacter*?

4. Why is ammonification essential to most functioning ecosystems?

5. Were there any significant differences in the amounts of ammonia produced by the pure-culture species? How did the amount of ammonification in the soil sample compare with ammonification by specific species?

6. Why are sugars (and other organic compounds) deliberately excluded from the nitrite formation broth?

7. Why did we incubate the soil-inoculated nitrite formation broth for a week instead of the 24–48-hour periods that were typical with other media used in this course?

8. Was there any evidence that nitrification occurred in the soil-inoculated nitrite formation broth? What was that evidence?

9. If both ammonium and nitrate can be absorbed by crop species (and other plants as well), does it matter if nitrification occurs in farm fields? Why or why not?

10. If denitrification converts nitrates to nitrogen and nitrous oxide gases, what is the purpose of inoculating a nitrate-free broth if one is testing for denitrification?

11. Was there any evidence that denitrification occurred in either the *Bacillus cereus* or the soil-inoculated nitrate broth? What was that evidence?

12. How might the activity of denitrifying bacteria in a farm field affect our ability to produce food?

13. What happens when there is more nitrogen in the soil than can be used by the available vegetation or converted by bacteria into atmospheric nitrogen? Why is this a problem?

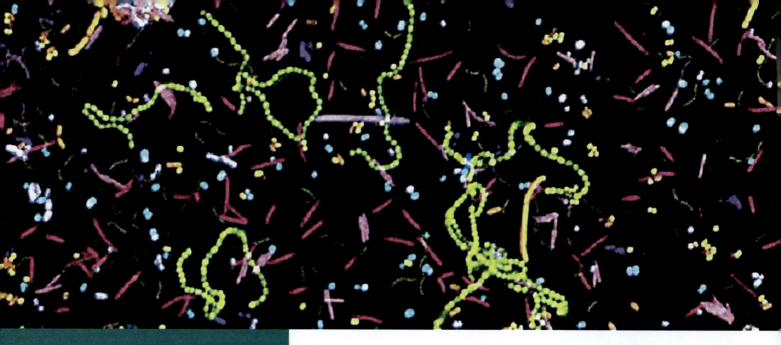

Although isolating bacteria from a sample is an essential skill for microbiologists, it is important to realize that bacteria rarely live alone. They exist in complex communities in which many species interact, often sharing resources and providing protection for one another. The striking image above is of fifteen species of bacteria (labeled with fluorescent molecules) living together in a sample of human dental plaque, a type of biofilm that we all battle with every time we brush our teeth.

LAB # 34

Winogradsky Columns

Learning Objectives

- Understand how Winogradsky columns favor specific types of microbes and model microbial ecosystems.

- Know the groups of microbes favored in the columns, including groups characterized by their interactions with oxygen and by how they obtain energy, carbon, and electrons and protons.

- Describe the chemical environment of the columns, including the roles of cellulose and phosphates, and of oxygen and sulfur gradients.

- Create Winogradsky columns with variable chemical conditions to favor different groups of microorganisms.

Winogradsky columns are used to model or simulate microbial ecosystems composed of many interacting species. They are named after Sergei Winogradsky, a Ukrainian-Russian soil microbiologist and ecologist whose research in the late 1800s and early 1900s focused on the role of microbes in nitrogen and sulfur cycles. Winogradsky understood that it was important to study multispecies systems because, in the natural world, species are not isolated from one another in pure cultures. Instead, a given species is one part of one or more complex **ecosystems** composed of many interacting and interdependent species.

Winogradsky columns also provide an opportunity to study the range of metabolic abilities that have evolved in the microbial world. **Metabolism** is the collection of biochemical reactions that sustain life inside of living cells, and these reactions require sources of energy, carbon, and protons and electrons. In the case of animal cell metabolism, it's organic molecules such as glucose that supply the required items. However, bacteria (as a group) are capable of a much greater variety of metabolic pathways than animal cells; that is, bacteria have evolved many different ways to skin the metabolic cat. Many bacterial solutions to basic metabolic needs can be seen in the species favored by the conditions created in Winogradsky columns. In particular, these columns promote the growth

FIGURE 34.1 A Winogradsky column just after construction.

FIGURE 34.2 A mature Winogradsky column.

of **sulfur cycle bacteria**, which convert sulfides to sulfates and sulfates to sulfides, and **anaerobic photosynthetic bacteria. Figures 34.1** and **34.2** show examples of just-built and mature Winogradsky columns. The size of these columns can vary considerably, but typical columns are about 10–12 inches (25–30 cm) tall and about 3–4 inches (8–10 cm) in diameter.

In summary, Winogradsky columns create selective environments that favor a diversity of interacting soil and aquatic microorganisms with a variety of metabolic abilities. These microbes can be gathered from a wide variety of environments, but the organisms are usually derived from the pond water and pond sediments used to build the columns. Had Shakespeare been a microbiologist, he might have put it this way in *As You Like It*, Act II, Scene VII:

> *All the column's a stage,*
> *And all the aerobes and anaerobes merely players.*
> *The species have their exits and their entrances,*
> *And one microbe in its time plays many parts.*

THE PLAYERS

As noted above, Winogradsky-column microbes demonstrate a wide range of metabolic abilities, so it's useful to sort species into **trophic groups** based on the sources of energy, carbon, and protons and electrons that a particular species uses.

1. What is the source of energy?

   Phototrophs use light as an energy source.
   Chemotrophs use reduced organic or inorganic molecules as energy sources.

2. What is the source of carbon used to build organic molecules and structures within cells?

   Autotrophs use carbon dioxide as their carbon source.
   Heterotrophs use organic molecules (glucose, organic acids, lipids, etc.) as carbon sources.

3. What is the source of electrons and protons needed for oxidation-reduction reactions such as those occurring in photosynthesis and in other metabolic pathways?

   Lithotrophs use inorganic molecules such as H_2O and H_2S as electron and proton sources.
   Organotrophs use organic molecules as electron and proton sources.

Because assignment of a species to each of these three different sets of trophic groups depends on the answers to three different questions, the trophic groups can be combined into eight different merged groups (e.g., photoautolithotroph or chemoheteroorganotroph). In reality, almost all autotrophs are also lithotrophs and almost all heterotrophs are also organotrophs, so in practice, those eight possible groups can be reduced to four groups; photoauto(litho)troph, chemoauto(litho)troph, photohetero(organo)troph and chemohetero(organo)troph (**Table 34.1**). Examples of all four groups can be found in many Winogradsky columns.

Table 34.1	TROPHIC GROUP CLASSIFICATION OF ORGANISMS		
TROPHIC GROUPS	**PHOTOTROPHS**	**CHEMOTROPHS**
AUTOTROPHS*	PHOTOAUTOTROPHS Light for energy CO_2 for carbon	CHEMOAUTOTROPHS Reduced inorganic molecules for energy CO2 for carbon
HETEROTROPHS†	PHOTOHETEROTROPHS Light for energy Organic molecules for carbon	CHEMOHETEROTROPHS Reduced organic molecules for energy Organic molecules for carbon

*Autotrophs are almost always also lithotrophs, so inorganic molecules are almost always their source of protons and electrons.

†Heterotrophs are almost always also organotrophs, so organic molecules are almost always their source of protons and electrons.

Photoautotrophs

Photoautotrophs use light, which they capture by photosynthesis, as a source of energy, and they use carbon dioxide (CO_2) as a source of carbon. So these organisms do not need any preformed organic molecules such as sugars, lipids, or amino acids. Instead, they use the process of carbon fixation to make organic carbon molecules from inorganic CO_2 via the Calvin cycle. Photoautotrophy is how multicellular plants make a living, but in the context of this lab, it's the microbial photoautotrophs that we care about. These microbes can be subdivided into two groups: those that produce oxygen as part of photosynthesis and those that do not.

Oxygenic (oxygen-generating) photoautotrophs

Oxygenic photoautotrophs use water (H_2O) molecules as a source of electrons and protons, so these microbes are also lithotrophs. Electrons are used for photosynthesis and protons are needed for reducing and fixing CO_2 in the Calvin cycle. The use of water for electrons and protons yields oxygen (O_2). In a Winogradsky column, oxygenic photoautotrophs that you might observe include eukaryotic algae, photosynthetic protozoa such as euglenoids, and prokaryotic cyanobacteria. They are found in the upper layer of the column, where oxygen is abundant (**Figure 34.3**).

Anoxygenic (non-oxygen-generating) photoautotrophs

Anoxygenic photoautotrophs live in anaerobic environments that receive at least some sunlight. For example, these microbes might be found in the lower water column or on top of aquatic sediments in a still-water pond (still water usually holds less oxygen than water mixed with air). In place of water (H_2O), anoxygenic photoautotrophs use hydrogen sulfide (H_2S) as their source of electrons for photosynthesis. This type of lithotrophy yields elemental sulfur, not oxygen, which accounts for the name "anoxygenic photoautotroph."

Winogradsky columns usually contain supplemental sulfur and may support two types of anoxygenic photoautotrophs, purple sulfur bacteria and green sulfur bacteria.

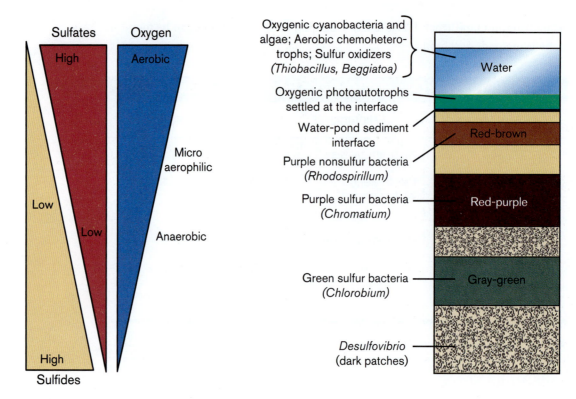

FIGURE 34.3 Typical layers and gradients in a mature Winogradsky column.

Purple sulfur bacteria include species in the genus *Chromatium*. These cells capture light energy using pigments such as bacteriochlorophyll *a* and carotenoids. The pigments give the cells their red-purple and brown colors, because bacteriochlorophyll *a* reflects more red light than other types of bacteriochlorophylls, whereas the carotenoids absorb green light. *Chromatium* species deposit the sulfur derived from sulfides within the cells in the form of sulfur granules. In Winogradsky columns, the red-purple-tinted *Chromatium* and related species are found in the upper part of the anaerobic zone (see Figure 34.3).

Green sulfur bacteria include species in the genus *Chlorobium*. These cells capture light energy with a variety of bacteriochlorophyll types, including bacteriochlorophyll *c*, a pigmented molecule that absorbs more red light than bacteriochlorophyll *a*. Like *Chromatium* cells, *Chlorobium* cells also contain carotenoids, but there is usually a greater concentration of green-reflecting bacteriochlorophyll than green-absorbing carotenoids. As a result of these differences in bacteriochlorophylls and relative amounts of carotenoids, *Chlorobium* cells are green instead of purple. In addition, unlike *Chromatium* and most other photoautotrophs, green sulfur bacteria do not fix carbon dioxide using the Calvin cycle. Instead, CO_2 is fixed by a reductive citric acid cycle, a process that adds inorganic CO_2 to intermediates in the citric acid (Krebs) cycle to create new organic molecules. In addition, these microbes deposit sulfur outside the cells. In Winogradsky columns, the green-tinted *Chlorobium* and related species are usually found in the lower part of the anaerobic zone (see Figure 34.3).

Chemoautotrophs

Like photoautotrophs, **chemoautotrophs** fix or use CO_2 via the Calvin cycle as a source of carbon instead of using preformed organic molecules such as sugars. However, unlike photoautotrophs, these microbes use simple reduced inorganic molecules such as ammonium (NH_4) and hydrogen sulfide (H_2S) as an energy source. The process of oxidizing reduced inorganic molecules generates the energy needed to fix CO_2. As a result, chemoautotrophs do not need any light or any preformed organic molecules to survive. Instead, they can thrive in the dark and on inorganic salts alone. This unusual way of life was discovered by Winogradsky and others in the 1880s.

Small reduced inorganic molecules do not yield much energy when oxidized, so it takes lots of ammonium or sulfide molecules to keep a given chemoautotroph cell going. Since these cells must process many nitrogen- or sulfur-based molecules to survive, chemoautotrophs may be quite important in the cycling of nitrogen and sulfur in soil and aquatic ecosystems. Here we will focus on the chemoautotrophic sulfide-oxidizing bacteria, because most Winogradsky columns are designed to favor sulfur cycle bacteria and because chemoautotrophic nitrifying bacteria are covered in Lab 33.

Thiobacillus and *Acidithiobacillus* are two genera of aerobic sulfur-oxidizing bacteria. Species in these genera oxidize elemental sulfur and the sulfides in hydrogen sulfide and metal sulfide to sulfuric acid (H_2SO_4). The production of sulfuric acid can significantly lower environmental pH, and not surprisingly, many *Thiobacillus* and *Acidithiobacillus* species are adapted to acidic habitats. Some species, such as *Acidithiobacillus ferroxidans*, can oxidize both sulfur and the iron released from iron sulfides using the reactions below:

Oxidation of sulfide to sulfuric acid:

$$4\ FeS_2 + 14\ O_2 + 4\ H_2O \rightarrow 4\ Fe^{2+} + 8\ SO_4^{2-} + 8\ H^+$$

Oxidation of ferrous iron to ferric iron:

$$4\ Fe^{2+} + O_2 + 4\ H^+ \rightarrow 4\ Fe^{3+} + 2\ H_2O$$

Ferric ions tend to form red-orange-colored precipitates of ferric hydroxide—$Fe(OH)_3$—and other insoluble ferric compounds. In an acidic Winogradsky column with plenty of iron in the pond sediments used to build it, these iron precipitates may appear as orange patches or as a rusty suspension in the upper or more aerobic parts of the column (see Figure 34.3). Out in the natural world, these oxidations produce acid mine drainage, a type of pollution associated with mining that acidifies creeks and rivers and leaves stream bottoms coated in red-orange iron precipitates.

Photoheterotrophs

Photoheterotrophs use light as their source of energy, just as photoautotrophs do. However, instead of CO_2 fixation, they rely on organic molecules, especially organic acids, as their main source of carbon for building other organic molecules and cell structures. These organic carbon sources can also provide electrons and protons, and so these heterotrophs are usually organotrophs, too.

One type of photoheterotroph that you may find in a Winogradsky column is the **purple nonsulfur bacteria** group. This group includes the genus *Rhodospirillum*, a genus of spiral-shaped, motile, microaerophilic bacteria. In Winogradsky columns, *Rhodospirillum* species are usually found in red to purple bands near the top of the anaerobic zone (see Figure 34.3). If the water column above the sediments becomes anaerobic, these cells may also be found in patches on the surface of the sediments.

Rhodospirillum cells can and will grow photoheterotrophically, but they also have very diverse metabolic abilities: Depending on conditions, they may grow chemoheterotrophically, photoheterotrophically, or photoautotrophically. Under aerobic conditions, these cells may grow as chemoheterotrophs by using organic molecules such as succinate, acetate, and malate as their source of carbon, energy, and electrons. If oxygen is available, then energy is released by aerobic respiration. If the environment becomes anaerobic, and if no light is available, the cells may still grow chemoheterotrophically by using organic molecules in fermentation pathways, but this is a much less efficient type of energy metabolism.

However, species in this genus have another trick up their tiny sleeves, because they also produce bacteriochlorophyll *a* and carotenoids, especially if there is little or no oxygen in the environment. So, if light is available in an anaerobic habitat, *Rhodospirillum* cells make a metabolic switch from fermentation to photosynthesis, using light as a source of energy. As long as organic molecules are present, the cells will continue to use these molecules as carbon and electron sources, and will now grow as reddish-purple photoheterotrophs.

Finally, if organic carbon supplies are exhausted, the cells can change their metabolism yet again and use CO_2 for carbon and H_2S for electrons for photosynthesis to grow as photoautotrophs (assuming light is still available). Since H_2S, and not H_2O, is the electron source, *Rhodospirillum* photosynthesis does not produce oxygen (it's anoxygenic). Despite occasionally using H_2S as an electron source, these species are considered to be "nonsulfur bacteria" because they do not oxidize elemental sulfur to sulfates, and because freshwater forms can't tolerate high concentrations of H_2S.

Freshwater *Rhodospirillum* metabolism is summarized in **Table 34.2**.

Table 34.2 | FRESHWATER *RHODOSPIRILLUM* METABOLISM

Oxygen	Light	Organic Molecules Available	Metabolism
Aerobic	—	Yes	Chemoheterotroph (aerobic respiration)
Anaerobic	No light	Yes	Chemoheterotroph (fermentation)
	Light	Yes	Photoheterotroph
		No	Photoautotroph

Chemoheterotrophs

Chemoheterotrophs use organic molecules for energy, for carbon, and for electrons and protons (so chemoheterotrophs are also almost always organotrophs). Humans and other animals are chemoheterotrophs, but in a Winogradsky column, it's the microbial chemoheterotrophs that count. A given column will be home to numerous species of chemoheterotrophs, mostly using organic carbon via fermentation and other anaerobic pathways. The anaerobic metabolism of chemoautotrophs generates a wide range of end products, including organic acids, ketones, alcohols, hydrogen gas, and carbon dioxide. Furthermore, organic acids produced by one species often serve as carbon and energy sources for other chemoheterotrophs in turn.

In Winogradsky columns, chemoheterotrophs include *Clostridium*, *Desulfovibrio*, and *Beggiatoa*.

Clostridium species are rod-shaped, spore-forming chemoheterotrophic bacteria that are common in soils and aquatic sediments. These species are strictly anaerobic, so they will be found in the lower anaerobic layers of the Winogradsky column. Some types of *Clostridium* are able to degrade cellulose, an organic carbon source that is routinely added to the columns. After cellulose has been broken down into its glucose subunits, the glucose molecules are fermented, a process that yields a variety of products such as organic acids, acetone, butanol, and gases. Gases may produce bubbles that can be seen against the walls of the column. Organic end products of *Clostridium* fermentation may be used as carbon and energy sources by other heterotrophs in the column, so the activity of *Clostridium* cells may generate food for other microbes.

Desulfovibrio species are curved rods that thrive in the anaerobic zones of the Winogradsky column. Like the many microbes capable of aerobic respiration, *Desulfovibrio* cells use electron transport systems driven by the oxidation of organic molecules to make ATP. However, *Desulfovibrio* cells do not use oxygen to remove electrons from the electron transport chain. Instead, they carry out anaerobic respiration, using oxidized forms of sulfur such as sulfates (SO_4^{2-}) as terminal electron acceptors in place of oxygen. When the sulfates accept the electrons from the electron transport chain, the sulfur in the sulfates is reduced to sulfides (S^{2-}).

The sulfides produced by *Desulfovibrio* can take a few forms. They may form hydrogen sulfide (H_2S), a compound that can be toxic to other microbes in aquatic environments. In contrast, the H_2S can diffuse through a Winogradsky column and can be used by green and purple sulfur bacteria as an electron source for photosynthesis, a process that converts sulfides to elemental sulfur. H_2S can also be used by *Acidithiobacillus* and *Thiobacillus* cells as an energy source, a process that oxidizes sulfides to sulfates and creates a simple sulfur cycle (sulfate to sulfide to sulfate to …). Finally, sulfides may also react with metal cations, such as ferrous ions (Fe^{2+}), to produce metal sulfides. The metal sulfides are gray to black in color, and they form very visible dark patches on the clear walls in the anaerobic zone of Winogradsky columns.

Beggiatoa species are aerobic bacteria that oxidize sulfides to elemental sulfur granules. Many marine *Beggiatoa* species are chemoautotrophic, oxidizing sulfides to acquire the energy needed to convert carbon dioxide into organic molecules; this process is similar to the metabolic strategy seen in *Thiobacillus* and *Acidithiobacillus*. However, most freshwater species, such as those common in ponds, appear to be primarily chemoheterotrophic, using organic acids and other organic carbon molecules for carbon and energy. If sulfides and other reduced forms of sulfur are not an energy source for heterotrophic *Beggiatoa*, then what is the value of sulfide oxidation to these cells? One possibility is that the sulfur granules produced by oxidation serve as electron acceptors in anaerobic respiration. This process would help *Beggiatoa* cells to survive periods of low oxygen concentrations and would also reduce elemental sulfur back to sulfides. In Winogradsky columns, the *Beggiatoa* species are usually found in zones where there is sufficient oxygen diffusing into the water from the air above the column to support aerobic metabolism and where there is sufficient sulfide diffusing up from the bottom of the column for sulfide oxidation (see Figure 34.3). At the oxygen-sulfide interface, the filamentous *Beggiatoa* may form whitish patches or layers.

THE STAGE

Winogradsky columns create environments that favor a diversity of interacting soil and aquatic microorganisms. Broadly speaking, two major components or features are needed to set the Winogradsky microbial stage: (1) chemical enrichment and (2) the establishment of gradients and cycles.

Chemical enrichment

Most Winogradsky columns are built using pond sediment, and since every pond is unique in its chemistry, columns built from sediments from different ponds will also be chemically different from one another. Some ponds may be naturally high in calcium, others may be high in iron or other metals, and yet others may be high in dissolved organic molecules. These pond-to-pond differences make it difficult to predict the outcome of Winogradsky experiments, but they also create opportunities to explore the effects of pond-specific variation in chemistry.

In addition to chemicals derived from collected sediments, it is standard procedure to enrich the column environment by adding additional chemicals to favor or select for certain groups of bacteria.

Sulfur All bacteria need a little sulfur for amino acids such as cysteine and methionine, and sulfur is typically added to the column in the form of calcium sulfate ($CaSO_4$) or egg yolks. While all species need sulfur, some need much more than others, because some species use various forms of sulfur to do more than synthesize a couple of amino acids and a few other types of molecules. As noted above, different forms of sulfur can be used as energy sources, as electron donors, and as electron acceptors. Enriching the environment by adding sources of sulfur to the column favors the growth of bacteria with a greater need for sulfur. Such bacteria include green and purple sulfur bacteria (*Chromatium*, *Chlorobium*) and purple nonsulfur bacteria (*Rhodospirillum*), as well as *Thiobacillus*, *Acidithiobacillus*, *Desulfovibrio*, and *Beggiatoa*. In fact, Winogradsky columns are notable for their ability to favor various sulfur cycle bacteria.

Phosphates Phosphate enrichment is achieved by adding K_2HPO_4 (dipotassium phosphate or potassium phosphate, dibasic) to the column. All bacteria need phosphates for ATP and nucleic acids (among other things), so adding K_2HPO_4 to the column will support the growth of all types of cells. K_2HPO_4 also acts as a buffer, slowing changes that would make the column too alkaline or too acidic for a wide range of microbes. For example, in the absence of a buffer, the pH drops faster when organic acids are produced by anaerobic fermentation or when sulfuric acid is generated by aerobic sulfide oxidation by *Thiobacillus* and *Acidithiobacillus*. However, *Thiobacillus* and *Acidithiobacillus* species prefer a low pH, so the buffering from phosphate enrichment may somewhat inhibit these species by maintaining a higher pH than is ideal for them.

Carbonates In most columns, carbonates (CO_3^{2-}) are provided by calcium carbonate ($CaCO_3$) or eggshells. Like K_2HPO_4, calcium carbonate can act as a buffer against rapid pH changes (calcium carbonate is the main ingredient in TUMS antacid tablets). In addition, in aqueous solutions, carbonates exist in a dynamic equilibrium with carbon dioxide (CO_2), so some carbonate can be converted to CO_2. As noted, CO_2 can be used as a carbon source by autotrophs, so carbonate enrichment favors and supports the growth of the many types of autotrophs found in the columns. These autotrophs include eukaryotic algae, cyanobacteria, green and purple sulfur bacteria, and *Thiobacillus* and *Acidithiobacillus*.

Cellulose Cellulose can be added to the Winogradsky column in the form of cellulose powder or shredded paper. Cellulose is a glucose polymer, so it represents a potential organic carbon source for a variety of heterotrophs. However, most bacteria lack the enzymes needed to break the bonds between the glucose molecules in cellulose, so they rely on certain *Clostridium* species to degrade it. Once cellulose has been broken down to glucose, heterotrophic bacteria can use the glucose for carbon and energy. Aerobic heterotrophs use glucose and oxygen in aerobic respiration, producing H_2O and CO_2 (which can be used by autotrophs). With time, this process helps to create anaerobic conditions, so enrichment with cellulose may promote the formation of anaerobic zones (see the discussion of dynamic gradients below). *Clostridium* and many other types of anaerobic bacteria can ferment the glucose, a process that yields a variety of products such as organic acids, acetone, butanol, and gases. As noted above, the organic products of fermentation may be used as carbon and energy sources by other heterotrophs, so the fermentation may generate food for other microbes.

Establishment of dynamic gradients

The environment in a Winogradsky column is dynamic and begins to change from the moment the column is filled. These changes are largely the product of microbial activity, so it's the players themselves that construct and maintain certain aspects of the Winogradsky stage. For example, microbial activity creates at least three significant gradients within the column: an oxygen gradient and two sulfur gradients (see Figure 34.3).

Oxygen gradient The oxygen gradient in a Winogradsky column is the result of the diffusion of oxygen from the air above the column and the metabolic activities of microbes that create and consume oxygen. Over time, an oxygen gradient is established, with aerobic zones at the top, microaerophilic zones in the middle, and anaerobic zones at the bottom of the column.

Within the column, aerobic chemotrophs use oxygen as they extract energy from reduced molecules. Numerous chemoheterotrophs use aerobic respiration to release the energy in organic molecules, while chemoautotrophs such as *Acidobacillus* species use oxygen to derive energy from reduced sulfides by oxidation. At or near the top of the column, the depletion of oxygen is countered to some degree by the diffusion of oxygen from the air and by the activity of oxygenic photoautotrophs, including cyanobacteria, eukaryotic algae, and photosynthetic protozoa (such as euglenoids). This addition of oxygen to upper levels may be sufficient to maintain aerobic regions that will continue to support the aerobic chemoheterotrophs and chemoautotrophs. Just below the aerobic zones, the microaerophilic areas with lower oxygen concentrations favor bacteria such as *Rhodospirillum*.

By contrast, in the lower layers, oxygen consumption exceeds oxygen replenishment from the upper layers. These layers soon become anaerobic, and these conditions favor strictly anaerobic heterotrophs such as *Clostridium* species. These species degrade the added cellulose and provide organic molecules for other heterotrophs. Some of the organic products of *Clostridium* metabolism may diffuse up to support aerobic heterotrophs, but in the anaerobic lower layers, those products are used by heterotrophs extracting energy by fermentation and anaerobic respiration (*Desulfovibrio*). The anaerobic environment also favors the green and purple sulfur bacteria, including *Chromatium* and *Chlorobium*.

Sulfur gradients Sulfur gradients, enhanced by sulfate enrichment, are the product of the interaction of sulfur cycle bacteria with the oxygen gradient and with one another. In anaerobic layers, heterotrophic *Desulfovibrio* cells carry out anaerobic respiration, reducing sulfates in place of oxygen and producing sulfides as end products (there may also already be sulfides in the pond sediment). These sulfides may react with metals to form gray-black metal sulfides, which appear as dark patches in the column. Sulfides will also diffuse throughout the column to the other sulfur cycle bacteria.

Where light is available in the anaerobic zones, green and purple sulfur bacteria will use sulfides as electron donors in anoxygenic photosynthesis, producing elemental sulfur as an end product. When sulfides diffuse up to aerobic areas of the column, they may be oxidized to elemental sulfur by *Beggiatoa*, or they may be oxidized to sulfuric acid by *Acidithiobacillus* and *Thiobacillus*. Sulfuric acid yields sulfates, and these sulfates may diffuse down into the anaerobic zone, where they will be reduced to sulfides by *Desulfovibrio*, thus completing a simple sulfur cycle.

PROTOCOL

Day 1
MATERIALS

- 22–24-oz (650–710-ml) clear plastic water bottle, about 2.75 inches (about 7 cm) in diameter
- Calcium sulfate ($CaSO_4$)
- Calcium carbonate ($CaCO_3$)
- Dipotassium phosphate (K_2HPO_4)
- Cellulose powder or shredded newspaper
- Spatulas
- Weigh boats
- Digital balances (accurate to 0.01 g)
- 500- or 1,000-ml graduated cylinder
- Large kitchen spoon
- Aluminum foil
- Aquatic sediment from nearby ponds
- Pond water from the same source as the sediment

Pond sediment can be collected from any point in the pond, but good results are usually produced by using sediment from areas that are underwater but right next to the shore. Sediments that have a visible surface coat of cyanobacteria or green algae will usually produce a good, strong green color at the top of the column. You can also mix in soils that may be just above the waterline but which are also saturated with water. Finally, the samples should contain as few rocks as possible, as this will make it easier to mix the mud with the various chemical additives. For variety, different groups of students can use sediment from different sources.

1. You will be assigned to build a Winogradsky column using one of five combinations of chemical additives (see list below). Chemicals for a column may be weighed out by instructors before the lab or by students during the lab. After weighing, all of the components for your column should be combined and mixed in a single large weigh boat prior to building the column.

   Chemical additive combinations:

   a. All chemicals
Calcium sulfate	3.0 grams
Calcium carbonate	3.0 grams
Dipotassium phosphate	3.0 grams
Cellulose	4.0 grams

   b. No sulfur
Calcium carbonate	3.0 grams
Dipotassium phosphate	3.0 grams
Cellulose	4.0 grams

   c. No phosphate
Calcium sulfate	3.0 grams
Calcium carbonate	3.0 grams
Cellulose	4.0 grams

   d. No carbonate
Calcium sulfate	3.0 grams
Dipotassium phosphate	3.0 grams
Cellulose	4.0 grams

   e. No cellulose
Calcium sulfate	3.0 grams
Calcium carbonate	3.0 grams
Dipotassium phosphate	3.0 grams

2. To build the column, begin by carefully cutting the top off of a 22–24-oz. clear plastic bottle. The cut should be made just above the point where the parallel sides of the bottle begin to curve inward to the opening at the top. When you are finished, the opening should be large enough to fit the large spoon that will be used to load sediment into the column.

3. After cutting off the top, use a 500- or 1,000-ml graduated cylinder to pour about 400–450 ml of water into the bottle, and draw a line at the water line. This line should be about 2–2.5 inches below the new rim of the bottle. After the line has been marked, pour the water out of the bottle.

4. Use the large spoon to begin filling the bottle with sediment. Fill the bottle until the sediment line is about 25% of the distance between the bottom of the bottle and the line drawn in step 3.

5. Add about 25% of the assigned chemical additive mix to the bottle sediment, and use the large spoon to thoroughly mix the chemicals into the sediment.

6. Use the large spoon to add more sediment until the sediment line is about halfway to the line drawn in step 3.

7. Add another 25% of the chemical mix to the bottle sediment, and use the large spoon to thoroughly mix the chemicals into the sediment.

8. Use the large spoon to add more sediment until the sediment line is about 75% of the way to the line drawn in step 3.

9. Add another 25% of the chemical mix to the bottle sediment, and use the large spoon to thoroughly mix the chemicals into the sediment.

10. Use the large spoon to add more sediment until the sediment line is about level with the line drawn in step 3.

11. Add the last 25% of the chemical mix to the bottle sediment, and use the large spoon to thoroughly mix the chemicals into the sediment.

12. Add pond water until the water level is about 0.5 inches above the sediment line, then use the large spoon to mix and pack the column from top to bottom. This process should distribute the chemical additives evenly throughout the column and remove as many air bubbles as possible.

13. Add more pond water until the water level is about 0.5 inches below the rim of the bottle. There should be a total of about 2 inches of water above the sediment line.

14. Cover the opening at the top of the column with aluminum foil to reduce the rate of evaporation. Be sure that the edges of the foil extend no more than about 0.5 inches down the sides of the column, because the water at the top of the column needs to be exposed to the light.

15. Incubate your column at room temperature near a window so that it receives plenty of light, but be sure that sunlight does not hit the column directly, as this could lead to overheating.

If windows are not available, use 60- or 75-watt incandescent bulbs placed at a distance of about 1–1.5 feet from the column to provide light for the various photosynthesizers.

Subsequent Days
MATERIALS

• Winogradsky columns built on day 1

1. Columns should be checked at least once or twice a week. If the water level is 1 inch or less from the top of the sediment layer, add more water to bring the water depth to about 2 inches.

2. Observe examples of all five chemical combinations by observing your fellow students' columns. That is, be sure that you observe all of the five different types of columns.

3. Record the appearance of each of the five column types, paying attention to the colors in the columns and the presence of gas bubbles in the bottom of the columns. If digital cameras are available, photos can be used to follow weekly changes.

4. It can take 1 to 2 weeks before the various colors can be seen, and then columns can be observed for as long as you'd like: 4–6 weeks is a typical length of time.

5. At the end of the lab, remove the aluminum foil cap and allow the column to dry out for several weeks before disposal. If a fume hood is available, columns should be placed in the hood to vent volatile sulfides during drying.

Name: _____

Section: _____

Course: _____

Date: _____

WEEK 1

Column Type	Appearance of Water Column and Surface of Sediments	Appearance, Thickness, and Height of any Colored Layers below Water/Sediment Interface (can show by drawing or photo)	Bubbles in Lower Layers? (yes/no)
All chemicals			
No sulfur			
No phosphate			
No carbonate			
No cellulose			

WEEK 2			
Column Type	Appearance of Water Column and Surface of Sediments	Appearance, Thickness, and Height of any Colored Layers below Water/Sediment Interface (can show by drawing or photo)	Bubbles in Lower Layers? (yes/no)
All chemicals			
No sulfur			
No phosphate			
No carbonate			
No cellulose			

WEEK 3

Column Type	Appearance of Water Column and Surface of Sediments	Appearance, Thickness, and Height of any Colored Layers below Water/Sediment Interface (can show by drawing or photo)	Bubbles in Lower Layers? (yes/no)
All chemicals			
No sulfur			
No phosphate			
No carbonate			
No cellulose			

WEEK 4

Column Type	Appearance of Water Column and Surface of Sediments	Appearance, Thickness, and Height of any Colored Layers below Water/Sediment Interface (can show by drawing or photo)	Bubbles in Lower Layers? (yes/no)
All chemicals			
No sulfur			
No phosphate			
No carbonate			
No cellulose			

WEEK 5

Column Type	Appearance of Water Column and Surface of Sediments	Appearance, Thickness, and Height of any Colored Layers below Water/Sediment Interface (can show by drawing or photo)	Bubbles in Lower Layers? (yes/no)
All chemicals			
No sulfur			
No phosphate			
No carbonate			
No cellulose			

WEEK 6			
Column Type	Appearance of Water Column and Surface of Sediments	Appearance, Thickness, and Height of any Colored Layers below Water/Sediment Interface (can show by drawing or photo)	Bubbles in Lower Layers? (yes/no)
All chemicals			
No sulfur			
No phosphate			
No carbonate			
No cellulose			

Name: _____   Section: _____

Course: _____   Date: _____

1. Taking all of the Winogradsky columns in this lab together, list all of the various groups or genera or microbes that you saw in one or more of the columns. For each group or genera:

   a. Describe what led you to conclude that the group or genus was present in one or more of the columns.

   b. Describe the conditions that favor the growth of that group or genus, and relate this to conditions in the particular columns in which these microbes are found.

   c. Describe how and why the amount of growth of these organisms changed over time, that is, describe how changing conditions in the columns affected the growth or appearance and disappearance of these microbes.

2. For each of the four column types that were missing one chemical (i.e., the "No sulfur" column, the "No phosphate" column, etc.), list one or two groups or genera that were absent from this column, but present in at least one other column. In each case, explain why the absent group or genus is missing from this column. That is, what may have worked against or inhibited the growth of those microbes?

3. Hydrogen sulfide is toxic to animals because it binds to iron atoms in cell respiration cytochromes and so inhibits respiration, and because it binds to iron atoms in hemoglobin, reducing the binding of oxygen to this blood protein. Hydrogen sulfide can accumulate in ponds and other enclosures where large numbers of fish and shellfish are reared, so it is a big problem in aquaculture. Are there any bacteria that might be isolated from a Winogradsky column (not just your columns) that could help with this problem? How or why could this Winogradsky-column resident be useful in the bioremediation of hydrogen sulfide?

4. Hydrogen sulfide is also a problem because it corrodes metals, including metals used to make water pipes. If your job was to control corrosion, which typical Winogradsky-column resident would be of greatest concern to you, and how might you use a Winogradsky column to test various ways of controlling these types of microbes?

5. In some parts of the United States, wetlands have been constructed to treat acid mine drainage from coal mines. These constructed wetlands receive drainage from the mines and then hold that drainage for an extended period with the goal of removing iron and sulfur from the water. If you were building such wetlands to treat acid mine drainage, which types of organisms found in the Winogradsky column could be useful to you? Why or how would these particular organisms be valuable?

6. Pick one variable that could be altered in this experiment, describe how you would alter this variable, and predict how altering this variable would change the outcome or what you see in the column.

7. The Winogradsky column enables us to see how mineral elements are cycled in natural environments by microorganisms. We focused mainly on sulfur, but there are also cycles in operation for carbon. Use the names of carbon-containing molecules mentioned in this lab (for example, cellulose, organic acids, carbon dioxide) to sketch out a closed-loop carbon cycle for these molecules. For each step in the cycle, be sure to list the name of a microorganism likely to be found in these columns that could carry out that step, and note what type of "troph" it would be (photoautotroph, etc.). You may want to review Figure 33.1 in Lab 33 as an example of how to sketch out a cycle, and you can use the back of this page if you need more room for your sketch.

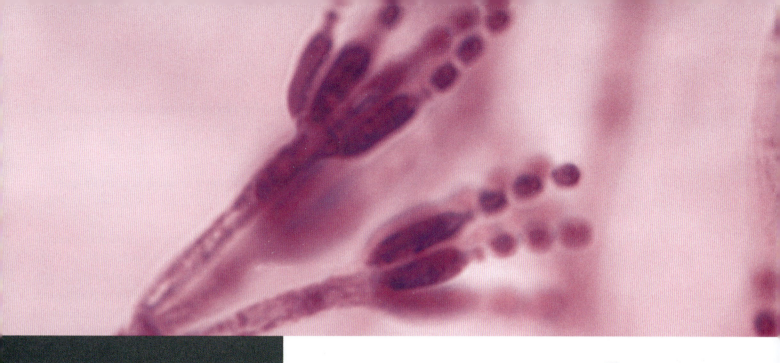

Many antibiotics are isolated from bacteria and fungi, which use them as defense molecules as they vie with other microbes for turf and food. The *Penicillium* hyphae in this micrograph send up stalks of conidiophores. The fungal spores (or conidia) are visible at the ends of the stalks.

Learning Objectives

- Understand how and why microbiologists look in the soil for antibiotic-producing microorganisms.

- Describe the taxonomy and characteristics of the *Streptomyces* genus.

- Explain how selective media can be used to find and isolate *Streptomyces* species.

- Isolate *Streptomyces* species and test isolates for their ability to inhibit a range of bacterial species.

LAB **35**

Antibiotic Producers in Soil and Selection for *Streptomyces*

Antibiotics, as we saw in Lab 17, are chemical control agents that have the very important property of selective toxicity; that is, these drugs are usually much more toxic to microorganisms than they are to us or our cells. Many of these chemicals are of microbial origin. In this lab, we're going to take a look at one way in which microbes capable of producing antibiotics have been discovered.

DISCOVERING ANTIBIOTICS: LOOKING FOR DRUGS IN ALL THE RIGHT PLACES

In 1928, Alexander Fleming, a bacteriologist, was working with *Staphylococcus aureus* cultures at St. Mary's Hospital in London. Fleming's crowded lab was filled with plates of *S. aureus*, and inevitably, some of these plates became contaminated with various fungi. There was certainly nothing new or special about contamination of bacterial plates by unwanted mold. This was a problem that had plagued microbiologists since Robert Koch's days, and Fleming himself was said to have observed:

> As soon as you uncover a culture dish, something tiresome is sure to happen. Things fall out of the air.

As a rule, when working with pure cultures, we ignore the need to prevent contamination at our peril. However, sometimes just a little bit of peril can lead to a Nobel Prize. In September 1928, Fleming spotted a gray-green mold growing on a plate of *S. aureus* colonies. He was about to discard the contaminated plate when he noticed that the area around the fungal colony was free of bacterial growth. Fleming reasoned that the mold must be producing and secreting an antibacterial chemical. And he was right.

Fleming saved and subcultured the mold. He quickly demonstrated that it secreted a yellow-pigmented chemical that killed many Gram-positive pathogens (it was ineffective against most Gram-negatives) **(Figure 35.1)**. The fungus was identified as a strain of *Penicillium notatum*, and the antibacterial chemical that it produced was dubbed "penicillin." Fleming also saved the original plate. It now resides in the British Museum.

Unfortunately, penicillin proved to be a somewhat unstable molecule, and the quantities produced by the original mold were too small to be of much clinical value. It would take the work of many doctors, pharmacologists, and biochemists in the late 1930s and early 1940s to turn penicillin into a wonder drug available to millions. In 1945, Fleming and two of these doctors and biochemists, Howard Florey and Ernst

A.

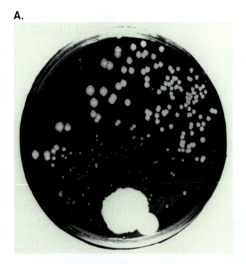

B.

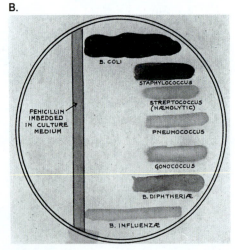

FIGURE 35.1 Fleming's culture plate.
A. A photo of Fleming's culture plate. **B.** Figure from Fleming's 1929 paper, "On the antibacterial action of cultures of a penicillium, with special reference to their use in the isolation of *B. influenza*." Fleming tested for inhibition by filling a furrow in a plate with a mix of agar and the yellowish broth from a *Penicillium* culture (labeled "penicillin imbedded in culture medium"). The figure shows inhibition of various pathogenic bacteria by penicillin. The inhibited species are all Gram-positive except for "gonococcus" (*Neiserria gonorrhoeae*), while the two resistant species, *E. coli* and *B. influenzae* (now called *Hemophilus influenzae*), are Gram-negative species.

Chain, shared the Nobel Prize in Physiology or Medicine for the discovery and development of penicillin.

Fleming's discovery in 1928, and the subsequent success in turning a microbial product into a life-saving drug, would inspire many others to turn to the fungal and bacterial worlds in the search for more antibiotics. Many hypothesized that the microbes were making antibiotics as a way to win the competition with other microbes for resources such as essential nutrients. In other words, antibiotic production by one microbial species might be a means of conducting chemical warfare against another microbial species. Given this thought, an obvious place to look for drug producers was in the soil, because soils hold large, diverse microbial communities, and competition in these environments is intense.

In the late 1930s, a soil microbiologist at Rutgers College (now Rutgers University) began to look for antibiotic producers among strains of soil-dwelling bacteria in the phylum Actinobacteria. This microbiologist, Dr. Selman Waksman, had already spent 25 years studying the actinobacteria. Based on his observations and the scientific literature, he had reason to think that these microbes would be an excellent source of antimicrobial drugs. Waksman established a working relationship with nearby Merck Pharmaceuticals so that any discoveries could be turned into products. Within a few years, he'd passed along a couple of potential antibiotics, some from fungi and some from actinobacteria. Unfortunately, these first microbial products were almost as effective at killing test animals as they were at controlling microorganisms. That is, these products were not selectively toxic enough to be of much clinical value.

By the early 1940s, Waksman's Rutgers lab team included several graduate students, including Albert Schatz, a student who was determined to find an antibiotic that would stop the many varieties of penicillin-resistant Gram-negative pathogens. In the summer of 1943, Schatz was searching for possible drug producers in soil samples taken from the leaf, straw, and manure compost found in a college barn. He mixed the samples with a little tap water, spread drops of the mix onto nutrient-rich agar plates, and then selected promising actinobacterial strains based on colony morphology and on the ability of the colonies to inhibit the growth of other microbes on the plates. In one case, he chose to subculture a green-gray colony of an actinobacterium called *Actinomyces griseus* for further study (*griseus* is Latin for "gray"). Around the same time, Schatz received a second *A. griseus* culture from another grad student, Doris Jones. Jones had isolated this strain of *A. griseus* from the throat of a sick chicken. (Yes, that's where she found it ... in the throat of a sick chicken. It was probably inhaled in barnyard dust.)

By October 1943, Schatz had clearly demonstrated that *A. griseus* strains produced an antibiotic, although the barn soil strain was a more potent producer than the chicken throat strain. This new antibiotic could inhibit or kill Gram-negatives, including *E. coli*, *Proteus*, *Salmonella*, and *Listeria*, as well as halting *Mycobacterium tuberculosis*, the cause of the lung disease tuberculosis. Later in 1943, *A. griseus* was reclassified as *Streptomyces griseus*, and its antimicrobial product was christened streptomycin (Figure 35.2).

Once it was clear that streptomycin inhibited a wide range of pathogens, most of the ensuing testing, developing, and producing of the drug was done at the Mayo Clinic and Merck Pharmaceuticals; Rutgers lacked the facilities to do this work. Within just a few years after its discovery, streptomycin was available to patients, and its impact in

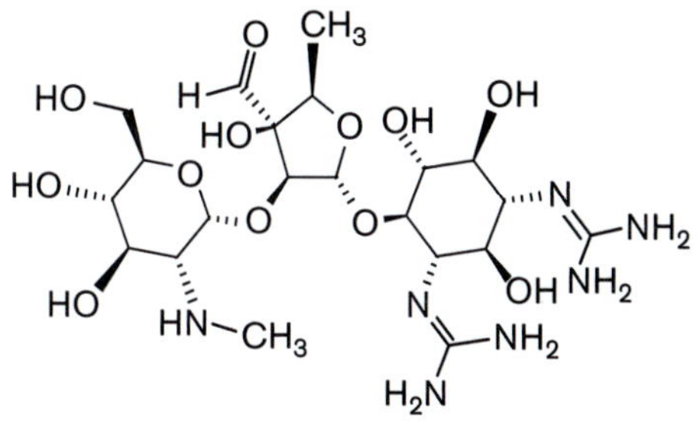

FIGURE 35.2 Streptomycin: An amino-modified trisaccharide that inhibits protein synthesis.

the treatment of deadly diseases such as tuberculosis was enormous. For the first time in history, TB could be effectively cured, and a disease that once killed millions began to fade into the background. (The emergence of drug-resistant strains, however, has led to a bit of a comeback for the "white plague.")

Dr. Selman Waksman went on to develop other *Streptomyces*-derived antibiotics, most notably neomycin in 1949. As a result of his drug discoveries, he won the Nobel Prize in Physiology or Medicine in 1952. However, despite his role in the discovery of streptomycin, Albert Schatz was not named as a co-recipient of this award. Waksman and Schatz would spend many years battling over how much credit each scientist should receive for streptomycin and over the appropriate allocation of royalties for the patents for this drug. Remember … scientists are human, too.

Streptomycin was the first broadly successful drug derived from *Streptomyces* species, and the work of Waksman and Schatz led numerous researchers to search around the world for more *Streptomyces* strains capable of producing antibiotics. Many of these quests were quite successful, and *Streptomyces* remains the most important genus in antibiotic production today. A partial list of the antibiotics derived from species in this one genus is given in **Table 35.1**. Most, though not all, of these drugs inhibit protein synthesis by binding to the bacterial 70S ribosome.

Given the significance of this genus, it's worth knowing a little bit about the nature and taxonomy of *Streptomyces*, and about how one might isolate these organisms from the soil.

Table 35.1	SOME ANTIBIOTICS DERIVED FROM *STREPTOMYCES*		
Antibiotic	***Streptomyces* Species**	**Original Source of *Streptomyces* Isolate**
Streptomycin	*S. griseus*	Soil from Rutgers (New Jersey) barn
Chloramphenicol	*S. venezuelae*	Soil from near Caracas, Venezuela
Daptomycin	*S. roseosporus*	Soil from Mt. Ararat, Turkey
Kanamycin	*S. kanamyceticus*	Soil from Japan
Lincomycin	*S. lincolnensis*	Soil from Lincoln, Nebraska
Neomycin	*S. fradiae*	Soil in Waksman's Rutgers lab, original source of soil unclear
Tetracycline	*S. rimosus* and *S. aureofaciens*	Soil from Missouri

Classification and description of *Streptomyces*

The important *Streptomyces* genus is a member of the phylum Actinobacteria, one of the largest of the bacterial phyla. The diverse species within this phylum are all Gram-positive, and they all have DNA with a high guanine + cytosine content. The cell shapes range from coccoids to fungus-like fragmenting hyphae. Actinobacterial species are found in the soil, in fresh and salt water, and living on plants and animals. In addition to *Streptomyces*, other genera in this phylum include the following:

Table 35.2 | ACTINOBACTERIA

Genus Name	Comments
Bifidobacterium	Part of normal intestinal flora
Corynebacterium	Includes species that cause diphtheria
Mycobacterium	Includes species that cause tuberculosis
Propionibacterium	Causes acne; used to make Swiss cheese
Micrococcus	Found on the skin; usually nonpathogenic
Brevibacterium	Causes foot odor; used to make Limburger cheese

Streptomyces is the largest genus in the phylum Actinobacteria, with hundreds of identified species. Members of this genus produce filamentous hyphae, exospores, and dry to leathery colonies, so when they were first studied and described, *Streptomyces* were mistaken for fungi. This error is reflected in the genus name, which is derived from two Greek words: *streptos*, meaning "twisted," and *mykes*, meaning "fungus" or "mushroom." Despite their confusing similarities to fungi, cells in this group lack nuclei, and their cell walls contain peptidoglycan, so they are clearly prokaryotic bacterial cells.

Species in the *Streptomyces* genus are very common in soil, where they play a critical role in improving soil fertility by decomposing large biological molecules such as starches and proteins. These bacteria also break down many compounds that are difficult to degrade, including cellulose, lignin, chitin, and even agar. In the process of decomposition, they produce and release a compound called **geosmin**, a volatile organic molecule that gives soil its distinctive earthy, musty odor, especially after it rains (**Figure 35.3**). The compound's name is derived from the Greek terms *geo*, meaning "earth," and *osmin*, meaning "smell"; literally translated, geosmin means "earth-smell." This odor can also be detected when *Streptomyces* colonies grow in petri dishes, and so it can help you to determine that you've isolated a *Streptomyces* strain in lab. (Note: Species in other actinobacterial genera can also produce geosmin.)

And *Streptomyces* species may also produce antibiotics, especially at certain key points in their life cycles.

Compared with most non-Actinobacteria, *Streptomyces* species have a relatively complex life cycle, which includes multicellular and spore production stages. Individual *Streptomyces* spores germinate to produce long, fungus-like filaments. These filaments, called vegetative hypthae, grow through the soil to form dense mycelial mats. As the concentration of nutrients in the immediate vicinity declines, the vegetative hyphae will grow up into the air to form aerial hyphae. This is the point in the life cycle when most of the antibiotic molecules are produced. So, while conditions are good and nutrients are abundant, the absorbed carbon and nitrogen are directed toward more vegetative hyphal growth. Then, as nutrients are depleted, cells divert more resources to producing the antimicrobial chemicals that may improve their survival in the crowded, tangled bank of the soil environment. As the aerial hyphae grow, they begin to twist and curl (remember, "twisted fungus"), and walls, or septa, appear within the filaments. These septa divide the hyphal filaments into chains of cellular compartments, and at the tip or distal end of a filament, the compartments develop into chains of exospores. The spores eventually separate from the "mother" mycelium and disperse to a better world to begin the cycle again. This cycle

FIGURE 35.3 Geosmin: A bicyclic (two ring) alcohol that gives soil its earthy aroma.

is similar to what is seen in many fungi, but again, these are prokaryotic organisms—they are bacteria.

Isolation of *Streptomyces*

To isolate *Streptomyces* species, we are going to plate a soil suspension on a slightly modified version of a starch-casein-nitrate agar that was developed in the 1960s for just this purpose. The composition of this medium is:

Soluble starch	10.0 grams
Casein	1.0 gram
Dipotassium phosphate	2.0 grams
Potassium nitrate	2.0 grams
Sodium chloride	2.0 grams
Magnesium sulfate ($MgSO_4 \cdot 7\ H_2O$)	0.05 grams
Calcium carbonate ($CaCO_3$)	0.02 grams
Ferrous sulfate ($FeSO_4 \cdot 7\ H_2O$)	0.02 grams
Agar	15.0 grams
Distilled water	1,000 ml

As mentioned above, *Streptomyces* species are notable for producing enzymes that degrade large biological molecules, and this medium favors and selects for such microbes in several ways.

Carbon and energy are supplied in the form of starch, rather than glucose or any other low-molecular-weight organic carbon molecules. Starch molecules are too large to be absorbed across cell membranes. However, if a cell can secrete amylase, then the starch can be converted into a pile of easily absorbed glucose molecules, and the cell will be provided with carbon and energy. Conversely, if a cell can't produce amylase, then it's going to go hungry. Thus, a medium with starch, but no glucose, favors amylase producers such as *Streptomyces* over numerous other bacterial species that lack amylases.

Furthermore, this medium uses casein, rather than peptones, as a nitrogen source. (Casein can also act as a carbon and energy source.) Recall that peptones are created by digesting proteins, so they supply easily absorbed amino acids. In contrast, like starch, casein molecules are too large to cross the cell membrane, and so can't be absorbed as is. If a cell doesn't secrete proteases capable of degrading casein, then casein cannot act as a nitrogen source for that cell. However, if a cell can secrete casein-degrading proteases, then the casein can be turned into a collection of absorbable amino acids. So a medium with casein, but no peptones, favors protease-producing *Streptomyces* over numerous other bacterial species that can't break down casein.

In addition, most *Streptomyces* are adapted to neutral or slightly alkaline soils, so potassium phosphate is added as a buffer to maintain a near-neutral pH; it can also act as a source of phosphorus. Potassium nitrate serves as a supplemental nitrogen source, and magnesium sulfate, calcium carbonate, and ferrous sulfate provide sulfur and trace metals. Beyond meeting basic nutritional needs, trace metals can promote formation of distinctive pigments.

There are many ways to use a starch-casein-nitrate agar to isolate *Streptomyces* species. In this lab, you will suspend fresh, rich soil in water, make a series of tenfold dilutions from this suspension, and then spread a 0.1-ml drop from each dilution on an agar plate. It can then take 3–5 days for the colonies to reach a usable size. On this medium, colonies of *Streptomyces* species are usually dry, leathery, or powdery in texture. In some cases, the dryness of the growth may make it a little difficult to transfer cells and spores from a colony to another plate. Different species produce colonies of different colors, and you may see colonies that resemble tiny white volcanoes or colonies that are gray, red, or yellow **(Figure 35.4)**. Some species also secrete yellow-brown or reddish pigments into the surrounding medium.

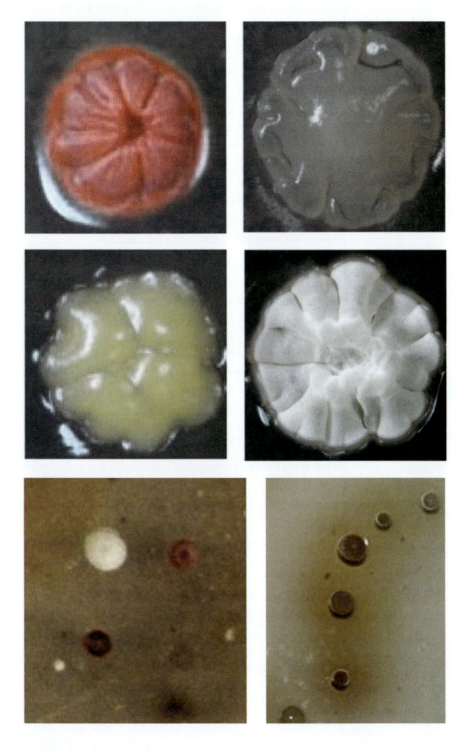

FIGURE 35.4 *Streptomyces* colonies.
Typical *Streptomyces* colonies may be red, gray, yellow, or white and may secrete yellow-brown pigment.

PROTOCOL

Day 1
MATERIALS

- 5 sterile 9.0-ml water blanks

- 4 starch-casein-nitrate agar plates

- 2 sterile 1.0-ml pipettes

- 2-ml (blue-barrel) pipette pump

- Small metal spatula or scoop

- Glass spreading rod

- Beaker of alcohol

- Fresh, rich soil (garden soil is good)

For steps 1–11, see **Figure 35.5**. You will prepare your tubes and plates, perform a serial dilution of the soil suspension, and then spread 0.1 ml from each dilution onto agar medium.

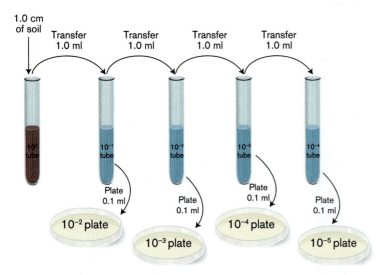

FIGURE 35.5 Preparing serial dilutions of the soil suspension and culturing the bacteria.

1. Label five sterile water blanks with your name and the date. The first tube should then also be labeled 10^0 (undiluted), the second tube should be labeled 10^{-1}, etc., until the fifth tube is labeled 10^{-4}.

2. Label four starch-casein-nitrate agar plates with your name and the date. The first plate should also be labeled 10^{-2}, the second plate 10^{-3}, the third plate 10^{-4}, and the final plate 10^{-5}.

3. Use the metal spatula to scoop about a cubic centimeter (a quarter-inch cube) of fresh, rich soil into the water blank labeled 10^0 (undiluted). Use a vortex mixer or flicks of your finger to suspend the soil in the water.

4. Use a sterile 1.0-ml pipette to transfer 1.0 ml of the undiluted soil suspension from the 10^0 tube to the 10^{-1} tube. Using

the same pipette, gently draw the water up and down to the 1.0-ml mark a couple of times to mix the suspension in the tube.

5. Use the same pipette as in step 4 to transfer 1.0 ml of the soil suspension from the 10^{-1} tube to the 10^{-2} tube. Again, gently draw the water up and down to the 1.0-ml mark a couple of times to mix the suspension in the tube. You don't have to change pipettes, because you're not doing a precise cell count.

6. Use the same pipette as in step 5 to transfer 1.0 ml of the soil suspension from the 10^{-2} tube to the 10^{-3} tube. Mix the suspension as in step 5.

7. Use the same pipette as in step 6 to transfer 1.0 ml of the soil suspension from the 10^{-3} tube to the 10^{-4} tube. Mix this suspension, too. Now you can discard the pipette.

8. With a new sterile 1.0-ml pipette, transfer 0.1 ml of soil suspension from the 10^{-4} tube to the 10^{-5} plate.

Recall that if you plate 0.1 ml, it reduces the cell population per milliliter by a factor of ten.

9. Using the same pipette, transfer 0.1 ml of soil suspension from the 10^{-3} tube to the 10^{-4} plate, then transfer 0.1 ml from the 10^{-2} tube to the 10^{-3} plate, and finally, transfer 0.1 ml from the 10^{-1} tube to the 10^{-2} plate.

As long as you move from more dilute to less dilute suspensions, you don't have to change pipettes between tubes.

10. Following the method demonstrated by your instructor, use an alcohol-sterilized spreader to spread the 0.1-ml drop to all areas of the 10^{-5} plate.

Instructions for using the spread plate method are also given in Lab 4.

11. Repeat step 10 with the 10^{-4} plate; then repeat with the 10^{-3} plate and the 10^{-2} plate.

As long as you move from the more dilute plates to the less dilute plates, you do not have to sterilize the spreader between plates.

12. Give the plates about 10 minutes to dry, then invert the plates and incubate all plates at 28°C–30°C for 2 days.

After 2 days, plates should be checked daily by students or instructors until the colonies are about 3–4 mm in size. Ideally, a few colonies will be pigmented or will be secreting pigments. At that point, plates can be refrigerated until used for day 2 procedures. (Total time required is typically 3–5 days.)

Day 2
MATERIALS

- 1 tube of sterile brain-heart infusion or glucose nutrient broth

- 3 starch-casein-nitrate agar plates

- Plates from day 1

1. Label three starch-casein-nitrate plates with your name and the date. Then label the first plate Isolate 1, the second Isolate 2, and the third Isolate 3.

2. Examine all of the plates inoculated with soil suspensions on day 1. You're looking for a plate with lots of well-separated colonies that are likely to be *Streptomyces* (see the introductory section for descriptions). Since the initial microbial population can't be predicted ahead of time, the best plate could be any of the four plates.

3. Dip a flame-sterilized inoculating loop into sterile broth to dampen the loop. This will help the loop pick up cells from the dry, powdery colonies.

4. Using the dampened sterile loop, touch one of the likely *Streptomyces* colonies a couple of times to transfer cells to the loop. In this step, your loop should **touch one and only one** isolated colony.

5. Streak the cells onto the starch-casein-nitrate agar plate labeled Isolate 1 in such a way as to produce more isolated colonies of the selected colony type (see Lab 3). Record a description of the chosen colony in the Results section. If a camera is available, you can also photograph the plates and then tape prints of the photos in the relevant locations in the Results section.

6. Repeat steps 3–5 using two other isolated colonies to streak the plates labeled Isolate 2 and Isolate 3. You now have a total of three plates with possible antibiotic producers.

7. Incubate all three plates at 28°C–30°C for 2 days.

   After 2 days, plates should be checked daily by students or instructors until the colonies are about 4–5 mm in size. Plates can then be refrigerated until used for day 3 procedures.

Day 3
MATERIALS

- 3 brain-heart infusion plates

- 3 glucose nutrient agar plates

- 1 tube of sterile brain-heart infusion or glucose nutrient broth

- Plates from day 2

1. Label three BHI plates and three GNA plates with your name and the date. Then label one BHI plate and one GNA plate Isolate 1, label another BHI plate and another GNA plate Isolate 2, and label a third BHI plate and a third GNA plate Isolate 3.

2. Dip a flame-sterilized inoculating loop into sterile broth to dampen the loop. This will help the loop pick up cells from the dry, powdery colonies.

3. Using the dampened sterile loop, touch an isolated colony on the Isolate 1 plate from day 2 to transfer cells to the loop.

Cells and spores of *Streptomyces* tend to cling to the dry colonies, so you should touch the colony a couple of times to ensure transfer of cells to the loop.

4. With the inoculated loop, streak in a single line across the diameter of the BHI plate labeled Isolate 1 (see **Figure 35.6** for the placement of the streak). Drag the loop back and forth along this line several times to deposit as many cells as possible. In this case, your goal is to generate a continuous line of growth across the diameter of the plate.

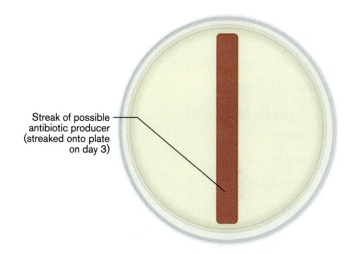

Streak of possible antibiotic producer (streaked onto plate on day 3)

FIGURE 35.6 Placement of streak of your antibiotic-producing microbe.

5. Return to the Isolate 1 plate from day 2 and look for another isolated colony that appears to be identical to the one you used for steps 3 and 4.

   If you were careful on day 2, every (or almost every) colony on the Isolate 1 plate should be of the same organism.

6. Using a flame-sterilized loop dipped into broth, touch this second colony with the loop a couple of times, and streak in a single line across the diameter of the GNA plate labeled Isolate 1. Again, drag the loop back and forth along this line several times to deposit as many cells as possible.

7. Repeat steps 2–6 using the colonies of Isolates 2 and 3 from day 2.

   When you're finished, you should have a total of six plates: three isolates streaked onto separate BHI plates and three isolates streaked onto separate GNA plates.

8. Incubate the six plates at 30°C for 2 days.

Day 4
MATERIALS

- Plates from day 3

1. After 2 days of incubation, examine the six plates with the soil isolates. If there is a solid line of colonial material across the plate, then return the plate to the incubator

without any further action. If the line of colonial growth is broken or is composed of isolated colonies, then sterilize a loop and drag the loop back and forth along the line to move cells into the spaces between the colonies.

2. Incubate all plates at 28°C–30°C for another 2–3 days. Plates should be used for day 5 procedures after a total of 4 or 5 days of incubation (4–5 days from the date of day 3).

Day 5
MATERIALS

- 4 10-ml BHI broths
- 4 sterile 1.0-ml pipettes
- 2-ml (blue-barrel) pipette pump
- Plates from day 4

Cultures (in BHI broth):

- *Bacillus cereus*
- *Escherichia coli*
- *Micrococcus luteus*
- *Staphylococcus epidermidis*

1. Mark all six plates from day 4 following the pattern shown in **Figure 35.7**. On the bottom of each plate, draw four lines perpendicular to the line of Isolate 1 growth (two lines on each side of the line of soil isolate growth), and label the lines *B. cereus*, *E. coli*, *M. luteus*, and *S. epidermidis*. The lines should stop 3 mm from the perpendicular line of soil isolate growth.

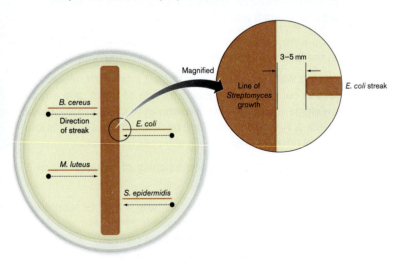

FIGURE 35.7 Streaking the test cultures.

2. For each of the four bacterial cultures, transfer 0.1 ml to a 10-ml BHI broth tube to dilute the culture approximately 1:100, or 10^{-2}. Use the diluted cultures for all subsequent steps.

3. Use a flame-sterilized inoculating loop to transfer a diluted bacterial culture from one of the BHI broths to the matching line on one of the six plates. Touch the loopful of cells to the end of the marked line farthest from the isolate growth. Carefully draw the loop along the marked line toward the soil isolate growth until the loop is about 3–5 mm (0.3–0.5 cm) away from the isolate growth. At this point, lift up the loop without dragging the loop back down the line.

The goal is to deposit bacterial cells up to the edge of the soil isolate growth without spreading cells from the isolate back along the line you just streaked.

4. Sterilize the loop. Repeat step 3 for all six plates and each of the four diluted bacterial cultures.

5. Incubate all six plates at 37°C for 24 hours.

Day 6
MATERIALS

- Plates from day 5
- Metric rulers

Without opening the plates, examine all plates and measure any zones of inhibition. That is, measure the gap or distance (in millimeters) from the edge of a soil isolate's growth to the edge of a given bacterial culture's growth. If it appears that there is no gap between the soil isolate and a given bacterium's growth, this could indicate that the bacterial species is resistant to any antibiotic produced by the isolate. However, you should take a close look at the colony morphology to be sure that the absence of the gap isn't due to the growth of the soil isolate (probable *streptomyces* isolate) into and along the line you followed when you streaked the bacteria. Record your observations in the Results section.

Name: _____   Section: _____

Course: _____   Date: _____

DESCRIPTION OF CHOSEN COLONIES (DAY 2), INCLUDING ANY DISTINCTIVE AROMAS:

Isolate 1:

Isolate 2:

Isolate 3:

INHIBITION ZONES FOR EACH ISOLATE AND BACTERIAL TEST SPECIES

MEASURE OF ZONES OF INHIBITION (DAY 6):		
BACTERIAL SPECIES	**ISOLATE 1**	
	BHI PLATE	GNA PLATE
B. cereus		
E. coli		
M. luteus		
S. epidermidis		

BACTERIAL SPECIES	ISOLATE 2	
	BHI PLATE	GNA PLATE
B. cereus		
E. coli		
M. luteus		
S. epidermidis		

BACTERIAL SPECIES	ISOLATE 3	
	BHI PLATE	GNA PLATE
B. cereus		
E. coli		
M. luteus		
S. epidermidis		

Name: _____   Section: _____

Course: _____   Date: _____

ANSWER QUESTIONS 1 AND 2 FOR EACH OF THE THREE SOIL ISOLATES

1. Did the isolate produce antibacterial chemicals? Cite data to support your conclusion.

2. If the isolate produced antibacterials, would you describe these products as being "narrow spectrum" or "broad spectrum" in their effects (see Lab 17)? Cite data to support your conclusions.

3. Overall, which of the three isolates appeared to be the most effective at controlling bacterial growth? Cite data to support your conclusions.

4. In this lab, we let the possible drug producers grow for about 5 days before adding the bacteria to check for secretion of inhibitory compounds. How might the results have been different if we'd added the bacteria within a day of streaking a line of possible drug producers? Would you expect to see more or less inhibition? Explain your reasoning.

5. How did the type of agar affect the results? Did the zones of inhibition for a given soil isolate and bacterial species differ between the BHI and GNA plates? If so, offer two or three possible explanations for this difference. That is, why should the medium matter?

6. Assuming that you've isolated a soil microbe capable of producing antibiotics, why aren't you rich now? A new antibiotic could be worth billions of dollars, so what stands between you and your dream mansion with its gold-plated lavatory fixtures? What hurdles stand between your observations in lab and a profitable medical product? Describe several things that can go wrong in the process of developing a new antibiotic.

7. Antibiotics from microbes such as *Penicillium* and *Streptomyces* species have saved millions of lives, but why have microbes been so nice to us? Could it be that *Penicillium notatum* has a fondness for *Homo sapiens*? Probably not. So, why do microbes produce antibiotics? Or, to put it another way: When, where, and why might natural selection favor antibiotic-producing microbial strains over non-antibiotic producers? Could there be a cost to antibiotic production, and if so, what types of environments might favor antibiotic producers over nonproducers? In short, how could antibiotic production have evolved?

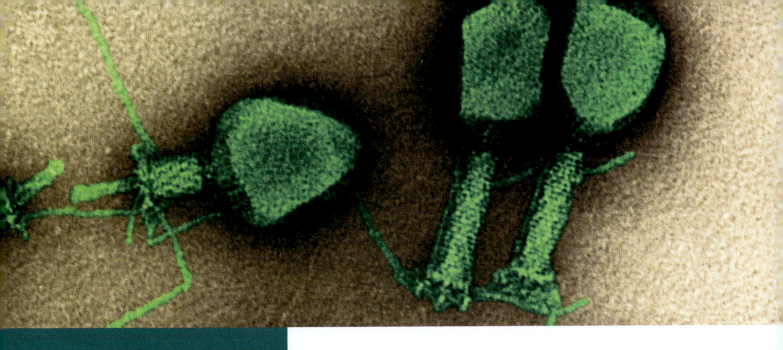

The T4 bacteriophage that infects *E. coli* is a complex virus made up of an icosahedral head packed full of DNA, a body, and six leglike structures.

LAB 36

Bacteriophages and Bacteriophage Titers

Learning Objectives

- Explain the differences between viruses and cellular microbes such as bacteria.

- Describe the host specificity and replication cycle of bacteriophages, particularly the T4 phage.

- Understand the steps used to determine a bacteriophage titer (the number of virus particles in a given virus suspension).

- Determine the titer of a bacteriophage suspension using a plaque assay.

THE NATURE OF VIRUSES

Viruses differ in several ways from the microbes we've examined up to this point:

1. **Viruses are acellular microorganisms.** They are not made of cells, they cannot be described as cellular life, and there is no such thing as a "viral cell." Viruses are particles that are composed of nucleic acid wrapped in a protein coat. Some types of viruses also have a phospholipid envelope with embedded proteins that surrounds the protein coat.

2. There are exceptions, but **most viruses are much smaller than even bacterial (prokaryotic) cells**. The majority of viruses range in size from about 10 to 500 nanometers (nm). A nanometer is 1/1,000 of a micrometer, so viruses are about 0.010–0.500 micrometers (μm) in size. By contrast, prokaryotic cells are about 1,000–3,000 nm (1 to 3 μm) in size, and a eukaryotic red blood cell is about 7000 nm (7 μm) in diameter. Objects smaller than about 200–400 nm in size cannot be seen with a light

microscope, so only the very largest viruses can be observed through a light microscope. Fortunately, viruses can be seen with electron microscopes.

3. **Viruses are obligate intracellular parasites.** Some types of bacteria are also obligate intracellular parasites, but this is an uncommon trait among cellular organisms. As obligate intracellular parasites, viruses must be inside of live host cells in order to copy their genes, replicate, and evolve; that is, they must be inside of host cells to carry out functions associated with being alive. Viruses lack certain structures, enzymes, and other components needed for independent existence, so they must enter host cells and use their genes to subvert the host cell machinery and direct it to make more viruses, (usually) at the expense of the health of the cell. By contrast, when viruses are outside of host cells, these "organisms" are completely inert or inactive collections of proteins, nucleic acids, and other biological molecules.

BACTERIOPHAGES

All viruses must infect host cells in order to replicate, but a given virus can typically infect only a narrow range of host species (host specificity will be explored in much more depth in Lab 37). This means, for example, that viruses that infect animal cells do not infect plant cells, and that viruses that infect plant cells do not infect animal cells. Still other viruses infect only bacterial cells.

We are familiar with the idea that viruses infect animal cells because we suffer the ill effects of animal viruses (colds, influenza, and measles, for example). However, the fact that even lowly single-celled bacteria may be afflicted with viruses is not as well known. The same tiny bacteria that may cause disease in humans are also vulnerable to attack from their own viral pathogens, offering support for Jonathan Swift's observation:

So naturalists observe, a flea …
Hath smaller fleas that on him prey;
And these have smaller fleas to bite 'em.
And so proceeds Ad infinitum.

On Poetry: A Rhapsody (1733)

Or as Augustus De Morgan later noted (probably with Swift's poem in mind):

Great fleas have little fleas,
Upon their backs to bite 'em,
And little fleas have lesser fleas,
and so ad infinitum.

A Budget of Paradoxes (1872)

Viruses that infect bacteria are called **bacteriophages**, or just **phages** for short. *Phage* is derived from the Greek word *phagein*, meaning "to eat" or "to have a share of food," and/or the Greek word *phagos*, meaning "eater of" or "glutton." So, bacteriophages are gluttons for bacterial cells. In the process of "eating the cell," or consuming the cell's nutrients and machinery to make more virus particles, the phages usually lyse and kill the cell.

One other note about the word *phage*. According to various dictionaries, the word may be pronounced two different ways. Some pronounce the word with the "a" sounding like the long "a" in "page," but others pronounce it with the "a" sounding like the ahh-sounding "a" in "far." Well, to quote the old George and Ira Gershwin tune, "you like po-tay-to, and I like po-tah-to, you like to-may-to, and I like to-mah-to; po-tay-to, po-tah-to, to-may-to, to-mah-to, let's call the whole thing off". [see Astaire and Rogers in *Shall We Dance* (1937)]

THE T4 BACTERIOPHAGE

There are numerous bacteriophage types, with a variety of shapes, host species, and life cycles, but here we're going to focus on the *E. coli*–infecting T4 phage, which we'll be working with in this lab. The T4 phage is constructed from many different proteins, which are assembled within a host cell into a particle with several distinct parts. These parts include a head, or **capsid**, and a **tail** consisting of a hollow core encircled by a contractile sheath, a spiked baseplate, and several tail fibers **(Figure 36.1)**. The viral genes are built of double-stranded DNA, which is stored and protected within the capsid.

T4 phages are **lytic** or **virulent** phages; this means that the activity of this virus in an *E. coli* host cell will result in the lysis and death of the host **(Figure 36.2)**. The phage life cycle begins with the attachment or adsorption of the virus to the cell wall of the host. Attachment or binding of the virus to a cell depends on a precise, shape-dependent fit between the tips of the tail fibers and a receptor on the surface of the host cell. Given the profile of the T4 phage, this process resembles the touchdown of the *Apollo* lunar module on a rod-shaped moon ("Tranquility Base here. The Eagle has landed"). In the case of the T4 phage, the host cell receptor is an outer-envelope porin protein, OmpC. If the OmpC receptor with its specific, T4-fitting shape is absent from the surface of a potential host cell, then the virus cannot bind to and infect the cell. This fact is part of the explanation for the host specificity of the virus (see Lab 37, for more details).

After docking with the host cell, the virus uses lysozyme stored in the base of the tail to weaken the cell wall. The sheath of the tail contracts, driving the tail's hollow

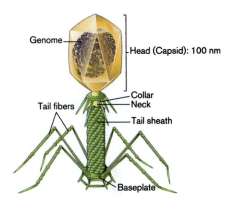

FIGURE 36.1 Structure of the T4 bacteriophage.

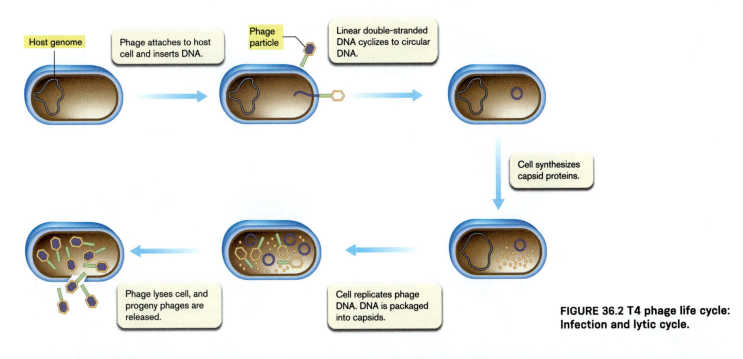

FIGURE 36.2 T4 phage life cycle: Infection and lytic cycle.

core through the cell wall. At this point, the viral DNA can move from the capsid into the cytoplasm of the host cell. In the cytoplasm, the viral genes will be transcribed and translated, and the resulting collection of viral proteins will (1) inhibit normal host cell function, (2) synthesize multiple copies of the viral genome, and (3) be assembled into new phage particles, with the new viral genomes packaged inside. As the new particles are being assembled, a viral gene for lysozyme is expressed, and the lysozyme acts to weaken the host cell wall. The combination of lysozyme activity and outward pressure from hundreds of new virus particles lyses the host cell, killing the cell and releasing the progeny virus. Total time for this viral life cycle? About 30 minutes under ideal conditions.

While the T4 phage is only capable of causing lytic, cell-killing infections, other types of bacteriophages carry out another type of infection, called a **lysogenic** or **temperate** infection. During lysogenic infections, viral genes are incorporated or integrated into the bacterial chromosome, new virus particles are not immediately produced, and host cells are not immediately lysed **(Figure 36.3)**. Instead, when the host cell divides, phage genes are replicated along with the host cell genes. While this process does not immediately produce new phage particles, the virus is still getting its genes copied, and in biology and evolution, that's how you win the game. The virus may be maintained in the DNA-only form for many bacterial generations, but eventually, something triggers the expression of genes involved in the lytic cycle. At that point, the virus reverts to the pathway that will ultimately produce hundreds of new particles and the death of the host cell.

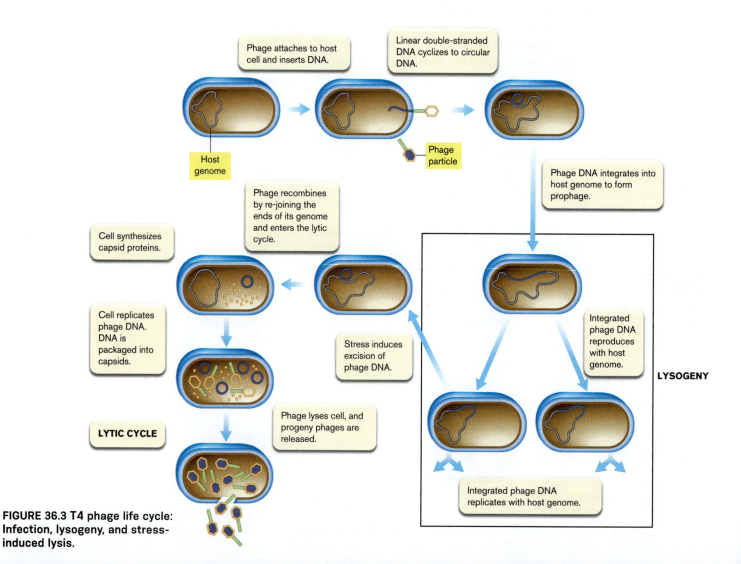

Phage attaches to host cell and inserts DNA.

Linear double-stranded DNA cyclizes to circular DNA.

Host genome

Phage particle

Phage DNA integrates into host genome to form prophage.

Phage recombines by re-joining the ends of its genome and enters the lytic cycle.

Cell synthesizes capsid proteins.

Cell replicates phage DNA. DNA is packaged into capsids.

Integrated phage DNA reproduces with host genome.

LYSOGENY

Stress induces excision of phage DNA.

LYTIC CYCLE

Phage lyses cell, and progeny phages are released.

Integrated phage DNA replicates with host genome.

FIGURE 36.3 T4 phage life cycle: Infection, lysogeny, and stress-induced lysis.

DETERMINING BACTERIOPHAGE TITERS

It's often quite useful to know how many virus particles are in a given suspension. This quantity or count is usually referred to as the **bacteriophage titer**. Such a measurement is analogous to a bacterial cell count, and the methods used to determine a virus titer are comparable to those we used in the quantitative plate count exercise (see Lab 4): A virus suspension is serially diluted and then plated to reveal the presence and activity of the individual virus particles. The revealed virus particles are then counted and multiplied by a dilution factor to produce a phage titer. This all seems simple and straightforward, but recall that viruses are completely inactive outside of host cells. So, how do we persuade the virus particles to reveal themselves?

The trick here is to provide a host for the virus. The T4 virus kills bacterial cells, so we can look for the loss or absence of host cells as evidence that a phage particle was present. In other words, if we don't see visible bacterial growth, then this reveals the presence of the invisible virus (invisible, that is, unless you have an electron microscope). It's like the old adage about jazz: It's not the notes you play, but the notes that you don't play that make the difference. When looking for phage, it's the bacteria that aren't there that matter.

Specifically, to determine the bacteriophage titer, we add a large number of *E. coli* cells to the various phage dilutions, then combine this mix with a soft agar and pour it onto a tryptic soy agar or BHI agar plate (either medium works well). Phage particles are nonmotile and are held in place once the soft agar solidifies, but *E. coli* cells can move a bit within the soft agar layer. After plating, the *E. coli* cells grow and multiply, and those cells form a visible, continuous haze or lawn of cells across the surface of the plate. However, if a cell in the lawn is infected by a phage particle, then that cell is killed, and hundreds of phage particles are released to infect adjacent cells. The infection process repeats itself for several rounds of virus replication, producing an expanding circle of phage particles and lysed *E. coli* cells.

The contents of the lysed cells are mostly water-soluble and dissolve into the water in the agar, and so **lysis creates an expanding clear spot in the lawn**. After thousands of cells have been killed, there will be a roughly circular cleared area, or **plaque**, in the lawn that is large enough to be seen with the unaided eye. Plaques can continue to expand in size for many hours, but the T4 phage replicates only in multiplying bacterial cells. With time, the *E. coli* cells fill the soft agar layer, exhaust the available nutrients, and stop multiplying. This stops virus replication and limits the size of the plaque.

Remember that the agar physically separated the particles in the initial virus population that were poured into the plate, and so it's likely that each individual plaque is the product of a single particle infecting a single bacterial cell. Therefore, the number of plaques in the lawn should be roughly equal to the number of phage particles that were applied to the plate. We can then multiply the number of plaques on a given plate by the dilution factor for that plate to estimate the number of particles in the initial stock suspension. Unfortunately, we can't be absolutely sure that each and every virus particle was completely separated from each and every other particle, and so phage titers are reported in **plaque-forming units** (PFU) per milliliter instead of particles per milliliter.

Phage titer calculations are very similar to those used in a quantitative plate count, except that we're counting plaques instead of colonies. As with a quantitative plate count, we need to generate "countable plates." There may be so many virus particles in some of our dilutions that all the plaques overlap because all the bacterial cells on the plate have been killed, and there is no bacterial lawn at all. Such dilutions are of little value to us. Furthermore, there may be other dilutions in which there is so little virus that there are no plaques; instead, there is a continuous, intact bacterial lawn. Again, such dilutions are of little value to us. For a bacteriophage titer, we're looking for dilutions that produce plates in which the plaque count falls between certain values. Specifically, we're looking for plates with 30 to 300 plaques. These are considered "countable plates" for the bacteriophage titer.

For the calculation of PFU/ml of the stock suspension, you will use (1) the same rounding rules, (2) scientific notation, and (3) equation that you used in the quantitative plate count experiment (in Lab 4):

When 1.0 ml of phage suspension produces countable plates, then the PFU/ml of the original, undiluted phage suspension equals:

$$\text{Number of plaques} \times \text{dilution factor}$$

- "Number of plaques" is number of plaques on countable plates after adjustment by rounding rules.
- "Dilution factor" is the total dilution factor for the tube that produced countable plates.

When 0.1 ml of phage suspension produces countable plates, then the PFU/ml of the original, undiluted phage suspension equals:

$$10 \times \text{number of plaques} \times \text{dilution factor}$$

- We multiply by 10 because number of plaques is actually plaques per 0.1 ml and final value must be in PFU per 1.0 milliliter.
- "Number of plaques" is the number of plaques on countable plates after adjustment by rounding rules. (see Lab 4).
- "Dilution factor" is the total dilution factor for the tube that produced countable plates.

PROTOCOL

Day 1
MATERIALS

- 5 tryptic soy agar or BHI agar plates
- 5 tubes of melted LB soft agar, 5.0 ml of agar per tube
- 1 9.9-ml water blank test tube
- 6 9-ml water blank test tubes
- 10 sterile 1.0-ml pipettes

- 2-ml (blue-barrel) pipette pump
- 48°C–50°C water bath

Cultures:

- *Escherichia coli* B: overnight culture grown in BHI broth
- T4 phage (10^6 to 10^8 PFU/ml)

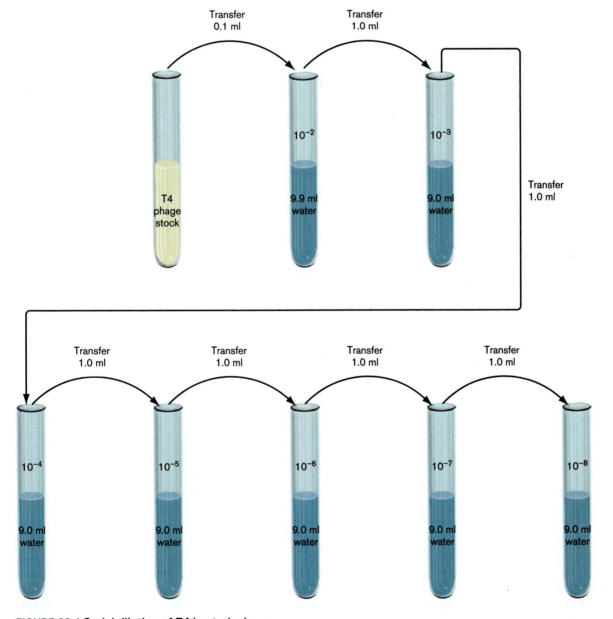

FIGURE 36.4 Serial dilution of T4 bacteriophage.

1. Label the tube containing 9.9 ml of water with your name, the date, and the label 10^{-2}.

2. Label the six tubes containing 9 ml of water with your name and the date. The first tube should also be labeled 10^{-3}, the second tube 10^{-4}, the third tube 10^{-5}, and so on until the last tube is labeled 10^{-8}.

3. Label the five agar plates with your name and the date. The first plate should be labeled 10^{-4}, the second plate 10^{-5}, and so on until the last plate is labeled 10^{-8}.

PREPARATION OF T4 PHAGE DILUTIONS (SEE FIGURE 36.4)

4. Gently swirl the T4 phage stock culture to evenly distribute the virus particles in the stock. Then use a sterile 1.0-ml pipette to transfer 0.1 ml of stock to the 10^{-2} tube (9.9-ml water blank). Using the same pipette, gently draw the water up and down to the 1.0-ml mark a couple of times to mix the phage into the water. Discard this pipette.

5. Using a second pipette, transfer 1.0 ml of suspended virus from the 10^{-2} tube to the 10^{-3} tube. Gently draw the water up and down to the 1.0-ml mark a couple of times to mix the phage into the water. Do not discard the pipette.

6. With the second pipette, transfer 1.0 ml of suspended virus from the 10^{-3} tube to the 10^{-4} tube. Gently draw the water up and down to the 1.0-ml mark a couple of times to mix the phage into the water. Now discard the second pipette.

7. Using a third pipette, transfer 1.0 ml of suspended virus from the 10^{-4} tube to the 10^{-5} tube. Gently draw the water up and down to the 1.0-ml mark a couple of times to mix the phage into the water. Do not discard the pipette.

8. With the third pipette, transfer 1.0 ml of suspended virus from the 10^{-5} tube to the 10^{-6} tube. Gently draw the water up and down to the 1.0-ml mark a couple of times to mix the phage into the water. Now discard the third pipette.

9. Using a fourth pipette, transfer 1.0 ml of suspended virus from the 10^{-6} tube to the 10^{-7} tube. Using the same pipette, gently draw the water up and down to the 1.0-ml mark a couple of times to mix the phage into the water. Do not discard the pipette.

10. With the fourth pipette, transfer 1.0 ml of suspended virus from the 10^{-7} tube to the 10^{-8} tube. Gently draw the water up and down to the 1.0-ml mark a couple of times to mix the phage into the water. Now discard the fourth pipette.

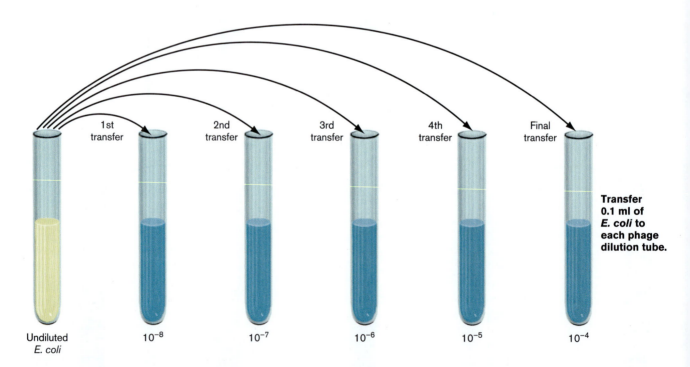

Transfer 0.1 ml of *E. coli* to each phage dilution tube.

Undiluted *E. coli* 10^{-8} 10^{-7} 10^{-6} 10^{-5} 10^{-4}

FIGURE 36.5 Addition of *E. coli* to bacteriophage dilutions.
Note that the addition of the cells proceeds in a series from the tube with the greatest phage dilution (10^{-8}) to the tube in which the phage is least diluted (10^{-4}).

ADDITION OF BACTERIA TO PHAGE (SEE FIGURE 36.5)

11. Using a new, sterile 1.0-ml pipette, draw up 0.5 ml of *E. coli* broth culture.

12. Starting with the 10^{-8} tube, add 0.1 ml of the *E. coli* culture to each of the T4 dilution suspensions in this order: 10^{-8}, 10^{-7}, 10^{-6}, 10^{-5}, and 10^{-4}. Do not add *E. coli* to the 10^{-3} and 10^{-2} tubes. When adding *E. coli* to each of the T4 dilutions, be careful to avoid touching anything inside the tube with the pipette tip. You do not want to transfer virus from tube to tube.

PLATING BACTERIA-PHAGE MIXTURES (SEE FIGURE 36.6)

13. Remove a tube containing 5.0 ml of soft agar from the water bath and use a 1.0-ml pipette to transfer 1.0 ml of virus-bacteria suspension from the 10^{-8} dilution tube to the soft agar tube.

14. Use the pipette to thoroughly mix and stir the suspension into the melted soft agar. Discard the pipette.

15. Pour the soft agar into the plate marked 10^{-8}. Gently swirl the plate to evenly distribute the melted agar across the surface of the agar in the plate.

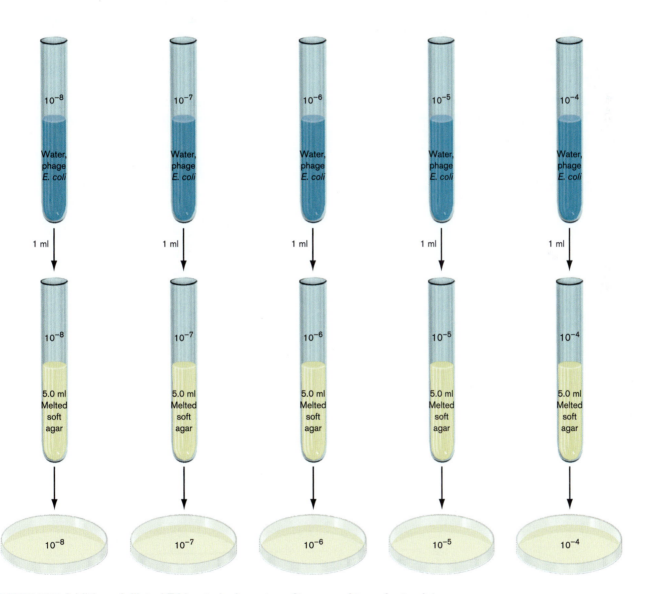

FIGURE 36.6 Addition of diluted T4 bacteriophage to soft agar and transfer to plates.

16. Wait about 15 minutes before inverting the plate. Soft agar contains a lower percentage of agar than most plate media and so takes a little longer to solidify.

17. Repeat steps 13–16 for the 10^{-7}, 10^{-6}, 10^{-5} and 10^{-4} dilution tubes.

18. Incubate all five plates at 37°C for 24 hours.

Day 2
MATERIALS

- Plates inoculated on day 1

1. Examine your plates for virus-created plaques (cleared areas) within the hazy, opaque lawns of *E. coli* (**Figure 36.7**). Count all of the plaques on all of your plates and record the values in the table in the Results section.

   If all of the plaques overlap, and all of the bacterial cells on the plate have been killed, there will be no bacterial lawn at all. If this is the case, you can record the count as TNTC ("too numerous to count").

   There may also be dilutions in which there was so little virus that there are no plaques, and instead,

Plaques

FIGURE 36.7 Bacteriophage plaques on a lawn of *E. coli*.

there will be a continuous, intact bacterial lawn. In this case, the plaque count is zero.

If all went well, there should also be plates with 30–300 plaques in the lawn. These are your countable plates.

2. After counting the plaques, dispose of all plates in the receptacle designated for trash to be autoclaved.

Name: _____   Section: _____

Course: _____   Date: _____

	Dilution Factor	Number of Plaques
10^{-4} plate		
10^{-5} plate		
10^{-6} plate		
10^{-7} plate		
10^{-8} plate		

Name: _____   Section: _____

Course: _____   Date: _____

1. Calculate the PFU/ml in the stock solution. Show your work.

2. Say you found a small *E. coli* colony in the middle of a plaque. What might explain the presence of this colony?

3. How might the initial concentration of *E. coli* (that is, the concentration of *E. coli* added to the phage dilutions) affect the outcome of the experiment? For example, what might you see if you'd used a 10^{-5} dilution of *E. coli* instead of a 10^{-2} dilution (0.1 ml of cells diluted in a 10-ml phage suspension)?

4. We can determine the number of *E. coli* cells per milliliter using a light microscope and a gridded slide to count the cells in a given volume. Why can't we determine the number of virus particles per milliliter using the same method?

5. We can also determine the number of *E. coli* cells in terms of CFU/ml by spreading the bacterial cells alone on a nutrient agar plate. Why can't we determine the number of virus particles in terms of PFU/ml by spreading virus particles alone on a nutrient agar plate?

6. Are viruses alive? Give one argument for concluding that viruses are living organisms and one argument for concluding that they are not living organisms.

7. The T4 bacteriophage always follows a lytic life cycle, but let's say you attempted to determine the phage titer of a temperate (lysogenic) phage instead of a lytic phage.

   a. What would you see on the plate if all of the temperate virus particles remained in a lysogenic state in a lawn of bacteria?

   b. What would be happening at the cellular level if some of the temperate virus alternated between rounds of lytic and lysogenic cycles within a lawn of bacteria? Explain how this might this change the appearance or number of plaques.

8. Say that you were in the yogurt production business. How or why might the phage titer procedure be useful to you?

9. How would you proceed if none of your plates fell within the guidelines of a countable plate?

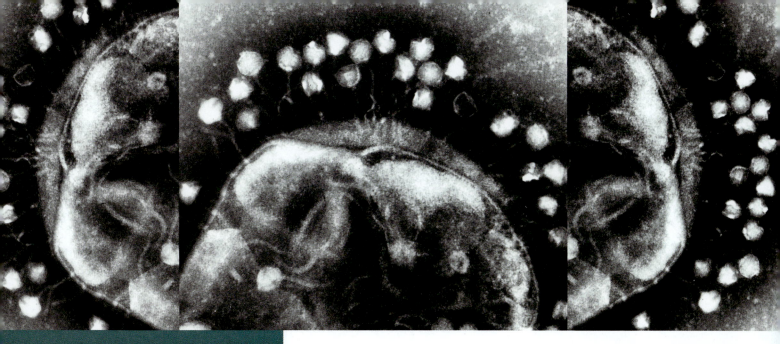

LAB 37

Bacteriophage Typing

Learning Objectives

- Understand the mechanism of bacteriophage attachment to the bacterial host cell surface.

- Describe how the specificity of attachment allows us to use phages to identify bacteria.

- Consider the history and potential future uses of phages in treating bacterial infections.

- Use T4 bacteriophage to identify an *E. coli* culture in a pool of unknown bacterial cultures.

THE IMPORTANCE OF FORMING ATTACHMENTS

The first step in infection of a bacterium by a phage is attachment or adsorption of the virus to the host cell. This step is the product of interactions between the following molecules found on the phage and the host:

Attachment proteins on the virus. These proteins are found on the surface of the virus. Like all proteins, each attachment protein has a specific three-dimensional shape, and in this case, it is a shape that will fit the host cell receptors.

Host cell receptors. In the case of Gram-negative species, receptors for phages are found on the surface of the outer phospholipid membrane or envelope. These receptors may be proteins, glycoproteins (proteins with sugars attached), or the lipopolysaccharides. To bind to the host cell, the proteins on the surface of the virus must interact with these receptors precisely in three-dimensional space. Without binding or attachment, there is no infection, so if a given cell lacks the receptor that fits a given virus's attachment proteins, that cell cannot be infected by that virus. This fact will become important when we get to phage typing.

If viruses are using host cell proteins, glycoproteins, and lipopolysaccharides to lock onto cells as part of the infection process, why do host cells have these receptors in the first place? Why leave the castle drawbridge down and the moat empty when there are barbarians in the neighborhood? The answer is that these receptors usually have important or essential roles to play in the normal function of the cell. That is, the cells often can't do without them. For example, the virus receptors may be bacterial cell transport proteins involved in the routine movement of essential molecules into and out of healthy cells. Unfortunately for cells, viruses have evolved attachment proteins that can bind to these important cell-surface components, and once they do, these cellular molecules also serve as virus receptors.

ATTACHMENT OR ADSORPTION OF THE T4 PHAGE TO SUSCEPTIBLE *E. COLI* CELLS

The infection of *E. coli* by the T4 bacteriophage begins with the attachment of the virus to the surface of the host cell. This process is initiated by the binding of a needlelike gp37 protein, found at the tips of the virus's long tail fibers, to two different receptors (Figure 37.1A):

An outer envelope protein called OmpC. OmpC is an integral membrane protein that normally functions as a porin or molecular channel. The OmpC porin, which is composed of three noncovalently bound amino acid chains, enables transport by passive diffusion of small, polar molecules across the nonpolar, phospholipid-based outer envelope or outer membrane of Gram-negative species. OmpC is a relatively nonspecific transporter, allowing the passage of several types of molecules, including glucose, ions, and other nutrients. The tip of the T4 phage's long tail fiber fits into a bowl-shaped cavity in the OmpC protein.

Specific lipopolysaccharide molecules. Lipopolysaccharide, or LPS, molecules are found in all Gram-negative outer cell membranes. However, there is considerable variation among Gram-negative strains and species in terms of the type and sequence of the sugars that are present at the ends of the LPS carbohydrate side chains. The T4 phage interacts with these sugars, but will bind to an LPS molecule only if certain specific sugars are present in a certain specific order. If mutations in the *E. coli* genome should change any of the enzymes required to add sugars to the LPS, this could change the type and order of sugars in the LPS side chain, which might alter the likelihood of T4 phage binding.

If either OmpC or the specific LPS molecule is absent due to mutation, then infection rates are greatly reduced, and if both are absent, then infection may be nearly impossible.

The binding of the phage's long tail fibers to the host cell's OmpC protein and LPS is initially reversible. However, if enough long tail fibers are attached to the cell surface, this

A.

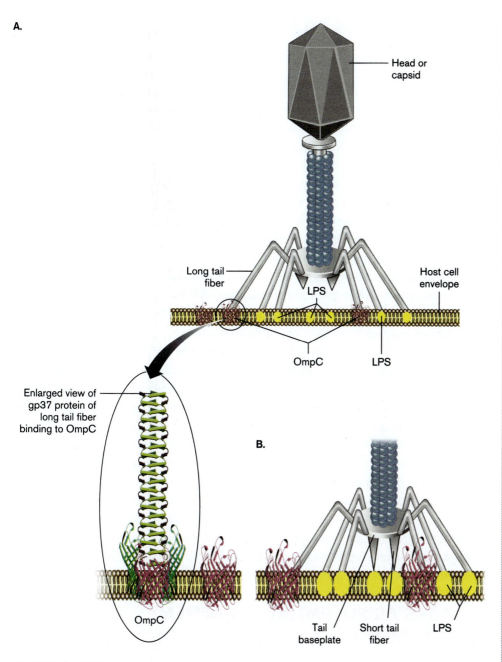

B.

FIGURE 37.1 T4 bacteriophage binding to *E. coli*.
A. First, the long tail fibers of T4 bacteriophage bind to *E. coli* OmpC protein and lipopolysaccharide (LPS) molecules, which are in the bacterial cell envelope. **B.** Second,
the phage's short tail fibers bind to more LPS molecules, strengthening the binding and
leading to the next steps of infection.

binding induces a change in the configuration (shape) of short tail fibers found on the
base plate of the tail. These short tail fibers rotate out and into a position from which
they can bind to other LPS molecules in the *E. coli* cell envelope **(Figure 37.1B)**. Once
the short tail fibers have attached to the lipopolysaccharides, the binding of T4 to *E. coli*
is irreversible. Now the tail sheath contracts, driving the hollow core of the tail through
the cell envelope and delivering the T4 phage genome into the cytoplasm of the host
cell. The remaining steps of the infection and replication cycle are described in Lab 36.

USE OF PHAGES TO TYPE OR IDENTIFY BACTERIAL STRAINS AND SPECIES

As we've seen, phage particles must bind to host cells in order to infect them, and this binding requires a precise, three-dimensional fit between virus proteins and host cell receptors. In addition, relatively small changes in host cell genes can lead to small changes in the shapes of the host cell receptors, and these changes can prevent phage binding, infection, and host cell death. So, two different bacterial strains or species could be quite similar genetically to each other, and yet one strain or species might be infected by a given phage while the other strain or species is unaffected by the same virus. (see Lab 27 for definitions of strain and species).

If you tried to differentiate closely related strains using biochemical assays such as phenol red carbohydrate assays, this approach might fail because closely related strains could give the same pattern of results for these tests. **However, if the strains differed in their susceptibility to a given type of phage, you could differentiate them by exposing each strain to that phage followed by examination of which cells survived and which died.** In broth cultures, this is as easy as observing which tubes are cloudy or turbid due to cell growth and which are clear due to lysis of the bacteria by the phage. The use of specific phages to identify different, specific bacterial strains and species is known as **phage typing**.

Phage typing is frequently used to identify the disease source in outbreaks due to bacterial pathogens, especially when the outbreak is caused by a widely distributed species with many strains. For example, an outbreak of food-borne illness might be linked to a *Salmonella* species, but there are a huge number of strains of *Salmonella*, and it's not unusual to find one or more of these different strains in several different types of food. In addition, on any given day, a large number of people nationwide could be suffering from salmonellosis, but these individuals would not necessarily have been infected with the same strains of *Salmonella*. Phage typing allows an investigator to show that a given set of cases is due to the same specific strain of *Salmonella*, and that this particular strain is present in a specific food item consumed by all the people with the illness, thus identifying the source of the outbreak.

PHAGES AS POTENTIAL TREATMENT FOR INFECTIONS

Bacteriophages were discovered about a hundred years ago, and almost from the moment of their discovery, doctors and microbiologists have thought about using phages to treat bacterial infections. In the 1920s and 1930s, there was much study of the possibility of using phages to treat infection by scientists such as Felix d'Herelle who studied the use of phage to treat intestinal diseases. However, this research produced mixed results. Then events overtook this line of inquiry. Interest in phage therapy faded with the development of sulfa drugs in the 1930s, large-scale production of penicillin in the 1940s, and the discovery of many other antibiotics in the 1950s. Antibiotics were easy to use and clearly very effective against a broad range of bacterial pathogens.

So, why waste time with phages? The short answer is that the spread of antibiotic-resistant strains of pathogenic bacteria has significantly reduced the effectiveness of antibiotic therapy in the treatment of certain bacterial diseases. When an infection is caused by a drug-resistant strain, an illness that was easily cured a few decades ago can be very difficult to control today. So, with the emergence of drug resistance, there is renewed and growing interest in the use of phages to prevent and treat disease. To paraphrase Paul Simon, "Hello, virus, my old friend, I've come to talk with you again." Phage therapy could have several advantages:

1. Bacteriophages do not infect human cells, and so these viruses do not directly harm our cells. Furthermore, there are fewer worries about toxicity or side effects than with the use of antibiotics and other chemicals. In short, phage therapy should be relatively safe.

2. Once in contact with host cells within the human body, phage particles can replicate in the host cells. As long as host cells are available, bacteriophages are an example of a "drug" that makes more of itself.

3. For reasons we've discussed, a given phage will have a very narrow, specific host strain or host species range; it will infect only a few strains or a few species of bacteria. So, a particular phage should also be harmless to the vast majority of bacterial species. As a result, a phage used to treat a specific infection should be safe for most species in the body's normal flora communities. These communities include hundreds of bacterial species that colonize various parts of healthy bodies, such as the skin and large intestine. Their presence actually protects us against disease, so we'd like to keep them happy, healthy, and present in our bodies. By contrast, antibiotics can greatly disrupt these communities, and this can lead to later, secondary infection.

However, there are also several significant problems and hurdles with the use of phages to treat bacterial infection:

1. The specificity, or narrow host species range, of phages is a double-edged sword. It has the benefits described above, but it also means that a given phage can be used to treat infections caused by only a few specific strains of bacteria. In contrast, most antibiotics can be used to treat many different illnesses caused by a wide range of bacterial species. With phages, to be effective, you would need to know the specific strain responsible for the infection so that you could match the phage to the bacterium. Or, alternatively, you would have to use a virus cocktail containing many different types of phages in the hope that one of the phage types in this "shotgun" would hit the target.

2. The human immune system doesn't distinguish between human viruses and bacteriophages. It doesn't differentiate between bad viruses and good viruses. The immune system will attempt to attack and eliminate phages, especially if they are injected into the bloodstream, where there is extensive immune surveillance. In addition to the destruction of the phage, there is also the possibility of a damaging overreaction by the immune system. This observation suggests that phage therapy would be most effective in treating diseases in which the bacteria are living in the intestines or on the surface of the skin.

3. Naturally occurring organisms that replicate without human intervention are considered "products of nature," and so they usually cannot be patented. In the absence of the commercial advantages provided by patent protection, companies are rarely

interested in developing a given treatment. However, it's possible that if a phage had been genetically altered by human engineers, it could then be protected by a patent.

4. In addition to evolving drug resistance, bacteria can evolve resistance to phages. For example, mutations could lead to changes in the shape of host cell receptors used by viruses, which could reduce the ability of the virus to bind to the host cell. On the bright side, viruses evolve, too, and a change in the shape of tail fiber proteins could restore the ability of a phage to attach to a host cell.

PROTOCOL

Day 1

MATERIALS

- 6 sterile 1.0-ml pipettes
- 2-ml (blue-barrel) pipette pumps
- 5 tubes of sterile BHI broth
- Safety glasses
- Disposable gloves

Cultures:

- T4 phage (10^6 to 10^8 PFU/ml)
- Stock cultures: BHI broth cultures labeled A, B, C, D, and E

Each of the five BHI broth cultures contains one of the following bacteria: *Escherichia coli* B, *Proteus vulgaris*, *Citrobacter freundii*, *Enterobacter aerogenes*, or *Staphylococcus epidermidis*.

IMPORTANT: *Proteus vulgaris* is a biosafety level 2 species, so gloves and safety glasses must be worn throughout this lab.

---

1. Label five sterile BHI broths with your name and the date. The first tube should also be labeled A, the second tube B, and so on, until the fifth tube is labeled E.

2. Use a sterile 1.0-ml pipette to draw up 0.5 ml of T4 phage suspension, and from this aliquot, add 0.1 ml of the phage suspension to each of the five sterile BHI broths.

   Be sure that you don't do anything that would introduce bacteria into these broths at this point in time. Follow practices that would maintain the virus as a pure culture of phages in each of the broth tubes.

3. Use a new, sterile 1.0-ml pipette to transfer 0.1 ml of Stock Culture A to the BHI broth labeled A. Use a second sterile 1.0-ml pipette to transfer 0.1 ml of Stock Culture B to the BHI broth labeled B. Continue this pattern until you've inoculated BHI Broth E with Stock Culture E using a different pipette for each culture.

4. To mix the bacteria and the phage particles, gently flick the bottom of each tube to swirl the broths in the tubes.

5. Incubate all broths at 37°C for 24 hours.

Day 2

1. Examine Broths A through E, and for each broth, describe what you see with terms such as "cloudy," "turbid," and "clear." Record your observations in the Results section.

2. After recording your results, dispose of all tubes in the receptacle designated for trash to be autoclaved.

Name: _____   Section: _____

Course: _____   Date: _____

DESCRIPTION OF GROWTH IN TUBES

Culture A:

Culture B:

Culture C:

Culture D:

Culture E:

Name: _____ Section: _____

Course: _____ Date: _____

1. Which tube contained *E. coli* B?

2. How do you know that this tube contained *E. coli* B? That is, what are the observations that support your conclusions? How is this evidence related to what you observed in Lab 36?

3. How do you know that the other tubes did not contain *E. coli* B? That is, what is the evidence that supports your conclusions?

4. How does our knowledge about the specificity of the attachment of phages to bacteria allow you to draw the conclusions that you did for Questions 1–3?

5. What are the specific roles of OmpC and LPS in the attachment of phage T4 to the bacterial cell envelope?

6. What if you examined a tube known to contain *E. coli* B, and that tube appeared to be cloudy after a 24-hour incubation? Give three possible explanations for this observation.

7. Do a little searching on the Internet and describe a specific example of the successful use of phages to treat a specific bacterial infection. What did you find interesting about this case, and what results or outcomes in this study led researchers to conclude that the phage treatment worked? Your example should include a reference to a research journal paper in which treatment effectiveness was assessed under controlled conditions.

8. Was there anything about the case described in Question 7 that made success more likely; that is, why do you suppose those who developed this treatment thought ... "Hey, this just might work this time?"

9. Human cases of salmonellosis caused by *Salmonella enteritidis* can often be traced to the consumption of *S. enteritidis*–contaminated eggs and egg products. In the 1990s, the USDA and CDC (Centers for Disease Control and Prevention) developed a traceback program designed to identify the particular chicken flock that was the source of the eggs causing a particular *S. enteritidis* outbreak. The agencies needed to test the effectiveness of this traceback program; specifically, they wanted to know if various salmonellosis outbreaks could be linked to eggs and if the traceback program could correctly identify the specfic *S. enteritidis* source.

   So, researchers for the USDA and the CDC turned to phage typing. They phage-typed numerous *S. enteritidis* strains isolated from several sources: human patients in several different egg-associated salmonellosis outbreaks; the environments of poultry operations implicated in the different outbreaks; and the organs of selected chickens from the same poultry operations.

The results of their study (S. Altekruse, et al., 1993, *Epidemiol. Infect.* 110:17–22) are shown in **Table 37.1** below, in which "PT" stands for phage type and the different phage types are identified by numbers (8, 13a, and so on). Examine the results and decide (a) if the evidence supports the hypothesis that eggs are linked to human cases of salmonellosis and (b) if it supports the conclusion that the traceback program worked. Explain your conclusions.

Table 37.1. PHAGE TYPE (PT) OF *SALMONELLA ENTERITIDIS* ISOLATES FROM HUMAN OUTBREAK AND THE EGG OPERATION IMPLICATED BY TRACEBACK

Flock	Outbreak Location	Month (1990)	Human Outbreak PT	Poultry Environment PT	Internal Organ PT
A	Nashville, TN	June	8	8	Negative
B	Delmar, DE	Sept.	8	8,23	8
C	Chicago, IL	Dec.	13a	13a,23,28	13
D	Versailles, KY	Aug.	8	8,13a.23,28	8,13a,13
E	Chicago, IL	Oct.	8	8,13,23	8
F	Jefferson Co., TX	Oct.	8	8,13a,23,28	8,23,28
G	Woburn, MA	Jan.	14b	14b	14b
G	Suffield, CT	Feb.	14b	14b	14b
H	Tarreyton, NY	Mar.	8	3,8,13,13a,23	8,23
J	Linthicum, MD	Mar.	8	8,23,14b,13,34	8
J	Nazareth, PA	May	8	8,23,14b,13,34	8
J	Fallston, MD	July	8	8,23,14b, 13,34	8
K	Bristol, CT	June	8	8,13a	2,8,13a,23
L	Eastern Pennsylvania	June	8	*	8,23
M	Eatonton, NJ	Sept.	8,34	8,23	†
X	Dauphin, PA	Dec. (1991)	8	2,8,23	8
P	Gaithersburg, MD	June	8	8,13a,22,23,34	8,13a,28
Q	Syracuse, NY	May	2	2	2

* Hens' organs were tested without environmental sampling because the flock was implicated in three outbreaks.

† Flock slaughtered before the hens were tested.

Using Glass and Plastic Serological Pipettes

Pipettes are used to transfer precise and accurately measured volumes of liquid. In this course, you will do several labs that require the use of serological pipettes, so it is important that you understand and apply basic principles for pipette use. Serological pipettes are also known as "blow-out pipettes" because **accurate delivery of a given volume of liquid depends on blowing out the last drop of liquid from the tip of the pipette** with a bulb, pipette pump, or pipette aid. In other words, if the exact volume that needs to be transferred has been drawn up to the appropriate line in the pipette, you should be certain that no liquid remains in the pipette at the end of the transfer.

PIPETTE VOLUMES AND GRADUATIONS

Serological pipettes come in a variety of volumes; in this course, you will be using 1.0-ml and 10-ml pipettes.

1.0-ml pipettes

1.0-ml pipettes are usually marked at the top with the words "1 ml in 1/100." This phrase indicates that the pipette holds a total volume of 1.0 ml and is subdivided into 1/100 or 0.01-ml graduations. The 0.01-ml graduations are the unlabeled hash marks that run one-eighth to one-quarter of the way around the pipette's circumference. The 0.1-ml graduations are marked with lines that wrap halfway around the circumference of the pipette, and these lines are labeled .1, .2, .3, and so forth (**Figure A1.1**).

A 1.0-ml pipette will have a line marked "0" near the top end (the top is the cotton-plugged wider end). When liquid is drawn into the pipette up to the 0 line, the pipette holds 1.0 ml of liquid (**Figure A1.2**). It may seem odd that drawing liquid up to a line marked 0 means that the pipette holds 1.0 ml of liquid, but when pipetting, the 0 refers to the fact that 0.0 milliliters have been dispensed from a "full" pipette (a pipette holding 1.0 ml).

If you start with the liquid drawn up to the 0 mark, and you then empty the pipette to the first 0.1-ml graduation, a line marked .1, then you have now dispensed or delivered 0.1 ml. Empty the pipette to the .2, and you've dispensed 0.2 ml; empty the pipette to the .5 line, and you've dispensed 0.5 ml; and so on. Empty the entire pipette, blow out the last drop, and you've dispensed 1.0 ml.

What if you need to measure out and dispense a volume of less than 1.0 ml? In this case, you need to remember that **the pipette scale is designed to tell you how much you've dispensed** rather than how much you've drawn into the pipette at the beginning. This means that the scale is inverted, so you will need to subtract the volume you wish to dispense from 1.0 to figure out which line you're looking for as you draw up the liquid. So, for example, if you wish to draw up and dispense exactly 0.9 ml, then you will draw the liquid up to the .1 or 0.1-ml line (1.0 − 0.9 = 0.1), because when the liquid is at the 0.1-ml line, the pipette is holding 0.9 ml (Figure A1.2). As another example,

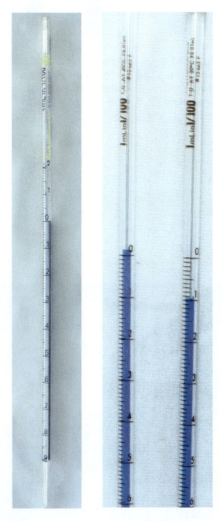

Figure A1.1
A 1-ml serological pipette.

Figure A1.2
1-ml serological pipettes holding 1.0 ml (left) and 0.9 ml (right) of liquid.

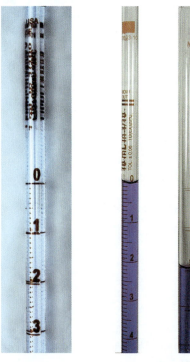

Figure A1.3 Detail of a 10-ml serological pipette.

Figure A1.4 10-ml serological pipettes holding 10 ml (left) and 8 ml (right) of liquid.

to draw up and dispense 0.1, you would draw the liquid up to the .9 or 0.9-ml line (1.0 − 0.1 = 0.9), because at that point, the pipette would be holding 0.1 ml.

10-ml pipettes

10-ml pipettes are usually marked at the top with the words "10 ml in 1/10." This indicates that the pipette holds a total volume of 10 ml and is subdivided into 1/10 or 0.1-ml graduations. The 0.1-ml graduations are the unlabeled hash marks that run one-eighth to one-quarter of the way around the pipette's circumference. The 1.0-ml graduations are marked with lines that wrap halfway around the circumference of the pipette, and these lines are labeled 1, 2, 3, and so forth (**Figure A1.3**).

A 10-ml pipette will have a line marked "0" near the top end. When liquid is drawn into the pipette up to the 0 line, the pipette holds 10 ml of liquid. If you start with the liquid drawn up to the 0 mark, and you then empty the pipette to the first 1.0-ml graduation, a line marked 1, then you have now dispensed or delivered 1.0 ml. Empty the pipette to the 2, and you've dispensed 2.0 ml (**Figure A1.4**); empty the pipette to the 5 line, and you've dispensed 5.0 ml; and so on. Empty the entire pipette, blow out the last drop, and you've dispensed 10 ml.

When measuring and dispensing volumes of less than 10 ml, you may have to subtract the volume you wish to dispense from 10 to find the appropriate draw line. For example, to dispense 1.0 ml, you would draw up the liquid to the 9-ml line. However, 10-ml pipettes have a greater diameter than 1.0-ml pipettes, and so there is room for more than one scale. Therefore, **many 10-ml pipettes have two scales**, one for dispensing and another drawing up. In this case, when you are drawing up a liquid to a given volume, you can use the scale that runs from 1 or 1 ml at the bottom to 10 ml at the top, thus skipping the math. The 10-ml line at the top will also be labeled 0 to mark the zero point when dispensing from a full pipette. For example, when using this second scale, if you wish to draw up and dispense 1.0 ml, you just draw the liquid up to the 1-ml line.

MAINTAINING STERILITY AND PREVENTING CONTAMINATION WHEN HANDLING PIPETTES

In almost every case in which you will use pipettes in the microbiology lab, it is very important to be aware of the need to maintain sterility and to use techniques that prevent contamination when handling pipettes and transferring liquids. Sterile glass pipettes are usually provided in metal canisters, while plastic pipettes are usually individually wrapped in sterilized paper sleeves.

Sterile glass pipettes in metal canisters

1. Lay the metal pipette canister <u>on its side</u> before you open it (**Figure A1.5**). If a canister is opened while it is standing on its end, it's much more likely that contaminating microbes will fall into the canister while the sterile pipettes are being removed.

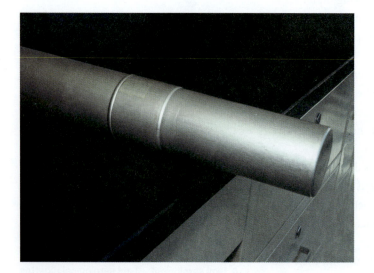

Figure A1.5 Pipette canister on its side prior to opening.

2. With the canister lying on its side, remove the metal canister top, and hold the top in one hand while withdrawing the pipette with your other hand **(Figure A1.6)**.

3. Remove a single pipette by gripping the <u>very top</u> of the pipette with your thumb and forefinger. Pull the pipette straight out of the canister, and avoid touching the remaining pipettes in the canister.

4. Replace the metal cap as soon as possible to reduce the probability of the remaining pipettes being contaminated.

5. **Hold the pipette you've selected at or near the top** to prevent contamination of the part of the pipette that will come into contact with liquids to be transferred **(Figure A1.7)**.

Figure A1.6 Withdrawing a single pipette without contaminating the tip.

6. Think of the lower end of the pipette as you would think of a flame-sterilized loop, and **never put the pipette down on the bench top!** If the lower end of the pipette touches any

nonsterile surface, discard the pipette and start over.

7. When you are finished with the pipette, place it in a used pipette canister filled with disinfectant. Used pipettes will be washed and sterilized before reuse.

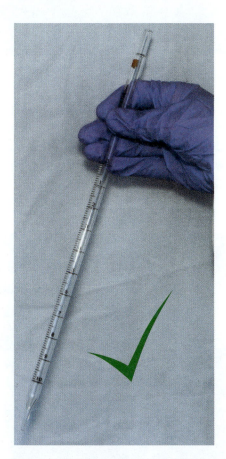
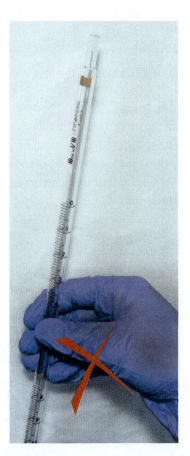

Figure A1.7 The right way to hold the pipette (left), and the wrong way to hold the pipette (right).

Sterile plastic pipettes in paper wrappers

1. Remove a sterile plastic pipette from its paper wrapper **(Figure A1.8)** in a manner that minimizes the probability that the <u>lower part</u> of the pipette will be contaminated.

2. Orient the wrapped pipette with the top pointing up, and use two hands to peel apart the two halves of the wrapper until the wrapper is peeled back **for about a quarter of the length of the pipette**.

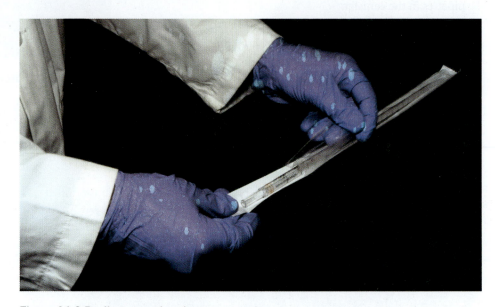

Figure A1.8 Peeling open the pipette wrapper.

3. Holding the <u>still-wrapped</u> bottom part of the pipette in one hand, grip the <u>very top</u> of the pipette with the thumb and forefinger of the other hand, and pull the pipette straight out of the wrapper **so that the bottom part of the pipette touches only the inside of the wrapper** as it's withdrawn **(Figure A1.9)**. That is, the pipette should be removed in such a way as to prevent the bottom half of the pipette from touching any nonsterile surfaces.

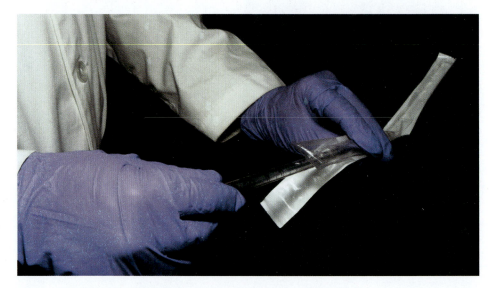

Figure A1.9 Removing the pipette without contaminating the tip.

4. **Hold the pipette you've selected at or near the top** to prevent contamination of the part of the pipette that will come into contact with liquids to be transferred **(Figure A1.10)**.

5. Think of the lower end of the pipette as you would think of a flame-sterilized loop, and **never put the pipette down on the bench top!**

6. When you are finished with the pipette, place it in the receptacle designated for trash to be autoclaved.

PIPETTING DEVICES AND THEIR USE

No one should ever use their mouths to pipette in the microbiology laboratory. There are several devices available that make mouth pipetting completely unnecessary. You will use one or more of these devices, which include rubber bulbs, plastic-barreled pipette pumps, and electric-powered pipette aids.

Pipette pumps are widely used in teaching labs. Here we describe the transfer of liquids using a sterile pipette held in a pipette pump, but most of this description also applies to the use of bulbs or powered pipette aids.

1. Remove a sterile pipette from a metal canister or paper wrapper as described previously.

2. Hold the top end of the pipette in one hand. **Always be very careful to keep the bottom end of the pipette from touching any nonsterile surface.**

3. Pick up the pipette pump with your free hand, and with a <u>slight</u> amount of pressure and a little bit of twisting, <u>gently</u> press the top end of the pipette into the pipette-holder opening of the pipette pump **(Figure A1.11)**.

   The pipette must be held securely by the pipette pump in such a manner as to create an airtight seal between the pipette and the pump. However, when seating a glass pipette in a pipette pump, **it is essential that you do not exert too much pressure**, because glass pipettes can snap, and the jagged edge of the broken pipette can inflict a deep cut in the palm of your hand.

4. Once the pipette is securely in place in the pump, hold the barrel of the pump between your palm and your first three fingers with the bottom end of the pipette pointing downward. Again, always be very careful to keep the bottom end of the pipette from touching any nonsterile surface.

5. With your free hand, pick up the test tube containing the liquid to be transferred, and remove the cap using the crook of the pinkie finger of the pipette-holding hand **(Figure A1.12)**. The act here is very similar to the act of removing the cap from a test tube while holding a sterilized loop, except that the pipette and pump are a little more difficult to handle than a sterilized loop. As with loop transfers, **you should hold the test-tube cap in your pinkie finger until it is time to replace the cap**.

Figure A1.10 Proper way to hold the pipette.

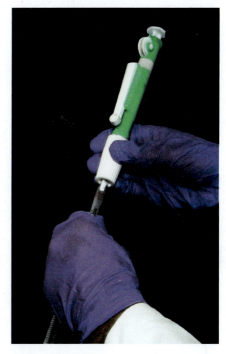

Figure A1.11 Inserting the pipette into the pipette pump.

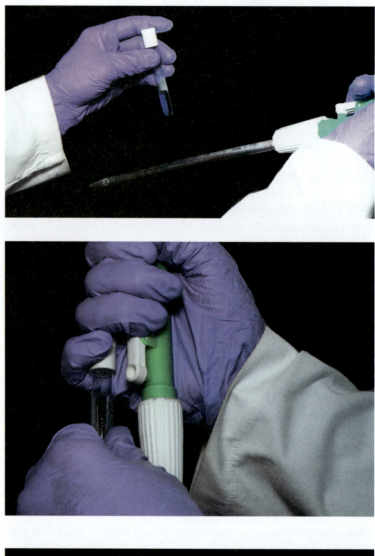

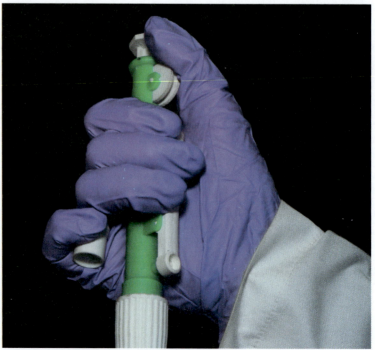

Figure A1.12 Preparing to transfer liquid from a test tube.

6. If the test tube is glass, then briefly flame the lip of the open test tube, and insert the pipette into the test tube without touching any of the outer surfaces of the tube **(Figure A1.13)**. <u>Do not</u> flame plastic test tubes; they will melt!!

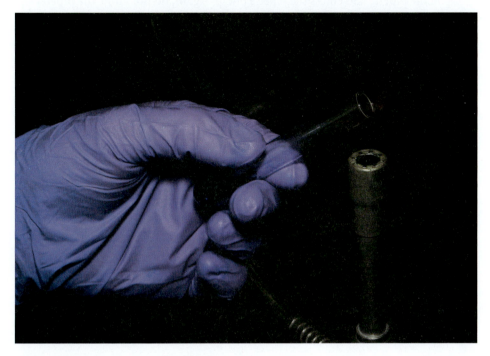

Figure A1.13 Flaming the test tube lip.

7. Lower the pipette into the liquid in the tube until the tip is below the surface of the liquid. Use your thumb to rotate the wheel on the side of the pump; this will draw liquid up into the pipette **(Figure A1.14)**.

 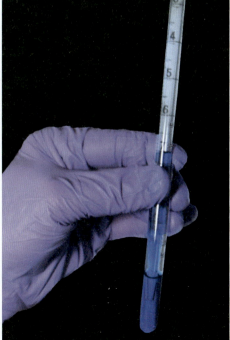

Figure A1.14 Drawing liquids.
Rotate the pump wheel (left) to draw the liquid out of the tube (right).

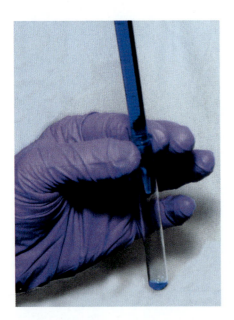

Figure A1.15 Dispensing liquid into a sterile test tube.

8. Draw the liquid into the pipette until it reaches the desired line, and then carefully withdraw the pipette from the tube **(Figure A1.15)**. <u>Never</u> draw so much liquid into a pipette that it soaks into the cotton plug at the top of a plastic pipette or exits the top of the pipette and enters the pipette pump. **Again, do not allow the pipette tip to touch the outside of the test tube or any other nonsterile surface.**

9. Briefly re-flame the lip of the test tube and replace the cap.

10. Pick up the test tube that is to receive the transferred liquid and remove the cap using the crook of the pinkie finger of your pipette-holding hand.

11. Flame the lip of the open test tube, and insert the pipette into the test tube **without touching any of the outer surfaces of the tube**.

12. Use your thumb to rotate the pump wheel in the opposite direction (opposite from the direction used to draw up the liquid) to dispense the liquid into the tube. **If the full volume is to be dispensed, rotate the wheel until all of the liquid is blown out of the tip of the pipette.**

13. Carefully withdraw the pipette, briefly re-flame the lip of the test tube, and replace the cap.

14. Dispose of the used pipette in the appropriate container.

USING DIGITAL MICROPIPETTORS

Digital micropipettors and micropipette tips are designed to accurately measure and transfer <u>much smaller volumes</u> than are normally measured and transferred by serological pipettes. These volumes are usually <u>expressed in microliters (µl)</u> instead of milliliters (ml); 1 microliter is 1/1,000 of 1 milliliter. For example, these devices might be used to measure and dispense 10 microliters (0.01 ml) or 20 microliters (0.02 ml). In this course, you will usually be able to do your work with serological pipettes. However, in some labs, such as the PCR lab (Lab 20), you will need to measure very small volumes of reagents accurately, and so will require digital micropipettors.

MICROPIPETTOR VOLUMES

A given micropipettor size covers a specific range of volumes. Most labs will have available three types or three sizes or micropipettors (**Figure A2.1**) covering three volume ranges:

Size	Range of volumes measured
Large volume	100–1,000 microliters
Medium volume	10–100 microliters **OR** 20–200 microliters
Small volume	0.5–10 microliters **OR** 0.5–20 microliters

Figure A2.1 Digital micropipettors.

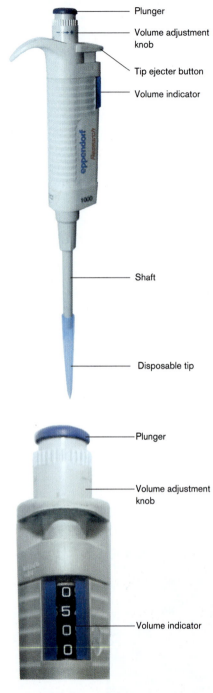

Figure A2.2 Features of a micropipettor.

In each case, the volume to be measured can be dialed in by **rotating a volume adjustment knob (Figure A2.2)**, which is usually located near the top of the micropipettor. The knob should be rotated until the desired volume is seen in the readout window.

You should never adjust the volume beyond the range of the micropipettor! Micropipettors are relatively delicate and expensive instruments, and forcing the adjustment knob to a volume either above or below the micropipettor's range can cause significant damage. This includes setting the volume to zero.

Depending on your specific type of micropipettor, the digital volume readout window will show either three or four digits. The digits may be in a horizontal line, or they may be stacked on top of each other. For pipettors in which the digits are stacked vertically, the digits should be read or interpreted as shown below:

100–1,000-µl pipettors

If the pipettor has a three-digit readout, the top digit stands for thousands, the second digit for hundreds, and the third digit for tens. If there is a fourth digit, that digit represents ones. So, in the example in the far left column of **Figure A2.3**, the dial on the pipettor is set to 420 microliters (0.42 ml) for both displays. Since the pipettor should never be set for any volume above 1,000 µl, the only time the top digit would be anything other than 0 would be if the pipettor were set to deliver exactly 1,000 µl; in that case, the top digit would be "1" and all other digits would be "0."

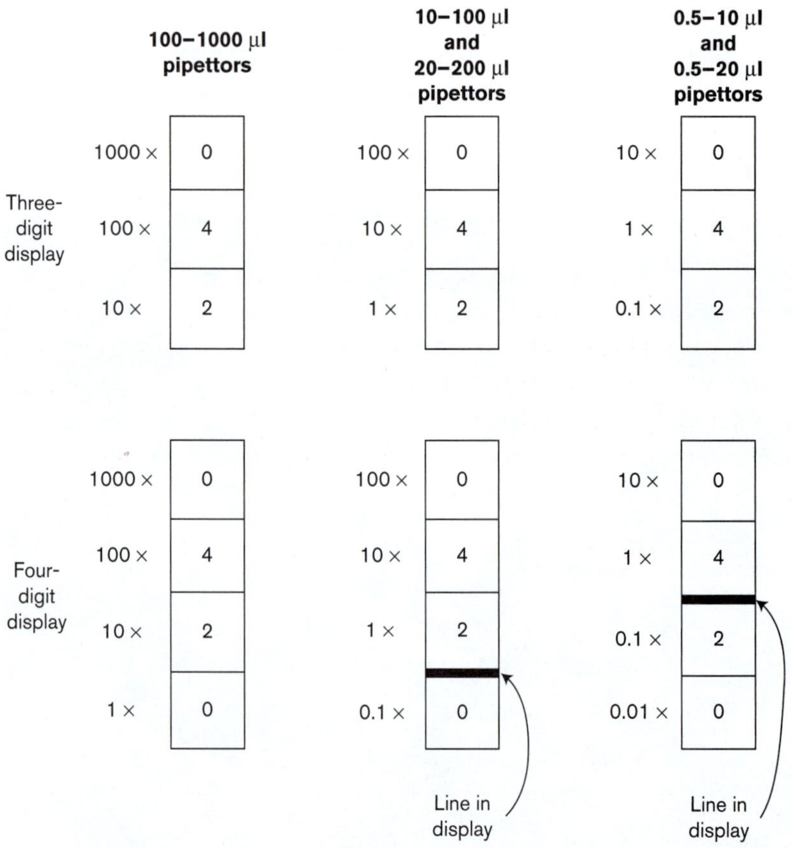

	100–1000 µl pipettors	10–100 µl and 20–200 µl pipettors	0.5–10 µl and 0.5–20 µl pipettors
Three-digit display	1000 × [0] 100 × [4] 10 × [2]	100 × [0] 10 × [4] 1 × [2]	10 × [0] 1 × [4] 0.1 × [2]
Four-digit display	1000 × [0] 100 × [4] 10 × [2] 1 × [0]	100 × [0] 10 × [4] 1 × [2] 0.1 × [0] — Line in display	10 × [0] 1 × [4] 0.1 × [2] 0.01 × [0] — Line in display

Figure A2.3 Reading the adjustable digital display on a micropipettor.

10–100-µl and 20–200-µl pipettors

If the pipettor has a three-digit readout, the top digit stands for hundreds, the second digit for tens, and the third digit for ones. If there is a fourth digit, that digit represents tenths of microliters, and there is often a dark or colored horizontal line above this digit to represent the decimal point. So, in the example in the middle column of Figure A2.3, the dial on the pipettor is set to 42 microliters (0.042 ml).

The pipettor should never be set for any volume above 100 µl for a 10–100-µl pipettor or above 200 µl for a 20–200-µl pipettor. Therefore, the only time the top digit would be anything other than 0 for a 10–100-µl pipettor would be if the pipettor were set to deliver exactly 100 µl. The only time the top digit would be anything other than 0 or 1 for a 20–200-µl pipettor would be if the pipettor were set to deliver exactly 200 µl.

0.5–10-µl and 0.5–20-µl pipettors

If the pipettor has a three-digit readout, the top digits stands for tens, the second digit for ones, and the third digit for tenths of a microliter. If there is a fourth digit, that digit would represent hundredths of microliters. There is often a dark or colored horizontal line above the tenths of a microliter digit to represent the decimal point. So, in the example in the far right column of Figure A2.3, the dial on the pipettor is set to 4.2 microliters (0.0042 ml) for both displays.

The pipettor should never be set for any volume above 10 µl for a 0.5–10-µl pipettor or above 20 µl for a 0.5–20-µl pipettor. Therefore, the only time the top digit would be anything other than 0 for a 0.5–10-µl pipettor would be if the pipettor were set to deliver exactly 10 µl. The only time the top digit would be anything other than 0 or 1 for a 0.5–20-µl pipettor would be if the pipettor were set to deliver exactly 20 µl.

MICROPIPETTORS AND PIPETTE TIPS

Liquids are never drawn directly into the shaft of the micropipettor! Instead, disposable tips are attached to the ends of the shafts, and liquids are drawn into the tips and only into the tips. When you are finished, the tips should always be placed in the receptacle designated for trash to be autoclaved.

The different sizes of micropipettors use different tips (**Figure A2.4**), which hold different volumes of liquids. In most cases, tips of different volumes also differ in color, and the colors are usually as follows:

Range of volumes measured	Tip color
100–1,000 microliters	Blue
10–100 microliters **OR** 20–200 microliters	Yellow
0.5–10 microliters **OR** 0.5–20 microliters	Colorless or yellow

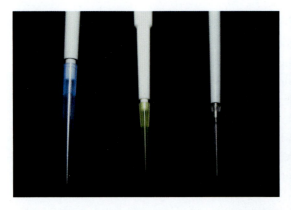
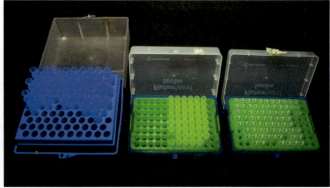

Figure A2.4 Micropipette tips.

In the microbiology lab, micropipette tips are sterilized before use and are stored in sterile boxes until used. It is critically important to prevent contamination of tips during micropipettor use, so when micropipetting, always follow these steps:

1. Select the correct tip size.
2. Open the pipette tip box, but do **not** touch any of the inner surfaces, including the inside of the lid and the tips themselves. **You should not touch the micropipette tip at any time during the use of the tip, nor should you put the lid of the box down on the bench.**
3. Gently insert the end of the micropipettor shaft into the top of the tip and press down with just enough force to attach the tip firmly and securely to the shaft of the micropipettor (**Figure A2.5**).

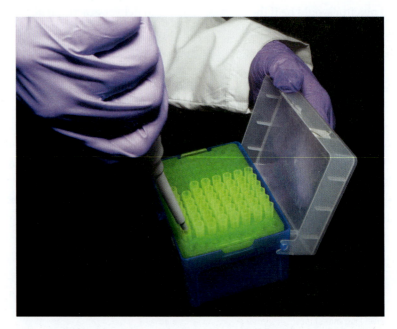

Figure A2.5 Inserting the micropipettor shaft into the pipette tip.

4. Pull the attached tip out of its slot in the box, **taking care not to touch any surface**. That is, at all times, treat the tip as you would treat a flame-sterilized wire loop or wire.

5. Close the box while touching the outer surface of the lid only. Again, do not touch the inside of the lid or any of the remaining pipette tips **(Figure A2.6)**.

Figure A2.6 Closing the box after the tip has been removed.

MEASURING AND DISPENSING LIQUIDS WITH A MICROPIPETTOR

Micropipettors draw up and dispense liquids using a plunger at the top of the device (see Figure A2.2), which displaces the air in the shaft and micropipette tip. **The plunger has two stops**, or two points at which you will feel a clear change in resistance as you depress the plunger **(Figure A2.7)**. When <u>drawing up</u> a desired volume of liquid, you

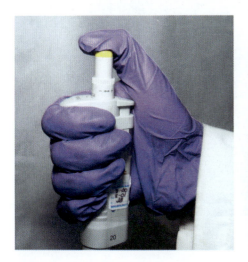
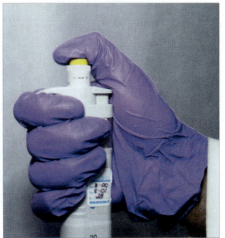
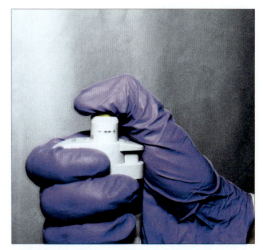

Figure A2.7 Preparing to depress the plunger (left); depressed to the first stop to draw up liquid (middle); depressed to the second stop to dispense liquids (right).

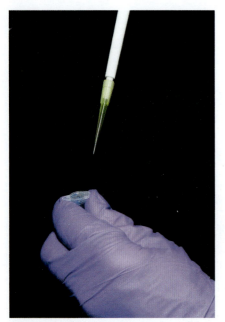

Figure A2.8 Holding the microcentrifuge tube prior to opening.

will depress the plunger to the <u>first stop</u>, the first point at which you feel a clear change in resistance in the plunger, and when <u>dispensing</u> liquids, you will depress the plunger to the <u>second stop</u> (see details below). Before using the micropipettor to transfer liquids for the first time, practice depressing the plunger until you are certain that you can feel the points at which the plunger reaches the first and second stops.

Micropipettors also have a feature that will allow you to remove the micropipette tip without touching the tip with your hand. Most micropipettors have a second plunger-like button on the side that will eject the tip when it is depressed (see Figure A2.2). Other micropipettors do not have a separate ejection button, but instead are designed to eject the tip when the dispensing plunger is depressed past the second stop. Before using a micropipettor, you should be certain that you understand how your particular instrument ejects micropipette tips.

For most procedures using microcentrifuge tubes, gloves should be worn throughout the procedure.

1. Select the micropipettor of the appropriate size, and dial in the volume to be transferred (in microliters).
2. Attach a micropipette tip to the micropipettor shaft as described in the previous section. **Remember, never touch the tip itself.**
3. Once the micropipette tip is securely in place in the shaft, hold the micropipettor between your palm and your last three fingers, with the micropipette tip pointing downward. Again, always be very careful to keep the micropipette tip from touching any nonsterile surface.
4. Pick up the microcentrifuge tube containing the liquid to be transferred with your free hand **(Figure A2.8)**. Use your thumb and index finger to open the cap of the microcentrifuge tube **(Figure A2.9)**. As you do so, **take care to avoid contact between your fingers and the rim of the microcentrifuge tube.**

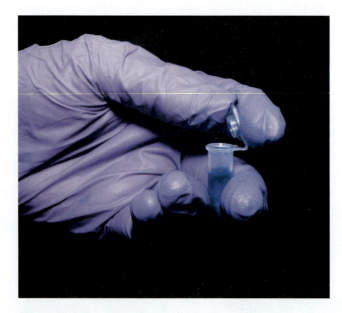

Figure A2.9 Opening the microcentrifuge tube.

5. Use your thumb to press the plunger of the micropipettor to the <u>first stop</u> **(Figure A2.10)**. Insert the micropipettor into the microcentrifuge tube **without allowing the sterile pipette tip to touch any of the outer surfaces of the tube**. If the micropipette tip touches an outer surface at any point, discard the tip and start over.

6. Lower the pipette tip into the liquid in the tube until the tip is a few millimeters below the surface of the liquid **(Figure A2.11)**. <u>Slowly and steadily</u> release the plunger to smoothly draw liquid up into the tip **(Figure A2.12)**.

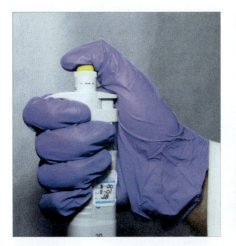

Figure A2.10 Depressing the plunger to the first stop prior to drawing liquids.

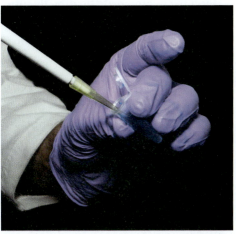

Figure A2.11 Inserting the tip into the liquid.

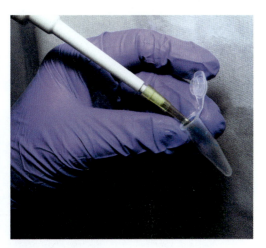

Figure A2.12 Drawing the liquid into the pipette tip by releasing the plunger.

7. Carefully withdraw the micropipettor, and again, **do not allow the pipette tip to touch the outside of the tube or any other nonsterile surfaces**.

8. Close the microcentrifuge tube by pushing the cap back into the tube.

9. Pick up the microcentrifuge tube that is to receive the transferred liquid, and remove the cap as before with your thumb and index finger.

10. Insert the pipette tip into the microcentrifuge tube **without touching any of the outer surfaces of the tube.**

11. To dispense the liquid, use your thumb to <u>slowly depress the plunger</u> to the first stop, pause, and then continue to slowly depress the plunger to the <u>second stop</u> to blow out all liquid from the pipette tip **(Figure A2.13)**.

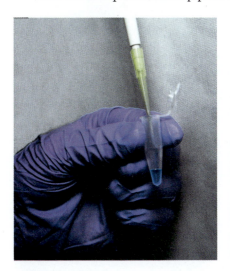
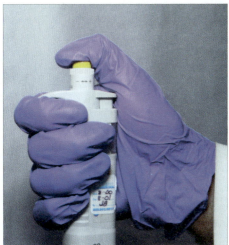
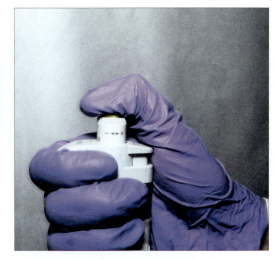

Figure A2.13 Dispensing liquid into a new tube (left); by depressing the plunger to the first stop (middle); and then to the second stop (right).

If the tip is below the level of the liquid in the tube when you empty the tip, do <u>not</u> release the plunger until the tip has been withdrawn from the liquid, or you will inadvertently draw liquid back into the tip.

If your micropipettor is designed to eject the tip when the plunger is depressed past the second stop, take care <u>not</u> to push beyond the second stop, as this will eject the tip into the test tube.

12. Carefully withdraw the micropipettor, and replace the cap on the tube.

13. Eject the micropipette tip **directly into the receptacle designated for trash to be autoclaved** using the appropriate ejection button on the micropipettor **(Figure A2.14)**. Do <u>not</u> remove the tip with your hands, and do <u>not</u> eject the tip onto the bench.

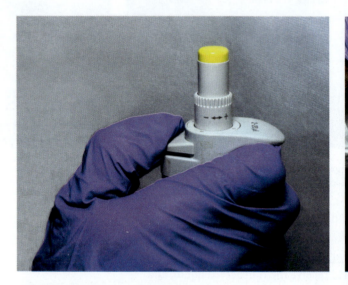 

Figure A2.14 Properly disposing the contaminated micropipette tip.
Use the eject button (left) to eject the micropipette directly into the receptacle designated for trash to be autoclaved (right).

acetone An organic solvent and decolorizing agent that may be used as part of the Gram stain. [Lab 6]

acid-fast bacteria Bacteria that are positive for the acid-fast stain; these bacteria are considered neither Gram-positive nor Gram-negative, and they include the medically important *Mycobacterium* and *Nocardia* genera. [Lab 7]

acid-fast stain A differential stain that distinguishes and identifies bacteria with high concentrations of waxy lipids in their cell walls. [Lab 7]

acid fuchsin An acidic dye used in Hektoen enteric (HE) agar as an inhibitor of Gram-positive species and as a pH indicator. [Lab 31]

acidic Having a pH value below 7.0. [Lab 13]

acidic stain See **indirect stain**.

acidophile A microbe that prefers and grows better in acidic environments, especially in environments with pH below 5.5. [Lab 13]

ADME An acronym for the four processes that must be understood in order to grasp what happens when a drug is used to treat disease: absorption, distribution, metabolism, and excretion. [Lab 18]

aerotolerant Of or referring to a microbe that is incapable of using or metabolizing oxygen but can tolerate its presence, usually growing best under low oxygen concentrations. [Lab 14]

agar A gelling agent extracted from a multicellular red-purple algal species and composed of a complex mix of polysaccharides; components include galactose and 3,6-anhydro-L-galactopyranose polymers (agarose) along with other galactose-based polymers. [Lab 2]

agarose A component of agar; a very large polysaccharide composed of repeating units of a disaccharide composed of galactose and the sugar derivative 3,6-anhydro-L-galactopyranose; used as a component of electrophoresis gels. [Lab 20]

alcohol Organic solvents used as disinfectants, antiseptics and decolorizing agents in the Gram stain, usually in the form of ethanol. [Lab 6]

alkaline See **basic**.

alkaliphile A microbe that prefers and does better in basic environments, especially in environments with pH above 8.5. [Lab 13]

alpha (α) hemolysis A hemolysis pattern resulting from damage to and partial breakdown of the red blood cells in a blood agar plate; this type of hemolysis produces partial clearing and a greenish brown color in the blood agar due to pigments such as biliveridin. [Lab 21]

ammonia (NH_3) A species of inorganic nitrogen; when dissolved in water, it will exist in an equilibrium with ammonium (NH_4^+). [Lab 9]

ammonification A process that releases NH_3/NH_4^+ (ammonia/ammonium) from organic nitrogen sources, such as amino acids. [Lab 33]

ammonium (NH_4^+) A species of inorganic nitrogen; when dissolved in water, it will exist in an equilibrium with ammonia, and at pH 7, it will account for 99% of the (NH_4^+/NH_3) molecules in solution. [Lab 9]

amylase A type of exoenzyme produced by many bacterial species that can hydrolyze large, coiled starch molecules into a collection of glucose monomers or into molecules containing two to about ten glucose subunits apiece. [Lab 10]

amylopectin A type of starch composed of loosely coiled and branching chains of glucose molecules. [Lab 10]

amylose A type of starch composed of glucose molecules that are linked together into long, tightly coiled, unbranched chains. [Lab 10]

anaerobe jar See **Brewer jar**.

anaerobic photosynthetic bacteria Bacteria living under anaerobic conditions and using light as an energy source; these bacteria use molecules other than water as electron and proton sources, and so photosynthesis does not generate oxygen. [Lab 34]

anaerobic respiration A type of metabolism, distinct from fermentation, in which inorganic molecules other than oxygen are used as terminal electron acceptors at the end of respiratory electron transport systems; for example, nitrates (NO_3) may be used in place of oxygen as terminal electron acceptors, generating nitrogen (N_2) and nitrous oxide (N_2O) gases as end products. [Lab 23]

anion A negatively charged ion. [Lab 5]

annealing A process in which two single strands of complementary DNA are joined by hydrogen bonding between complementary bases to form double-stranded DNA; often used to describe the binding of primer DNA to a target sequence during PCR. [Lab 20]

anoxygenic photoautotroph A photoautotroph that uses hydrogen sulfide (H_2S) as its source of electrons for photosynthesis, thereby producing elemental sulfur. [Lab 34]

anthrax A deadly human disease caused by *Bacillus anthracis* that usually involves skin infection (cutaneous anthrax), digestive tract infection (gastrointestinal anthrax), or lung infection (pulmonary anthrax). [Lab 8]

antibiotic A chemical control agent, often of microbial origin, that is selectively toxic to microbes (usually bacteria). [Lab 17]

antibody An antimicrobial immune system protein, produced and secreted into the blood plasma by B lymphocytes, that binds to an antigen with a specific shape. [Lab 28]

antigen A molecule that stimulates antibody production and that fits a particular antibody's antigen-binding sites; includes molecules that characterize the cell surface of a particular bacterial species or strain. [Lab 28]

antigen-binding site The region with a specific amino acid sequence at the tips of the arms of the "Y" shape of an antibody that binds to a specific antigen. [Lab 28]

antiseptic A chemical applied to living tissue to reduce the microbial population and the risk of infection; usually less toxic than a disinfectant. [Lab 15]

atmospheric oxygen (O_2) Oxygen molecule formed of two atoms of oxygen; most common form of oxygen in the atmosphere. [Lab 14]

autoclave A device that uses high-pressure steam or moist heat to kill cells and sterilize media and equipment; typically operated at temperatures of around 120°C and pressures of around 15 PSI for 15 to 20 minutes. [Lab 9]

autoinfection Infection caused by pathogens, such as *Staphylococcus aureus*, that had been inhabiting the body without causing harm until a break in the skin, a weakening of the immune system, or some other event gave them an opportunity to cause an infection. [Lab 22]

autotroph An organism that is able to use carbon dioxide as its sole carbon source for the synthesis of all of its organic molecules such as sugars. [Lab 34]

autotrophic Able to use carbon dioxide as the sole carbon source for the synthesis of all required organic molecules such as sugars. [Lab 33]

bacillus A rod-shaped cell; derived from a Latin word meaning "staff," "walking stick," or "little rod." [Lab 5]

bacterial species A collection of different bacterial strains that, though not genetically identical, share many genetically stable traits and differ significantly from other collections of strains. [Lab 27]

bacterial strain A bacterial population derived from a single cell whose members are genetically identical, or very nearly so. [Lab 27]

bacteriophage A virus that infects bacteria, usually host specific; derived from the word *phagein*, "to eat" or "to have a share of food," and/or *phagos*, "eater of" or "glutton." Also called a **phage**. [Lab 36]

bacteriophage titer The quantity or count of bacteriophage particles in a given suspension. [Lab 36]

bacteriostasis A condition in which bacterial cells are alive, but inactive and nonreproducing. [Lab 3]

basic Having a pH value above 7.0; also referred to as **alkaline**. [Lab 13]

basic fuchsin See **carbolfuchsin**.

basic stain See **direct stain**.

basophilic Having an affinity for basic stains; literally, "base-loving." [Lab 5]

beta-fructosidase An enzyme that hydrolyzes sucrose, producing glucose and fructose as end products; required for the fermentation of sucrose. [Lab 24]

beta-galactosidase An enzyme that splits lactose into the monosaccharides glucose and galactose; required for the utilization of lactose. [Lab 24]

beta (β) hemolysis A hemolysis pattern resulting from the complete lysis or destruction of the red blood cells in a blood agar plate by small toxic polypeptides called streptolysins; this type of hemolysis produces completely transparent and nearly colorless zones around the colonies. [Lab 21]

biliveridin A greenish brown–tinted pigment generated as a result of the breakdown of hemoglobin by alpha (α) hemolytic bacteria. [Lab 21]

biochemical fingerprint A pattern of pluses and minuses, based on the presence or absence of enzymes or the ability or inability to use certain substrates and produce certain products, that characterizes a particular bacterial species or strain. [Lab 19]

biocode A five-digit numeral that summarizes the results of the EnteroPluri assays and can be used to identify the test bacterium. [Lab 27]

blood agar A nutrient-rich medium composed of peptones, yeast extract, and liver or heart extracts with 5% sheep's blood added to the medium after autoclaving; differentiates bacterial species on the basis of hemolysis patterns or how they break apart, or fail to break apart, red blood cells. [Lab 21]

botulism A deadly type of food poisoning that occurs when *Clostridium botulinum* spores germinate under anaerobic conditions, such as in sealed cans of food, producing a protein neurotoxin that causes paralysis by blocking transmission of signals from nerve cells to skeletal muscles. [Lab 8]

bound coagulase A coagulase on the cell surface of *Staphylococcus aureus* that acts as a receptor for fibrinogen; causes clumping of blood plasma proteins and promote the attachment of *S. aureus* cells to blood clots, traumatized tissue, and implanted medical devices such as catheters. Also called **clumping factor**. [Lab 22]

Brewer jar A small container from which the free oxygen can be removed by chemical reactions after the jar has been sealed, used in the cultivation of strict anaerobes; also referred to as an **anaerobe jar**. [Lab 14]

broad-spectrum antibiotic An antibiotic that inhibits or kills a greater range of bacterial species than a narrow-spectrum antibiotic, typically inhibiting or killing a large number of both Gram-positive and Gram-negative species. [Lab 17]

bromcresol purple A pH indicator that has a structure similar to phenol red, but is yellow when uncharged at lower pH values (pH < 5.8) and purple when it loses protons at higher pH values (pH > 6.8). [Lab 10]

bromthymol blue A pH indicator that is yellow in color when it is uncharged at lower pH values (pH < 6.0) and blue in color when it loses a proton (H^+) and becomes negatively charged at higher pH values (pH > 8.0); the indicator is green around pH 7.0. [Lab 25]

buffer A compound that slows the rate of change in pH value as more acid or more base is added to a solution by binding H^+ or OH^- ions and preventing them from contributing to the pH of the solution. [Lab 13]

cadaverine A diamine produced by the decarboxylation of lysine; in a few species of anaerobic Gram-negative bacteria, cadaverine is an essential component in peptidoglycan, and thus in cell wall synthesis. In many other species, including *E. coli* and other Gram-negative intestinal bacteria, cadaverine interacts with protein channels found in cell membranes and in the outer phospholipid envelopes of the Gram-negative cell walls, altering membrane permeability and reducing the rate at which molecules move through and across the membrane. This, in turn, can protect the cell from chemicals such as antibiotics and prevent an excess inflow of protons (H^+) in acidic environments. [Lab 10]

calcium dipicolinate (calcium salt of dipicolinic acid) A compound found in the core of endospores, thought to confer resistance to heat and radiation. [Lab 8]

capsid The protein case or capsule of a bacteriophage that contains the viral DNA; resembles a head. [Lab 36]

carbolfuchsin A direct stain used in the acid-fast stain that is soluble in phenol. Also called **basic fuchsin**. [Lab 7]

carbon The element that forms the backbones of the rings and chains of organic molecules. [Lab 9]

carbon skeletons Rings or chains of carbon used to build organic molecules. [Lab 9]

casein One of a family of milk proteins that make milk white and opaque; these proteins have phosphates attached to some of the amino acids, which cause them to clump or curdle under acidic conditions. [Lab 29]

casein agar See **skim milk agar**.

caseinase A protease or peptidase that can specifically degrade casein to small, soluble peptides and amino acids; produced by many soil bacteria, such as *Bacillus* species. [Lab 10]

casein hydrolysis assay An assay designed to differentiate between microbial species that can produce a protease that hydrolyzes a milk protein called casein and those that cannot. [Lab 10]

catalase An oxygen-detoxifying enzyme produced by some microbial species that converts hydrogen peroxide (H_2O_2) to molecular oxygen (O_2) and water (H_2O). [Lab 14]

cation A positively charged ion. [Lab 5]

cell membrane The semi-permeable, phospholipid-based membrane that surrounds and separates the cytoplasm of the cell from the outside environment; in bacterial cells, it is located inside of the cell wall. [Lab 11]

cell morphology Traits of a cell such as size, shape, color, Gram status, and the arrangement of connected cells. [Lab 5]

CFU/ml See **colony-forming units per milliliter**.

chemoautotroph An organism that can fix CO_2 via the Calvin cycle as the sole source of carbon and can use simple reduced inorganic molecules as energy sources. [Lab 34]

chemoheterotroph An organism that uses organic molecules as a source of carbon and energy; most also use organic molecules as a source of electrons and protons. [Lab 9, 34]

chemotherapeutic Of or referring to an agent, such as an antibiotic, whose selective toxicity allows it to be used to treat a microbial infection, including internal infections. [Lab 17]

chemotroph An organism that uses reduced molecules as energy sources. [Lab 34]

chemotrophic Able to oxidize reduced molecules to acquire the energy needed for metabolism. [Lab 33]

chromogen The complete colored ion in a stain or dye, including the chromophore or chromophores. [Lab 5]

chromophore The chemical group or region of a dye molecule that absorbs and reflects particular wavelengths of light; often added to a colorless organic molecule to create a chromogen. [Lab 5]

chymotrypsin A protease found in the mammalian pancreas that can be used to hydrolyze proteins to make peptones or tryptones. [Lab 9]

-cidal A suffix added to prefixes such as "germi-," "bacteri-," or "fungi-," as in "germicidal," "bactericidal," and "fungicidal," to indicate a cidal agent. [Lab 15]

cidal agent An agent that kills microbes. [Lab 15]

citrate assay An assay that detects the ability of bacteria to use the organic compound citrate (citric acid) as its sole source of carbon and energy. [Lab 25]

clumping factor See **bound coagulase**.

coagulases Virulence factors produced by *Staphylococcus aureus* that cause coagulation of blood proteins. [Lab 22]

coccobacillus A short rod-shaped cell, for example, only about twice as long as it is wide. [Lab 5]

coccus A spherical cell; derived from the Greek *kokkos*, meaning "berry" or "grain." [Lab 5]

coliform species Members of the Gram-negative enteric family that are capable of fermenting lactose to organic acids and gas; includes almost all species and strains in the genera *Escherichia* (including *E. coli*), *Enterobacter*, and *Klebsiella*, and most of the common species and strains in the genus *Citrobacter*. [Lab 24]

colony-forming units per milliliter (CFU/ml) A measure of the viable cell density of microbial suspensions; equal to the number of cells or groups of cells producing separate colonies on plate media per milliliter. [Lab 4]

colony morphology Characteristics of the growth of populations of microbes on solid and semisolid media that can be seen without a microscope, including the color, size, shape, elevation, and surface features of the isolated colonies of a given species. [Lab 2]

completed test Last of a series of three assays used to measure total coliforms in water, in which a colony from the confirmed test is used to inoculate (1) a lactose broth with a Durham tube and (2) a nutrient agar slant to produce cells to be Gram-stained. If this isolate is (1) positive for gas from lactose fermentation and (2) a Gram-negative rod, it is assumed to be a coliform. [Lab 30]

complex medium A medium containing one or more components in which the exact chemical composition is unknown or undefined, such as a meat extract, yeast extract, or protein digest. [Lab 9]

confirmed test Second of a series of three assays used to measure total coliforms in water; involves growing the bacteria identified in the presumptive test as possible coliforms on EMB agar to confirm that they are coliforms and not some other type of lactose-fermenting microbes. [Lab 30]

copper (Cu) A metallic element required in trace amounts by bacteria. [Lab 9]

countable plate In quantitative plate counting, a plate containing a number of colonies within the range required to accurately calculate CFU/ml. For spread plates, countable plates are those with 20 to 200 colonies; for pour plates, countable plates are those with 30 to 300 colonies. [Lab 4]

crystal violet A direct, basic stain, used in the first step of the Gram stain, that colors cells of all types a blue-purple color. [Lab 6]

curds Clumps formed in milk products when casein molecules stick together. [Lab 29]

cutaneous anthrax An infection of the skin caused by *B. anthracis*. [Lab 19]

cyanobacteria A group of photosynthetic bacteria that are able to convert carbon dioxide gas into carbohydrates (carbon fixation) and create ammonium for use in amino acids and other organic nitrogen compounds from nitrogen gas (nitrogen fixation). [Lab 33]

cysteine An amino acid that contains sulfur; used in growth media to supply sulfur and as a reducing agent in the cultivation of strict anaerobes. [Lab 9, 14]

cystic fibrosis A genetic disease that, among other symptoms, causes a buildup of thick mucus in the lungs that leaves affected individuals vulnerable to opportunistic infections. [Lab 23]

cytochrome *c* oxidase A specific enzyme involved in the electron transport system of aerobic metabolism; produced by some groups of Gram-negative bacteria, particularly *Pseudomonas aeruginosa*. [Lab 23]

cytochrome *c* oxidase assay A chromogenic oxidation-reduction assay used to differentiate bacterial species that produce cytochrome *c* oxidase from those that do not produce that enzyme. [Lab 23]

dark repair system An enzyme system that can locate and repair damaged and mutated sections of DNA and does not require light for activation; replaces thymine dimers with new, separated thymines. [Lab 16]

decarboxylase An enzyme that removes the carboxyl group (–COOH) from a particular amino acid. [Lab 10]

decimal reduction time The time required to reduce a microbial population by 90% or by one log unit less than the starting count; reduction is equivalent to moving the decimal point for the initial population count one digit to the left. [Lab 29]

defined medium A growth medium in which the quantity and identity of each specific chemical component is known or defined; a medium in which the nutritional needs of a bacterial species are met by providing known quantities of specific pure chemicals. [Lab 9]

degrees Brix (°Bx) A measure of the sugar concentration of a liquid in which one degree Brix is equal to 1.0 g of sucrose in 100 g of solution. The Brix scale yields sucrose concentration as a percentage by weight (% w/w), or the percentage of total grams of solution that is composed of sucrose. [Lab 32]

denature (1) To change the shape or structure of a protein, usually as a result of an environmental factor such as heat or acidity; (2) to separate the two strands of double-stranded DNA by heating it. [Lab 11, 20]

denitrification A process in which nitrates (NO_3^-) are used in anaerobic respiration and are reduced to either nitrogen (N_2) gas or nitrous oxide (N_2O) gas; also called **dissimilatory nitrate reduction**. [Lab 33]

depth of field (1) The thickness of the slice of a specimen that will be in focus with a given lens, (2) the distance above and below the subject that appears to be in focus, or (3) the distance between the highest point and the lowest point of a three-dimensional object that is in sharp focus. [Lab 1]

diarrheal form A form of food poisoning characterized by cramps and diarrhea. [Lab 8]

differential medium A medium designed to aid in the identification of various microorganisms by differentiating, distinguishing, or separating different types of microbes based on their ability to perform particular biochemical reactions. [Lab 19]

differential stain A stain that aids in the identification of microbes by revealing distinguishing traits of different types of bacteria. [Lab 5]

dimethyl-aminobenzaldehyde The active ingredient in Kovac's reagent; combines with indole to produce a cherry-red end product, a di-dimethyl ammonium salt. [Lab 25]

diplo- A prefix used to describe cells that remain connected as pairs after cell division; derived from the Greek *diplous*, meaning "double." [Lab 5]

direct stain A stain in which the chromogen binds directly to negatively charged cells and cell components because it uses dye salts in which the chromogen is a cation, and thus is positively charged. Also called a **basic stain**. [Lab 5]

disinfectant A chemical applied to nonliving or inanimate surfaces and objects to reduce the microbial population; disinfectants are not intended for use on living tissue and tend to be more toxic than antiseptics. [Lab 15]

dissimilatory nitrate reduction See **denitrification**.

DNA polymerase An enzyme that links individual nucleotides to the 3′ end of an existing DNA strand to form longer strands of DNA. [Lab 20]

dose The concentration or the amount of antibiotic that is to be consumed, applied, or injected into the body. [Lab 18]

Durham tube A small glass tube that is inverted in a tube of broth to capture any gases that might be produced by fermentation at the anaerobic bottom of the broth. [Lab 10]

E. coli O157:H7 A strain of *Escherichia coli* with extra genes that can produce a toxin capable of destroying red blood cells and causing potentially fatal kidney damage, resulting in a condition known as hemolytic-uremic syndrome. [Lab 31]

ecosystem A complex community composed of many interacting and interdependent species along with the nonliving components of their environment. [Lab 34]

electrophoresis A technique for separating macromolecules such as DNA on the basis of the size of the molecules; uses an electric current to draw macromolecules through molecular sieves. In DNA electrophoresis, negatively charged DNA molecules are separated by size when they are drawn through an agarose gel sieve toward the positively charged electrode. [Lab 20]

EMB agar See **eosin methylene blue (EMB) agar**.

emetic form A form of food poisoning characterized by cramps and vomiting (emesis). [Lab 8]

endocarditis Inflammation of the inner lining of the heart; may be caused by *Enterococcus faecalis* infection. [Lab 21]

endoenzyme An intracellular enzyme; an enzyme that acts within the cell that produces it on absorbed nutritional substrates and on other molecules created internally by various metabolic processes. Endoenzymes may be involved either in energy-releasing catabolism or in macromolecule-building anabolic pathways. [Lab 10]

endospore A tough, resistant, and dormant structure or form produced by a few genera of bacteria, including *Bacillus* and *Clostridium*, when environmental conditions become unfavorable. [Lab 8]

energy A requirement for building organic molecules, transporting atoms and molecules across the cell membrane, moving the cell, and for many other tasks as well. [Lab 9]

enteric family See **Enterobacteriaceae**.

Enterobacteriaceae A family of rod-shaped, oxidase-negative, Gram-negative bacteria capable of fermenting glucose to organic acids and gases that includes the genera *Escherichia*, *Enterobacter*, *Citrobacter*, *Klebsiella*, *Proteus*, *Salmonella*, *Shigella*, *Serratia*, and *Providencia*; often found as part of the normal flora in the human large intestine, but may also act as pathogens; also called the **enteric family**. [Lab 24]

EnteroPluri system A rapid, multi-test system used in the identification of members of the enteric family; consists of a tube with twelve chambers containing different media for fifteen assays, all of which can be inoculated at once. [Lab 27]

enterotoxin A toxin secreted by bacteria that acts in the digestive tract to cause food poisoning. [Lab 12]

enzyme A biological molecule that catalyzes a biochemical or metabolic reaction that converts a specific substrate to a specific end product. [Lab 10]

eosin methylene blue (EMB) agar A medium containing eosin and methylene blue dyes; designed to select for Gram-negative enterics and other Gram-negative species and to inhibit or prevent the growth of Gram-positive species; also differentiates lactose-fermenting cells from non-lactose-fermenting cells. [Lab 24]

essential amino acids Amino acids that must be supplied in an organism's food because it cannot synthesize these compounds from other biological molecules. [Lab 9]

ethidium bromide A molecule used to stain DNA in electrophoresis gels; it intercalates, or inserts itself, between the stacks of bases in double-stranded DNA and fluoresces under UV light, but it is also mutagenic. [Lab 20]

exoenzyme An extracellular enzyme; that is, an enzyme released by a cell into the surrounding environment, where it acts on substrates outside the cell, usually breaking down large macromolecules such as starches and proteins. [Lab 10]

exponential response In microbial control, a response in which the same fraction of the remaining microbial population is eliminated in any given time interval. [Lab 29]

extension In PCR, the step in which individual nucleotides pair up with complementary nucleotides along the template DNA, and a DNA polymerase links them together to form a new, complementary strand. [Lab 20]

facultative Of or referring to a facultative anaerobe. [Lab 14]

facultative anaerobe A microorganism that will use oxygen if it is available in aerobic respiration, but does not require oxygen to grow and multiply; can grow in either the presence or absence of oxygen. [Lab 14]

fastidious species Microbial species that have complex nutritional requirements; organisms that need specific organic molecules or nutrients to grow. [Lab 9]

fibrin An insoluble, stringy protein involved in blood clotting; normal blood clotting mechanisms involve a number of clotting factors found in the bloodstream that interact sequentially to produce fibrin from fibrinogen. [Lab 22]

fibrinogen A soluble blood plasma protein that will be converted to fibrin during normal blood clotting. [Lab 22]

flaccid paralysis A symptom of botulism caused by a protein neurotoxin that blocks the transmission of signals from nerve cells to skeletal muscles, preventing muscle contraction; causes death when muscles essential to breathing fail to contract. [Lab 8]

flagella Protein-based filaments possessed by many species of bacteria that rotate to enable the cells to move from one location to another. [Lab 25]

flagella stain A staining method that makes it possible to visualize flagella by "thickening" them with stain in a process akin to applying mascara to eyelashes. [Lab 25]

food poisoning Intestinal illness caused by an enterotoxin, characterized by symptoms of cramps, diarrhea, and vomiting. [Lab 8]

free coagulase See **secreted coagulase**.

freeze-drying See **lyophilizing**.

freezing The use of low temperatures to control microbial growth; effective freezing requires temperatures of at least −20°C, and many labs use freezers set at −70°C for long-term culture storage. Temperatures this low will stop all metabolic activity. [Lab 3]

galactose A monosaccharide important in carbohydrate metabolism, found in the disaccharide-based polymers in agar; for example, D-galactose and 3,6-anhydro-L-galactopyranose form the repeating disaccharide found in the agar component agarobiose. [Lab 2]

gamma (γ) hemolysis The absence of hemolysis of the red blood cells in a blood agar plate, characterized by the absence of any change in the color or opacity of the blood agar medium. [Lab 21]

gases End products of carbohydrate fermentation, typically carbon dioxide (CO_2) and hydrogen (H_2) gas. [Lab 10]

gas gangrene A type of wound infection caused by *Clostridium perfringens* when these bacteria multiply in dead, anaerobic tissue and produce toxins such as a toxic phospholipase; characterized by the production of fermentation gases in the dead tissue. [Lab 8]

geosmin A volatile organic molecule produced by bacteria in the genus *Streptomyces* that gives soil its distinctive earthy, musty odor, especially after it rains; derived from the Greek *geo*, "earth," and *osmin*, "smell." [Lab 35]

glucose A monosaccharide that is very important in carbohydrate metabolism and is the subunit in polysaccharides such as starch and glycogen. [Lab 2]

glutamate An amino acid that can be synthesized from inorganic nitrogen and carbon skeletons in cells. [Lab 9]

glutamine An amino acid that can be synthesized from inorganic nitrogen and carbon skeletons in cells. [Lab 9]

Gram morphology See **Gram status**.

Gram-negative Having a cell wall with an outer phospholipid envelope or membrane and a thin layer of peptidoglycan that accounts for only about 10% to 15% of the dry weight of the wall; does not retain crystal violet when rinsed with alcohol in the Gram stain. [Lab 6]

Gram-positive Having a cell wall with a thick layer of peptidoglycan; retains crystal violet when rinsed with alcohol in the Gram stain. [Lab 6]

Gram's iodine An aqueous solution of iodine and potassium iodide (KI), used in the Gram stain as a mordant to fix crystal violet stain to the cells. [Lab 6]

Gram stain A differential stain that allows the bacterial world to be divided into two groups, Gram-positive species and Gram-negative species, based on differences in cell wall structure. [Lab 6]

Gram status A distinction based on cell wall structure, determined by Gram staining; also referred to as **Gram morphology**. [Lab 6]

Green sulfur bacteria Anoxygenic photoautotrophic bacteria that capture light energy, fix CO_2 by a reductive citric acid cycle, and deposit sulfur outside the cells; include species in the genus *Chlorobium*. [Lab 34]

halophilic Requiring, or able to thrive in, high environmental salt concentrations of 3% salinity or higher. [Lab 12]

halotolerant Able to grow at low to moderate environmental salt concentrations. [Lab 12]

Hansen's disease A human disease caused by *Mycobacterium leprae*; also called **leprosy**. [Lab 7]

heat-labile Of or referring to molecules that are chemically altered or denatured and become biologically nonfunctional if exposed to high temperatures, including autoclave temperatures. [Lab 9]

Hektoen enteric (HE) agar A medium used primarily for the isolation and identification of two enteric family genera, *Salmonella* and *Shigella*. [Lab 31]

hemocytometer slide A type of microscope slide marked with grid lines; used to determine the number of cells per milliliter. [Lab 4]

hemolysis The act of breaking apart and destroying red blood cells. [Lab 21]

hemolytic-uremic syndrome A condition caused by the *E. coli* O157:H7 strain, characterized by destruction of red blood cells (hemolytic) and fatal kidney damage (uremic). [Lab 31]

heterotroph An organism that uses organic molecules as carbon sources. [Lab 34]

high-level disinfectant A chemical capable of sterilization, such as formaldehyde. [Lab 15]

HTST (high temperature, short time) A method of pasteurization in which raw milk is heated to about 72°C (161°F) for 15–20 seconds. [Lab 29]

hydrogen peroxide (H_2O_2) A highly reactive and toxic form of oxygen; often created in cells during oxygen-using metabolism. [Lab 14]

hydrometer A sealed glass tube, weighted on one end, that is used to measure the specific gravity of a liquid; sinks to a certain depth in a liquid depending on its specific gravity. [Lab 32]

hyperthermophiles Microbes with temperature growth ranges between 65°C–70°C and 90°C–105°C, or even higher, with optimum growth temperatures at around 75°C–90°C; some texts combine thermophiles and hyperthermophiles into a single class labeled "thermophiles." [Lab 11]

hypertonic Of or referring to an environment in which the solute concentration is higher than the solute concentration of the cytoplasm of a cell, in which water tends to diffuse out of the cell. [Lab 12]

hypotonic Of or referring to an environment in which the solute concentration is lower than the solute concentration of the cytoplasm of a cell, in which water tends to diffuse into the cell. [Lab 12]

immunoassay An assay that relies on specific antibodies that bind to specific antigen molecules; used to identify particular bacterial strains or species on the basis of strain- or species-specific antigens. [Lab 28]

indirect stain A stain in which the chromogen is repelled by negatively charged cells and cell components because it uses dye salts in which the chromogen is an anion, and thus is negatively charged; a stain that colors the background of a slide but not the cell itself. Also called an **acidic stain.** [Lab 5]

indole An intermediate in tryptophan metabolism, produced by some bacteria during the breakdown of tryptophan and detected by the Kovac's reagent. [Lab 25]

indole production assay An assay designed to detect the breakdown of the amino acid tryptophan to indole, pyruvate, and ammonia; one of the SIM assays. [Lab 25]

inorganic ammonium salts Small inorganic molecules used to supply nitrogen in growth media in the form of ammonium; includes $(NH_4)_2SO_4$ (ammonium sulfate) and $NH_4H_2PO_4$ (ammonium dihydrogen phosphate). [Lab 9]

inorganic nitrogen salts Small inorganic molecules used to supply nitrogen in growth media in the form of ammonium, nitrite, or nitrate. [Lab 9]

iron (Fe) A metallic element required in trace amounts by bacteria. [Lab 9]

isotonic Of or referring to an environment in which the solute concentration is about equal to the solute concentration of the cytoplasm of a cell, in which there is no net diffusion of water into or out of the cell. [Lab 12]

keratin A tough, resilient protein that is abundant in the outer spore coat of endospores; the same protein that makes up human skin, hair, and fingernails. [Lab 8]

Kinyoun stain A differential acid-fast stain technique that relies on high concentrations of phenol and dye, rather than heat, to partially dissolve and stain the waxes in bacterial cell walls. [Lab 7]

Kirby-Bauer assay A method for determining the spectrum of an antibiotic or the response of a particular bacterial strain to an antibiotic, in which a standard number of cells are spread across the surface of a Mueller-Hinton agar plate and a paper disk impregnated with a standard amount of the antibiotic is applied to the plate. [Lab 17]

Kovac's reagent A reagent composed of dimethyl-aminobenzaldehyde dissolved in butanol; in the indole production assay, it combines with indole to produce a bright red color. [Lab 25]

lactic acid bacteria A collection of Gram-positive species in genera such as *Lactococcus*, *Lactobacillus*, *Leuconostoc*, and *Streptococcus* that produce large amounts of lactic acid by fermentation. [Lab 29, 30]

leaching Removal of nitrate and other nutrients by water flowing through soils. [Lab 33]

leprosy See **Hansen's disease**.

limiting element An element essential for cell growth that is in short supply relative to other elements and is often the first nutrient to be exhausted, limiting or stopping growth; also called a **limiting nutrient**. [Lab 33]

limiting nutrient See **limiting element**.

lipopolysaccharide (LPS) A molecule composed of fatty acids and chains of sugars and found in the outer cell membranes of Gram-negative bacteria; varies considerably among Gram-negative strains and species in terms of the type and sequence of the sugars that are present at the ends of the polysaccharide side chains. [Lab 37]

lithotroph An organism that uses inorganic molecules as electron and proton sources. [Lab 34]

loop dilution technique A serial dilution technique in which one to two inoculating loopfuls of cells are transferred from one tube to the next in a series of three or four tubes. Each subsequent tube in the series will have a lower density of cells than the tube before it. [Lab 3]

LTLT (low temperature, long time) A method of pasteurization in which raw milk is heated to about 63°C (145°F) for 30 minutes. [Lab 29]

lyophilizing A dehydration process that requires freezing of the material to be preserved, followed by the removal of water by sublimation under low atmospheric pressure; also called **freeze-drying**. [Lab 3]

lysine An amino acid. [Lab 10]

lysine decarboxylase An enzyme that removes the carboxyl group from lysine under acidic and anaerobic conditions, producing carbon dioxide and cadaverine as end products. [Lab 10]

lysine decarboxylase assay An assay designed to detect the production of cadaverine and carbon dioxide that results from the removal of the carboxyl group from lysine. Specifically, the assay reveals a rise in pH due to the accumulation of cadaverine in the medium. [Lab 10]

lysogenic Of or referring to a bacteriophage that produces an infection in which the viral genes are incorporated into the chromosome of the host cell, new virus particles are not immediately produced, and the host cell is not immediately lysed; instead, when the host cell divides, phage genes are replicated along with the host cell genes. Also referred to as **temperate**. [Lab 36]

lytic Of or referring to a bacteriophage whose activity in a host bacterial cell results in the lysis and death of the host. Also referred to as **virulent**. [Lab 36]

MacConkey agar A medium containing crystal violet and bile salts, specifically designed to select for Gram-negative enterics and other Gram-negative species and to inhibit or prevent the growth of Gram-positive species; also differentiates lactose-fermenting cells from non-lactose-fermenting cells. [Lab 24]

magnification The relative enlargement of a specimen when it is seen through the microscope. [Lab 1]

malachite green A green stain used to stain endospores with the application of heat. [Lab 8]

manganese (Mn) A metallic element required in trace amounts by bacteria. [Lab 9]

mannitol A sugar alcohol used in mannitol salt agar and mannitol salts broth. [Lab 22]

mannitol salt agar A highly selective medium that favors the growth of *Staphylococcus* and *Micrococcus* and tests for their ability to ferment mannitol. [Lab 22]

mannitol salts broth A medium that contains mannitol as a carbon source and no nitrogen in any form, either organic or inorganic, and is therefore selective for nitrogen fixers. [Lab 33]

maximum effective dilution In the phenol coefficient assay, the greatest dilution of the test chemical that kills the test bacteria. [Lab 15]

maximum growth temperature The highest temperature at which growth and cell division can occur, and above which the heat begins to denature proteins. [Lab 11]

McFarland turbidity standards References used to adjust the turbidity of bacterial cell suspensions; the turbidity, or cloudiness, of a broth culture is directly related to the density of the cells in that culture, so adjusting turbidity allows the setting of a suspension to a standard cell density. [Lab 17]

mesophile A microbe with a temperature growth range between 10°C–15°C and 45°C–50°C; those associated with soils and plants have

optimum growth temperatures at 25°C–30°C, whereas those associated with warm-blooded animals, including humans, have optimum growth temperatures at 35°C–40°C. [Lab 11]

metabolism The collection of biochemical reactions that sustain life inside of living cells. [Lab 34]

metabolite A substrate or product in biological metabolism. [Lab 9]

methionine An amino acid that contains sulfur. [Lab 9]

methylene blue A basic stain that can be used as a direct stain or as a counterstain in a number of staining techniques. [Lab 7]

methyl red A pH indicator that is yellow in color when it is uncharged at higher pH values (pH > 6.1) and red in color when it gains a proton (H^+) and becomes positively charged under highly acidic conditions (pH < 4.1). [Lab 25]

methyl red assay An assay designed to differentiate glucose-fermenting bacterial species that produce small amounts of organic acids from those that generate high concentrations of acids. [Lab 25]

Methyl Red Voges Proskauer (MRVP) broth A medium used for the methyl red (MR) assay and Voges-Proskauer (VP) assays. In almost all cases, methyl red–negative species will be positive for the VP assay, and vice versa, because the VP assay tests for the presence of a neutral product, acetoin, which is produced by fermentation via an organic acid intermediate; therefore, more acetoin (organism will be VP-positive) usually means correspondingly lower concentrations of organic acids (organism will be methyl red–negative). [Lab 25]

microaerophile A microorganism that requires oxygen, but grows best at oxygen concentrations below 20% (most commonly 2%–10%). [Lab 14]

minimum growth temperature The lowest temperature at which growth and cell division can occur. [Lab 11]

minimum inhibitory concentration (MIC) The lowest dose of an antibiotic that will inhibit the growth of a given bacterium. [Lab 18]

minimum inhibitory concentration (MIC) assay An assay that uses serial dilution of an antibiotic to determine the minimum inhibitory concentration of that drug. [Lab 18]

mordant A chemical that increases the affinity of a stain for the object to be stained; derived from the French *mordre*, meaning "to bite." [Lab 6]

most probable number (MPN) An estimate of coliform population density that is based on probability theory—specifically, on the probability of getting a certain number of positive tubes in each of three sets of dilution tubes when the initial population is X. [Lab 30]

motile Able to move through a medium; in the case of bacteria, movement is by rotating, corkscrew-shaped flagella. [Lab 25]

motility assay An assay used to assess the ability of cells to move; one of the SIM assays. [Lab 25]

motility protein B (MotB) A motor protein found in the bacterial flagellum. [Lab 20]

MRSA (pronounced "mer-sa") Methicillin-resistant *Staphylococcus aureus*; an antibiotic-resistant strain of *S. aureus* that has become a life-threatening problem in many clinical settings. [Lab 22]

MRVP broth See **Methyl Red Voges Proskauer broth**.

mycolic acids Strongly hydrophobic, waxy molecules composed of very long hydrocarbon chains that account for a substantial portion of the dry weight of the cell walls of acid-fast bacteria of the genus *Mycobacterium* and significantly slow the movement of nutrients and stains into these bacteria. [Lab 7]

N-acetylglucosamine An amino sugar that is part of the streptococcal group A antigens. [Lab 21]

narrow-spectrum antibiotic An antibiotic that inhibits or kills fewer bacterial species than a broad-spectrum antibiotic; typically, the drug will inhibit or kill either Gram-positive or Gram-negative species but will not be equally effective against both groups. [Lab 17]

neutral red A pH indicator that passes through a transition from red to orange to yellow as the pH rises from 6.4 to 7.4 to 8.4. [Lab 24]

nitrate broth A medium designed to select for denitrifying bacteria; contains potassium nitrate salt as a source of NO_3^- for denitrification. [Lab 33]

nitrate-free broth Nutrient broth lacking the nitrates required for denitrification. [Lab 33]

nitrate salts Compounds such as $NaNO_3$ (sodium nitrate) that can be used in growth media as a source of nitrogen; the nitrates (NO_3^-) must be reduced to ammonium or amino nitrogen before they can be incorporated into amino acids, and this reduction is energetically expensive. [Lab 9]

nitrification A process that oxidizes ammonium (NH_4^+) to nitrite (NO_2^-) and then oxidizes nitrite to nitrate (NO_3^-). [Lab 33]

nitrite formation broth A medium that selects for nitrifying bacteria that oxidize ammonium to nitrites. [Lab 33]

nitrogen One of the five or six elements in amino acids, proteins, nucleotides, DNA, and RNA; required all by organisms. [Lab 9]

nitrogen cycle The series of steps or transitions from one form of nitrogen to another as this element moves through ecosystems from one type of organism to another. [Lab 33]

nitrogen fixation A process that reduces atmospheric nitrogen (N_2) to ammonia (NH_3) or ammonium (NH_4^+). [Lab 33]

non-coliform enterics Members of the enteric family that are unable to ferment lactose; include almost all the species and strains in the genera *Proteus, Salmonella, Shigella, Serratia,* and *Providencia*. [Lab 24]

normal flora Microbes that colonize the large intestine and protect their host against disease by manufacturing vitamins that can be absorbed into the bloodstream and by competing with pathogenic species. [Lab 29]

Novobiocin An antibiotic that is effective against some, but not all, *Staphylococcus* species and most *Micrococcus* species; used to differentiate *S. saprophyticus* from other *Staphylococcus* species. [Lab 22]

oligonucleotide primer In PCR, a DNA sequence, typically composed of 15–25 nucleotides. chosen for its ability to complement and bind to the DNA at the beginning or at the end of the section of the DNA one wishes to copy. PCR usese two oligonucleotide primers, which flank the stretch of DNA to be amplified. [Lab 20]

OmpC An integral membrane protein that normally functions as a porin or molecular channel, used by T4 bacteriophage to bind the virus to its host cell. [Lab 37]

opportunistic infection An infection that typically occurs after an individual has been weakened, injured, or made vulnerable by a preexisting condition or decline in the immune system. [Lab 23]

optimum growth temperature The relatively narrow range of temperatures at which cell division, or the rate of reproduction, for a given species is at a maximum. [Lab 11]

organic acids End products of carbohydrate fermentation; depending on the microbial species, fermentation may produce any number of different organic acids, including lactic, acetic, formic, and butyric acids. [Lab 10]

organic nitrogen Nitrogen that is a part of large organic molecules such as amino acids, proteins, and nucleic acids. [Lab 33]

organotroph An organism that uses organic molecules as electron and proton sources. [Lab 34]

osmophilic Able to thrive in a very hypertonic environment. [Lab 12]

osmosis The net diffusion or movement of water from the side of a semipermeable cell membrane with the lower solute concentration to the side with the higher solute concentration; derived from the Greek *osmos*, meaning "push" or "thrust." [Lab 12]

osmotolerant Able to tolerate a very hypertonic environment. [Lab 12]

outer envelope A structure found in Gram-negative bacteria, composed primarily of alcohol-soluble phospholipids, that is similar in structure to the inner cell membrane; considered by some to be a part of the cell wall, but by others to be a structure separate from the cell wall itself. Also referred to as the **outer membrane**. [Lab 6]

outer membrane See **outer envelope**.

oxidation/reduction potential (ORP) A measure of the tendency of a molecule or compound to gain or lose electrons, often a product of the relative amount of oxygen in a microbial environment. [Lab 14]

oxidative/fermentative (O/F) glucose assay An assay that differentiates bacteria, particularly *Pseudomonas* species, on the basis of the bacterium's ability produce organic acids from glucose oxidatively. [Lab 23]

oxyduric Of or referring to a strict or obligate anaerobe that can endure or survive exposure to oxygen and can grow later if oxygen is removed. [Lab 14]

oxygenic photoautotroph A photoautotroph that uses water (H_2O) molecules as a source of electrons and protons, thereby producing oxygen. [Lab 34]

oxylabile Of or referring to a strict or obligate anaerobe that is quickly killed by contact with oxygen. [Lab 14]

papain A papaya juice protease that can be used to digest proteins to make peptones or tryptones. [Lab 9]

parfocal Of or referring to an arrangement in which the different objective lenses in a microscope with different focal lengths are set in a rotating nosepiece at distances above the stage such that the image should remain nearly in focus when the user switches from one lens to another. [Lab 1]

pasteurization A method of microbial control, best known from its use with milk, that involves heating food items and beverages for a brief period of time; this method reduces microbial populations, extends shelf life, and increases safety. [Lab 29]

pasteurized milk Milk that has been subjected to pasteurization. [Lab 29]

PCR See **polymerase chain reaction.**

pepsin A protease found in the human stomach that can be used to digest proteins to make peptones or tryptones. [Lab 9]

peptidoglycan A latticelike biological polymer found in bacterial cell walls. [Lab 6]

peptones Products used in growth media; made by digesting animal and plant proteins with hot acidic solutions or proteolytic enzymes, producing crude mixtures of individual amino acids and short peptide (amino acid) chains in which the exact concentration of any given amino acid is unknown. [Lab 9]

peritrichous flagella Multiple flagella distributed uniformly over the cell surface; characteristic of enteric bacteria. [Lab 25]

PFU See **plaque-forming unit.**

phage See **bacteriophage.**

phage typing The use of specific phages to identify different, specific bacterial strains and species. [Lab 37]

phenol coefficient A measure of the effectiveness of a disinfectant that compares its effectiveness against that of phenol under standard conditions; determined by dividing the reciprocal of the maximum effective dilution for the test chemical by the reciprocal of the maximum effective dilution for phenol. [Lab 15]

phenol red carbohydrate assay An assay used to assess a microorganism's ability to produce the endoenzymes needed to ferment monosaccharides and disaccharides, including sugars such as glucose, lactose, and sucrose, by detecting a variety of organic acid and gas end products of fermentation. [Lab 10]

phosphorus An element required in growth media, needed for the phosphate (PO_4) groups of ATP, for cell membrane phospholipids, and for the nucleotides used to synthesize the nucleic acids RNA and DNA. [Lab 9]

photoactivated repair system An enzyme system activated by light that can locate and repair damaged sections of DNA; cleaves the covalent bonds that create thymine dimers, restoring the DNA to its original state without physically replacing the thymines. [Lab 16]

photoautotroph An organism that uses light, captured by photosynthesis, as a source of energy and carbon dioxide as a source of carbon. [Lab 34]

photoheterotroph An organism that uses light as a source of energy and organic molecules, especially organic acids, as its main source of carbon and usually as a source of electrons and protons as well. [Lab 34]

phototroph An organism that uses light as an energy source. [Lab 34]

plaque A roughly circular cleared area in a bacterial lawn that is large enough to be seen with the unaided eye, representing an area from which bacterial cells are absent. [Lab 36]

plaque-forming unit (PFU) A term referring to the virus particle or tight cluster of particles that produces a single plaque, or cleared area, in a bacterial lawn; when measuring the titer of a bacteriophage suspension, phage titers are reported in plaque-forming units per milliliter instead of particles per milliliter. [Lab 36]

plasmolysis A process by which water flows out of the cytoplasm of a cell in a hypertonic environment, causing the cell membrane to shrink away from the cell wall. [Lab 12]

pneumonia A lung disease characterized by the accumulation of fluid in the bronchi and alveoli and caused by a variety of microorganisms. [Lab 22]

polymerase chain reaction (PCR) A widely used technique for copying DNA outside of cells that uses high temperatures to repeatedly separate the two strands in double-stranded DNA and uses DNA polymerase to produce new strands; depends on DNA polymerase enzymes that retain their shape and remain functional at temperatures between 70°C and 95°C. [Lab 11]

potable Safe for human consumption. [Lab 30]

pour plate counting A plate counting technique in which the density of a cell suspension is determined by mixing serially diluted aliquots of the suspension with melted ager; the agar is then poured into plates, the colonies are counted, and the cell density is determined by multiplying colony counts by the dilution factor for a counted sample. [Lab 4]

pour plating A technique for producing colonies in which cells are added to melted agar, the agar is poured into a plate, and colonies appear at points where the cells were trapped in the agar. [Lab 3]

presumptive test One of a series of three assays used to measure total coliforms in water; uses a lactose broth to favor the growth of coliforms and a Durham tube to detect the ability of microbes to rapidly produce gas as one end product of the fermentation of lactose. [Lab 30]

probiotic Of or relating to normal flora or microorganisms believed to be beneficial to health. [Lab 29]

protease An enzyme that degrades proteins, producing a mix of short peptide chains and free amino acids as end products. [Lab 10]

protein A high-molecular-weight biomolecule composed of chains of amino acids. [Lab 10]

psychrophile A "cold-loving" microbe with a temperature growth range between −5°C and 15°C–20°C; derived from the Greek *psukhros* and *philos*, meaning "cold" and "love," respectively. [Lab 11]

psychrotolerant microbe See **psychrotroph.**

psychrotroph A microbe with a temperature growth range between 0°C–5°C at the low end and 30°C–37°C at the high end, and with an optimum growth temperature at around 20°C–30°C. These broad temperature growth ranges mean that some of these species can grow at both refrigerator (2°C–4°C) and human body (37°C) temperatures Also called a **psychrotolerant microbe.** [Lab 11]

pulmonary anthrax An infection of the lungs caused by *B. anthracis.* [Lab 19]

pure culture A collection of cells of a single strain or species of microorganism growing in an environment free from contamination by any other living forms. [Lab 2]

purple nonsulfur bacteria A group of bacteria with diverse metabolic abilities that do not use or produce sulfur during photosynthesis; includes the genus *Rhodospirillum.* [Lab 34]

purple sulfur bacteria Anoxygenic photoautotrophic bacteria that capture light energy using pigments such as bacteriochlorophyll *a* and carotenoids and produce elemental sulfur during photosynthesis; includes species in the genus *Chromatium*. [Lab 34]

pyrimidine dimer A covalent link between adjacent pyrimidine bases on the same DNA strand; may result from exposure of DNA to UV light, especially at wavelengths of 260–265 nm. [Lab 16]

raw milk Milk that has not been pasteurized. [Lab 29]

redox indicator A molecule whose color reflects the oxidation/reduction potential of a growth medium. [Lab 14]

redox reaction A reaction that involves the transfer of electrons from one molecule to another; the molecule that gains the electrons becomes reduced, while the molecule that loses the electrons becomes oxidized. [Lab 23]

reducing agent A molecule that (1) binds free oxygen or (2) donates electrons and protons to keep other molecules in a reduced state; used in the cultivation of strict anaerobes. [Lab 14]

refrigeration The use of temperatures of 0°C–4°C (32°F–40°F) to significantly slow cell activity. [Lab 3]

resazurin A redox indicator; when used in thioglycollate medium, it is colorless when in the fully reduced state and turns pink as it becomes less reduced and more oxidized. [Lab 14]

residual sugar The sugar that remains at the end of fermentation of wine. [Lab 32]

resistant Of or referring to a bacterial strain that is not effectively inhibited or killed by an antibiotic at the concentration or dose that can usually be tolerated by a patient. [Lab 17]

resolution The ability to see objects that are close together as multiple, distinct, and separate objects; the minimum distance between two objects that reveals them as separate objects. [Lab 1]

resolving power The ability of an optical instrument to resolve microscopic objects; it is a function of the wavelength of the light (λ) that forms the image and the numerical aperture of the lens (a measure of the lens's ability to capture light). [Lab 1]

rhamnose A sugar that is part of the streptococcal group A antigens. [Lab 21]

rheumatic fever A serious complication of strep throat that can cause permanent damage to the heart and may be fatal. [Lab 21]

safranin A direct, basic, red-colored stain used as a counterstain in Gram staining and spore staining. [Lab 8]

salmonellosis A type of gastroenteritis that features diarrhea, cramps, vomiting, and a low fever, caused by intestinal infection by *Salmonella*. [Lab 28]

scarlet fever A serious complication of strep throat that includes a bright red rash on the chest and neck. [Lab 21]

secreted coagulase A coagulase secreted by *Staphylococcus aureus* that catalyzes a reaction that converts fibrinogen into fibrin, forming clots. Also called **free coagulase**. [Lab 22]

selective medium A medium designed to select for and favor the growth of specific microbial groups or species of interest, or to prevent, inhibit, or reduce the growth of interfering or undesired species. [Lab 19]

selective toxicity The property of being much more toxic to one particular group of organisms, such as bacteria, than to others, such as humans. [Lab 17]

semi-log paper Graph paper with a log scale on the *y*-axis. [Lab 29]

sensitive Of or referring to a bacterial strain that can be inhibited or killed by an antibiotic at the concentration or dose that human patients can usually tolerate; a drug that a bacterial strain is sensitive to should be effective in treating an infection caused by that strain. [Lab 17]

septicemia An infection of the blood. [Lab 22]

serial dilution A series of stepwise dilutions in which the aliquot for a given subsequent dilution step is taken from the tube created in the previous step; used to create a suspension with the right number of cells to produce a plate with tens to hundreds of isolated colonies per plate. [Lab 3]

serovar A bacterial strain; specifically, a strain of *Salmonella enterica*. [Lab 28]

SIM agar A semisolid medium designed for carrying out the SIM assays. [Lab 25]

SIM assays A set of tests for three independent traits—sulfide production, indole production, and motility—which can be done simultaneously on a single tube of SIM agar. [Lab 25]

simple stain A method of staining cells that uses only one type of stain. [Lab 5]

Singlepath *Salmonella* Rapid Test A test that relies on antibodies that bind to bacterial antigens specific for, or unique to, the numerous *Salmonella* strains capable of causing intestinal diseases in humans. [Lab 28]

skim milk agar An agar-based medium developed to test a microbe's ability to produce and secrete caseinases; includes 10% to 20% skim milk by volume; also called **casein agar**. [Lab 10]

spastic paralysis Uncontrollable paralytic contraction of muscles; a symptom of tetanus. [Lab 8]

specific gravity A measure of the density of a liquid; defined as the ratio of the density of a test substance to the density of a reference solution, usually water. [Lab 32]

spectrophotometer A device that measures the amount of light passing through a medium; used to measure the absorbance of a suspension of cells, which can be correlated with cell density. [Lab 4]

spectrum The range of bacterial species that can be inhibited or killed by a particular antibiotic. [Lab 17]

spirillum A bacterial cell with a spiraled or coiled form, usually with two or three twists or turns; derived from *spira* (Latin) and *speira* (Greek), meaning "coiling" or "winding." [Lab 5]

spirochete A bacterial cell with a spiraled or coiled form, usually a very long, thin, threadlike or hairlike cell with multiple (four or more) twists along the length of the cell; derived from *chaeta*, meaning "hair" or "bristle." [Lab 5]

spore stain A structural stain designed to visualize endospores. [Lab 8]

spread plate counting A plate counting technique in which a starting cell suspension is serially diluted, the dilutions are spread on the surfaces of plates of various media with a sterile glass rod, and then the number of viable cells in each dilution is determined by counting colonies on the plate. [Lab 4]

staphylo- A prefix used to describe cells that form clusters after cell division; derived from the Greek *staphyle*, meaning "bunches of grapes." [Lab 5]

starch A very large, plant-produced polysaccharide composed of hundreds or thousands of glucose molecules. [Lab 10]

starch agar A medium used in the starch hydrolysis assay that contains soluble starch as a substrate for amylase activity. [Lab 10]

starch hydrolysis assay An assay designed to detect amylases. [Lab 10]

-static A suffix added to prefixes such as "germi-," "bacteri-," or "fungi-," as in "germistatic," "bacteristatic," and "fungistatic," to indicate a static agent. [Lab 15]

static agent An agent that inhibits or stops microbial growth, but may not kill the microbes; if the agent is removed from the environment, the microorganisms that it held in check can resume growth and cell division. [Lab 15]

sterile Of or referring to an object or medium on or in which there are no viable organisms of any kind or in any form, active or inactive. [Lab 2]

streak plating A technique for physically separating cells to produce isolated colonies in which an inoculating loop is used to drag and physically separate cells on the surface of an agar plate. [Lab 3]

strep throat A type of pharyngitis caused by *Streptococcus pyogenes*, characterized by the formation and appearance of pus in the throat and on the tonsils. [Lab 21]

strepto- A prefix used to describe cells that are arranged in chains; derived from the Greek *streptos*, meaning "twisted" or "twisted chain." [Lab 5]

streptobacillus A rod-shaped cell that is arranged in chains. [Lab 5]

strict aerobe A microorganism that depends completely on aerobic respiration for the production of ATP. [Lab 14]

strict anaerobe A microorganism that lacks oxygen-detoxifying enzymes and is therefore unable to tolerate or grow in the presence of oxygen. [Lab 14]

structural stain A stain designed to visualize particular structures in or on a bacterial cell. [Lab 5]

sulfates Anions with the empirical formula SO_4^{2-}; sulfate salts such as Na_2SO_4 may be added to bacterial media to provide a source of sulfur for the microbes and may be used in place of oxygen by *Desulfovibrio* in anaerobic respiration, a process that reduces sulfates to sulfides. [Lab 25]

sulfhydryl group A functional group consisting of a sulfur bonded to a hydrogen atom; this group makes an effective reducing agent because the –SH group readily donates electrons and protons to other molecules, thus reducing them. In some cases, the sulfhydryl-bearing compounds react directly with certain reactive forms of oxygen, removing the reactive oxygen species by reducing them to water. [Lab 14]

sulfide production assay An assay designed to detect the generation of hydrogen sulfide (H_2S) from inorganic salts containing sulfates (SO_4^{2-}) or thiosulfates ($S_2O_3^{2-}$) or from sulfur-containing amino acids such as cysteine; one of the SIM assays. [Lab 25]

sulfur An element required in growth media, used by microbes to synthesize the amino acids cysteine and methionine. [Lab 9]

sulfur cycle bacteria Bacteria that convert sulfides to sulfates and sulfates to sulfides. [Lab 34]

superoxide (O_2^-) A highly reactive and toxic form assumed by oxygen in cells. [Lab 14]

superoxide dismutase An oxygen-detoxifying enzyme produced by some microbial species that catalyzes the reduction of superoxide to hydrogen peroxide. [Lab 14]

symbiotic Of or referring to a mutually beneficial relationship between two or more species. [Lab 33]

symbiotic nitrogen fixation Nitrogen fixation carried out by *Rhizobium* species, as well as by a few other genera, which establish a symbiotic relationship with the roots of plants in the legume family (clover, alfalfa, peas, soybeans). [Lab 33]

tail The part of a bacteriophage consisting of a hollow core encircled by a contractile sheath, a spiked baseplate, and several tail fibers. [Lab 36]

tannin One of a group of chemicals formed in plant cells by a combination of flavonoids, phenolics, and sugars; contributes to the astringency of wine. [Lab 32]

Taq polymerase A heat-stable DNA polymerase derived from the thermophilic bacterium *Thermus aquaticus*, used in PCR. [Lab 20]

TB See **tuberculosis**.

temperate See **lysogenic**.

temperature growth range The range of temperatures within which a given microbial species can carry out metabolic tasks, grow, and divide. [Lab 11]

template DNA A strand of DNA that acts as a template for DNA replication. [Lab 20]

tetanus A human disease, caused by *Clostridium tetani*, that develops after the bacteria multiply in oxygen-free tissue and produce a neurotoxin that causes muscles to contract uncontrollably. [Lab 8]

tetrad An arrangement in which four cells are joined together in a symmetrical packet of cells; derived from the Greek *tetra*, meaning "four." [Lab 5]

therapeutic index A quantity that compares the toxicity of an antibiotic to a given microbe with the toxicity of that antibiotic to humans; calculated as the dose of the drug that is harmful to humans divided by the dose of the drug that is harmful to the microbe. [Lab 18]

thermocycler Apparatus that cyclically raises and lowers the temperature of a solution in which PCR is occurring. [Lab 20]

thermoduric Of or referring to microbes that can outlast or endure heating; specifically, those that can survive temperatures of up to 100°C for up to 10 minutes, often as endospores; derived from the Latin *durare*, meaning "to last" or "to make hard." [Lab 11]

thermophile A microbe with a temperature growth range between 40°C–45°C and 65°C–70°C, with optimum growth temperatures at around 55°C 65°C; derived from the Greek *thermos* and *therme*, meaning "hot" or "heat"; these species are often found in hot springs. [Lab 11]

thioglycollate A reducing agent used in the cultivation of strict anaerobes, usually in the form of sodium thioglycollate. [Lab 14]

3× phenol red glucose agar A medium in which the pH indicator phenol red is present at three times the concentration found in phenol red broths. [Lab 29]

thymine dimer A covalent link between adjacent thymine bases on the same DNA strand; distorts the shape of the DNA molecule and prevents the accurate copying or replication of DNA, resulting in mutations. [Lab 16]

total dilution factor A value that is a part of the CFU/ml formula; the value for a given dilution tube can be determined by multiplying together all of the individual dilution factors for all tubes up to the given tube. [Lab 4]

triglyceride A lipid or fat composed of a glycerol molecule bonded to three fatty acid chains. [Lab 10]

trophic group Group or category of organisms that share traits based on the sources of energy, carbon, and protons and electrons that they use. [Lab 34]

trypsin A protease found in the human pancreas that can be used to digest proteins to make peptones or tryptones. [Lab 9]

tryptones Products used in growth media, made by degrading animal and plant proteins with hot acidic solutions or proteolytic enzymes, producing crude mixtures of individual amino acids and short peptide (amino acid) chains in which the exact concentration of any given amino acid is unknown. [Lab 9]

tryptophan An amino acid. [Lab 25]

tryptophanase L-tryptophan indole-lyase; an enzyme required for the breakdown of tryptophan to indole. [Lab 25]

TSST-1 toxin Toxic shock syndrome toxin; a toxin produced by *Staphylococcus aureus* that triggers a fever and causes a potentially fatal drop in blood pressure. [Lab 22]

tuberculosis (TB) A human disease caused by *Mycobacterium tuberculosis*. [Lab 7]

turbidity Cloudiness; can be used as a measure of the number of cells in a solution: The more cells there are, the greater the turbidity of the solution. [Lab 13]

ultraviolet (UV) light The segment of the electromagnetic spectrum between about 100 and 400 nm. [Lab 16]

urea Organic nitrogen compound with the formula CH_4N_2O; broken down into ammonia (NH_3) and carbon dioxide (CO_2) by ureases. [Lab 25]

urease An enzyme that breaks down urea into ammonia (NH_3) and carbon dioxide (CO_2). [Lab 25]

urease assay An assay designed to detect the ability of bacteria to produce urease. [Lab 25]

virulence factor A trait of a pathogen that enhances its ability to cause disease. [Lab 22]

virulent See **lytic**.

vitamin A compound needed as a cofactor for certain metabolic enzymes; not usually needed in growth media because most bacteria can synthesize these compounds. [Lab 9]

water blank A test tube containing pure water. [Lab 4]

wavelength Distance between successive peaks of a wave; for wavelengths of visible light, the distance between wave peaks is between 400 and 750 nm. [Lab 1]

Winogradsky column An artificial ecosystem used to model or simulate microbial ecosystems composed of many interacting species. [Lab 34]

working distance Working distance is related to focal length or focal distance. Working distance is defined as the distance between the front edge of the lens and the specimen (focal length is measured from the center of the lens); working distance decreases with increasing magnification. [Lab 1]

yeast extract An excellent source of B vitamins that can be added to growth media. [Lab 9]

Ziehl-Neelsen stain An acid-fast stain technique that relies on heat to partially liquefy the waxes in bacterial cell walls. [Lab 7]

zone of inhibition A circular area of clearing around an antibiotic disk placed on plate media seeded with a lawn of bacteria; the edge of the zone marks the point at which the concentration of the antibiotic is sufficient to inhibit or kill the bacteria. [Lab 17]

TEXT PERMISSIONS

Lab 37, Table 1: Table 1 from S. Altekruse, J. Koehler, F. Hickman-Brenner, R. V. Tauxe and K. Ferris, "A comparison of *Salmonella enteritidis* phage types from egg-associated outbreaks and implicated laying flocks," *Epidemiology and Infection* 110(1): pp. 17–22. Copyright © Cambridge University Press 1993. Reproduced with permission.

PHOTO CREDITS

Chapter 1
1 CDC/ Jeff Hageman, M.H.S. **2** W. W. Norton & Co. **7** W. W. Norton & Co.

Chapter 2
13 SPL / Science Source **14 (all)** W. W. Norton & Co. **15 left** AFP /Getty Images **15 right** Private Property/Courtesy of the Hesse Family **19 top** W. W. Norton & Co. **19 bottom** W. W. Norton & Co. **20 left** W. W. Norton & Co. **20 right** W. W. Norton & Co. **21 top left** W. W. Norton & Co. **21 top right** W. W. Norton & Co. **21 bottom** W. W. Norton & Co.

Chapter 3
27 Courtesy of Dr. Roger Tsien, University of California, San Diego. Artwork by Nathan Shaner, photography by Paul Steinbach, created in the lab of Roger Tsien. © 2006 The Regents of the University of California. All Rights Reserved. **28** W. W. Norton & Co. **30 top** W. W. Norton & Co. **30 bottom** W. W. Norton & Co.

Chapter 4
43 Leungchopan/Shutterstock **44** W. W. Norton & Co. **52** Courtesy of John Foster, University of South Alabama **55 top left** W. W. Norton & Co. **55 top right** W. W. Norton & Co. **55 bottom left** W. W. Norton & Co. **55 bottom right** W. W. Norton & Co.

Chapter 5
65 Yen Teoh/Getty Images **67** BSIP/UIG/Getty Images **69 top left** Ed Reschke/Getty Images **69 top right** Dr. Gladden Willis/Visuals Unlimited/Corbis **69 bottom left** CDC/Dr. Leanor Haley **69 bottom right** Dr. John D. Cunningham/Getty Images **73** Joseph Duris and Silvia Rossbach, Western Michigan University

Chapter 6
82 (both) Dr. Gladden Willis/Visuals Unlimited **86 left** CDC/Dr. Mike Miller **86 right** Dr. Gladden Willis/Visuals Unlimited/Corbis

Chapter 7
81 toetoey/Shutterstock **82 (both)** SPL/Science Source **95** CDC **97** SSPL/Getty Images **98** Stained M. tuberculosis from Koch, R. 1884. DIE AETIOLOGIE der TUBERKULOSE. Mittheilungen aus dem Laiserlichen Gesundheitsampte. 2: 1-88. **100** CDC/Dr. George P. Kubica

Chapter 8
109 VisitBritain/Jason Hawke/Getty Images **110** Dr. Tony Brain/Science Source **111 top** Michael Abbey/Getty Images **111 center** BSIP/UIG/Getty Images **111 bottom** CDC/Dr. Holdeman **112 left** Oxford Science Archive/ Print Collector/Getty Images **112 right** A. Loeffler/Corbis **114** Wikimedia Commons

Chapter 9
121 Bill Branson/National Cancer Institute **126** W. W. Norton & Co. **129** Javier Larrea/AgeFotostock **133** Ron Nichols/National Resources Conservation Services **136** W. W. Norton & Co. **140** W. W. Norton & Co. **142** W. W. Norton & Co. **144** Photogravure after M. J. A. Mercié. Courtesy Wellcome Images, operated by Wellcome Trust

Chapter 10
146 W. W. Norton & Co. **155** AF Archive/Alamy **158** Penn State University/ Science Source **160 left** Thermal Biology Institute (TBI), BioScience Montana Program, Montana State University **160 right** FotoFeeling/AgeFotostock

Chapter 11
167 Antonin Vinter/Shutterstock **171** Winnie The Pooh and the Honey Tree, 1966. The Everett Collection **173 top left** Eye of Science/Photo Researchers, Inc. **173 top right** David Scharf Photography **173 bottom left** Wayne P. Armstrong **173 bottom right** Courtesy of S. Dassarma, U. of Maryland BIiotechnology Institute

Chapter 12
179 Shane W. Thompson/Shutterstock **180 left** SONNY X. LI **180 right** THOMAS D. BROCK, UNIVERSITY OF WISCONSIN, MADISON **181 top** Birute Vijeikiene | Dreamstime **181 center** DSMZ/GBF, ROHDE **181 bottom** Guy Edwardes Photography / Alamy **184 left** W. W. Norton & Co. **184 top right** W. W. Norton & Co. **184 bottom right** W. W. Norton & Co.

Chapter 13
185 W. W. Norton & Co. **191** Joseph Sohm/Shutterstock

Chapter 14
195 left BSIP SA/Alamy **195 right** Courtesy of and © Becton, Dickinson and Company

Chapter 15
203 Trye-Maison, G. de/Wellcome Library, London

Chapter 16
215 Floris van Breugel/Nature Picture Library/Corbis

Chapter 17
225 CDC/Don Stalons **227 top (all)** Courtesy and © Becton, Dickinson and Company **227 center** © Hardy Diagnostics, www.HardyDiagnostics.com **227 center bottom** W. W. Norton & Co. **227 bottom** CDC/Gilda Jones **229 (both)** © Hardy Diagnostics, www.HardyDiagnostics.com

Chapter 18
235 Mikecphoto/Shutterstock

Chapter 19
247 © Daniela Beckmann **248 top** W. W. Norton & Co. **248 bottom** W. W. Norton & Co. **251** Science Photo Library/Getty Images

Chapter 20
261 Richard Wheeler **263** W. W. Norton & Co. **274** Paul J. Jackson Et Al. 1998. PNAS 95: 1224 **278** John Lindquist, University of Wisconsin-Madison